Modeling, Data Processing and Remote Sensing in Atmospheric Sciences

Modeling, Data Processing and Remote Sensing in Atmospheric Sciences

Edited by **Smith Paul**

SYRAWOOD
PUBLISHING HOUSE
New York

Published by Syrawood Publishing House,
750 Third Avenue, 9th Floor,
New York, NY 10017, USA
www.syrawoodpublishinghouse.com

Modeling, Data Processing and Remote Sensing in Atmospheric Sciences
Edited by Smith Paul

International Standard Book Number: 978-1-68286-362-6 (Hardback)

Printed in China.

Contents

Preface

Atmospheric sciences is an umbrella discipline comprising sub-fields such as meteorology, climatology, aeronomy, etc. Advances in science and technology have led to the expansion and advancement of this discipline. Some of the topics that have been covered include remote sensing of the atmosphere, measurement of wind, precipitation, etc. This book is an essential guide for both professionals and those who wish to pursue this discipline further. It attempts to assist those with a goal of delving deeper into the progress of atmospheric sciences.

Significant researches are present in this book. Intensive efforts have been employed by authors to make this book an outstanding discourse. This book contains the enlightening chapters which have been written on the basis of significant researches done by the experts.

Finally, I would also like to thank all the members involved in this book for being a team and meeting all the deadlines for the submission of their respective works. I would also like to thank my friends and family for being supportive in my efforts.

Editor

An evaluation of COSMIC radio occultation data in the lower atmosphere over the Southern Ocean

L. B. Hande[1], **S. T. Siems**[1], **M. J. Manton**[1], **and D. H. Lenschow**[2]

[1]School of Earth, Atmosphere and Environment, ARC Centre of Excellence for Climate System Science, Monash University, Melbourne, Australia

[2]National Center for Atmospheric Research, Boulder, Colorado, USA

Correspondence to: L. B. Hande (luke.hande@gmail.com)

Abstract. The global positioning system (GPS) radio occultation (RO) method is a relatively new technique for taking atmospheric measurements for use in both weather and climate studies. As such, this technique needs to be evaluated for all parts of the globe. Here, we present an extensive evaluation of the performance of the Constellation Observing System for Meteorology, Ionosphere, and Climate (COSMIC) GPS RO observations of the Southern Ocean boundary layer. The two COSMIC products used here are the "wetPrf" product, which is based on 1-D variational analysis with European Centre for Medium-Range Weather Forecasts (ECMWF), and the "atmPrf" product, which contains the raw measurements from COSMIC. A direct comparison of temporally and spatially co-located COSMIC profiles and high resolution radiosonde profiles from Macquarie Island (54.62° S, 158.85° E) highlights weaknesses in the ability of both COSMIC products to identify the boundary layer structure, as identified by break points in the refractivity profile. In terms of reproducing the temperature and moisture profile in the lowest 2.5 km, the "wetPrf" COSMIC product does not perform as well as an analysis product from the ECMWF. A further statistical analysis is performed on a large number of COSMIC profiles in a region surrounding Macquarie Island. This indicates that, statistically, COSMIC performs well at capturing the heights of main and secondary break points. However, the frequency of break points detected is lower than the radiosonde profiles suggest, but this could be simply due to the long horizontal averaging in the COSMIC measurements. There is also a weak seasonal cycle in the boundary layer height similar to that observed in the radiosonde data, providing some confidence in the ability of COSMIC to detect an important boundary layer variable.

1 Introduction

The structure and dynamics of the atmospheric boundary layer (ABL) not only directly impact the weather, through the transport of scalars such as water vapour, but also the climate, most obviously through their role in the formation and dissipation of clouds. The ABL height is an important variable, which is controlled by a balance of large-scale subsidence tending to decrease the ABL height and turbulent processes tending to increase the ABL height (Stull, 1988). The exact definition of the ABL height is ambiguous, making it difficult to quantify and study, particularly as routine observations of the depth of the boundary layer are generally not available over much of the globe. This problem is exacerbated over remote locations, such as the Southern Ocean, where in situ observations are sparse. Given that clouds over the Southern Ocean is responsible for large biases in modelled net radiation (Trenberth and Fasullo, 2010), it is important to understand the fundamental processes at work in this region, which is dominated by the presence of boundary layer cloud year round (Huang et al., 2012a).

Huang et al. (2012a) present a climatology of Southern Ocean clouds from *CloudSat*, and identify weaknesses in the ability of CloudSat to identify low clouds due to interference with the surface. This finding was reinforced by Huang et al. (2012b), who found similar difficulties with other satellite products. The height of clouds is strongly correlated with the height and structure of the boundary layer. Hande et al. (2012b) present a detailed study of the structure of the ABL over the Southern Ocean from in situ measurements taken from Macquarie Island (54.62° S, 158.85° E). The observations from radiosonde data suggest that the ABL is shallow and often decoupled, a feature manifested as multiple layers in the lowest few kilometres, and these features are not well captured in a state-of-the-art reanalysis data set from the ECMWF. It was shown that the reanalysis data set had a median primary inversion about 200 m lower than the radiosonde data. An analysis of proxy cloud fields from the radiosondes indicated that the low-level clouds are not typically capping a well-mixed boundary layer, in stark contrast to the well-studied subtropical stratocumulus in the Eastern Pacific and Atlantic (e.g. Albrecht et al., 1995; Bretherton et al., 2004). Furthermore, multiple cloud layers were observed to exist both within, and above the ABL. This supports earlier field observations of boundary layer decoupling in regions with less subsidence as typified in the First Aerosol Characterization Experiment (ACE-1) (Boers and Krummel, 1998; Russell et al., 1998). During ACE-1, as well as the Southern Ocean Cloud Experiments (SOCEX) (Jensen et al., 2000), a main inversion in virtual potential temperature was observed below 2 km from aircraft data, with a weaker inversion below the main inversion. Cloud was observed throughout this decoupled boundary layer.

Another distinctive feature of the Southern Ocean is high wind speeds and wind shear (Hande et al., 2012b; Xiaojun, 2004), producing some of the largest observed wave heights on the globe (Vinoth and Young, 2011). These conditions lead to the possibility of significant sea spray being injected into the boundary layer. This could have an, as yet, unaccounted for effect on the thermodynamics of the boundary layer, as well as the cloud microphysics (Andreas, 1998). With observations supporting an increase in the strength of the Southern Hemisphere westerlies over recent decades (Young et al., 2011; Hande et al., 2012a), the potential long-term impact of these effects on the Southern Ocean boundary layer, as well as the associated clouds, could be significant.

Sokolovskiy et al. (2006) demonstrated the usefulness of using global positioning system (GPS) radio occultation (RO) data to study the ABL height. They found that estimating the ABL height from the refractivity profile provided good agreement with radiosonde and reanalysis data sets. Most commonly used methods for determining the ABL height from GPS RO data involve identifying large gradients in the refractivity profile (Ao et al., 2012; Basha and Ratnam, 2009). A global analysis of ABL heights was performed by Guo et al. (2011) from GPS RO data. Their technique involved looking for a break point, or first-order discontinuities, in the refractivity profile, which served as an indicator of the ABL top. It was shown to agree well with boundary layer heights estimated from high-resolution radiosonde observations, particularly in the subtropical high-pressure regions where there is often a well-defined decrease in moisture above the main inversion. Chan and Wood (2013) use a similar technique to study the global variability of the ABL height. The authors find good agreement of ABL heights between COSMIC and radiosonde data on seasonal timescales.

Similarly, Seidel et al. (2010) present a climatology of ABL heights using several methods to estimate the heights from a number of different measurement systems. That study concludes that ABL heights based on the profile of refractivity do not agree with those based on gradients of other meteorological variables, particularly in the presence of clouds. There are large differences between different methods to calculate the PBL top in the presence of clouds, due to large humidity lapses over cloud top which can commonly lead to higher PBL heights when calculated from the humidity gradient. This conclusion is quite poignant, particularly for a cloud-dominated region such as the Southern Ocean.

The aim of this study is to evaluate the performance of the GPS RO technique of measuring boundary layer height, and other significant inversions, as well as the ability of the "wetPrf" COSMIC product to reproduce the temperature and moisture profiles within the boundary layer. An evaluation of this COSMIC data product is needed in this region because the structure of the boundary layer is unlike that of the tropical and subtropical high-pressure region, where the break point method for determining the ABL height has been shown to perform well.

2 Observational and analysis data

2.1 COSMIC

The Constellation Observing System for Meteorology, Ionosphere, and Climate (COSMIC)/Formosa Satellite 3 (FORMOSAT-3) is a constellation of six identical microsatellites each carrying a GPS RO receiver (Anthes et al., 2008). This allows for around 1500–2000 soundings per day around the globe, from 2006 onwards. The COSMIC measurement process is outlined by Kursinski et al. (1997), and summarized below. The primary measurement is of the Doppler shift of a radio signal that is emitted from the GPS satellite, occulted by Earth's atmosphere, and received by the Low Earth Orbiting satellite on the opposite side of the atmospheric limb. From this, a bending angle is derived, then the refractivity can be computed as a vertical integral of the bending angle from the top of the atmosphere down. The limb scanning geometry, either rising or setting occultations, is produced by the relative motion of the two satellites.

The raw measurements of the refractivity are used to identify the ABL height using the common technique of identifying break points in the refractivity profile, to be discussed in Sect. 3. This technique amounts to computing the vertical derivative of the refractive index, essentially reverting back to the bending angle measurement. This "atmPrf" data set contains approximately 802 data points in the lowest 3 km, and will be referred to as the $COSMIC_{raw}$ data set, where appropriate.

In the neutrally charged atmosphere, the refractive index (N) can be related to the pressure (p), temperature (T) and water vapour pressure (e) by

$$N = 77.6\frac{p}{T} + 3.73 \times 10^5 \frac{e}{T^2}, \quad (1)$$

where T is in Kelvin, and p and e are in hPa. With the aid of the hydrostatic equation, Eq. (1) can be used to estimate vapour pressure if temperature is specified, and vice versa. In this study, the "wetPrf" product was also used, which combines the raw observations with moisture information from the ECMWF TOGA 2.5 analysis using 1-D variational analysis. The ECMWF analysis product is used as a first guess for moisture below 10 km. The resulting profile is interpolated onto 100 m levels to produce the "wetPrf" profiles. Therefore in this data set, information on moisture and temperature is available at the expense of the high vertical resolution available in the raw refractivity measurements. The refractivity used in the "wetPrf" data set is the analysed refractivity, not the raw measurements. There are, on average, 31 data points in the lowest 3 km of the profiles over the Southern Ocean, and 400 levels available for the whole sounding. For the sake of clarity, this data product will be referred to as the $\mathrm{COSMIC_{wet}}$ data set. The profiles are constructed from 50 km long horizontal transects which cover the lowest 3 km. Therefore, the profiles from both COSMIC products would represent an average of the conditions over this line.

2.2 ECMWF analysis

The European Centre for Medium-Range Weather Forecasts (ECMWF) has a number of data products providing global coverage of various atmospheric variables. The ECMWF TOGA 2.5 global Upper Air Analysis data set (ECMWF, 1990) is used here to understand the influence of the background data in the 1-D variational data assimilation. The process of 1-D variational analysis uses the estimates of moisture from the ECMWF data to produce the $\mathrm{COSMIC_{wet}}$ measurements. This analysis data set is not independent of the other data sets considered here. The Macquarie Island radiosonde profiles are used in the assimilation process, however their contribution to the reanalysis data is weighted depending on the error characteristics of this data set. These error characteristics are determined by comparing these observations to others in this regions, mostly from satellite-based platforms over the Southern Ocean. When there is agreement between the various observations, the radiosonde data receive a higher weighting, and vice versa. These radiosonde data will be used for the evaluation of the thermodynamics for a limited number of cases. In addition to the radiosonde data, the ECMWF also assimilates the GPS RO data used in this study. Atmospheric profiles from ECMWF represent box-averaged quantities. Thus, these profiles represent the regional conditions over a larger portion of the ocean than the radiosonde profiles.

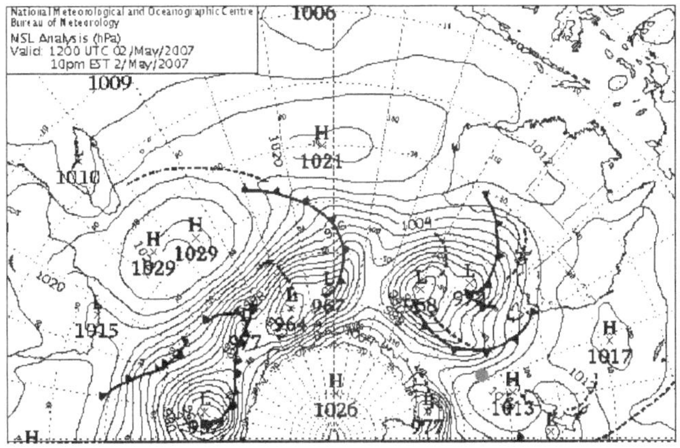

Figure 1. MSLP chart for 2 May 2007 at 12 Z. The approximate location of Macquarie Island is shown by the red dot. (Source: Australian Bureau of Meteorology.)

2.3 Macquarie Island

Macquarie Island, located at 54.62° S, 158.85° E, is one of the few Southern Ocean islands with a dedicated meteorological station. Here, radiosondes are released twice daily from an altitude of 8 m, with direct exposure to the prevailing westerly winds. The data set used here (MAC) consists of the 10 s vertical resolution soundings covering the period 1995–2014, consisting of 13 396 soundings. On average, there are 171 measurements in the lowest 3 km of the atmosphere. The profiles from the radiosonde represent a point measurement, and as such would be much more representative of the local conditions around Macquarie Island, rather than the regional conditions in the Southern Ocean. The difference in measurement techniques between the three data sets will contribute to differences in their respective representations of the atmospheric conditions.

The lack of observations at these high latitudes compared to other regions with better coverage of observations reduces the accuracy of numerical weather prediction (Adams, 1997). As a result, the location and extent of cold fronts, for example, can be inaccurate. The frequency of fronts passing over Cape Grim, Tasmania, was determined by Jimi et al. (1997) to be up to twice a week.

Figure 1 shows an example of a mean sea level pressure (MSLP) chart over much of the Southern Ocean, with the approximate location of Macquarie Island shown by the red dot. It is interesting to note the complicated frontal structures associated with the low-pressure systems over the Southern Ocean south of Australia. Hence, determining the exact synoptic conditions during each COSMIC or MAC profile may be prone to error.

3 Determining the ABL depth

A method for determining the height of the atmospheric boundary layer similar to Guo et al. (2011) is implemented for this study, and outlined below. We look for a break point in the $N(z)$ profile between 100 and 2500 m by defining an approximately 300 m sliding window and calculating the gradient of a linear regression of the form $Az+B$ within the window. The break point, which we use as an estimate of ABL height, is defined as the maximum difference of A from the windows immediately above and below the break point. A minimum value of $50\,\mathrm{km}^{-1}$ for the gradient of the window immediately below the break point of the main inversion is required. A detailed description of the method is outlined in Guo et al. (2011), including an example of how the method identifies inversions in sounding profiles. A secondary break point, or inversion, is defined in the same way, with a maximum height of 80 % of the height of the main inversion, and a minimum value of $40\,\mathrm{km}^{-1}$ for the gradient of the linear regression immediately below the secondary inversion. The requirement for having a weaker inversion below a main inversion ensures consistency with previous observations of the Southern Ocean boundary layer (Russell et al., 1998; Jensen et al., 2000), and is also commonly observed in other marine environments (Rémillard et al., 2012).

The magnitude of the gradient defining the secondary inversion is weaker than the main inversion, and hence differs from Guo et al. (2011). This value was chosen so the mean and standard deviation for the height of the secondary inversions, and also the frequency of occurrence for the MAC data set, were roughly the same as Hande et al. (2012b), who defined the ABL height based on the virtual potential temperature profile from soundings. Changing the value of the gradient below the break point by $\pm 5\,\mathrm{km}^{-1}$ had little effect on the height of inversions, with changes less than 15 m, but changed the frequency of occurrence by approximately ±16 % for secondary inversions.

The above method for determining the ABL height relies on identifying changes in the refractivity profile. However, it is changes in the temperature and moisture which are often used to define the top of the ABL. So at this point it is worthwhile considering how changes in the refractivity relate to changes in temperature and moisture. Differentiating Eq. (1) with height gives

$$\frac{\mathrm{d}N}{\mathrm{d}z} = 77.6\frac{1}{T}\frac{\mathrm{d}p}{\mathrm{d}z} - 77.6\frac{p}{T^2}\frac{\mathrm{d}T}{\mathrm{d}z} + 3.73\times 10^5\frac{1}{T^2}\frac{\mathrm{d}e}{\mathrm{d}z} - 7.46\times 10^5\frac{e}{T^3}\frac{\mathrm{d}T}{\mathrm{d}z}. \quad (2)$$

This shows that gradients in $N(z)$ are linked to gradients in temperature and vapour pressure, which are of the opposite sign. Therefore, the break point technique is most sensitive to a temperature increase and a moisture decrease occurring together. This is typical of a sub-tropical marine boundary layer. However, Hande et al. (2012b) showed that this structure only occurs in around 18 % of radiosonde data analysed over the Southern Ocean. Both temperature and vapour pressure can increase above the ABL, and in this case the refractivity may not necessarily change across the ABL top, and hence no ABL top detected. This was confirmed by tests on idealized profiles which were constructed from a series of straight-line segments with either a temperature or moisture increase or decrease inserted into the profile within the lowest 2 km. The break point method in the refractivity profile often failed to detect an ABL top in the presence of an increase in vapour pressure, even if this occurred at the same level as a strong temperature increase. Obviously this depends on the magnitude of the changes in temperature and moisture; however, the respective magnitudes were chosen to be consistent with observations from MAC.

Chan and Wood (2013) show that the changes in moisture contribute about an order of magnitude more than changes in temperature to the refractivity profile. This has the potential to complicate the ABL top detection over the Southern Ocean, where multiple cloud layers are common, meaning that multiple significant gradients in moisture may be present in the lowest few kilometres.

4 Case study evaluation of COSMIC

4.1 Boundary layer height

Here we use the MAC data set to make a direct comparison with spatially and temporally co-located profiles from COSMIC to evaluate the performance of the RO technique over the Southern Ocean. COSMIC profiles from within a 2×2 degree box around Macquarie Island occurring within ±1 h of the radiosonde launches were considered to be co-located. In addition to this, the COSMIC soundings were required to penetrate to within 500 m of the surface. For the period 2006–2013, only 35 soundings were found to coincide. The small sample is a result of three constraints: the time window, the size of the box, and the requirement to penetrate to below 500 m. Relaxing these constraints makes the sample bigger, but less relevant to the local conditions around Macquarie Island. The conclusions presented in this evaluation were drawn from the analysis of all the co-located profiles. However, to emphasize the typical ABL structures encountered over the Southern Ocean, only the results from four profiles will be shown. These four examples are shown in Figs. 2 to 5.

From the MAC data, Figs. 2 and 4 represent the case where multiple layers are found in the lowest few kilometres. These two cases have similar meteorological conditions. Both profiles represent pre-frontal conditions with northerly winds, and high values of relative humidity indicates that cloud exists throughout the boundary layer in both cases. Figure 3 represents a well-mixed boundary layer with a strong temperature inversion, and a decrease in the vapour pressure oc-

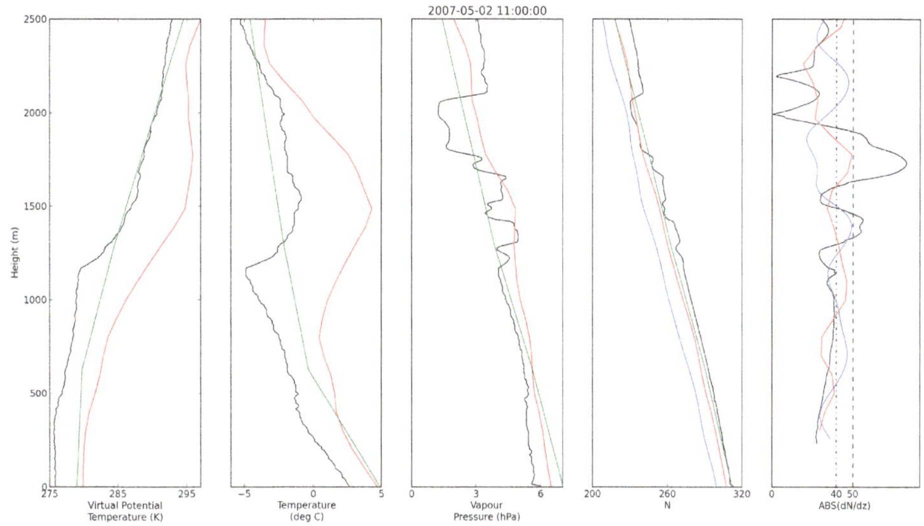

Figure 2. Profiles for 2 May 2007 at 11 Z for MAC (black), $COSMIC_{wet}$ (red), $COSMIC_{raw}$ (blue), and ECMWF (green). Vertical dashed (dash-dotted) line represents the threshold defining main (secondary) break point.

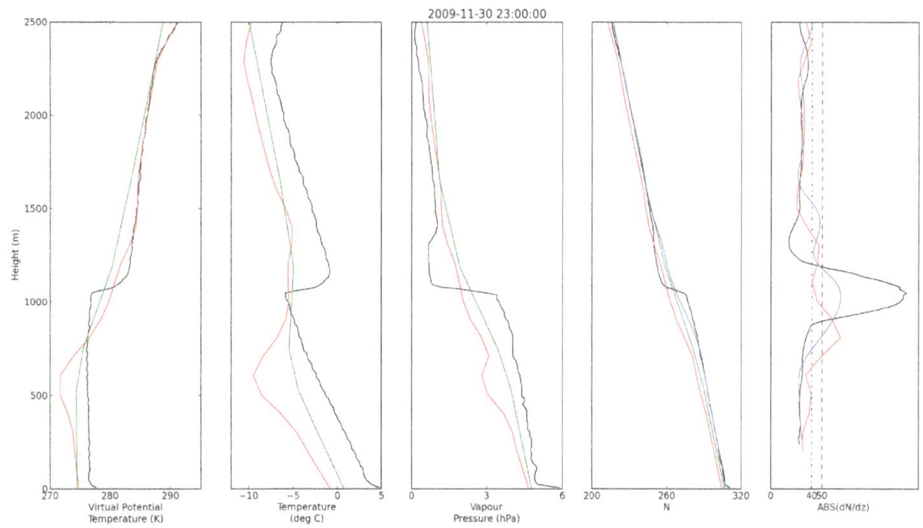

Figure 3. Profiles for 30 November 2009 at 23 Z for MAC (black), $COSMIC_{wet}$ (red), $COSMIC_{raw}$ (blue), and ECMWF (green). Vertical dashed (dash-dotted) line represents the threshold defining main (secondary) break point.

curring at the same height. The MSLP chart corresponding to Fig. 3 shows that Macquarie Island is under the influence of a high-pressure system, centred just south of Tasmania. There is a sharp drop in the relative humidity just over 1000 m, indicating a well-defined cloud top at the same height. This structure represents a typical boundary layer, not unlike those found in the sub-tropics where the break-point method on the refractivity profile has been found to work well (Guo et al., 2011). Finally, Fig. 5 shows no distinctive features, and a more-or-less stably stratified layer extending from about 500 m to above 2.5 km. This final profile is also under the influence of a high-pressure system to the north which produces westerly surface winds. In all the profiles, the black lines represent the MAC profile, the red is the $COSMIC_{wet}$ product, the green is the background ECMWF analysis profile, and the blue refractivity profile is the raw refractivity measurements from the $COSMIC_{raw}$ product.

The heights of any main and secondary break points detected in the various data sets are shown in Table 1. The height of the break points in the refractivity profile (N-bp) are shown for all data sets. However, the heights from the ECMWF data are not shown because the vertical resolution is too low for a 300 m sliding window. As a more conventional measure of any main and secondary inversions, the gradient in virtual potential temperature ($\frac{d\theta_v}{dz}$) in the MAC data set is also shown.

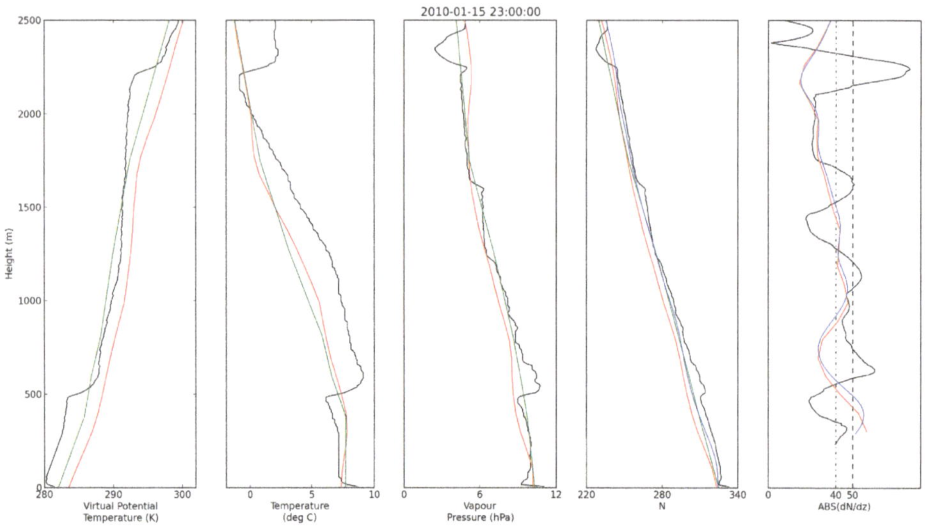

Figure 4. Profiles for 15 January 2010 at 23 Z for MAC (black), $COSMIC_{wet}$ (red), $COSMIC_{raw}$ (blue), and ECMWF (green). Vertical dashed (dash-dotted) line represents the threshold defining main (secondary) break point.

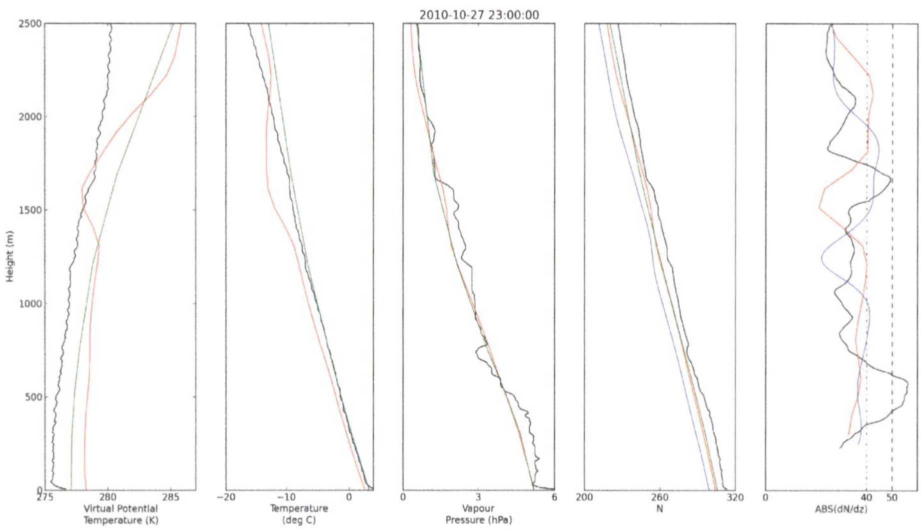

Figure 5. Profiles for 27 October 2010 at 23 Z for MAC (black), $COSMIC_{wet}$ (red), $COSMIC_{raw}$ (blue), and ECMWF (green). Vertical dashed (dash-dotted) line represents the threshold defining main (secondary) break point.

Figure 2 shows two clear inversions in virtual potential temperature in the MAC data set, however only one break point in the refractivity profile is detected. The vapour pressure increases above both temperature inversions at about 500 and 1200 m, so that the vertical change in refractivity is mostly cancelled out by the coincident increases in temperature and vapour pressure. The red profile shows that the $COSMIC_{wet}$ product has the same qualitative behaviour as MAC, however it is consistently slightly warmer and more moist. There are no break points detected here. Interestingly, the higher vertical resolution of the $COSMIC_{raw}$ product, shown as the blue refractivity profile, also fails to detect any break points. The final panel shows the absolute magnitude of the gradient in refractivity within a 300 m window. The dashed line and the dash-dotted line indicate the thresholds for the main and secondary inversion respectively. Gradients in the refractivity of the $COSMIC_{wet}$ and $COSMIC_{raw}$ products do not agree in this example.

An example of a well-mixed boundary layer is shown in Fig. 3. There is a strong temperature increase and a corresponding moisture decrease around 1100 m. This feature is identified by both the break point method, as well as the virtual potential temperature gradient method in the MAC data. Here, the break points in the COSMIC profiles show very good agreement with the MAC data set. The $COSMIC_{raw}$ product identifies the break point at the same location as the

Table 1. Main and secondary inversion heights for MAC and COSMIC using the refractivity profile break point method (N-bp) and the $\frac{d\theta_v}{dz}$ method for MAC.

	$COSMIC_{wet}$ N-bp		$COSMIC_{raw}$ N-bp		MAC N-bp		MAC $\frac{d\theta_v}{dz}$	
Profile	Main	Sec	Main	Sec	Main	Sec	Main	Sec
Figure 2	–	–	–	–	1909	–	1188	494
Figure 3	907	–	1187	–	1178	–	1069	–
Figure 4	495	–	595	–	2353	1262	499	178
Figure 5	–	–	–	–	663	520	–	–

main inversion in the MAC data. However, the gradient in the refractivity is much smaller. The $COSMIC_{wet}$ product identifies a break point about 200 m lower than the other data sets.

Another decoupled boundary layer is shown in Fig. 4. Here the strongest inversion is at 500 m, with a weaker inversion aloft at 2200 m. The higher inversion is detected by the break point method. However, since there is an increase in both moisture and temperature at 500 m, the break point method fails to detect the lower one. The $COSMIC_{wet}$ product has a break point in the refractivity profile at 495 m, nicely coinciding with the main inversion in MAC. Similarly, the $COSMIC_{raw}$ product identifies a break point at 595 m, in approximately the same location. Notice that there is a strong gradient in the refractivity profile from MAC around 600–700 m. This is not identified as a break point since the difference in the gradients between adjacent windows is not larger than those associated with the feature around 1200 m. Also notice that the refractivity gradients of the two different COSMIC products agree very well.

According to the MAC profile in Fig. 5, there are no significant features in either the temperature or vapour pressure. One would expect no break points or virtual temperature inversions to be detected, which is true for all but the MAC data set. Here there are two break points in the refractivity profile associated with some slight variability in the vapour pressure between about 300 and 700 m.

As a general observation from the 35 co-located profiles, the break point in the refractivity profile from either COSMIC products rarely aligns with a virtual potential temperature inversion from the high-resolution MAC soundings. The $COSMIC_{raw}$ product identifies marginally more break points (16 main and 9 secondary) than the $COSMIC_{wet}$ product (15 main and 7 secondary), and there appears to be no systematic difference in height between the two COSMIC data products, or the MAC data set. The common method for identifying the ABL top by identifying large gradients in virtual potential temperature would be inappropriate to use on the $COSMIC_{wet}$ data. The diagnostic variables from $COSMIC_{wet}$, such as virtual potential temperature, which are derived from a combination of the ECMWF analysis and the raw refractivity (which is in turn derived from the bending angle, which is derived from the first-order measurement of a Doppler shift) are not always consistent with in situ soundings. The $COSMIC_{wet}$ variables can sometimes be unphysical, such as having superadiabatic layers, as shown in Figs. 3 and 5.

4.2 Thermodynamics

Here we present an evaluation of the ability of the $COSMIC_{wet}$ product and the ECMWF data to reproduce the temperature and moisture profiles of the MAC soundings. The root mean squared error (RMSE) for temperature and vapour pressure, calculated using the differences between MAC and $COSMIC_{wet}$, and MAC and ECMWF at three levels in the atmosphere are shown in Table 2. In order to reduce interpolation errors of the data sets, the closest level to 500, 1500 and 2500 m were used as the three levels. The RMSE for each co-located profile and each variable is defined as

$$\mathrm{RMSE} = \sqrt{\sum_{l=1}^{3}(\mathrm{MAC}_l - \mathrm{COSMIC}_{\mathrm{wet},l})^2}, \tag{3}$$

where MAC_l is the observed variable from the MAC profile at level l, and $COSMIC_{wet,l}$ is the same for the co-located $COSMIC_{wet}$ profile. The $RMSE_{total}$ for each variable is just the root mean squared error of each co-located RMSE for that variable. That is,

$$\mathrm{RMSE}_{\mathrm{total}} = \sqrt{\sum_{p=1}^{35}\mathrm{RMSE}_{\mathrm{p}}^2}, \tag{4}$$

where $RMSE_p$ is the RMSE for each co-located profile.

The quantitative analysis shown in Table 2 indicates that overall, the $COSMIC_{wet}$ product is poorer than the ECMWF data at reproducing the MAC temperature and vapour pressure. Out of the four profiles shown earlier, only the temperature in Fig. 4 and the vapour pressure in Figs. 2 and 5 from $COSMIC_{wet}$ perform better than ECMWF. Over the 35 co-located profiles, the ECMWF data outperform $COSMIC_{wet}$ for temperature and vapour pressure, showing the lowest $RMSE_{total}$ for both these variables. This is likely due to the fact that the MAC soundings have been assimilated into the ECMWF data set used here.

Qualitatively, it can be seen that the background ECMWF profiles fail to capture any of the virtual potential tempera-

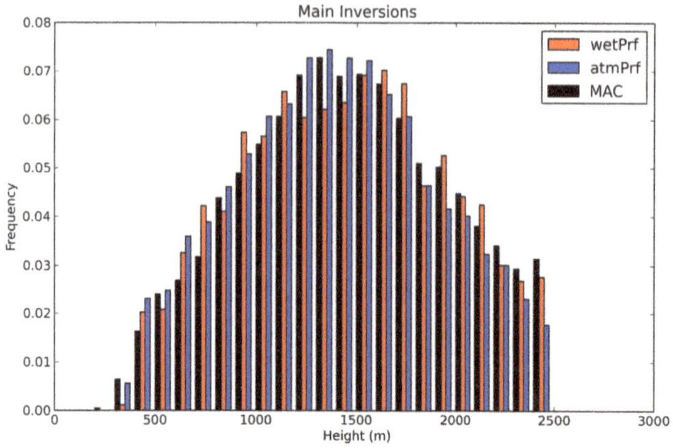

Figure 6. Main refractivity break point for MAC (black), $COSMIC_{wet}$ (red), $COSMIC_{raw}$ (blue).

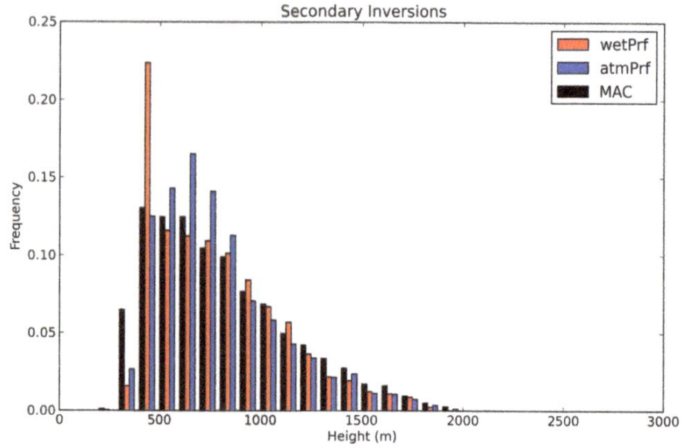

Figure 7. Secondary refractivity break point for MAC (black), $COSMIC_{wet}$ (red), $COSMIC_{raw}$ (blue).

ture inversions which would typically define the ABL top. The $COSMIC_{wet}$ product shows the same behaviour as the background ECMWF data. $COSMIC_{wet}$ does appear to have slightly more variability in most of the 35 co-located profiles, however it often produces unphysical superadiabatic layers – for example, in Figs. 3 and 5. Figure 2 is interesting in that the $COSMIC_{wet}$ profile appears to show the same features in the virtual potential temperature as the MAC data, only offset by several degrees. This raises the possibility that, given a higher resolution, and more accurate representation of either the temperature or moisture for the 1D-Var process, the performance of the $COSMIC_{wet}$ data in reproducing the thermodynamics and ABL height could be improved.

Schreiner et al. (2007) identify horizontal heterogeneity as a potential source of error in the GPS RO profiles in the neutrally charged atmosphere. In order to estimate this effect, we consider the impact of a front on the results. An approximate distance to the nearest front, based on MSLP charts, is given in Table 2. The distance to the nearest front is given in 5 degree increments in order to reflect the inherent uncertainty in the precise location of the frontal system. Figure 1 is the MSLP chart for the profile shown in Fig. 2, and shows Macquarie Island under the influence of a northerly pre-frontal air mass.

There does appear to be a relationship between the performance of $COSMIC_{wet}$ in reproducing the thermodynamics, and the structure of the ABL, as well as a weak relationship to the synoptic meteorology. Analysis of the four best and four worst performing $COSMIC_{wet}$ profiles, as judged by the RMSE, showed that the best performance of $COSMIC_{wet}$ in reproducing the thermodynamics occurred when the profile was stably stratified with no, or only weak inversions and little variability in the profile. These four profiles were typically closer to a cold front, suggesting that the influence of a front is to produce a stably stratified ABL with no strong inversions. On the other hand, the four worst performing profiles in reproducing the moisture and temperature mostly represented well-mixed ABLs of between 500 and 1200 m depth, with strong temperature inversions and a decrease in vapour pressure occurring together. Figure 3 is one of the four worst performing profiles, which shows a well-defined temperature inversion along with a decrease in moisture around 1000 m. These profiles with strong inversions tended to be further away from a front. Hande et al. (2012b) showed that reanalysis products typically have difficulty capturing strong changes in temperature or moisture near Macquarie Island. Hence, a possible explanation for this is the difficulty of the background ECMWF data to identify large gradients in temperature or moisture, and the influence this has on the $COSMIC_{wet}$ product.

5 Local statistical evaluation

To gain a broader appreciation of the performance of both the COSMIC products, we present a statistical analysis of the height and occurrence of main and secondary inversions. COSMIC RO data from a 10×10 degree box around Macquarie Island are used, in order to gain a good statistical sample size. In total, this amounts to 7768 profiles for the $COSMIC_{wet}$ product, and 7469 for the $COSMIC_{raw}$ product over the period 2006–2013. Previous studies (Huang et al., 2012a; Hande et al., 2012b) show that there are only very weak diurnal and seasonal cycles in thermodynamic and cloud properties over the Southern Ocean. Hence, any difference in the temporal distribution of the two data sets should not affect the statistics. Figures 6 and 7 show the distribution of main inversions and secondary inversions from both COSMIC products and MAC.

A main inversion was detected in 3559 of the $COSMIC_{wet}$ soundings, and a secondary inversion in 2002 cases. For the $COSMIC_{raw}$ product, a main inversion was detected in 3622 soundings, and a secondary inversion in 2210 profiles. Finally, for the MAC data set, a main inversion occurred in 8944 soundings, and a secondary inversion in 5136 sound-

Table 2. Root mean squared error for $COSMIC_{wet}$ and ECMWF data sets for the temperature (Temp) and vapour pressure (VPres). The approximate distance to a front is also shown in degrees. The letter indicates whether the front is to the west (W) or east (E) of Macquarie Island. The bottom row shows the $RMSE_{total}$ for all the 35 co-located profiles.

	$COSMIC_{wet}$ Temp (°C)	ECMWF Temp (°C)	$COSMIC_{wet}$ VPres (hPa)	ECMWF VPres (hPa)	Dist to Front (deg)
Figure 2	6.0	2.51	1.05	1.18	10–15 W
Figure 3	7.45	5.06	1.12	0.66	20+ W
Figure 4	3.15	4.11	1.63	1.05	20+ W
Figure 5	6.07	4.26	1.3	1.35	5–10 E
$RMSE_{total}$	27.37	18.5	9.89	8.56	

ings. The statistics for the main and secondary break points are shown in Table 3.

From Table 3, it can be seen that the relative frequency of the heights of the inversions are well represented in both COSMIC data sets. This can also be seen in Figs. 6 and 7, where the distribution of the inversions is very similar between the three data sets. The $COSMIC_{wet}$ product tends to have slightly higher and less frequent main and secondary break points than the $COSMIC_{raw}$ product; however, the difference between the two COSMIC products is small. There is an anomalously high frequency of secondary inversions below 500 m in the $COSMIC_{wet}$ data which is often associated with large gradients in moisture near the surface. According to Table 3, the frequency of occurrence of the two break points is significantly less in both COSMIC products compared to MAC. Therefore, the difference in vertical resolution between the two COSMIC products does not influence the frequency of main and secondary break points significantly. The statistics for the MAC data set are close to that of (Hande et al., 2012b), who use the gradient in virtual potential temperature to define the inversion heights. This implies that, statistically, the method of attributing break points in the refractivity profile to ABL interfaces is appropriate. However, there are intrinsic uncertainties in the measurement of the refractivity profile from the COSMIC products, for example the superadiabatic layers mentioned in the previous section. In terms of representing the height of the ABL, both COSMIC products offer an improvement over a high-resolution reanalysis product from the ECMWF, which was found to underestimate the height of virtual potential temperature inversions by about 200 m (Hande et al., 2012b). Note that the ECMWF product is not included here because the vertical resolution is too coarse to reliably compute break points.

We also investigated seasonal and diurnal cycles in the height of main and secondary break points; the statistics are presented in Fig. 8. It is encouraging to note that there is a weak seasonal cycle in the heights of the break points from all data sets, with slightly higher interfaces in Southern Hemisphere summer than winter. The cycle is less notable in the secondary break points. December is a notable exception for main break points in both COSMIC products, but nevertheless, the 3-month average still preserves the seasonal cycle. However, there is no clear diurnal cycle in any of the data sets.

Table 3. Statistics for the main and secondary break point in the refractivity profile for the MAC, $COSMIC_{wet}$ and $COSMIC_{raw}$ products.

	MAC	$COSMIC_{wet}$	$COSMIC_{raw}$
Main break point			
Frequency (%)	66.7	45.8	48.5
Mean height (m)	1481	1469	1430
Median height (m)	1471	1482	1413
Standard deviation (m)	521	502	517
Secondary break point			
Frequency (%)	38.3	25.7	29.6
Mean height (m)	821	809	795
Median height (m)	750	784	733
Standard deviation (m)	349	312	308

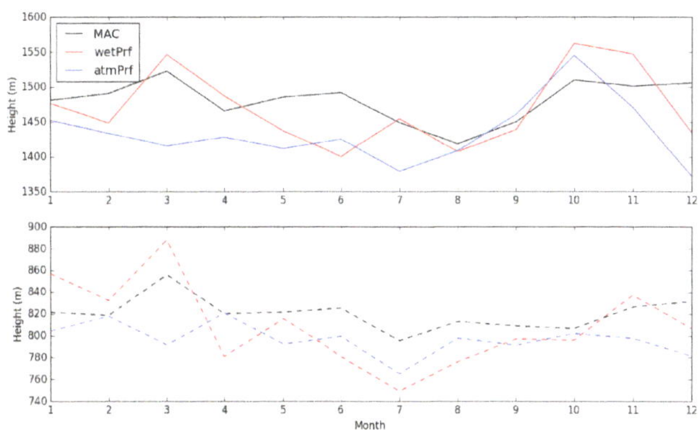

Figure 8. Seasonal cycle of main (top) and secondary (bottom) break points for MAC (black), $COSMIC_{wet}$ (red), and $COSMIC_{raw}$ (blue).

6 Conclusions

The performance of the GPS RO technique of estimating boundary layer heights over the Southern Ocean has been evaluated. A direct comparison between co-located COSMIC RO soundings and high resolution radiosonde profiles from Macquarie Island show that both COSMIC data products identify fewer boundary layer interfaces, identified as break points in the refractivity profile. The method of identifying break points in the refractivity profile as the ABL top has merit. The tests on idealized profiles indicated that this method can reliably identify increases in temperature and/or decreases in moisture, but can have difficulties when both temperature and vapour pressure increase across the interface. This type of boundary layer structure is commonplace over the Southern Ocean. The full analysis of the 35 co-located profiles shows that there are fewer break points detected in both COSMIC products, compared to the MAC data set. However, when co-located profiles from COSMIC and MAC both identify break points, the heights of the break points mostly agree.

The ability of $COSMIC_{wet}$ to reproduce the temperature and moisture profiles within the boundary layer was evaluated by comparison with the MAC soundings. It is shown that the $COSMIC_{wet}$ product does not agree with the MAC soundings as well as the background ECMWF data, even though this product is tied to the background ECMWF data. This illustrates the potential problems of deriving thermodynamic information from COSMIC data. However, the $COSMIC_{wet}$ product does show more variability in the vertical profiles, and the ECMWF data often fail to reproduce large temperature or moisture changes. This suggests that, given a more accurate and high-resolution background data set for the 1-D variational data assimilation, in may be possible to improve upon COSMIC data products.

GPS RO profiles from within a 10×10 degree box around Macquarie Island were used to perform a statistical analysis of the ability of COSMIC to reproduce the main and secondary break point heights. This indicates that, statistically, $COSMIC_{wet}$ reproduces the heights of break points in the refractivity profile well, as compared to MAC profiles. However, the frequency is much less. This is true for the raw refractivity measurements as well, which have much higher vertical resolution, indicating that the difference in the frequency of boundary layer interfaces detected is not due to differences in vertical resolution. The favourable agreement in terms of height is likely due to the break point detection algorithm using the vertical gradient of refractivity to identify break points. This essentially reverts back to the change in the bending angle, which is the fundamental COSMIC measurement. Differences between the MAC soundings and the COSMIC products should be expected due to differences in the measurement techniques between the different data sets used in this study. The different methods of averaging used to produce each data set would contribute to some of the differences in the RMSE values presented in the evaluation of the thermodynamics, as well differences in estimating the ABL height and frequency.

Finally, all data sets produce a weak seasonal cycle in the heights of the interfaces, as one would expect, which is gratifying. The ability of the COSMIC products to reproduce this ABL feature indicates there is some merit in using the break point of the refractivity profile to identify boundary layer interfaces. This analysis shows that the COSMIC data product is most useful when analysed statistically on seasonal, or longer, timescales.

Acknowledgements. The radiosonde data were obtained from the Australian Bureau of Meteorology. COSMIC data and ECMWF data were obtained from the COSMIC Data Analysis and Archive Center (CDAAC). This research was supported by Monash University, Australia. The National Center for Atmospheric Research is sponsored by the National Science Foundation.

Edited by: T. F. Hanisco

References

Adams, N.: Model prediction performance over the southern ocean and coastal region around east antarctica, Aust. Meteor. Mag, 46, 287–296, 1997.

Albrecht, B. A., Bretherton, C. S., Johnson, D., Scubert, W. H., and Frisch, A. S.: The atlantic stratocumulus transition experiment–astex, Bull. Amer. Meteorol. Soc., 76, 889–904, 1995.

Andreas, L. E.: A new sea spray generation function for wind speeds up to 32 ms^{-1}, J. Phys. Oceanogr., 28, 2175–2184, 1998.

Anthes, R. A., Ector, D., Hunt, D. C., Kuo, Y.-H., Rocken, C., Schreiner, W. S., Sokolovskiy, S. V., Syndergaard, S., Wee, T.-K., Zeng, Z., Bernhardt, P. A., Dymond, K. F., Chen, Y., Liu, H., Manning, K., Randel, W. J., Trenberth, K. E., Cucurull, L., Healy, S. B., Ho, S.-P., McCormick, C., Meehan, T. K., Thompson, D. C., and Yen, N. L.: The cosmic/formosat-3 mission: Early results, Bull. Am. Meteorol. Soc., 89, 313–333, 2008.

Ao, C., Waliser, D., Chan, S., Li, J., Tian, B., Xie, F., and Mannucci, A.: Planetary boundary layer heights from gps radio occultation refractivity and humidity profiles, J. Geophys. Res., 117, D16117, doi:10.1029/2012JD017598, 2012.

Basha, G. and Ratnam, M. V.: Identification of atmospheric boundary layer height over a tropical station using high-resolution radiosonde refractivity profiles: Comparison with gps radio occultation measurements, J. Geophys. Res.-Atmos., 114, D16101, doi:10.1029/2008JD011692, 2009.

Boers, R. and Krummel, P. B.: Microphysical properties of boundary layer clouds over the southern ocean during ace 1, J. Geophys. Res.-Atmos., 103, 16651–16663, 1998.

Bretherton, C. S., Uttal, T., Fairall, C. W., Yuter, S. E., Weller, R. A., Baumgardner, D., Comstock, K., Wood, R., and Raga, G. B.: The epic 2001 stratocumulus study, Bull. Am. Meteorol. Soc., 85, 967–978, 2004.

Chan, K. M. and Wood, R.: The seasonal cycle of planetary boundary layer depth determined using cosmic radio occultation data, J. Geophys. Res.-Atmos., 118, 12–422, 2013.

ECMWF: Ecmwf toga 2.5 degree global surface and upper air analyses, daily 1985-continuing. Research Data Archive at the National Center for Atmospheric Research, Computational and Information Systems Laboratory, available at: http://rda.ucar.edu/datasets/ds111.2/ (last access: 15 October 2013), 1990.

Guo, P., Kuo, Y. H., Sokolovskiy, S. V., and Lenschow, D. H.: Estimating atmospheric boundary layer depth using cosmic radio occulation data, J. Atmos. Sci., 68, 1703–1713, 2011.

Hande, L. B., Siems, S. T., and Manton, M. J.: Observed trends in wind speed over the southern ocean, Geophys. Res. Lett., 39, L11802, doi:10.1029/2012GL051734, 2012a.

Hande, L. B., Siems, S. T., Manton, M. J., and Belusic, D.: Observations of wind shear over the southern ocean, J. Geophys. Res., 117, D12206, doi:10.1029/2012JD017488, 2012b.

Huang, Y., Siems, T., Manton, S. J., Hande, M. B., and Haynes, M. J.: The structure of low altitude clouds over the southern ocean as seen by cloudsat, J. Climate, 25, 2535–2546, 2012a.

Huang, Y., Siems, T., Manton, S. J., Protat, M. A., and Delanoe, J.: A study on the low-altitude clouds over the southern ocean using the dardar-mask, J. Geophys. Res., 117, D18204, doi:10.1029/2012JD017800, 2012b.

Jensen, J. B., Lee, S., Krummel, P. B., Katzfey, J., and Gogoasa, D.: Precipitation in marine cumulus and stratocumulus. part i: Thermodynamic and dynamic observations of closed cell circulations and cumulus bands, Atmos. Res., 54, 117–155, doi:10.1016/S0169-8095(00)00040-5, 2000.

Jimi, S. I., Gras, J. L., Siems, S. T., and Krummel, P. B.: A short climatology of nanoparticles at the cape grim baseline air pollution station, tasmania, Environ. Chem., 4, 301–309, 1997.

Kursinski, E. R., Hajj, G. A., Schofield, J. T., Linfield, R. P., and Hardy, K. R.: Observing earths atmosphere with radio occultation measurements using global positioning system, Geophys. Res. Lett., 102, 23429–23465, 1997.

Rémillard, J., Kollias, P., Luke, E., and Wood, R.: Marine boundary layer cloud observations in the azores, J. Climate, 25, 7381–7398, 2012.

Russell, L. M., Lenschow, D. H., Laursen, K. K., Krummel, P. B., Siems, S. T., Bandy, A. R., Thornton, D. C., and Bates, T. S.: Bidirectional mixing in an ace 1 marine boundary layer overlain by a second turbulent layer, J. Geophys. Res., 111, 391–426, 1998.

Schreiner, W., Rocken, C., Sokolovskiy, S., Syndergaard, S., and Hunt, D.: Estimates of the precision of gps radio occultations from the cosmic/formosat-3 mission, Geophys. Res. Lett., 34, L04808, doi:10.1029/2006GL027557, 2007.

Seidel, D. J., Ao, C. O., and Li, K.: Estimating climatological planetary boundary layer heights from radiosonde observations: Comparison of methods and uncertainty analysis, J. Geophys. Res.-Atmos., 115, D16113, doi:10.1029/2009JD013680, 2010.

Sokolovskiy, S., Kuo, Y. H., Rocken, C., Schreiner, W. S., Hunt, D., and Anthes, R. A.: Monitoring the atmospheric boundary layer by gps radio occultation signals recorded in the open-loop mode, Geophys. Res. Lett., 33, L12813, doi:10.1029/2006GL025955, 2006.

Stull, R.: An introduction to boundary layer meteorology, Vol. 13, Springer, 1988.

Trenberth, K. E. and Fasullo, J. T.: Simulation of present-day and twenty-first-century energy budgets of the southern oceans. Journal of Climate, 23, 440–454, 2010.

Vinoth, J. and Young, I. R.: Global estimates of extreme wind speed and wind height, J. Climate, 24, 1647–1665, 2011.

Xiaojun, Y.: High-wind-speed evaluation in the southern ocean, J. Geophys. Res., 119, D13101, doi:10.1029/2003JD004179, 2004.

Young, I. R., Zieger, S., and Babanin, A. V.: Global trends in wind speed and wave height, Science, 332, 451–455, 2011.

2

Ionospheric assimilation of radio occultation and ground-based GPS data using non-stationary background model error covariance

C. Y. Lin[1], T. Matsuo[2,3], J. Y. Liu[1,4], C. H. Lin[5], H. F. Tsai[5], and E. A. Araujo-Pradere[2]

[1]Institute of Space Science, National Central University, Chungli, Taiwan
[2]Cooperative Institute for Research in Environmental Sciences, University of Colorado, Boulder, Colorado, USA
[3]Space Weather Prediction Center, National Oceanic and Atmospheric Administration, Boulder, Colorado, USA
[4]National Space Organization, HsinChu, Taiwan
[5]Department of Earth Sciences, National Cheng Kung University, Tainan, Taiwan

Correspondence to: J. Y. Liu (jyliu@jupiter.ss.ncu.edu.tw)

Abstract. Ionospheric data assimilation is a powerful approach to reconstruct the 3-D distribution of the ionospheric electron density from various types of observations. We present a data assimilation model for the ionosphere, based on the Gauss–Markov Kalman filter with the International Reference Ionosphere (IRI) as the background model, to assimilate two different types of slant total electron content (TEC) observations from ground-based GPS and space-based FORMOSAT-3/COSMIC (F3/C) radio occultation. Covariance models for the background model error and observational error play important roles in data assimilation. The objective of this study is to investigate impacts of stationary (location-independent) and non-stationary (location-dependent) classes of the background model error covariance on the quality of assimilation analyses. Location-dependent correlations are modeled using empirical orthogonal functions computed from an ensemble of the IRI outputs, while location-independent correlations are modeled using a Gaussian function. Observing system simulation experiments suggest that assimilation of slant TEC data facilitated by the location-dependent background model error covariance yields considerably higher quality assimilation analyses. Results from assimilation of real ground-based GPS and F3/C radio occultation observations over the continental United States are presented as TEC and electron density profiles. Validation with the Millstone Hill incoherent scatter radar data and comparison with the Abel inversion results are also presented. Our new ionospheric data assimilation model that employs the location-dependent background model error covariance outperforms the earlier assimilation model with the location-independent background model error covariance, and can reconstruct the 3-D ionospheric electron density distribution satisfactorily from both ground- and space-based GPS observations.

1 Introduction

The ionosphere is becoming more relevant to human society, with its reliance on modern technology, since the accuracy of navigation and quality of telecommunication is influenced by ionospheric conditions. Disruption of communications and navigation systems can have severe societal consequences. Even though the ionospheric observational techniques and the ionospheric models have gone through considerable development sustained over many decades, accurate monitoring and forecasting of the ionosphere conditions still presents obstinate challenges. Data assimilation procedures have recently come to play an important role in ionospheric research to overcome limitations of observations and models, and is an active area of research and development.

Some of the most comprehensive operational physics-based ionospheric data assimilation systems have been developed by Utah State University (e.g., Schunk et al., 2005; Scherliess et al., 2006) and the University of Southern California and Jet Propulsion Laboratory (e.g., Wang et al., 2004; Pi et al., 2009; Komjathy et al., 2010). Lee et al. (2012) recently assimilated the retrieved FORMOSAT-3/COSMIC

(F3/C) electron density profiles into a general circulation model of the thermosphere and ionosphere. Extensive variational data assimilation procedures have been developed with an empirical ionospheric model (e.g., Bust et al., 2004, 2007). Nicolls et al. (2009) applied this methodology to improve the E-region electron density retrieval from the F3/C data. Yue et al. (2011a, 2012) have assimilated F3/C and other radio occultation (RO) data into an empirical ionospheric model to obtain an improved global specification of the ionosphere.

The NOAA United States Total Electron Content (US-TEC) model (Fuller-Rowell et al., 2006) is designed to specify TEC over the continental United States (CONUS) from ground-based GPS slant TEC data. US-TEC uses the empirical ionosphere model, International Reference Ionosphere 95 (IRI95) (Bilitza, 1997), as a background model, and uses the empirical orthogonal functions (EOFs) and their coefficients (Spencer et al., 2004) to represent states in the Kalman filter (Kalman, 1960; Kalman and Bucy, 1961). The average rms (root mean square) error of US-TEC is 2.4 TEC units (Araujo-Pradere et al., 2007). The Kalman filter is one of the most commonly used approaches to ionospheric data assimilation. The Kalman filter update equation is given as

$$\boldsymbol{x}_{\mathrm{a}} = \boldsymbol{x}_{\mathrm{b}} + \mathbf{K}(\boldsymbol{y} - \mathbf{H}\boldsymbol{x}_{\mathrm{b}}), \tag{1}$$

$$\mathbf{K} = \frac{\mathbf{P}\mathbf{H}^T}{\mathbf{H}\mathbf{P}\mathbf{H}^T + \mathbf{R}}, \tag{2}$$

where $\boldsymbol{x}_{\mathrm{a}}$ is the state vector after an update, $\boldsymbol{x}_{\mathrm{b}}$ is the state vector of the background model, $\boldsymbol{y}$ is a vector of observations, $\mathbf{H}$ relates the state to the observations, $\mathbf{H}^T$ is transpose matrix of $\mathbf{H}$, $\mathbf{P}$ is the background model error covariance, $\mathbf{R}$ is the observation error covariance, and $\mathbf{K}$ is the Kalman gain (e.g., Welch and Bishop, 2000). In US-TEC, the Kalman filter state $\boldsymbol{x}_{\mathrm{b}}$ is composed of the coefficients of EOFs that characterize the principal modes of electron density vertical profiles. If Ψ_{v} and $\boldsymbol{a}_{\mathrm{v}}$ represent vertical EOFs and their coefficients, the electron density at a given altitude location v and longitude/latitude location h is expressed as $N_{\mathrm{e}}(v,h) = \sum_{m=1}^{6} \boldsymbol{a}_{\mathrm{v}_m}(h)\Psi_{\mathrm{v}_m}(v)$, where N_{e} is the grid-point electron density itself and m is an order of EOF. $\boldsymbol{x}$ in Eq. (1) is given by a vector of $\boldsymbol{a}_{\mathrm{v}}$ in US-TEC. The six EOFs are computed by using singular value decomposition; EOF1 is the dominant term of EOFs to represent the mean ionospheric profile. The higher-order EOFs gradually decrease their significance for allowing the reconstructed profile to depart from the mean (Spencer et al., 2004).

Mariner 4 first applied the RO observation technique to observe the Mars atmosphere and ionosphere in 1965 (Kliore et al., 1965). MicroLab-1 GPS/MET was launched in 1995 and applied to monitor the Earth's atmosphere and ionosphere by using the GPS RO technique (Hajj and Romans, 1998; Kursinski et al., 1997). The integrated content of ionospheric electron density, namely total electron content (TEC), has been measured between the GPS satellites and MicroLab-1. This is a way to observe the ionosphere horizontally at different altitudes and obtain the electron density profile using the Abel inversion technique. The F3/C mission was launched in April 2006, and has six micro-satellites in the different orbital planes. The GPS radio occultation experiment (GOX) is one of the satellite mission objectives, and observes the ionosphere and atmosphere vertical structure by using the RO observation technique. RO observations, particularly from F3/C, have significantly improved our capability of monitoring the global ionosphere (Cheng et al., 2006; Schreiner et al., 2007).

The overarching goal of our study is to assimilate not only the ground-based GPS slant TEC data but also F3/C RO slant TEC data into US-TEC. However, the original scheme was designed for assimilation of slant TEC between ground-based stations and GPS satellites, and is not ideal for assimilation of the RO slant TEC data. New data assimilation approaches that can better accommodate the RO data need to be developed. To this end, our study attempts to make the scheme more flexible; thus the Kalman filter equations will now be solved with respect to the grid-point electron density values (i.e., $\boldsymbol{x}$ is given by a vector of grid-point electron density itself, $N_{\mathrm{e}}(v,h)$) so that more realistic characteristics of the background model error can be incorporated. A new type of the background model error covariance is derived from an ensemble of IRI model outputs by representing location-dependent vertical and horizontal correlations using EOFs. Its performance is compared with a case where a simple stationary (i.e., location-independent) correlation is used in the Kalman filter. It should be noted that this study primarily focuses on implementation of the Kalman filter update step. The Kalman filter forecast step is outside of the scope of this study.

We use observing simulation system experiments (OSSEs), designed to mimic realistic observing systems of the ground-based GPS, and F3/C RO data to assess the impact of different types of the background model error covariance and the quality of assimilation analyses. Assimilation results are further validated with Millstone Hill incoherent scatter radar (ISR) electron density data, and compared with the F3/C GOX electron density profiles retrieved by the Abel inversion technique. Assimilation analyses are also presented in the form of TEC maps during both the daytime and nighttime. The use of F3/C RO data in data assimilation procedures is shown to improve both the vertical and horizontal structure of the ionospheric electron density distribution. The model domain extends from -140 to $-60°$ with $2°$ resolution in longitude, from 10 to $60°$ with $2°$ resolution in latitude, and from 80 to 1000 km with 15 km resolution in altitude.

2 Data error and data thinning

When ground-based GPS and F3/C RO observation data are ingested into the data assimilation model, the observational errors need to be taken into account. Data from 200 ground-based GPS stations on the CONUS and F3/C RO slant TEC are used in examples shown later. More than 2000 GPS stations exist on the ground all over the world, and can yield the high-resolution TEC observations to determine horizontal structures of the ionosphere for monitoring various ionospheric phenomena and diurnal variations (Liu et al., 1996). However, most stations do not provide the differential code bias (DCB) of GPS receivers. Kakinami et al. (2009) used the Global Ionospheric Map (GIM) developed by the Jet Propulsion Laboratory (Mannucci et al., 1998) as a reference to eliminate the instrumental biases. The differential instrumental biases of GPS receivers are estimated by the GIM minimum TEC value during 04:00 and 06:00 local time at GPS receivers, while the DCB of GPS satellite is calibrated by using the Center for Orbit Determination in Europe (CODE) values (Schaer, 1999). The ground-based GPS slant TEC in this study is treated as DCB-free TEC after these DCB calibration processes. For space-based RO, the main error sources of the absolute TEC value obtained from RO include multipath effects, DCBs from GPS and low Earth orbit (LEO) satellites, and errors in leveling of the phase to the pseudorange TEC (Yue et al., 2011b). This study uses the F3/C podTec data produced by the COSMIC Data Analysis and Archival Center (CDAAC), and assumes that the absolute TEC values are free of DCB and multipath errors because these major errors are eliminated by CDAAC. In ground-based GPS observations with elevation angles of 20° or greater, the rms errors of the multipath effect on the phase are typically less than 1 mm (Bevis et al., 1992), which is insignificant; however those on the pseudorange are often in excess of 50 cm (Mannucci et al., 1998), which cannot be ignored.

Every GPS satellite can transmit signals at dual-frequency carrier phases (L1: 1575.42 MHz; L2: 1227.60 MHz). Coarse/acquisition (C/A: 1.023 MHz) and precision codes (P1: 10.23 MHz) are coded on the L1 carrier phase, and another precision code (P2: 10.23 MHz) is coded on the L2 carrier phase. The absolute and relative TEC can be calculated from the pseudorange and carrier phase, respectively (Sardón et al., 1994; Liu et al., 1996). Since the satellite and ground-based receivers error sources influence the precision of TEC calculated from pseudorange (TEC_P), the precision of TEC_P is much lower than the TEC calculated from carrier phase (TEC_L). The leveled TEC is obtained by leveling the carrier phase TEC to the pseudorange TEC to benefit from the precision of TEC_L and the accuracy of TEC_P. The leveling error formula of each GPS observation arc is given by $\frac{\sqrt{\sum_{i=1}^{n}(TEC_{L_i} - TEC_{P_i})^2}}{n}$ (Yue et al., 2011b), where n is the number of observations in each GPS arc and i is the ith observation data point in each GPS arc, where the cycle slip has been resolved. Thus, the leveling error of ground-based GPS data is influenced by the pseudorange multipath error and some unknown errors, which is difficultly estimated. Nevertheless, the leveling error of F3/C RO and ground-based GPS data would be taken into consideration when modeling the observation error covariance.

Data thinning is required because sampling space and time frequency of the GPS technique is significantly higher than the temporal and spatial resolution of data assimilation analysis. In this study, the grid-point resolution is 2° for longitude and latitude, and 15 km for altitude. To thin the data, 1 out of 31 continuous observation data points for a given GPS arc is selected for ingestion into the assimilation model. The original data rates of ground-based GPS and F3/C RO are 30 and 1 s, respectively. After data thinning, the data rates of ground-based GPS and F3/C RO are 15.5 min and 31 s, respectively. This number for the data thinning is determined by trial and error. The standard deviation from 31 continuous samples is calculated to determine the representativeness error resulting from the fact that scales resolved by the data assimilation model are lower than what is observed.

The main sources of GPS data errors have been described in above. Errors resulting from the DCB of GPS satellites, F3/C satellites, and ground-based GPS stations have been reduced before ingestion through the various calibration processes mentioned above. The leveling TEC error and representativeness error are the two types of errors that are combined to model the observation error covariance.

3 Covariance models

We develop two types of the background model error covariance in this study (**P** in Eq. 2): location-independent covariance and location-dependent covariance, and compare their effects on the quality of assimilation analysis. Location-independent (i.e., stationary) covariance refers to a covariance whose correlation structure is homogeneous and isotropic (in other words, the vertical and horizontal correlations at every grid location are identical). A Gaussian function is used to describe the stationary correlation in examples shown later. The length scale of longitude and latitude is 3° and the length scale for altitude is 50 km. These length scales are determined by a number of length scale tests. The variance of the background model error covariance is calculated from 62 different IRI electron density distributions, the equivalent of 2 months' worth of data, obtained by using randomized values of IG (ionosphere global) index and sunspot number IRI input parameters centered at the actual value of IG index and sunspot number on the experiment date. Here, we conduct experiments with synthetic data and real observation on 5 February 2008 and 21 October 2008, respectively. The IG index affects the *Nm*F2 and sunspot number changes the *hm*F2; the IG index and sunspot number are specified by the random values sampled from a uniform distribution bounded by the given maximum and minimum value. In reality, spatial correlation of the electron density varies with location (i.e., inhomogeneous) and direction (i.e., anisotropic) in the ionosphere. Assuming the same model error correlation is likely to result in unexpected errors when assimilating observations, particularly from satellites. This study will examine how important this can be for determining ionospheric electron density distribution.

Here, the location-dependent correlation is designed by separating it into two parts: the vertical correlation and the horizontal correlation, and by using the EOFs to represent these correlations. First, we compute EOFs from the 62 different IRI ionospheric electron density distributions, N_e, by using the singular value decomposition. This is the same ensemble of IRI outputs used for calculating the variance. The electron density profiles are decomposed to three EOFs and their coefficients at each longitude and latitude location as in Eq. (3):

$$N_e^{(k)}(v) = \sum_{m=1}^{3} \boldsymbol{a}_{v_m}^{(k)} \Psi_{v_m}(v), \tag{3}$$

where $N_e^{(k)}$ is kth sample of the IRI electron density profiles ($k = 1, \ldots, 62$), v is the given altitude location, Ψ_v is a matrix that contains the EOFs for the vertical direction, and $\boldsymbol{a}_v$ is a vector of the vertical EOF coefficients. After determining the EOFs at each longitude and latitude location, the covariance of the EOF coefficients is computed from 62 sets, and then multiplied by the EOFs as in Eq. (4) to obtain the vertical covariance function $\mathbf{C}_v$:

$$\mathbf{C}_v = \Psi_v \mathrm{cov}(\boldsymbol{a}_v, \boldsymbol{a}_v^T)\Psi_v^T, \tag{4}$$

where cov denotes the covariance, and $\boldsymbol{a}_v^T$ and Ψ_v^T are the transpose of $\boldsymbol{a}_v$ and Ψ_v. The vertical correlation $\mathbf{Corr}_v$ can then be obtained after normalization and localization as in Eq. (5):

$$\mathbf{Corr}_{v_{i,j}} = \left(\frac{\mathbf{C}_{v_{i,j}}}{\sqrt{\mathbf{C}_{v_{i,j}}}\sqrt{\mathbf{C}_{v_{i,j}}}} \right) \circ \xi_{i,j} \tag{5}$$

where i and j are the element indices in the covariance matrix, ξ is the localization (i.e., tapering) function, and $\circ$ refers to an operation for the element by element multiplication. Localization is used to ameliorate effects of spurious correlations due to sampling errors. The Gaspari and Cohn function (Gaspari and Cohn, 1999) is used for the localization.

Figure 1a shows the vertical EOFs at a given longitude and latitude location. The location-independent vertical correlation (Fig. 1b) illustrates that the correlation is identical at every altitude, as calculated by using the same Gaussian function, while the location-dependent vertical correlation (Fig. 1c) shows that the vertical correlation is different at different altitudes. The vertical correlation at low altitude (below 150 km) is higher with lower-altitude locations but lower with higher-altitude locations. The vertical correlation at high altitude (above 150 km) is, on the other hand, higher with higher-altitude locations but lower with lower altitudes. This result suggests that the vertical correlation is strongly dependent on altitude locations. Since the vertical correlation is determined from the IRI empirical model outputs, rather than real observations, the vertical correlation used here is not free of errors when assimilating the real data. The length scale of localization function for altitude changes at different altitudes to reflect the vertical correlation length scale, which varies considerably with altitude.

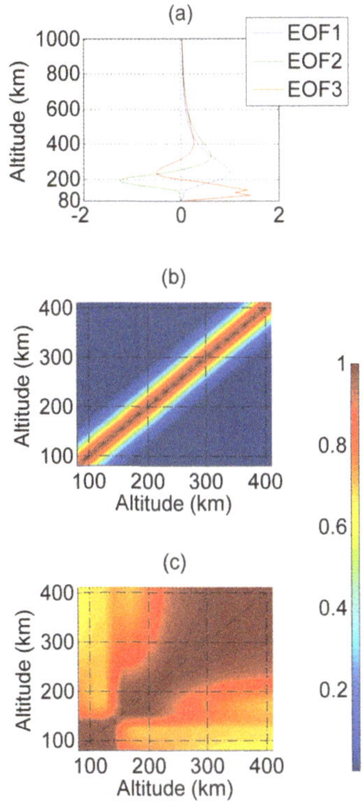

Figure 1. The vertical correlation at long −122° and lat 28°. **(a)** The EOFs calculated from 62 IRI N_e profiles with different sunspot numbers and IG indexes. **(b)** The location-independent vertical correlation modeled with a Gaussian function. **(c)** The location-dependent vertical correlation obtained from EOFs and its coefficients.

The idea of calculating the horizontal correlation, the second part of location-dependent correlation, is similar to the vertical correlation. The kth horizontal map of IRI electron density at a given altitude, v, is decomposed to three EOFs and its coefficients as shown by Eq. (6).

$$N_e^{(k)}(h) = \sum_{m=1}^{3} \boldsymbol{a}_{h_m}^{(k)} \Psi_{h_m}(h), \tag{6}$$

where h is the given horizontal (longitude and latitude) location, Ψ_h is a matrix of the EOFs for the electron density on horizontal surface, and $\boldsymbol{a}_h$ is a vector of the horizontal

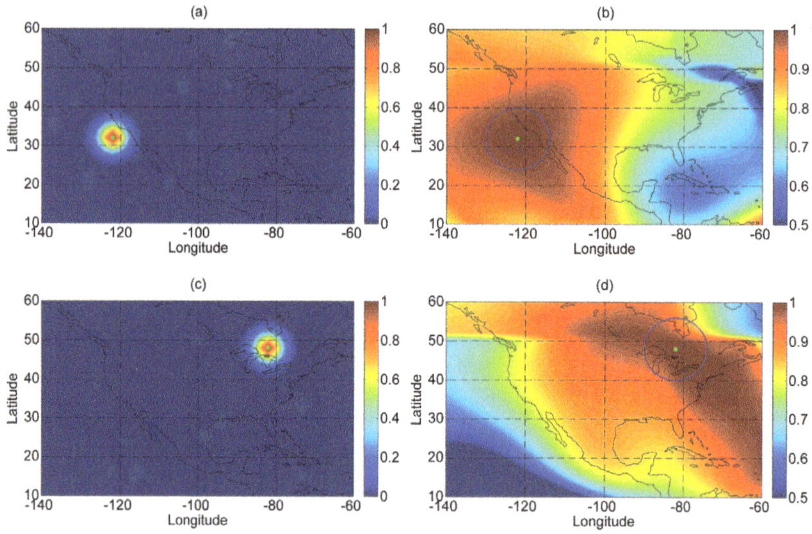

Figure 2. The horizontal correlation at 00:00 UTC. **(a)** The location-independent horizontal correlation modeled using a Gaussian function at altitude 200 km with respect to a grid point at long $-122°$ and lat $32°$ marked by the green dot. **(b)** The location-dependent horizontal correlation modeled with the horizontal EOF profiles and its EOF coefficients' covariance for the same reference point as **(a)**. The blue circle surrounding the dot is the full length of the Gaspari and Cohn localization function, with which the correlation is tapered towards zero from the center (green dot) to the rim of the blue circle. **(c)** The location-independent horizontal correlation at altitude 200 km with respect to a point at long $-82°$ and lat $48°$. **(d)** The location-dependent horizontal correlation for the same reference point as **(c)**.

EOF coefficients. After decomposing the total of 62 maps into 3 EOFs and their coefficient, the covariance of the EOF coefficient is computed and then multiplied by the EOFs to convert it to the covariance of electron density on a grid, as in Eq. (7). The horizontal covariance $\mathbf{C}_{\mathrm{h}}$ is given by

$$\mathbf{C}_{\mathrm{h}} = \Psi_{\mathrm{h}}\mathrm{cov}(\boldsymbol{a}_{\mathrm{h}}, \boldsymbol{a}_{\mathrm{h}}^{T})\Psi_{\mathrm{h}}^{T}, \tag{7}$$

where the $\boldsymbol{a}_{\mathrm{h}}^{T}$ and Ψ_{h}^{T} are the transpose matrixes of $\boldsymbol{a}_{\mathrm{h}}$ and Ψ_{h}.

The horizontal correlation is obtained after normalization and localization as in Eq. (8); this step is the same as calculating the vertical correlation.

$$\mathbf{Corr}_{\mathrm{h}_{i,j}} = \left(\frac{\mathbf{C}_{\mathrm{h}_{i,j}}}{\sqrt{\mathbf{C}_{\mathrm{h}_{i,j}}}\sqrt{\mathbf{C}_{\mathrm{h}_{i,j}}}}\right) \circ \xi_{i,j} \tag{8}$$

The length scale of localization for longitude and latitude is 3°. Figure 2a and b display the location-independent and dependent horizontal correlations at long $-122°$, lat $32°$, and Fig. 2c and d display the location-independent and dependent horizontal correlations at another location, long $-82°$, lat $48°$. As suggested by the fact that the horizontal correlation obtained from IRI empirical model outputs vary with location significantly, the location-dependent correlation information needs to be accounted for in the data assimilation scheme to successfully assimilate the real observations. As indicated by the blue circle in Fig. 2b and d, the extent of horizontal correlation is localized by the Gaspari and Cohn function to minimize the effect of spurious correlations resulting from the sampling error. Note that the correlation between the green point and surrounding locations outside of the blue circle is enforced to be zero. The location-dependent type of the background model error covariance is modeled by multiplying the location-dependent vertical correlation, location-dependent horizontal correlation, and model variance, as shown in Eq. (9).

$$\mathbf{P} = \mathbf{Var}_{i,j} \circ \mathbf{Corr}_{\mathrm{h}_{i,j}} \circ \mathbf{Corr}_{\mathrm{v}_{i,j}} \tag{9}$$

The advantage of modeling the background model error covariance in this manner is not only obtaining the variance but also the vertical and horizontal correlation of ionospheric electron density distribution varying according to its locations and geophysical conditions (e.g., day and night). The disadvantage is that it requires more computing time and memory than utilizing location-independent (i.e., stationary) correlation in the assimilation scheme. This issue can be dealt with by pre-computing the EOFs and the EOF coefficients covariance in advance and by adopting space matrix methods.

4 Assimilation experiments with synthetic data

The OSSEs, in which synthetically generated observations sampled from the IRI model output (i.e., the simulation truth) are assimilated, provide the assessment of the data assimilation (DA) procedure under idealistic conditions. A total of six different OSSEs have been conducted to examine the effects of different types of background model error covariance on the assimilation analysis quality. The simulation results are shown for the North American continent at 00:00 UTC on

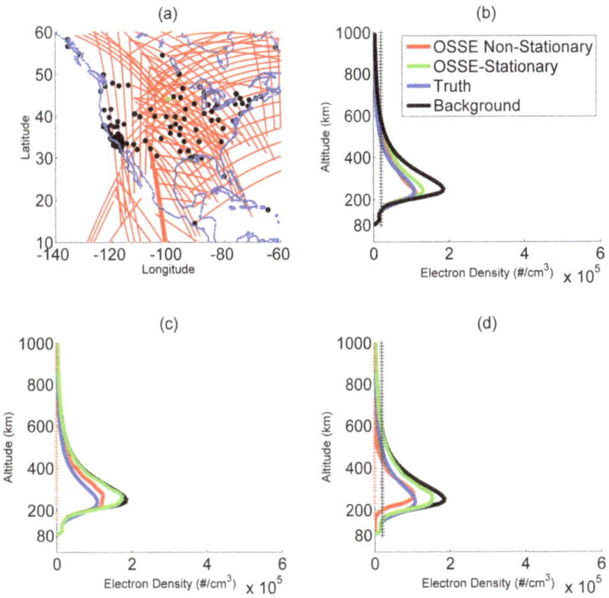

Figure 3. The OSSE result of the data assimilation model for 00:00 UTC on 5 February 2008. **(a)** The F3/C RO slant TEC paths of assimilated observations are shown as the red line, the black points are the ground-based GPS stations used in the OSSE. The data are accumulated over 1 h. The green dot indicates a location where electron density profiles are sampled for comparison and validation. **(b)** The electron density profiles are sampled from OSSE DA results obtained by assimilating the synthetic ground-based GPS data, which is integrated from the OSSE truth with different covariance models. The red line indicates the OSSE DA result with a location-dependent model error covariance (computed from EOFs), the green line is the OSSE DA result with a location-independent model error covariance (Gaussian), the blue line is the OSSE truth, and the black line is the background model. **(c)** The same as **(b)** but OSSE DA results are obtained by assimilating the synthetic F3/C data. **(d)** The same as **(b)** but OSSE DA results assimilated the synthetic data obtained from ground-based GPS and F3/C data. Note: the marks on the *y* axis indicate altitudes where slant TEC paths pass through this validation location, where red is used for F3/C and black is for ground-based GPS data.

5 February 2008 (Fig. 3). Note that the simulation truth and the background model states are different in the OSSEs presented here. The simulation truth is calculated from IRI with the lower IG index rather than the real IG index. This parameter setting makes synthetically generated observations of electron density lower than the climatological prediction by IRI in an effort to account for the tendency of IRI to overestimate the TEC during the extreme solar minimum conditions. This makes the simulation truth different from the background state. Two different kinds of slant TEC data are considered in the OSSEs: ground-based GPS data and F3/C occultation data.

Figure 3a displays the distributions of observations used in these experiments: ground-based GPS station (black point) and F3/C RO slant TEC path (red line). Six different experiment settings are as follows: (1) location-independent background model error covariance with synthetic ground-based GPS data (Fig. 3b), (2) location-independent background model error covariance with synthetic F3/C RO data (Fig. 3c), (3) location-independent background model error covariance with both ground-based and F3/C RO data (Fig. 3d), (4) location-dependent background model error covariance with synthetic ground-based GPS data (Fig. 3b), (5) location-dependent background model error covariance with synthetic F3/C RO data (Fig. 3c), and (6) location-dependent background model error covariance with both ground-based and F3/C RO data (Fig. 3d). The OSSE results are shown at long $-100°$ and at lat 44° to compare the vertical structures of assimilation analyses. Figure 3b and c compare electron density profiles from the IRI background model, the simulation truth, and data assimilation results for ground-based GPS data and F3/C RO data, respectively. The comparison suggests that assimilation results obtained using location-independent and location-dependent background model error covariance are both close to the simulation truth when assimilating ground-based GPS data. For assimilation of the F3/C RO data, on the other hand, a choice of the background model error covariance type makes a considerable difference. The electron density profile obtained using the location-independent covariance lies significantly away from the simulation truth and almost next to the background electron density. Figure 3d illustrates the OSSE electron density profile results for assimilation of both ground-based GPS data and F3/C RO data. The updated electron density profile using the location-dependent background model error covariance compares with the simulation truth far better than the one using location-independent model error covariance. The OSSE results suggest that the location-dependent background model error covariance effectively facilitates the assimilation of ground-based GPS and F3/C RO data independently, and of both simultaneously. In assimilation examples shown in the following sections, the location-dependent background model error covariance is used.

5 Assimilation results: TEC maps

We have conducted several data assimilation experiments over the CONUS with real ground-based GPS and F3/C RO slant TEC data on 21 October 2008. Assimilation results are presented as TEC, which is the column-integrated electron density from 80 to 1000 km. Results are compared to the background model TEC to examine the impact of F3/C RO data as well as ground-based GPS data. Figures 4 and 5 show TEC maps for the daytime (14:00–21:00 UTC) and nighttime (02:00–09:00 UTC) on 21 October 2008. At each hour UTC, three TEC maps are shown: a DA TEC map using

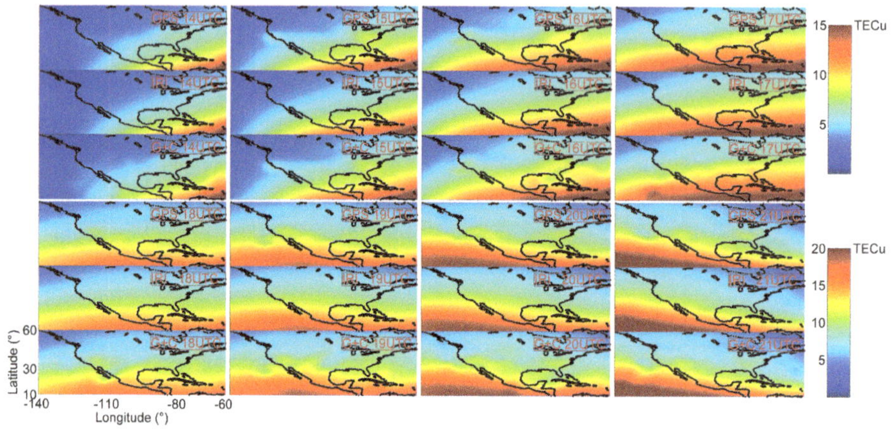

Figure 4. The DA and background model IRI TEC map of the North American Continent at daytime (14:00 to 17:00 UTC, top three rows) and 18:00 to 21:00 UTC (bottom three rows). Each time period has three different TEC maps, including a DA TEC map using ground-based GPS data, an IRI TEC map, and a DA TEC map using ground-based GPS and F3/C RO data.

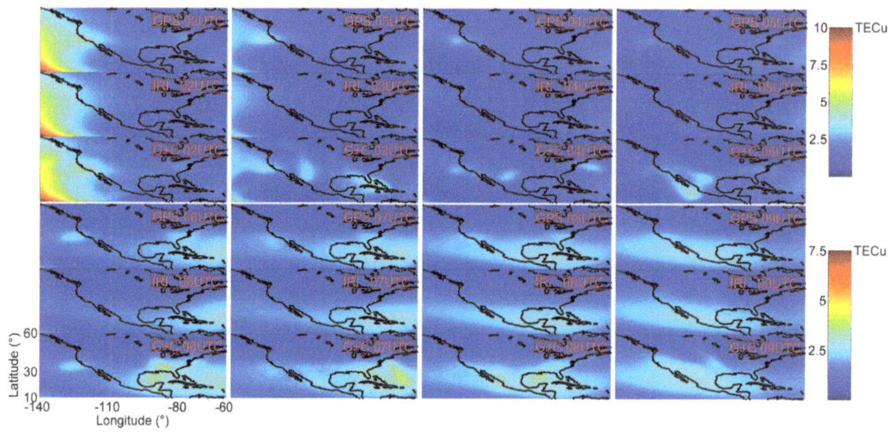

Figure 5. The DA and background model IRI TEC map of the North American Continent at nighttime (02:00 to 05:00 UTC, top three rows) and 06:00 to 09:00 UTC (bottom three rows). Each time period has three different TEC maps, including a DA TEC map using ground-based GPS data, an IRI TEC map, and a DA TEC map using ground-based GPS and F3/C RO data.

ground-based GPS data, an IRI background TEC map, and a DA TEC map using both ground-based GPS and F3/C RO data. It is clear that both types of data can modify the climatological TEC distribution predicted by the IRI successfully. Through comparison of two DA TEC maps, the influence of F3/C RO data is evident, particularly over the ocean, where ground-based GPS stations are absent. For instance, see the TEC maps at 02:00 UTC in Fig. 5, in which TEC have been updated over the ocean thanks to F3/C RO data. Our data assimilation system can effectively assimilate both space-based and ground-based GPS observations in the daytime and nighttime.

6 Validation with ISR data

The use of the location-dependent background model error covariance in the data assimilation scheme is essential when assimilating both ground-based GPS and F3/C RO data as suggested by the OSSE experiments. The accuracy of the data assimilation model must now be validated in terms of the model's ability to assimilate the real observation data. Here, the Millstone Hill incoherent scatter radar data are used to validate electron density profiles obtained from the assimilation of real ground-based GPS and F3/C RO data. Figure 6 displays the profiles at each 15 min interval between 16:00 and 17:45 UTC on 21 October 2008. Shown are the background model IRI (black), the data assimilation result (red is the DA electron density located at Millstone Hill; the green line is the DA electron density assimilating only ground-based GPS data), and the Millstone Hill data and related error bar (blue). The data assimilation results are obtained every 15 min. The displayed incoherent scatter radar profile is the median value over 15 min. There are three F3/C RO observation events over the region during this period. The alti-

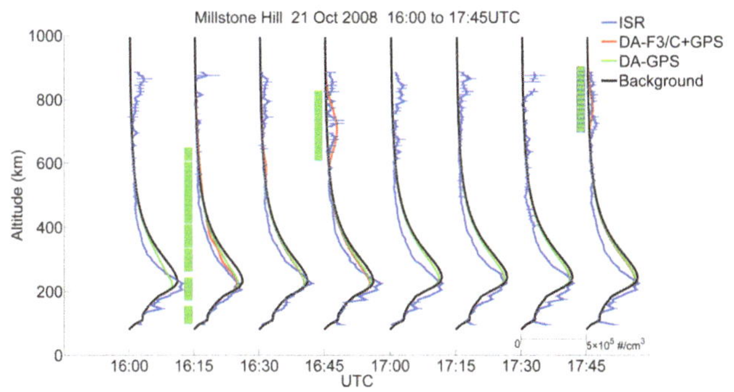

Figure 6. Comparison of data assimilation results and Millstone Hill incoherent scatter radar observations from 16:00 to 17:45 UTC, with a time period of 15 min. The red line is the DA electron density located at Millstone Hill and the green line is the DA electron density assimilating only ground-based GPS data, while the black line is the background of the DA model. The blue line is the observation and error bars of Millstone Hill incoherent scatter radar. The green boxes indicate that the altitude had F3/C RO passing through.

tude grids where F3/C RO paths intersect with the ISR line of sight are indicated by green boxes at 16:15, 16:45, and 17:45 UTC. The F3/C RO paths pass through at different altitude regions over Millstone Hill over this period; they pass through the *E* region and *F* region at 16:15 UTC, and between 500 and 800 km altitude at 16:45 and 17:45 UTC. In these three time periods the data assimilation result agrees well with ISR electron density; the rms deviation between DA results and ISR data is $4.06 \times 10^4\,\#\,cm^{-3}$. When no F3/C RO data are assimilated in addition to ground-based GPS data, the agreement of DA and ISR electron density profiles is considerably poorer; the rms deviation between DA results and ISR data is $4.72 \times 10^4\,\#\,cm^{-3}$.

The F3/C electron density profile calculated from the Abel inversion (obtained from CDAAC) can also be compared to our data assimilation analysis. The comparison is presented for 16:15 UTC on 21 October 2008, when the RO path is located close to Millstone Hill. Figure 7a displays the 3-D observing geometry of Millstone Hill ISR (blue) and F3/C RO (red), and the TEC map from the DA result is shown on the longitude–latitude plane at the bottom. Since the ISR location and F3/C RO tangent point location are different, we validated our DA result with ISR and F3/C RO electron density separately. Figure 7b compares vertical electron density profiles from the ISR and data assimilation analysis at the Millstone Hill location. The ISR error bar plotted along the ISR profile suggests high measurement uncertainty in the upper portion of the profile. The data assimilation electron density profile agrees well with ISR from altitude 200 to 400 km, where the ISR error is relatively small, including *hm*F2 and *Nm*F2 values. Figure 7c compares the Abel inversion retrieval of electron density profile (black) with the data assimilation result (red) along the RO tangent point (i.e., red line in Fig. 7a). The *hm*F2 from data assimilation results is

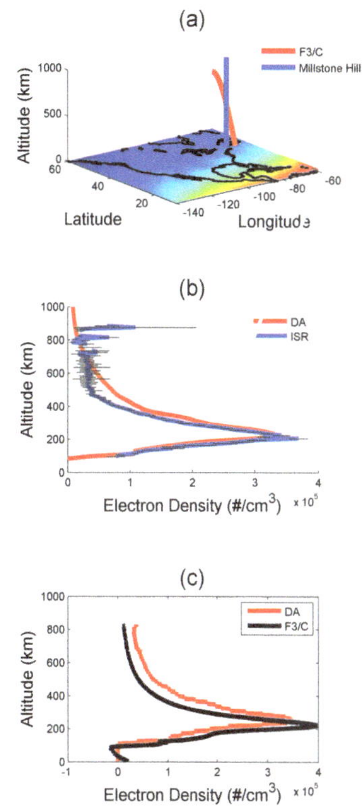

Figure 7. The comparison of the data assimilation results, Millstone Hill incoherent scatter radar observation, and F3/C electron density profile from the Abel inversion. **(a)** The location of F3/C RO tangent point electron density profile (red) and the location of Millstone Hill observation electron density profile (blue). **(b)** The electron density profile of DA (red), ISR (blue), and ISR's error bars (black) over Millstone Hill. **(c)** The electron density profile of DA (red) and F3/C (black) along the RO tangent point.

estimated to be slightly higher than the Abel inversion result, while the *Nm*F2 is estimated to be lower.

7 Discussion

In this study, we develop a new background model error covariance model to account for spatially variable correlation patterns of the model errors. Even though the location-dependent covariance model reflects more realistic correlations, the OSSE results indicate that both location-independent and location-dependent covariance work equally well for assimilation of ground-based GPS data. (See Fig. 3b, where electron density profiles obtained by using both types of the covariance models are shown to produce similar results and compare well with the simulation truth.) This result suggests that the vertical and horizontal error correlation might not have significant importance when assimilating the ground-based GPS slant TEC. Since the radio signal is transmitted from GPS satellites to ground-based receivers, the ground-based GPS slant TEC data tend to contain more information about the horizontal ionospheric structure and less about the vertical ionospheric structure. Vertical correlation in the background model error covariance is rather irrelevant in inverting the ground-based GPS slant TEC data since they are a nearly vertically integrated quantity. Moreover, the horizontal extent of ground-based GPS observations is mostly limited in the vicinity of ground-based stations, and therefore the long-range horizontal correlation does not play a significant role either.

RO slant TEC values are computed from radio signals transmitted from GPS satellites to LEO satellites that pass through the ionosphere at different altitudes over extensively large horizontal areas, and F3/C RO data can thus provide vertically resolved ionospheric information. To invert RO slant TEC data accurately, the vertical and horizontal correlation in the background model error covariance needs to be realistic. The OSSE results presented in Fig. 3c demonstrate this point, showing that electron density profiles obtained using a more realistic covariance model are closer to the simulation truth. When assimilating the F3/C RO data, it is therefore important that the background model error covariance reflects realistic location-dependent correlation structures.

The accuracy of the background model error covariance affects the quality of data assimilation analyses, especially when assimilating different kinds of GPS data. The OSSE results shown in Fig. 3d, obtained from assimilation of ground-based and RO synthetic data, illustrate this point clearly. Electron density profiles estimated by using the location-dependent error covariance are closer to the simulation truth in comparison to the cases with the location-independent error covariance. Even though the OSSE electron density profiles obtained by using the location-independent error covariance is reasonable for assimilating ground-based GPS data alone, the results are deteriorated when assimilating the same data in addition to the RO data. These detailed comparisons suggest that the choice of the background model error covariance needs to be suitable for all types of GPS data being assimilated; otherwise the accuracy of data assimilation analysis would not be optimal.

It should be noted that the current study has not considered the Kalman filter forecast step, so our approach is closer to the optimal interpolation for the moment. In the Kalman filter, the background model error covariance is expected to evolve and become more realistic over time through a recursive application of update and forecast steps in theory. In reality, it is difficult to fully incorporate nonlinear ionospheric dynamics into the Kalman filter forecast step. The model dynamics are often simplified, and the evolution of background (forecast) model error covariance is therefore approximated. The current study represents a low-dimensional modeling of the background model error covariance, which can facilitate the future implementation of the Kalman filter in the global domain. This point will be demonstrated in our future work. As part of our future work, the Taiwan Ionospheric Model (TWIM), which is constructed from F3/C RO electron density profiles (Tsai et al., 2009), can be used in the place of the IRI model in our data assimilation study to increase the accuracy of background model vertical structure.

To validate the electron density profiles, the DA results are evaluated and compared to the Millstone Hill ISR data from 16:00 to 17:45 UTC at every 15 min interval on 21 October 2008. Both ground-based GPS and F3/C observations are located closed to the ISR at three intervals: 16:15, 16:45, and 17:45 UTC. In these three time periods, the rms deviation between ISR data and DA results obtained from assimilation of RO and ground-based GPS data is $4.06 \times 10^4\,\#\,cm^{-3}$. At other time intervals, only ground-based GPS data are available. DA results agree poorly with ISR data when only ground-based GPS data are assimilated. The rms deviation between ISR data and DA results obtained from assimilation of only ground-based GPS data is $4.72 \times 10^4\,\#\,cm^{-3}$. Since the ground-based GPS data contain little information about the vertical ionospheric structure, assimilating ground-based GPS data alone will not modify the vertical structure predicted by the background model. On the other hand, the results from assimilation of both ground-based GPS and F3/C RO slant TEC agree better with the ISR electron density profile in the *F* region at (16:15 UTC) and at high altitude from 500 to 800 km (16:45 and 17:45 UTC). The validation with the ISR data supports our premise that our data assimilation procedure with the location-dependent background model error covariance effectively reconstructs the 3-D electron density from both ground-based GPS and F3/C RO data.

The tangent points of the F3/C are aligned with the Millstone Hill ISR line of sight at 16:15 UTC on 21 October 2008, when the DA results compare well with the ISR electron density profile. The comparison of our DA results with the ISR electron density profile and with the profile retrieved from the F3/C using the Abel inversion is shown in Fig. 7b and c, respectively. This comparison reflects the respective observing geometry of ISR and F3/C RO techniques, since the later one is slightly veered from vertical. The DA electron density profile agrees well with the ISR data from altitude 200 to 400 km, where the ISR error is relatively small, including *hm*F2 and *Nm*F2 values. In comparison with the Abel inversion result, the *hm*F2 is estimated to be slightly higher, while the *Nm*F2 is estimated to be lower. Below altitude 150 km, the Abel inversion yields unphysical negative electron density value. Liu et al. (2010) have shown that electron density profiles obtained by the Abel inversion are prone to a large amount of error below about 200 km. With the help of regularization through the background model error covariance in the Kalman filter, electron density profiles resulting from our data assimilation procedure are more robust than the Abel inversion results.

The ground-based GPS technique can provide valuable continual monitoring of the ionosphere, but its spatial coverage is regional. To achieve global monitoring of the ionosphere, including the oceans and remote regions where it is difficult to deploy ground-based stations, space-based observing techniques like the F3/C are crucial. TEC maps, shown in Figs. 4 and 5 for both daytime and nighttime, demonstrate this point. Because of its spatially limited observing geometry, the ground-based GPS observation can modify the background ionosphere horizontal structure only in the vicinity of station locations. On the other hand, the space-based GPS technique can provide greater spatial coverage because RO paths pass through the ionosphere rather horizontally. Data assimilation approaches, such as the one presented here, can take advantage of both observing techniques to improve the global ionospheric specification.

In addition, the TEC maps display the feature akin to the plasmaspheric tail (ionospheric plume) at 20:00 UTC (Fig. 4) (Foster et al., 2002). The similar plasmaspheric tail signature is shown in both assimilation results but is not seen in the background model. Our data assimilation procedure has the capability to reconstruct realistic transient ionospheric features, such as plasmaspheric tails, that are absent in IRI. Resulting assimilation analyses can be used for storm time studies to identify the plasmaspheric foot points in the ionosphere.

8 Conclusions

In this study, we presented a new approach to assimilate F3/C RO data as well as ground-based GPS data into the background model IRI and examined the impact of RO data on the reconstructed 3-D ionospheric election density structure. Unlike the current US-TEC model, which solves Kalman filter equations with respect to EOF coefficients, we developed a new data assimilation procedure in which the Kalman filter equations are solved in terms of grid-point electron density in order to accommodate a more flexible regularization strategy. Four main conclusions were drawn from our results.

The first conclusion is, as illustrated by the OSSE experiment results, that the location-dependent model error covariance facilitated inverting both ground-based GPS and F3/C RO slant TEC data independently, and that it is a key to successfully assimilating both data at the same time. This conclusion was drawn from demonstrating the limitation of the location-independent covariance model (based on the Gaussian function) for the assimilation of RO observational data, and developing a new covariance model whose correlation structures depend on longitude, latitude, and altitude location. The location-dependent covariance model was constructed by representing vertical and horizontal correlations with EOFs and their coefficients as shown in Figs. 1 and 2. OSSE experiments with the location-independent model error covariance showed that results were satisfactory for ground-based GPS data assimilation, but the OSSE result assimilating F3/C RO slant TEC does not agree well with the simulation truth. Therefore, the OSSE result illustrated that the location-dependent model error covariance worked well with both datasets separately, and even better assimilating both simultaneously.

Second, we validated assimilation analyses using Millstone Hill ISR data and concluded that assimilation of ground-based and RO data using our new procedure improved the agreement with ISR electron density profiles. This was true particularly when F3/C RO data are present in the vicinity of Millstone Hill, since the DA scheme is able to correct altitude structures quite well.

The third conclusion is that our scheme can produce a more robust profile than the Abel inversion thanks to regularization through the use of more realistic background model error covariance in the Kalman filter update equation. This point was demonstrated by sampling our DA results along the RO tangent points, and comparing them with F3/C RO electron density profiles calculated from the Abel inversion, as shown in Fig. 7 for the Millstone Hill location. Our DA results are generally similar to the Abel inversion results, but, importantly, our scheme was able to produce more robust profiles due to an improved regularization in inversion methods.

Our final conclusion is that the F3/C RO could improve the electron density results over greater areas, including oceans and remote regions where no ground-based stations exist. Furthermore, the F3/C RO data have the potential to improve the altitude structures of DA analysis. For the F3/C mission, 1000–2500 RO events per day (3000 at maximum) have occurred; however the current RO global coverage is not complete at a given time. This should be greatly improved when FORMOSAT-7/COSMIC-2 (F7/C2), with 12 micro-satellites, is launched in 2016 and 2018. Each micro-satellite will be equipped with the appropriate satellite signal receiver to conduct the RO experiment with GNSS and GLONASS. This new mission is expected to yield more than 14 000 RO events per day. By assimilating F7/C2 RO data into our data assimilation model, the global ionosphere specification will be significantly improved. Our new DA procedure can also be incorporated into the NOAA/SWPC ionospheric assimilation model US-TEC to increase the assimilation analysis accuracy. With the expanded availability of RO data, the electron density distribution analysis will be improved over greater regions, including areas where ground-based GPS stations are absent, and the approach presented in this paper supports such future prospects.

Acknowledgements. Chi-Yen Lin sincerely thanks Karen Fay O'Loughlin for her helpful assistance with the paper. Support for this study is provided through NASA award NNX09AJ83G to the University of Colorado at Boulder, and the Taiwan National Science Council (NSC) grant NSC 102-2628-M-008-001. The authors gratefully acknowledge the International GNSS Service (IGS) for providing GPS data and COSMIC Data Analysis and Archival Center (CDAAC) and the Taiwan Analysis Center for COSMIC (TACC) for the FORMOSAT-3/COSMIC data. Radar observations at Millstone Hill are supported by a cooperative agreement between the US National Science Foundation and the Massachusetts Institute of Technology.

Edited by: U. Foelsche

References

Araujo-Pradere, E. A., Fuller-Rowell, T. J., Spencer, P. S. J., and Minter, C. F.: Differential validation of the USTEC model, Radio Sci., 42, RS3016, doi:10.1029/2006RS003459, 2007.

Bevis, M., Businger, S., Herring, T., Rocken, C., Anthes, R., and Ware, R., GPS meteorology: Remote sensing of atmospheric water vapor using the Global Positioning System, J. Geophys. Res., 97, 15787–15801, 1992.

Bilitza, D.: International Reference Ionosphere – Status 1995/96, Adv. Space Res., 20, 1751–1754, 1997.

Bust, G. S., Garner, T. W., and Gaussiran II, T. L.: Ionospheric Data Assimilation Three-Dimensional (IDA3D): A global, multisensor, electron density specification algorithm, J. Geophys. Res., 109, A11312, doi:10.1029/2003JA010234, 2004.

Bust, G. S., Crowley, G., Garner, T. W., Gaussiran II, T. L., Meggs, R. W., Mitchell, C. N., Spencer, P. S. J., Yin, P., and Zapfe, B.: Four-dimensional GPS imaging of space weather storms, Space Weather, 5, S02003, doi:10.1029/2006SW000237, 2007.

Cheng, C. Z. F., Kuo, Y. H., Anthes, R. A., and Wu, L.: Satellite constellation monitors global and space weather, Eos T. Am. Geophys. Un., 87, 166, doi:10.1029/2006EO170003, 2006.

Foster, J. C., Coster, A. J., Erickson, P. J., Goldstein, J., and Rich, F. J.: Ionospheric signatures of plasmaspheric tails, Geophys. Res. Lett., 29, 1623, doi:10.1029/2002GL015067, 2002.

Fuller-Rowell, T., Araujo-Pradere, E., Minter, C., Codrescu, M., Spencer, P., Robertson, D., and Jacobson, A. R.: US-TEC: A new data assimilation product from the Space Environment Center characterizing the ionospheric total electron content using real-time GPS data, Radio Sci., 41, RS6003, doi:10.1029/2005RS003393, 2006.

Gaspari, G. and Cohn, S. E.: Construction of correlation functions in two and three dimensions, Q. J. Roy. Meteor. Soc., 125, 723–757, doi:10.1002/qj.49712555417, 1999.

Hajj, G. A. and Romans, L. J.: Ionospheric electron density profiles obtained with the global positioning system: results from the GPS/MET experiment, Radio Sci., 33, 175–190, doi:10.1029/97RS03183, 1998.

Kakinami, Y., Chen, C. H., Liu, J. Y., Oyama, K.-I., Yang, W. H., and Abe, S.: Empirical models of Total Electron Content based on functional fitting over Taiwan during geomagnetic quiet condition, Ann. Geophys., 27, 3321–3333, doi:10.5194/angeo-27-3321-2009, 2009.

Kalman, R. E.: A new approach to linear filtering and prediction problems, J. Basic Eng., 82, 34–45, doi:10.1115/1.3662552, 1960.

Kalman, R. E. and Bucy, R.: New results in linear filtering and prediction theory, J. Basic Eng., 83, 95–108, doi:10.1115/1.3658902, 1961.

Kliore, A. J., Cain, D. L., Levy, G. S., Eshleman, V. R., Fjeldbo, G., and Drake, E. D.: Occultation experiment: Results of the first direct measurement of Mars' atmosphere and ionosphere, Science, 149, 1243–1248, doi:10.1126/science.149.3689.1243, 1965.

Komjathy, A., Wilson, B., Pi, X., Akopian, V., Dumett, M., Iijima, B., Verkhoglyadova, O., and Mannucci, A. J., JPL/USC GAIM: On the impact of using COSMIC and ground-based GPS measurements to estimate ionospheric parameters, J. Geophys. Res., 115, A02307, doi:10.1029/2009JA014420, 2010.

Kursinski, E. R., Hajj, G. A., Schofield, J. T., Linfield, R., and Hardy, K. R.: Observing Earth's atmosphere with radio occultation measurements using the Global Positioning System, J. Geophys. Res., 102, 23429–23465, doi:10.1029/97JD01569, 1997.

Lee, I. T., Matsuo, T., Richmond, A. D., Liu, J. Y., Wang, W., Lin, C. H., Anderson, J. L., and Chen, M. Q.: Assimilation of FORMOSAT-3/COSMIC electron density profiles into a coupled thermosphere/ionosphere model using ensemble Kalman filtering, J. Geophys. Res., 117, A10318, doi:10.1029/2012JA017700, 2012.

Liu, J. Y., Tsai, H. F., and Jung, T. K., Total electron content obtained by using the global positioning system, Terr. Atmos. Ocean. Sci., 7, 107–117, 1996.

Liu, J. Y., Lin, C. Y., Lin, C. H., Tsai, H. F., Solomon, S. C., Sun, Y. Y., Lee, I. T., Schreiner, W. S., and Kuo, Y. H.: Artificial plasma cave in the low-latitude ionosphere results from the radio occultation inversion of the FORMOSAT-3/COSMIC, J. Geophys. Res., 115, A07319, doi:10.1029/2009JA015079, 2010.

Mannucci, A. J., Wilson, B. D., Yuan, D. N., Ho, C. H., Lindqwister, U. J., and Runge, T. F.: A global mapping technique for GPS-derived ionospheric total electron content measurements, Radio Sci., 33, 565–582, doi:10.1029/97RS02707, 1998.

Nicolls, M. J., Rodrigues, F. S., Bust, G. S., and Chau, J. L.: Estimating *E* region density profiles from radio occultation measurements assisted by IDA4D, J. Geophys. Res., 114, A10316, doi:10.1029/2009JA014399, 2009.

Pi, X., Mannucci, A. J., Iijima, B. A., Wilson, B. D., Komjathy, A., Runge, T. F., and Akopian, V.: Assimilative modeling of ionospheric disturbances with FORMOSAT-3/COSMIC and ground-based GPS measurements, Terr. Atmos. Ocean. Sci., 20, 273–285, doi:10.3319/TAO.2008.01.04.01(F3C), 2009.

Sardón, E., Rius, A., and Zarraoa, N.: Estimation of the transmitter and receiver differential biased and the ionospheric total electron content from Global Positioning System Observations, Radio Sci., 29, 577–586, doi:10.1029/94RS00449, 1994.

Schaer, S.: Mapping and predicting the Earth's ionosphere using the Global Positioning System, Ph.D. dissertation, Astron. Inst., Univ. of Bern, Bern, 1999.

Scherliess, L., Schunk, R. W., Sojka, J. J., Thompson, D. C., and Zhu, L.: The USU GAIM Gauss-Markov Kalman filter model of the ionosphere: Model description and validation, J. Geophys. Res., 111, A11315, doi:10.1029/2006JA011712, 2006.

Schreiner, W., Rocken, C., Sokolovskiy, S., Syndergaard, S., and Hunt, D.: Estimates of the precision of GPS radio occultations from the COSMIC/FORMOSAT-3 mission, Geophys. Res. Lett., 34, L04808, doi:10.1029/2006GL027557, 2007.

Schunk, R. W., Scherliess, L., Sojka, J. J., Thompson, D., and Zhu, L.: Ionospheric weather forecasting on the horizon, Space Weather, 3, S08007, doi:10.1029/2004SW000138, 2005.

Spencer, P. S. J., Robertson, D. S., and Mader, G. L.: Ionospheric data assimilation methods for geodetic applications, paper presented at IEEE PLANS 2004, Inst. Electr. Electron. Eng., Monterey, Calif., 26–29 April, 2004.

Tsai, L. C., Liu, C. H., Hsiao, T. Y., and Huang, J. Y.: A near real-time phenomenological model of ionospheric electron density based on GPS radio occultation data, Radio Sci., 44, RS5002, doi:10.1029/2009RS004154, 2009.

Wang, C., Hajj, G., Pi, X., Rosen, I. G., and Wilson, B.: Development of the global assimilative ionospheric model, Radio Sci., 39, RS1S06, doi:10.1029/2002RS002854, 2004.

Welch, G. and Bishop, G.: An introduction to the Kalman filter, Dept. Comput. Sci., Univ. North Carolina, Chapel Hill, Tech. Rep. TR95041, 2000.

Yue, X., Schreiner, W. S., Lin, Y. C., Rocken, C., Kuo, Y. H., and Zhao, B.: Data assimilation retrieval of electron density profiles from radio occultation measurements, J. Geophys. Res., 116, A03317, doi:10.1029/2010JA015980, 2011a.

Yue, X., Schreiner, W. S., Hunt, D. C., Rocken, C., and Kuo, Y. H.: Quantitative evaluation of the low Earth orbit satellite based slant total electron content determination, Space Weather, 9, S09001, doi:10.1029/2011SW000687, 2011b.

Yue, X., Schreiner, W. S., Kuo, Y.-H., Hunt, D. C., Wang, W., Solomon, S. C., Burns, A. G., Bilitza, D., Liu, J.-Y., Wan, W., and Wickert, J.: Global 3-D ionospheric electron density reanalysis based on multisource data assimilation, J. Geophys. Res., 117, A09325, doi:10.1029/2012JA017968, 2012.

3

A depolarisation lidar-based method for the determination of liquid-cloud microphysical properties

D. P. Donovan[1], H. Klein Baltink[1], J. S. Henzing[2], S. R. de Roode[3], and A. P. Siebesma[1,3]

[1]Royal Netherlands Meteorological Institute (KNMI), P.O. Box 201, 3730 AE, De Bilt, the Netherlands
[2]Netherlands Organisation for Applied Scientific Research (TNO), Princetonlaan 6, Utrecht, the Netherlands
[3]Technical University of Delft (TUD), Delft, the Netherlands

Correspondence to: D. P. Donovan (donovan@knmi.nl)

Abstract. The fact that polarisation lidars measure a depolarisation signal in liquid clouds due to the occurrence of multiple scattering is well known. The degree of measured depolarisation depends on the lidar characteristics (e.g. wavelength and receiver field of view) as well as the cloud macrophysical (e.g. cloud-base altitude) and microphysical (e.g. effective radius, liquid water content) properties. Efforts seeking to use depolarisation information in a quantitative manner to retrieve cloud properties have been undertaken with, arguably, limited practical success. In this work we present a retrieval procedure applicable to clouds with (quasi-)linear liquid water content (LWC) profiles and (quasi-)constant cloud-droplet number density in the cloud-base region. Thus limiting the applicability of the procedure allows us to reduce the cloud variables to two parameters (namely the derivative of the liquid water content with height and the extinction at a fixed distance above cloud base). This simplification, in turn, allows us to employ a fast and robust optimal-estimation inversion using pre-computed look-up tables produced using extensive lidar Monte Carlo (MC) multiple-scattering simulations. In this paper, we describe the theory behind the inversion procedure and successfully apply it to simulated observations based on large-eddy simulation (LES) model output. The inversion procedure is then applied to actual depolarisation lidar data corresponding to a range of cases taken from the Cabauw measurement site in the central Netherlands. The lidar results were then used to predict the corresponding cloud-base region radar reflectivities. In non-drizzling condition, it was found that the lidar inversion results can be used to predict the observed radar reflectivities with an accuracy within the radar calibration uncertainty (2–3 dBZ). This result strongly supports the accuracy of the lidar inversion results. Results of a comparison between ground-based aerosol number concentration and lidar-derived cloud-droplet number densities are also presented and discussed. The observed relationship between the two quantities is seen to be consistent with the results of previous studies based on aircraft-based in situ measurements.

1 Introduction

The fact that a linear polarisation lidar will detect a cross-polarised signal due to the occurrence of multiple-scattering in liquid water clouds has been recognised since at least 1970 (Liou and Schotland, 1971). Extensive field and laboratory observations (Sassen, 2005) of the depolarisation of laser radiation in water clouds have been made and various theoretical approaches have been developed ranging from Monte Carlo-based (MC-based) models to semi-analytic approaches; see Chaikovskaya (2008) for a review.

The penetration depth of lidars into water clouds (100–300 m) is limited to what may be considered the cloud-base region; thus limiting the region of the cloud where information can be directly retrieved. However for semi-adiabatic cloud layers, number concentration at cloud base and the rate of increase of the liquid water content (LWC) strongly constrain the structure of the cloud as a whole. The region of maximum supersaturation (above which no new cloud condensation nuclei (CCN) are activated) is typically only a few tens of centimetres to a few tens of metres above cloud base (Pinsky et al., 2012) and thus accessible, in general, to probing by lidars. Thus any microphysical information potentially provided by lidar observations will be of value for e.g. process studies involving the quantification of aerosol–cloud interactions (Lohmann and Feichter, 2005; McComiskey et al., 2009).

The general idea of using the depolarised return as a means to determine water cloud microphysical properties, such as number density, is not new and has been raised by several authors. Liou and Schotland (1971) briefly raised the possibility and presented the results of a double-scattering lidar model applied to homogeneous (i.e. constant LWC and number density) clouds. More recently, Roy et al. (1999) developed an inversion procedure based on the constrained linear inversion of a double-scattering model of the cross-polarised return applied to homogeneous clouds. Using observations and MC models which include higher-order scattering, it has also been noted that a tight correspondence exists between the layer accumulated depolarisation ratio, layer integrated backscatter (Cao et al., 2009) and multiple-scattering factor (Roy and Cao, 2010). An approach using (single field of view) depolarisation lidar has been suggested by Kim et al. (2010) who, based on MC model results, noted that for homogeneous water clouds that the depolarisation observed by a lidar with a suitably large field of view (FOV) [1] is expected to be, to a good approximation, only a function of the optical depth.

In spite of the long history and the increasing understanding of the relevant phenomenon, the use of depolarisation measurements to retrieve cloud extinction and microphysical information appears to not have seen widespread implementation. This may be due to the fact that much of the theoretical work has focused on homogeneous clouds (i.e. LWC and effective radius being constant with height) which are not necessarily suitable models of actual clouds (Sassen and Zhao, 1995). Another reason is the fact that while fast models limited to second-order scattering are well established (Roy et al., 1999), highly accurate general approaches taking into account higher-order scattering and applicable to inhomogeneous clouds are mainly limited to computationally costly MC approaches (although some exceptions may exist e.g. Chaikovskaya and Zege, 2004). Yet another, perhaps primary, reason may be the shift in attention towards using multiple FOV lidar observations (e.g. Bissonnette et al., 2005; Pounder et al., 2012; Schmidt et al., 2013) for which fast and accurate forward models that treat scattering orders above 2 have emerged in the past few years (e.g. Bissonnette et al., 2005; Hogan, 2006; Malinka and Zege, 2007). In spite of the their apparent under-utilisation, the potential advantages of using the depolarised lidar return in the context of water cloud lidar sensing have been previously noted (Roy et al., 1999; Veselovskii et al., 2006) and it should be noted that (single-view) depolarisation lidars, being of generally simpler design, are much more common than multiple-FOV systems. Thus a practical accurate depolarisation lidar water cloud microphysical inversion scheme could potentially yield a large amount of valuable data.

In this work we present a retrieval procedure using single FOV depolarisation lidars. The retrieval is based on assuming that the cloud-base region can be characterised by a quasi-linear (with height) LWC profile (i.e. constant LWC lapse rate) and constant cloud particle number density. This set of assumptions allows us to reduce the cloud variables to two parameters. In turn, this enables the development of a fast and robust inversion procedure using a look-up-table approach based on stored results from lidar MC simulations.

The outline of the remainder of this paper is as follows. In Sect. 2 we present the cloud representation (model); we employ and present and discuss the results of lidar multiple-scattering MC calculations applied to our cloud model. The lidar MC model is discussed in more detail in Appendix A. In Sect. 3 we first describe the basic inversion scheme based on the MC calculations and then describe the extension of the scheme in order to include non-ideal effects such as imperfect knowledge of lidar polarisation cross-talk. We then proceed to demonstrate the function of the inversion scheme using simulated lidar data based on large-eddy simulation (LES) cloud fields which include areas of drizzle (Sect. 3.1) and exhibit realistic (e.g. variable) cloud structure. In Sect. 4 we demonstrate the application of the inversion scheme to various case studies. The measurements in question were obtained at the Cabauw Experimental Site for Atmospheric Research (CESAR) multi-sensor atmospheric observatory in the central Netherlands (www.cesar-observatory.nl). In particular, we present evidence to support the accuracy of the inversion results by demonstrating the consistency between observed values of cloud-base region radar reflectivity compared with values of the reflectivity forward modelled using the corresponding lidar-derived cloud parameters (Sect. 4.3). In Sect. 4.4, we examine the values of the LWC produced by the lidar inversion procedure and compare them with the corresponding adiabatic values. Further, the results of a preliminary comparison between lidar-derived cloud-base droplet number densities and ground-based aerosol number density values are presented and discussed in Sect. 4.5. The paper concludes with a summary of the main points and findings.

2 Theory

2.1 Cloud model

The cloud model (i.e. representation) used in this work is a simple but still useful model of cloud-base conditions (de Roode and Los, 2008). In particular, we assume that cloud-droplet number density is constant as is the altitude derivative (or lapse rate) of the liquid water content[2] (Γ_l):

[1] Note that in this work FOV refers to the *full angle*.

[2] Lapse rates, in general, are usually defined to be the negative of derivatives of different quantities with respect to height. Note that

$$N(z) = N : z \geq z_\mathrm{b} \tag{1}$$

and

$$\mathrm{LWC}(z) = \frac{\mathrm{dLWC}}{\mathrm{d}z}(z - z_\mathrm{b}) = \Gamma_\mathrm{l}(z - z_\mathrm{b}) : z \geq z_\mathrm{b}, \tag{2}$$

where z is altitude and z_b is the cloud-base altitude. Noting that for droplets whose size is large compared to the wavelength of light involved, $\alpha = 2\pi\langle r^2\rangle$ where α is the extinction coefficient and we have

$$R_\mathrm{eff} = \frac{\langle r^3\rangle}{\langle r^2\rangle} = \frac{3}{2\rho_\mathrm{l}}\frac{\mathrm{LWC}}{\alpha}, \tag{3}$$

where ρ_l is the density of liquid water and the brackets denote averaging over the cloud particle size distribution.

If the LWC increases linearly with height above cloud base while the number density remains constant, then the cloud-droplet effective radius profile has the following form

$$R_\mathrm{eff}(z) = R_\mathrm{eff}(z_\mathrm{ref})\left(\frac{z - z_\mathrm{b}}{z_\mathrm{ref} - z_\mathrm{b}}\right)^{1/3}, \tag{4}$$

where z is the altitude and z_ref is some reference altitude The extinction coefficient profile can be found using Eqs. (2)–(4) leading to

$$\alpha(z) = \frac{3}{2\rho_\mathrm{l}}(z - z_\mathrm{ref})^{1/3}\frac{\Gamma_\mathrm{l}}{R_\mathrm{eff}(z_\mathrm{ref})}(z - z_\mathrm{b})^{2/3}. \tag{5}$$

In this work, z_ref is set, somewhat arbitrarily, to be 100 m above cloud base. Further in this paper, $R_\mathrm{eff,100}$ will be used to denote the value of the effective radius at the reference altitude (i.e. $z - z_\mathrm{b} = 100$ m).

In order to link the effective radius and liquid water content to cloud-number concentration it is necessary to specify the droplet size distribution (DSD). Here we model the size distribution of the droplets using a single-mode modified-gamma distribution (Miles et al., 2000):

$$\frac{\mathrm{d}N(r)}{\mathrm{d}r} = \frac{N_\mathrm{o}}{R_\mathrm{m}}\frac{1}{(\gamma - 1)!}\left(\frac{r}{R_\mathrm{m}}\right)^{\gamma-1}\exp\left[-r/R_\mathrm{m}\right], \tag{6}$$

where R_m is the so-called mode radius, N_o is the total number of particles in the distribution and γ is the shape parameter. For this type of distribution

$$\langle r^\mathrm{n}\rangle = \frac{(\gamma + n - 1)!}{(\gamma - 1)!}R_\mathrm{m}^\mathrm{n}, \tag{7}$$

where the brackets denote averaging over the size distribution. Thus the relationship between the effective radius (R_eff) and R_m is given by

$$R_\mathrm{eff} = \frac{\langle r^3\rangle}{\langle r^2\rangle} = R_\mathrm{m}(\gamma + 2), \tag{8}$$

in this work, for convenience, the LWC lapse rate (Γ_l) is defined to be positive.

and the LWC is given by

$$\mathrm{LWC} = N_\mathrm{o}\frac{4}{3}\pi\rho_\mathrm{l}\frac{(\gamma + 2)!}{(\gamma - 1)!}R_\mathrm{m}^3 = N_\mathrm{o}\frac{4}{3}\pi\rho_\mathrm{l}R_\mathrm{v}^3, \tag{9}$$

leading to

$$R_\mathrm{v} = \left(\frac{(\gamma + 2)!}{(\gamma - 1)!}\right)^{1/3}R_\mathrm{m}, \tag{10}$$

where R_v is the volume mean radius.

The ratio between the volume mean radius and the effective radius (k) is an important parameter for linking the cloud physical and optical properties Martin et al. (1994). From the preceding equations it can be seen that

$$k = \frac{R_\mathrm{v}^3}{R_\mathrm{eff}^3} = \frac{(\gamma + 1)\gamma}{(\gamma + 2)^2}. \tag{11}$$

Based on the results of LES modelling of stratocumulus (Lu and Seinfeld, 2006) in this work we adopt a value of k equal to 0.75 ± 0.15. Using Eq. (11) this corresponds to a range of γ values between 5 and 14 with a $k = 0.75$ corresponding to $\gamma = 9$. Once k has been specified N_o can be then be predicted from Γ_l and $R_\mathrm{eff,100}$ using Eqs. (2) and (9):

$$N_\mathrm{o} = \alpha_{100}\frac{1}{2\pi}R_\mathrm{eff,100}^{-2}\frac{1}{k}, \tag{12}$$

where α_{100} is the extinction 100 m from cloud base.

2.2 Lidar MC calculations

Lidars (like radars) are time-of-flight active measurement techniques. Pulses of laser light are transmitted into the atmosphere and the backscattered signal is detected as a function of time after each pulse has been launched. If only single scattering is considered, the relationship between the detected linearly polarised backscattered powers can be written directly as

$$P_\parallel\left(z = \frac{ct}{2}\right) = \frac{C_{l,\parallel}}{z^2}\beta_\parallel(z)\exp\left[-2\int_0^z \alpha(z')\mathrm{d}z'\right] \tag{13}$$

and

$$P_\perp\left(z = \frac{ct}{2}\right) = \frac{C_{l,\perp}}{z^2}\beta_\perp(z)\exp\left[-2\int_0^z \alpha(z')\mathrm{d}z'\right], \tag{14}$$

where z is the range from the lidar, c is the speed of light, t is the time-of-flight (so that $z = ct/2$), $\beta_\parallel$ is the range-dependent total (molecular + cloud + aerosol) parallel polarised backscatter coefficient, $\beta_\perp$ is the corresponding coefficient for the perpendicular polarisation state, $C_{l,\parallel}$ and $C_{l,\perp}$ the polarisation channel-dependent lidar instrument constants and α is the range-dependent extinction coefficient.

Table 1. Range of parameters used in the MC calculations.

Parameter	Values
Cloud base (km)	0.5, 1.0, 2.0, 4.0
FOV (mrad)	0.5, 1.0, 2.0, 4.0
$R_{\mathrm{eff},100}$ (μm)	2.0, 2.6, 3.3, 4.3, 5.6, 7.2, 9.3, 12.0
Γ_l ($\mathrm{g\,m^{-3}\,km^{-1}}$)	0.1, 0.2, 0.4, 0.6, 0.8, 1.0, 1.2, 1.4, 1.6, 1.8, 2.0

The backscatter coefficients can be further decomposed into the components corresponding to the molecular, aerosol and cloud components:

$$\beta_\parallel = \beta_{\parallel,\mathrm{m}} + \beta_{\parallel,\mathrm{a}} + \beta_{\parallel,\mathrm{c}} \quad (15)$$

and

$$\beta_\perp = \beta_{\perp,\mathrm{m}} + \beta_{\perp,\mathrm{a}} + \beta_{\perp,\mathrm{c}}, \quad (16)$$

where the m subscripts denote the molecular contribution, a denotes the aerosol contribution and c is used for the cloud contribution. If the aerosols and cloud droplets being probed are spherical then $\beta_{\perp,\mathrm{a}} = \beta_{\perp,\mathrm{c}} = 0$ and

$$\beta_\perp = \beta_{\perp,\mathrm{m}} = \delta_\mathrm{m}\beta_{\parallel,\mathrm{m}}, \quad (17)$$

where δ_m is the molecular scattering linear depolarisation ratio which mainly depends on the wavelength and spectral passband of the lidar and is on the order of 0.2–0.4 % for typical passband widths (Behrendt and Nakamura, 2002) in the UV–VIS–NIR (ultra-violet–visible–near-inferred) wavelength range. Thus under single-scattering conditions in water clouds, $\beta_\perp \ll \beta_\parallel$. However with respect to lidar cloud measurements, the multiple-scattering (MS) contribution to the signal can be many times the single-scattering contribution. The occurrence of multiple-scattering, in turn, may give rise to a perpendicularly polarised return from clouds which is many order of magnitude greater than that predicted from single-scattering theory (Sassen, 2005; Chaikovskaya, 2008).

In order to calculate the polarised lidar backscatter, the Earth Clouds and Aerosol Radiation Explorer (EarthCARE) simulator (ECSIM) lidar-specific MC forward model was used. ECSIM is a modular multi-sensor simulation framework original developed in support of the EarthCARE but is flexible enough to be applied to other instruments and platforms (Voors et al., 2007) including upward looking ground-based simulations. More information regarding the ECSIM lidar MC model is given in Appendix A.

Using our cloud model, MC runs were performed for various values of Γ_l, $R_{\mathrm{eff},100}$, different cloud-base heights and different lidar field of views. The range of parameters used is given in Table 1. Example results are shown in Figs. 1 and 2 for a lidar receiver FOV of 0.5 mrad and 2.0 mrad, respectively. The laser divergence was fixed at 0.1 mrad and the wavelength is 355 nm. The results were not found to be sensitive (above the 1–2 % level) to the laser divergence so long as the laser divergence was less than about half the receiver FOV. The MC calculations were run until the estimated error level in the calculated depolarisation ratio was below 5 % for ranges below where attenuation has reduced the normalised parallel return to a value below about 0.01, which, for a homogeneous cloud, corresponds to an apparent OD of 2.3. Beyond this point it was judged that the signal-to-noise (SNR) ratios of practical lidar measurements would be too unfavourable to be exploitable. Results are shown for both the parallel and perpendicular attenuated backscatters (ATB):

$$\mathrm{ATB}_\parallel(z) = z^2 P_\parallel(z) \quad (18)$$

and

$$\mathrm{ATB}_\perp(z) = z^2 P_\perp(z), \quad (19)$$

where $P_\parallel$ and $P_\parallel$ are the parallel and perpendicular received powers, respectively.

In this work, we fix the lidar wavelength at 355 nm (tripled Nd:YAG wavelength) since this corresponds to the wavelength of the depolarisation lidar measurements we will eventually apply the theory presented in this section to. We expect the results shown here to be indicative of the behaviour at other wavelengths for the same FOV if the $R_{\mathrm{eff},100}$ variable is rescaled by the ratio of the wavelengths and the LWC correspondingly adjusted to keep the extinction the same (see Eq. 5). This is due to the fact that cloud extinction does not vary appreciably between 355 and 1064 nm and multiple-scattering effects generally scale with the effective angular width of the forward-scattering lobe which, in turn, depends on the λ/R_eff ratio.

In Fig. 1 it can be seen that for a FOV of 0.5 mrad that the maximum depolarisation reached in the $R_\mathrm{eff}(100\,\mathrm{m}) = 2\,\mu$m cases is less than 0.2, while values of 0.4 are reached in the case with R_eff $(100\,\mathrm{m}) = 8\,\mu$m and $\Gamma_\mathrm{l} = 1\,\mathrm{g\,m^{-3}\,km^{-1}}$. In Fig. 1 the general pattern remains similar with depolarisation increasing with increasing Γ_l and effective radius but, as expected, the depolarisation ratios are correspondingly larger with the larger FOV. More example results of the MC calculations are shown in continuous form in Fig. 3. In all these examples the lidar laser divergence was modelled as being Gaussian with a $1/e$ full width of 0.1 mrad.

The MC calculations predict depolarisation profiles similar to those observed in previous investigations (e.g. Pal and Carswell, 1973). Note here that the clouds are effectively semi-infinite, that is, they have a cloud top at infinity, this leads to the prediction of a generally increasing depolarisation ratio profile with penetration into the cloud. Observations in thin water clouds often reveal that the depolarisation ratio may exhibit a peak (Sassen and Petrilla, 1986) which is associated with the penetration of the lidar signal to the cloud-top region or beyond (Sun and Li, 1989).

Figures 1–3 are informative and show that the shape of the return signals and the associated depolarisation ratio is

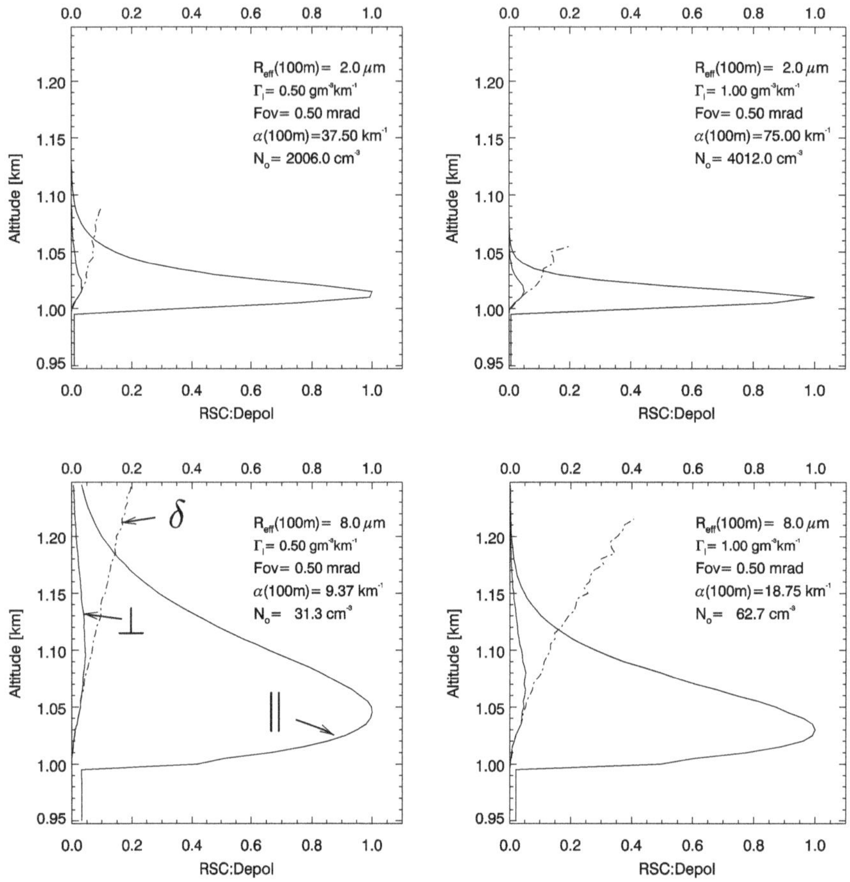

Figure 1. Example results MC calculations for a lidar wavelength of 355 nm corresponding to semi-infinite clouds with a cloud base of 1.0 km with the values of Γ_l and R_{eff} at 100 m as indicated in the top-right of each panel while $\gamma = 9$. Here the lidar FOV is 0.5 mrad. As labelled in the bottom-left panel, the right-most solid line in each plot shows the parallel (para) range-corrected signal (RCS). The other solid line shows the corresponding perpendicular (perp) RCS and the dashed-dotted line shows the depolarisation ratio. Both the para- and perp-RCS profiles have been normalised by the maximum para return.

a well-defined function of the LWC and effective radius profile. However since the extinction profile itself is a function of both the LWC and R_{eff} profiles, the variations shown in Figs. 1–3 are the result of changes in both the single-scattering return and the associated multiple-scattering contributions. Using Eq. (5) it is possible to interpolate between the MC look-up-table entries to examine how the signal and depolarisation ratio profiles behave as a function of $R_{\text{eff},100}$, while the extinction profile is held constant, thus isolating the effects of MS. Such an example is shown in Fig. 4 where the para, perp and depolarisation profiles are shown for values of $\alpha_{100} = 5\,\text{km}^{-1}$ and $10\,\text{km}^{-1}$ (the extinction coefficient at 100 m from cloud-base) as a function of $R_{\text{eff},100}$. If MS was not occurring, there would be no variation present in the para profile as $R_{\text{eff},100}$ changes and practically no perp signal would exist at all. As it is, a clear dependence on $R_{\text{eff},100}$ is present in the para and perp attenuated backscatters and in the depolarisation ratios.

A fixed value of $\gamma = 9$ was used to generate the results shown in Figs. 1–4. Other simulations (not shown) conducted with $\gamma = 2$ indicate that, for FOVs ranging from 0.5 to 2.0 mrad, the values of the para and perp signals and the associated depolarisation ratio change less than 10 % so long as $R_{\text{eff},100}$ is greater than 3 μm. For values of $R_{\text{eff},100}$ of 2 μm the depolarisation ratio profile remains the same within better than 10 %; however the shape of the normalised para and perp returns past the peak para signal altitude can change by up to 0.1 in absolute terms. This is likely not entirely due to changes in the relative MS contribution but more to do with the fact that for small effective radius values that the details of the phase function itself becomes sensitive to the width of the distribution and that even the approximate that $\alpha = 2\pi\langle r^2\rangle$ itself starts to break down.

2.3 Information content: towards an inversion scheme

Figures 1–4 strongly suggest (within the confines of our simplified model of cloud structure) the possibility that microphysical information can indeed be extracted from depolarisation lidar measurements. However it is necessary first to

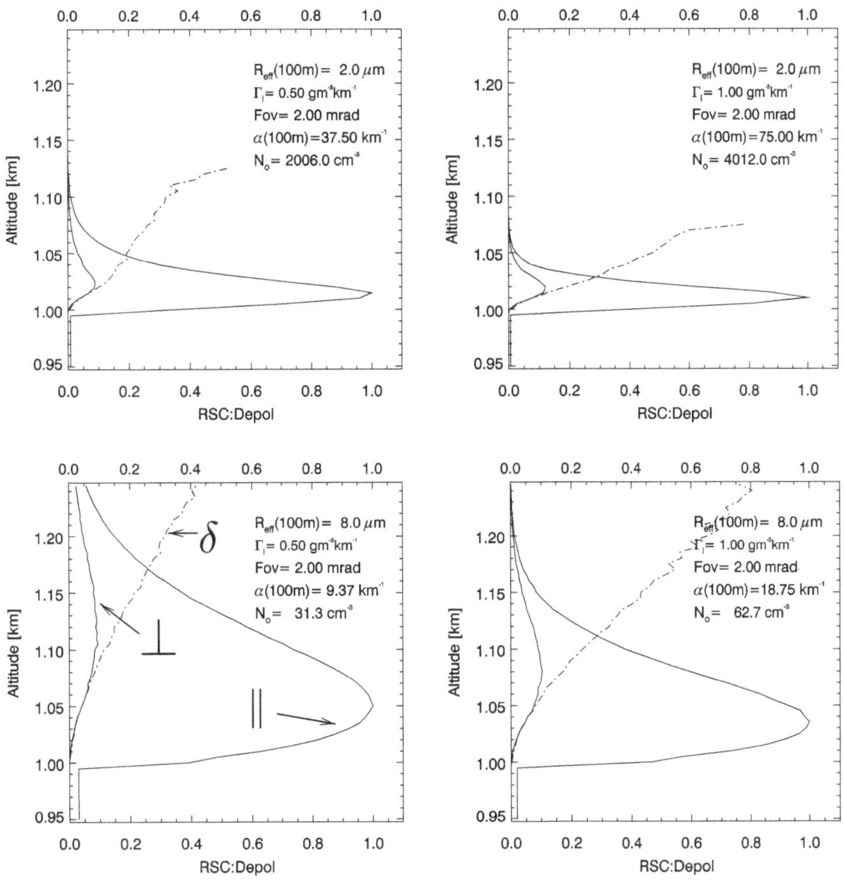

Figure 2. As Fig. 1 except for a lidar FOV of 2 mrad.

examine the degree of uniqueness of the information, i.e. how distinct are the signals corresponding to one distinct $(\alpha_{100}, R_{\mathrm{eff},100})$ pair from the set of all possible observed signals corresponding to other $(\alpha_{100}, R_{\mathrm{eff},100})$ pairs. In order to do this, here we make use of the following simple prototype *cost function* applied to our look-up-table results

$$\chi^2(\alpha_{100,j}, R_{\mathrm{eff},100,k}; \alpha_{100}, R_{\mathrm{eff},100}) = \sum_{i=i_\mathrm{b}}^{i_\mathrm{t}} \left[\left(\frac{B_\parallel(z_i;\alpha_{100}, R_{\mathrm{eff},100}) - B_\parallel(z_i;\alpha_{100,j}, R_{\mathrm{eff},100,k})}{\sigma_{B_\parallel}(z_i;\alpha_{100}, R_{\mathrm{eff},100})} \right)^2 + \left(\frac{B_\perp(z_i;\alpha_{100}, R_{\mathrm{eff},100}) - B_\perp(z_i;\alpha_{100,j}, R_{\mathrm{eff},100,k})}{\sigma_{B_\perp}(z_i;\alpha_{100}, R_{\mathrm{eff},100})} \right)^2 \right], \tag{20}$$

where i is the altitude index with i_b being the bottom and i_t the effective layer top indices. The indices j and k refer to the entries in the extinction and effective radius dimensions of the look-up tables. Keeping in mind our goal of developing a practical inversion algorithm and noting the fact that lidars are usually not well calibrated in an absolute sense, Eq. (20) makes use of the backscatters normalised by the maximum value of the parallel attenuated backscatter on a profile-by-profile basis, that is, $B_\parallel$ and $B_\perp$ where

$$B_\parallel(z) = \frac{\mathrm{ATB}_\parallel(z)}{\max(\mathrm{ATB}_\parallel(z))}, \tag{21}$$

$$B_\perp(z) = \frac{\mathrm{ATB}_\perp(z)}{\max(\mathrm{ATB}_\parallel(z))}, \tag{22}$$

and the σ terms in Eq. (20) represent the respective uncertainties, which in relative terms for actual measurements, will be in the range of a few percent above the immediate cloud bottom region and increasing to a few tens of a percent with increasing penetration into the cloud past the altitude of maximum return. Note also that there is a implicit dependence of χ^2, $B_\parallel$ and $B_\perp$ on the cloud-base altitude and the lidar FOV.

Using $B_\parallel$ and $B_\perp$ avoids the need for an absolute calibration of the lidar and accounting for the transmission between the lidar and cloud base. It is also useful to consider the altitude range of the signals to treat. Taking into account that, with actual observations, the below-cloud return will vary according to the possible presence of below-cloud drizzle and varying aerosol loads together with the finite SNR levels achievable as the lidar signal becomes attenuated as it penetrates into the cloud, we limit the altitude range to consider $(z_{i_\mathrm{b}} - z_{i_\mathrm{t}})$ according to the following criteria:

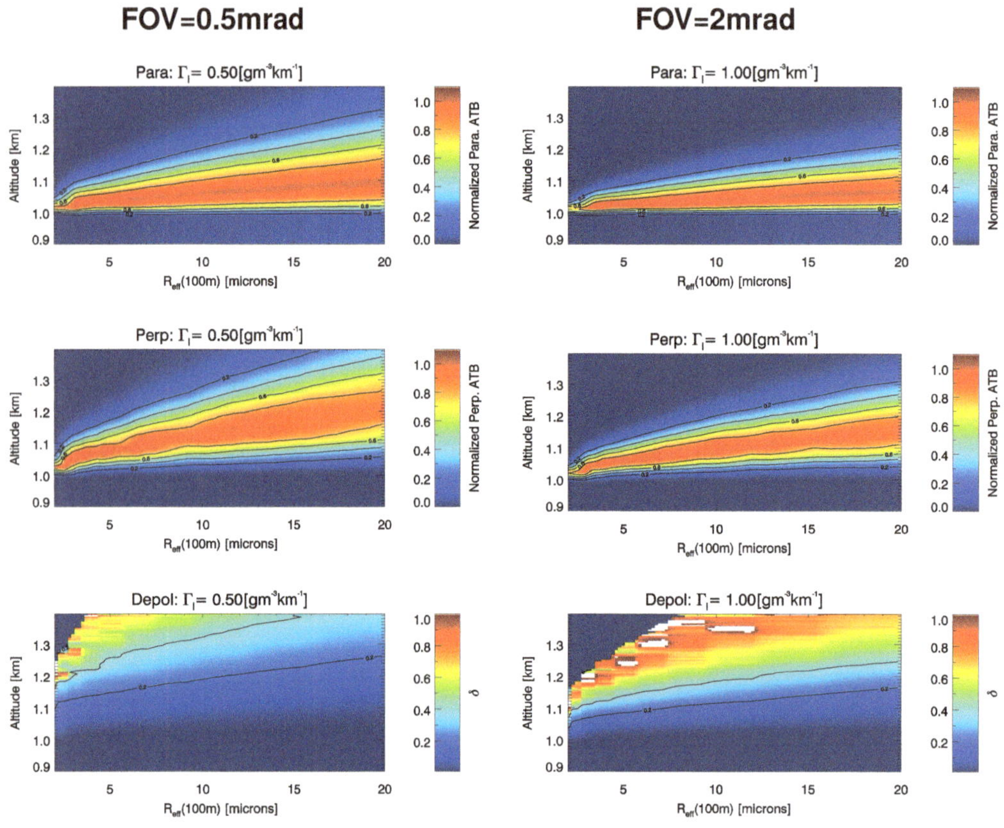

Figure 3. Example results of MC calculations for a lidar wavelength of 355 nm corresponding to semi-infinite clouds with a cloud-base of 1.0 km for two values of Γ_l as a function of $R_{\mathrm{eff},100}$ for lidar FOVs of 0.5 and 2.0 mrad. Here, for each value of $R_{\mathrm{eff},100}$, the para and perp attenuated backscatter (ATB) values have been normalised by the maximum para return.

- $z_i \geq z_b$

- $B_{\parallel}(z_i) \geq 10^{-2}$,

where here the value of the cloud base (z_b) is known precisely.

Clearly Eq. (20) achieves its global minimum at the point where the tabulated extinction and effective radius values match the specified values of α_{100} and $R_{\mathrm{eff},100}$ (i.e. $\chi^2 = 0$ at the point where $\alpha_{100,j} = \alpha_{100}$ and $R_{\mathrm{eff},100,k} = R_{\mathrm{eff},100}$). How well defined the global minimum Eq. (20) is and if other local minima exist strongly indicates the accuracy and precision we may expect in any inversion procedure based on minimising Eq. (20) or similar function. Normalised values of Eq. (20) (with $\sigma_{B_\parallel}$ and $\sigma_{B_\perp}$ proportional to $B_\parallel$ and $B_\perp$, respectively) for a lidar FOV of 1 mrad are presented in Fig. 5. Here it can be seen that, as expected, a well-defined global minimum exists where $\alpha_{100,j} = \alpha_{100}$ and $R_{\mathrm{eff},100,k} = R_{\mathrm{eff},100}$ and that the minimum is sharper for the smaller particle size cases. It can also be seen that, in spite of the unique global minimum, that the topology is complicated and that local minima exist. This indicates that in any eventual practical inversion scheme care must be taken so that the inversion scheme converges to the global minimum rather than one of the local minima. It can also be seen that the minima are less elongated along the effective radius axis for the $R_{\mathrm{eff},100} = 3\ \mu$m than the $R_{\mathrm{eff},100} = 9\ \mu$m cases. This is expected, since as the particle sizes increase, the associated forward-scattering lobe (which in the large particle limit contains 1/2 the scattered energy) will eventually become much smaller than the lidar FOV, leading to the decreased ability to distinguish between different particle sizes since practically all the forward scattered light will remain within the FOV. Results similar to those shown in Fig. 5 for a different FOV of 0.5 mrad are shown in Fig. 6. Here it can be seen that, when compared to the 1.0 mrad case, the minima associated with the $R_{\mathrm{eff}} = 9.0\ \mu$m cases are less elongated along the effective radius axis when compared to the FOV = 1.0 mrad case. This is a demonstration of the fact that smaller field of views allow more sensitivity at larger particle sizes. The reason for this is similar to the reason for the reduced sensitivity to larger particle sizes discussed above in relation to Fig. 5.

2.3.1 Effect of depolarisation calibration and FOV uncertainty

Our prototype cost function (Eq. 20) does not depend on the lidar backscatter signals being calibrated in a absolute sense;

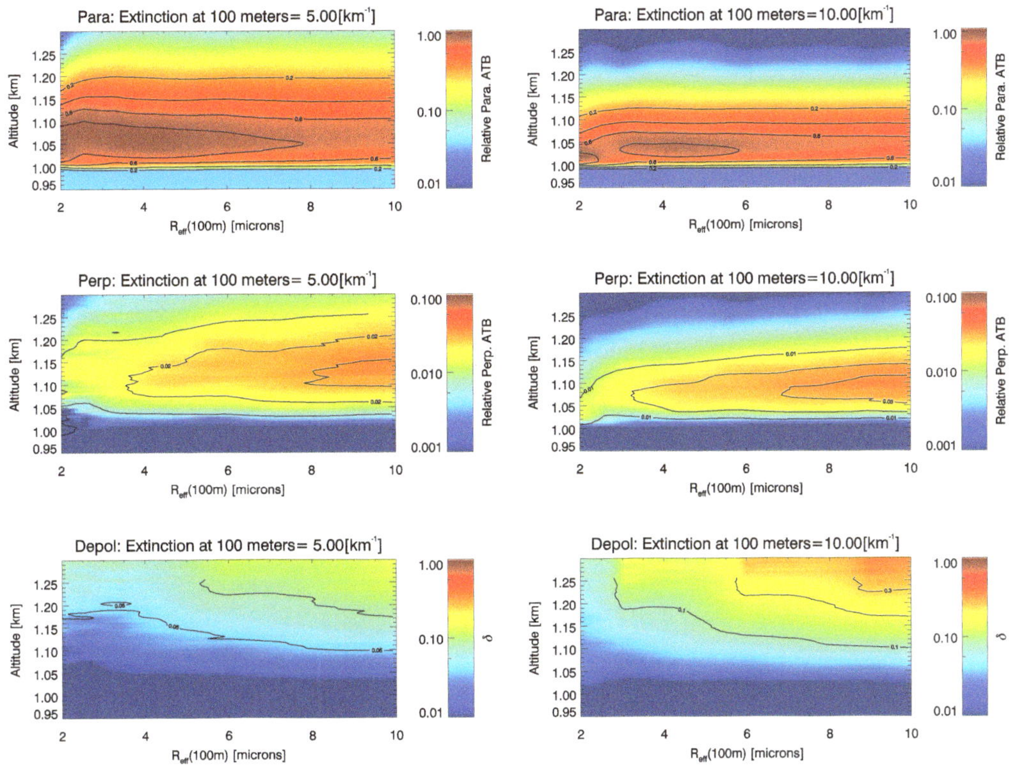

Figure 4. Example results of MC calculations for a lidar wavelength of 355 nm corresponding to semi-infinite clouds with a cloud-base of 1.0 km for two values of α_{100} as a function of $R_{\mathrm{eff},100}$ for a lidar FOV of 1.0 mrad.

however the para and perp channels must be calibrated in a relative sense. Further, for many practical depolarisation lidars, a degree of cross-talk between the two channels exist so that in practice one can write

$$\mathrm{ATB}_{\parallel} = z^2\left[\left(1-\delta^{\mathrm{C}}\right)P_{\parallel} + \delta^{\mathrm{C}}P_{\perp}\right] \tag{23}$$

and

$$\mathrm{ATB}_{\perp} = C_{\mathrm{r}}z^2\left[\left(1-\delta^{\mathrm{C}}\right)P_{\perp} + \delta^{\mathrm{C}}P_{\parallel}\right], \tag{24}$$

where δ^{C} is the polarisation cross-talk parameter and C_{r} is the inter-channel depolarisation calibration constant (Donovan and Apituley, 2013a, b).

Example results of a simulated 20 % error in the value of C_{r} are shown in Fig. 7. Here it can be seen that the location of the minimum can be shifted substantially by an error in C_{r} with the effect being generally felt more by the effective radius values. For practical lidar systems δ_{c} may be on the order of a few percent or less; thus even a 50 % error in the value of δ_{c} only produces a much smaller relative effect than a 20 % misspecification of C_{r}. Roughly speaking, we conclude that in order to be able retrieve $R_{\mathrm{eff},100}$ to within 10 % C_{r} should be known to better than 5 %, while, for typical cross-talk values, δ^{c} should be known to within about 50 %.

A similar exercise was carried out to examine the sensitivity of the results to the lidar FOV. It was found that, in general, a 15 % error in the assumed lidar FOV leads to less than a 5 % error in the extinction and/or effective radius. Since the FOV of lidar systems are generally known better than a few percent, we consider this error will generally be a secondary source of error in comparison with the errors associated with the depolarisation calibration.

3 Inversion scheme

On the basis of the results presented in the previous section we conclude that a practical inversion scheme is possible. That is, given a measurement of $B_{\parallel}$ and $B_{\perp}$, useful estimates of α_{100} and $R_{\mathrm{eff},100}$ can, in principle, be produced by finding the global minimum of Eq. (20) or similar function. However on a practical level care should be taken in the initialisation of the inversion scheme (due to the presence of multiple local minima) and errors in the depolarisation calibration (e.g. C_{r} and δ_{c}) should be taken into account. Further, since the prototype cost function uses normalised attenuated backscatters, the error in the normalisation should also be accounted for. Accordingly, our practical inversion scheme requires a more flexible functional form for our cost function.

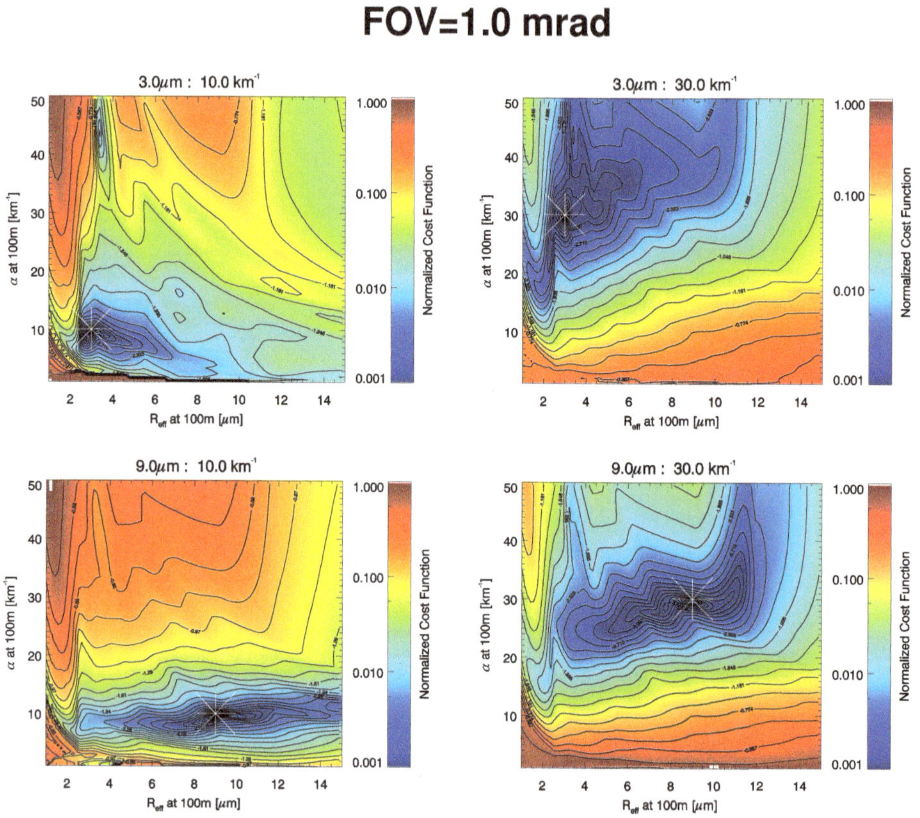

Figure 5. Normalised results from the application of Eq. (20) for a lidar wavelength of 355 nm as a function of $\alpha_{100,k}$ and $R_{\text{eff},100,\text{k}}$ for two true values of α_{100} and $R_{\text{eff},100}$ (as indicated) for a lidar FOV of 1.0 mrad and a cloud base of 1 km. The symbol is used to mark the location of the minimum of the cost function.

In particular, we will use the following optimal-estimation (Rodgers, 2000) cost function:

$$\chi^2(\boldsymbol{x};\boldsymbol{x}_\text{a},\mathbf{S}_\text{e},\mathbf{S}_\text{a}) = \left[\boldsymbol{y}-\boldsymbol{F}(\boldsymbol{x})\right]^\top \mathbf{S}_\text{e}{}^{-1}\left[\boldsymbol{y}-\boldsymbol{F}(\boldsymbol{x})\right] + \left[\boldsymbol{x}-\boldsymbol{x}_\text{a}\right]^\top \mathbf{S}_\text{a}{}^{-1}\left[\boldsymbol{x}-\boldsymbol{x}_\text{a}\right], \quad (25)$$

where $\boldsymbol{x}$ is the state vector, $\boldsymbol{y}$ is the observation vector and $\boldsymbol{F}(\boldsymbol{x})$ is the forward model estimate of the observations; $\mathbf{S}_\text{e}$ is the observational error covariance matrix, $\boldsymbol{x}_\text{a}$ is a vector containing an a priori estimate of the state vector and the a priori error covariance matrix is denoted by $\mathbf{S}_\text{a}$. As with the case with Eq. (20), the altitude limits of the summation are subject to the same conditions as listed with Eq. (20), with the additional constraint that altitudes past the maximum of the observed depolarisation profile are not considered. This is due to the fact that a sustained drop in the depolarisation profile is expected to be associated with penetration into the cloud-top region or beyond.

The observation vector ($\boldsymbol{y}$) is composed of the observed elements of $B_\parallel$ and $B_\perp$ as defined by Eqs. (21) and (22):

$$\boldsymbol{y} = [B_{\parallel,1}, B_{\parallel,2}\ldots B_{\parallel,n_z}, \ldots B_{\perp,1}\ldots B_{\perp,n_z}]. \quad (26)$$

The state vector ($\boldsymbol{x}$) is defined as

$$\boldsymbol{x} = \left[\ln(C_\text{N}), \Delta z \sin(\phi_{z_\text{p}}), \ln(C_\text{r}), \ln(\delta_\text{c}), \ln(R_{\text{eff},100}), \ln(\alpha_{100})\right], \quad (27)$$

where C_N is a factor introduced to account for any error in the signal normalisation process, Δz is the range resolution of the observations and $\sin(\phi_{z_\text{p}})$ is a factor (constrained by the use of the sine function to be between -1 and 1) used to account for the uncertainty in assigning the altitude of the peak return (see **Step 1** in Sect. 3.1.1) due to the finite vertical resolution of the measurements.

The forward model vector ($\boldsymbol{F}(\boldsymbol{x})$) is defined as

$$\boldsymbol{F}(\boldsymbol{x}) = \frac{C_\text{N}}{\max(\text{ATB}_\parallel)} \left[\text{ATB}_{\parallel,1}\ldots\text{ATB}_{\parallel,nz}, \text{ATB}_{\perp,1}\ldots\text{ATB}_{\perp,n_z}\right], \quad (28)$$

where $\text{ATB}_\parallel$ and $\text{ATB}_\perp$ given by Eqs. (23) and (24), respectively, with $P_\parallel$ and $P_\perp$ determined by interpolation using the pre-computed look-up tables. Before interpolation in α_{100} and $R_{\text{eff},100}$, the profiles are shifted in altitude by an distance given by $\Delta z \sin(\phi_{z_\text{p}})$ and then binned to a vertical resolution matching the observations. The look-up tables have been computed at a resolution of 5 m while, in this work, the observations we will consider are at a resolution of 15 m.

The elements of the error covariance matrix for the observations (S_e) can be found by calculating the expectation

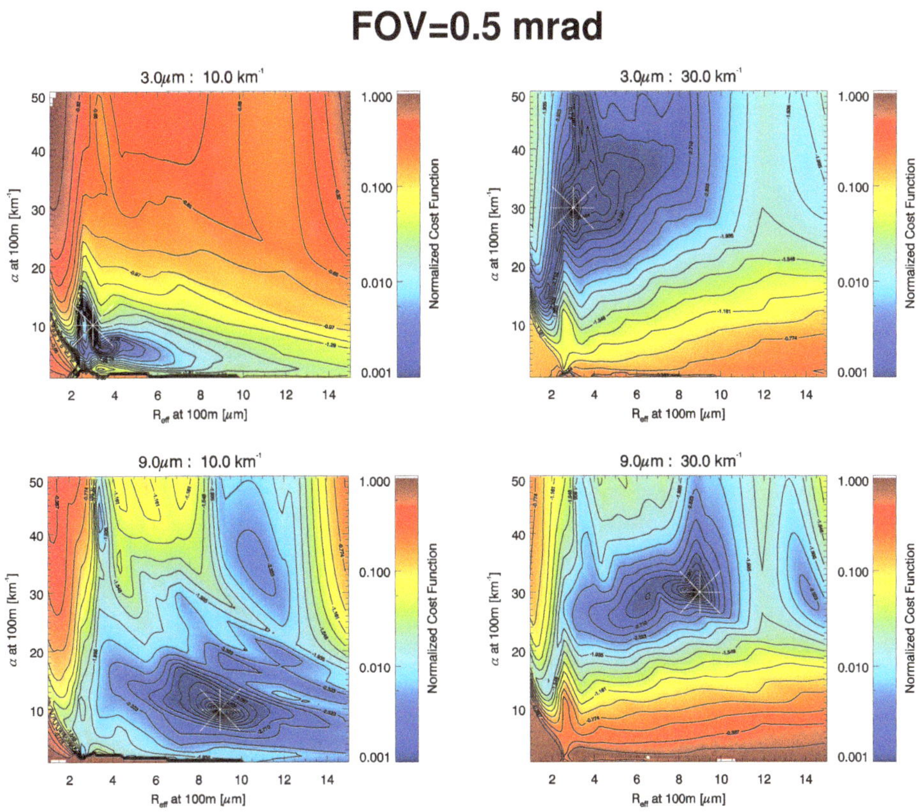

Figure 6. As Fig. 5 except for a receiver FOV of 0.5 mrad.

value of the difference between the observations and the optimal forward model fits:

$$S_{\mathrm{e},i,j} = E\left[(y_i - F_i)(y_j - F_j)\right]. \tag{29}$$

Accordingly, for simplicity if we ignore the correlation in the para and perp signals due to δ^{C}, it can be shown that

$$\begin{aligned} S_{\mathrm{e},i,j} &= \sigma_{C_\mathrm{N}}^2 y_i y_j : 1 \le i \le n_z, 1 \le j \le n_z, i \ne j \\ &= \sigma_{y_i}^2 + \sigma_{C_\mathrm{N}}^2 y_i^2 : 1 \le i \le n_z, i = j \\ &= \left(\sigma_{C_\mathrm{r}}^2 + \sigma_{C_\mathrm{N}}^2\right) y_i y_j : n_z < i \le 2n_z, n_z < j \le 2n_z, i \ne j \\ &= \sigma_{y_i}^2 + \left(\sigma_{C_\mathrm{r}}^2 + \sigma_{C_\mathrm{N}}^2\right) y_i^2 : n_z < i \le 2n_z, i = j, \end{aligned} \tag{30}$$

where $\sigma_{y_i}^2$ is the variance assigned to y_i which is estimated by averaging the observations themselves in time as a function of altitude and $\sigma_{C_\mathrm{N}}^2$ is the estimated variance of C_N which is similarly estimated from the observations. σ_{C_r} is the a priori uncertainty in the depolarisation inter-channel calibration factor.

In our procedure, we assign a priori estimates to the depolarisation calibration parameters (C_r and δ^{C}) and the normalisation factor C_N, all other factors are unconstrained by any explicit a priori. Thus non-zero elements of the inverse of the a priori error covariance matrix are given by

$$\begin{aligned} S_{\mathrm{a},1,1}^{-1} &= \left(\frac{\sigma_{C_\mathrm{N}}}{C_\mathrm{N}}\right)^{-2} \\ S_{\mathrm{a},3,3}^{-1} &= \left(\frac{\sigma_{C_\mathrm{r}}}{C_\mathrm{r}}\right)^{-2} \\ S_{\mathrm{a},4,4}^{-1} &= \left(\frac{\sigma_{\delta^\mathrm{C}}}{\delta^\mathrm{C}}\right)^{-2}, \end{aligned} \tag{31}$$

where we have assumed that the a priori estimates are all uncorrelated. Here $\sigma_{\delta^\mathrm{C}}$ is the assumed a priori uncertainty in the depolarisation cross-talk factor. The $S_{\mathrm{a},2,2}$ term is zero since no a priori knowledge is assumed for the ϕ_{z_p} term in the state vector; however the term $\sin(\phi_{z_\mathrm{p}})$ is still constrained by its very nature to be between -1 and $+1$.

Once the cost function is minimised, the retrieved values of α_{100} and $R_{\mathrm{eff},100}$ can be used along with Eq. (5) to find Γ_l, while N_o can be found via Eq. (12). The covariance matrix of the retrieved parameters (C_N, C_r, δ^C, α_{100}, $R_{\mathrm{eff},100}$) are found using standard approaches (e.g. Press et al., 2007), and standard error propagation techniques are then used to find the resulting error estimates for Γ_l and N_o including the effects of the uncertainty in k.

The form of the cost function and state vector presented here was found to be lead to rapid and reliable convergence, but should not be regarded as definitive. The reader should

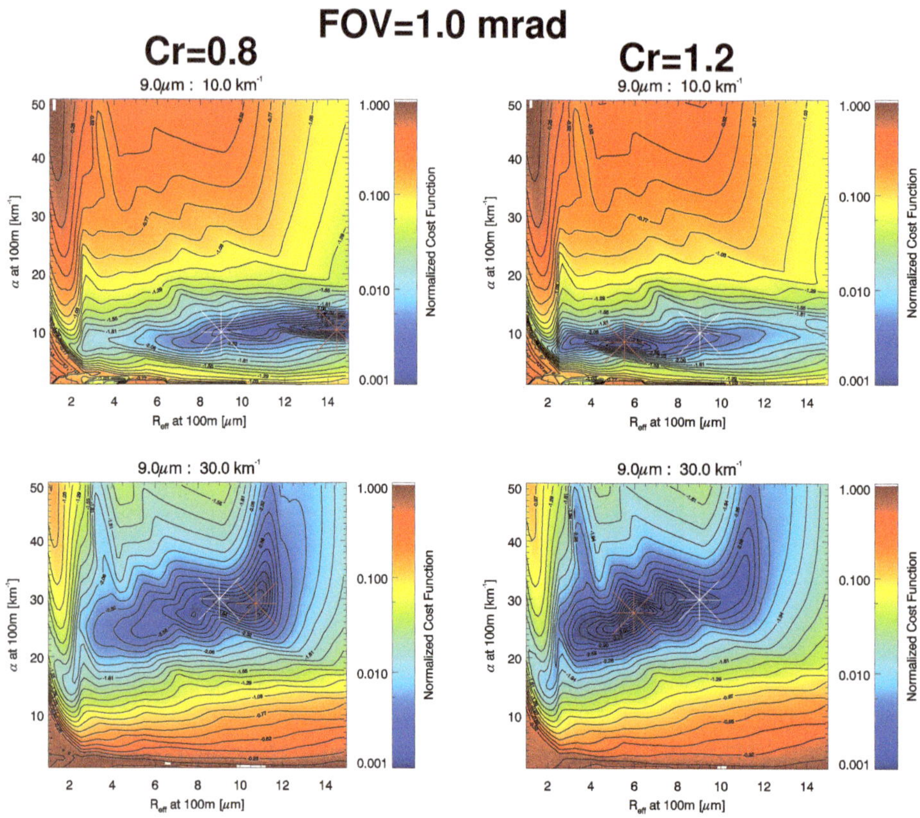

Figure 7. Normalised results from the application of Eq. (20) as a function of $\alpha_{100,k}$ and $R_{\mathrm{eff},100,\mathrm{k}}$ for true values of $\alpha_{100} = 10$ and $30\,\mathrm{km}^{-1}$ and true values of $R_{\mathrm{eff},100} = 9\,\mu\mathrm{m}$ with perturbed values of C_{r}. For the left panels C_{r} has been set 20 % too low while in the right panels C_{r} has been set 20 % too high. The white symbols show the location of the true values of $R_{\mathrm{eff},100}$ and α_{100} while the red symbols mark the position of the actual cost-function minimum in each case.

be aware that other strategies may be more appropriate, depending on the SNR of the observations and the availability (or lack thereof) of useful a priori information. For example, N_{o} and Γ_{l} could be used instead of α_{100} and $R_{\mathrm{eff},100}$. This would enable a priori estimates of both N_{o} and Γ_{l} to be taken into account as well as physical constraints, such as that the gradient of LWC, should not be steeper than adiabatic. In our formulation, however it was found not to be necessary to include a priori constrains on any state variables beyond C_{r}, δ_{c} and C_{N}.

3.1 Simulations: application to LES data fields

In order to further develop and test the inversion procedure in a manner which includes the effects of realistic cloud structure, end-to-end simulations were conduced based on results from LES model runs. In particular, output from the Dutch Atmospheric LES model (DALES) (Heus et al., 2010) was used. DALES uses a bulk scheme for precipitating liquid-phase clouds. Condensed water is separated into cloud water and precipitation. Cloud-droplet number density is a prescribed parameter, while a two-moment bulk scheme is used to treat precipitation (Khairoutdinov and Kogan, 2000). Temperature, pressure, non-precipitable cloud water, precipitation water content and precipitation droplet number density extracted from DALES snapshots were used to create ECSIM scenes. ECSIM requires the explicit specification of the cloud (DSDs). The bulk scheme used in DALES does not provide explicit DSDs; thus in order to build an ECSIM scene, it was necessary to impose DSDs based on the LES output fields. For the precipitation mode droplets the size distribution function embedded in the scheme of Khairoutdinov and Kogan (2000) was used. For the cloud droplets, modified-gamma distributions (Eq. 6) with a fixed value of γ were assumed. Using the LES cloud LWC along with an imposed value of N_{o} (which could be different from that assumed internally in the LES model which in this case was $100\,\mathrm{cm}^{-3}$) together with the assumed functional form of the size distribution then allows the DSDs to be fully defined.

Once a scene was created, the ECSIM lidar and radar forward models were applied to generate time series of simulated observations. The ECSIM lidar and radar forward models both simulate the effects of the respective virtual instrument footprints, sampling rate and instrument noise levels (for more information see the ECSIM Models and Algorithm Document, Donovan et al., 2010). An example of a DALES-

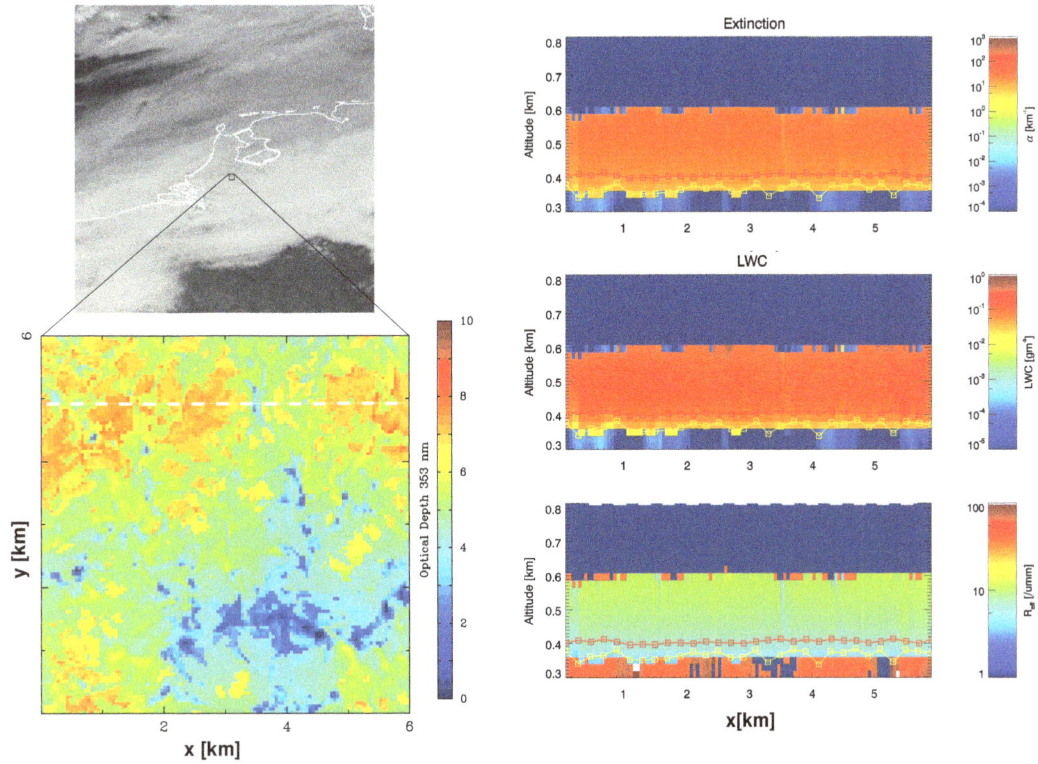

Figure 8. Visible MetoSat-SG Satellite image (top left) cloud optical thickness (COT) field for an DALES simulation for the Cabauw measurement site (bottom left) for the afternoon of 30 January 2011. Vertical extinction, LWC and effective radius slices corresponding to the "y" = 5 km line indicated on the COT plot (right panels). The red lines in the right panels indicate the peak of the simulated lidar parallel attenuated backscatter while the yellow lines indicate the cloud base returned by the retrieval procedure.

derived cloud optical thickness (COT) field along with vertical slices corresponding to the 355 nm extinction, LWC and R_{eff} fields taken along the indicated path is shown in Fig. 8. The scene corresponds to a overcast stratocumulus deck with a degree of drizzle present. The LES model was driven with boundary conditions corresponding to the meteorological situation surrounding the CESAR measurement site in the central Netherlands (52° N, 5° W) (Leijnse et al., 2010) on 30 January–1 February 2011. More details concerning the meteorological context of this scene can be found in Jacobs et al. (2012). The scene shown here corresponds to a snapshot at 16:00 UTC on the 31 January.

Here the cloud-droplet number density was fixed to a value of 85 cm^{-3}. The scene has a horizontal resolution 50 m and a vertical resolution of 10 m. The LWC panel shows the total (cloud + precipitation/drizzle) water. Here the drizzle water content is much lower the cloud water content and contributes little to the extinction. However the presence of drizzle is clear in the effective radius panel, particularly below the cloud base.

Virtual lidar and radar measurements corresponding to the track shown in Fig. 8 are shown in Fig. 9. Here a 355 nm depolarisation lidar with a field of view of 1 mrad was simulated along with the observed radar reflectivity corresponding to a 35 GHz cloud-profiling radar with a pulse length of 20 m and a simulated antenna diameter of 1.25 m. It can be seen that the depolarisation ratio increases from cloud-base and decreases sharply above cloud top, although it is quite noisy in this region. It can also be seen that while the lidar measurements are apparently not strongly influenced by the presence of drizzle the simulated radar signals are. This is, of course, expected since the radar reflectivity is proportional to the sixth moment of the hydrometer size distribution so that the radar reflectivity is strongly impacted by the presence of even small numbers of drizzle-sized droplets (see Eq. 35).

3.1.1 Inversion procedure

An inversion procedure based on the minimisation of Eq. (25) was developed and tested using the scene described above and other similar DALES-derived scenes. The steps of the full procedure are outlined below.

Step 1: Averaging and binning of data

The altitude of the peak observed parallel lidar attenuated backscatter is found for each profile. Each profile is shifted in altitude so that the peaks match and then the desired num-

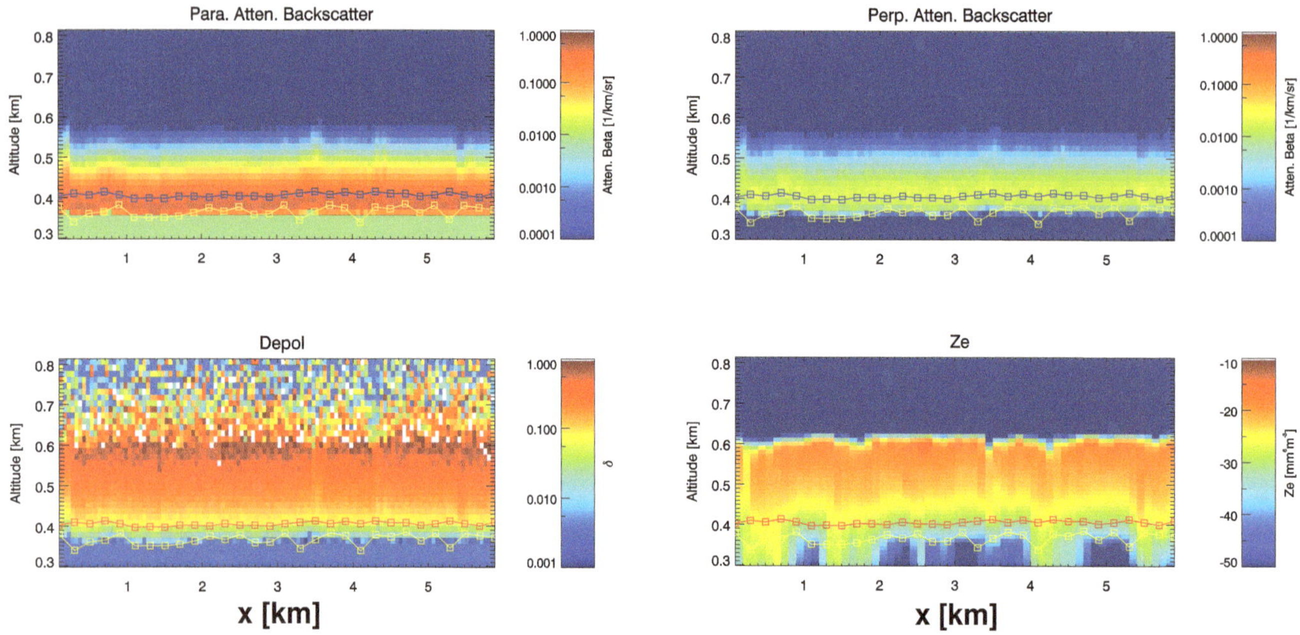

Figure 9. Simulated parallel and perpendicular attenuated backscatter signals for a 355 nm depolarisation lidar with a FOV of 1 mrad. Also shown are the corresponding linear depolarisation ratio and the radar reflectivity (Z_e).

ber of profiles are averaged. The uncertainties (the $\sigma_{y_i}^2$s) are estimated by evaluating the corresponding variance profiles.

The logic behind this averaging strategy can be illustrated as follows. In Fig. 9 it can be seen that the altitude of the peak return is not constant. Further, even in these simulations the cloud base can be difficult to unambiguously define due to the variations in cloud altitude and the presence of sub-cloud drizzle. When dealing with real observation the additional complicating factor of the presence of growing aerosol particles may also complicate the determination of the effective cloud base. In our procedure, we largely avoid the need to very accurately identify cloud base directly from the observation by using the peak of the observed parallel lidar attenuated backscatter as our reference. The minimum altitude considered in the inversion procedure is based on a threshold of $B_{\text{para}} = 0.05$ (which likely overestimates the true cloud base but largely avoids drizzle and aerosol effects), while an estimate of the true cloud base can be produced as a by-product of the fitting procedure determined by the optimal fit to the observations.

Step 2: Initialisation of minimisation procedure

From the investigations into the structure of Eq. (20) we know that spurious local minima in our cost function likely exist. For this reason it is necessary to specify an appropriately close initial guess when numerically minimising Eq. (25). It was found that a simple grid search of 10–15 values of α_{100} between 1 and 30 km^{-1} and $R_{\text{eff},100}$ between 3 and 12 microns with the values of C_N, C_r, δ^C set to their respective a priori values was appropriate in order to find a suitable initial guess for the minimisation procedure.

Step 3: Minimisation of Eq. (25)

A two step method to minimise the cost function was implemented in a robust manner. First we apply the gradient-free Nelder–Mead simplex algorithm (Press et al., 2007). Then, as an additional convergence check and to improve the accuracy of the minimisation, the simplex algorithm results were followed by an application of Powell's algorithm (Press et al., 2007). Finally, as described in Press et al. (2007), after convergence the curvature matrix around the minimisation point was numerically evaluated and the resulting covariance matrix of the retrieved parameters was found.

3.1.2 Inversion results

Two sample inversion results corresponding to x equal to 2.0 and 2.5 km are shown in Fig. 10. Here it can be seen that a very good match between the simulated observations and the results of the retrieval procedure are obtained. The results shown here correspond to a horizontal averaging of 0.2 km which corresponds to averaging across five consecutive simulated lidar profiles. It is interesting to note that that the simulated signals bear a striking similarly to actual observations extending even to the qualitative appearance of the signals above cloud top (Sassen and Petrilla, 1986).

Time series of inversion results as well as the true model values are shown in Fig. 11. In this set of trials (which contain the results presented in Fig. 10), the assumed error in C_r

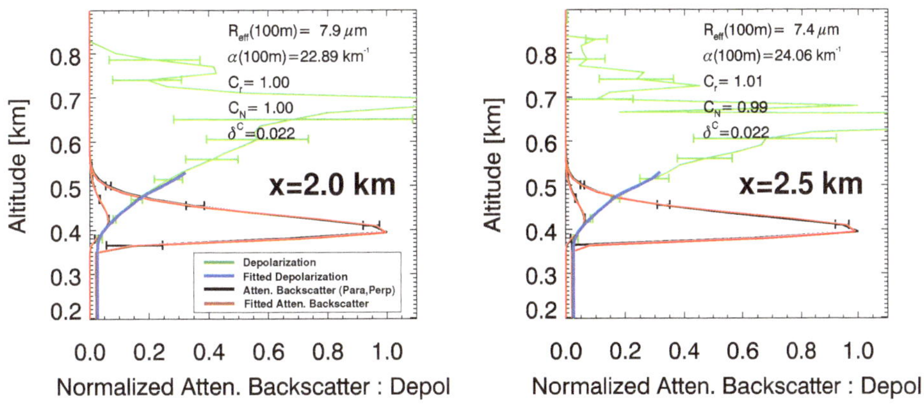

Figure 10. Results of the retrieval applied to the simulated lidar data along for two columns (corresponding to $x = 2.0$ and 2.5 km). Here the black lines are the simulated observations at a vertical resolution of 15 m and a horizontal resolution of 400 m while the corresponding depolarisation ratio is given by the green line. Here C_r was set to 1.0 and the depolarisation cross-talk parameter (δ^C) was set to 0.3. The red lines are the fits to the parallel and perpendicular attenuated backscatter and the blue line is the corresponding fit depolarisation ratio.

was set to 5 %, and for δ^C 20 % and the a priori values were set to match the true values. The SNR of the lidar signals themselves are functions of the signal strength but are generally in the range of 20 to 40 for the case depicted here. It can be seen that the agreement between the retrieval results for α_{100} and $R_{\mathrm{eff},100}$ as well as the derived variables Γ_l and N is generally within 10 % or better on a profile-by-profile basis.

The bottom panel of Fig. 11 shows the radar reflectivity corresponding to a level 100 m above the retrieved cloud base. In order to predict the radar reflectivities corresponding to the lidar retrieval results we note that the the relationship between radar equivalent reflectivity (Z_e) and LWC can, by rearranging Eq. (22) of Donovan and van Lammeren (2001), be written as

$$Z_e = \frac{\mathrm{LWC}}{\rho_l}\frac{48}{\pi}\left(\frac{|K|}{|K_w|}\right)^2 \frac{R_{\mathrm{eff}}}{R'^{4}_{\mathrm{eff}}}, \tag{32}$$

where $|K|$ is the dielectric factor for water which is temperature and frequency dependent and $|K_w|$ is a reference value of $|K|$ corresponding to a fixed reference temperature. For our purposes at 35 GHz, $|K_w|$ is fixed to a value 0.964. R'_{eff} is the so-called lidar–radar effective radius and for spheres is defined as

$$R'_{\mathrm{eff}} = \left(\frac{\langle r^6\rangle}{\langle r^2\rangle}\right)^{1/4}. \tag{33}$$

Equation (32) can be re-written to emphasise the role played by the ratio of R'_{eff} to R_{eff}. If we define

$$R_r \equiv \frac{R'_{\mathrm{eff}}}{R_{\mathrm{eff}}}, \tag{34}$$

then we have

$$Z_e = \frac{\mathrm{LWC}}{\rho_l}\frac{48}{\pi}\left(\frac{|K|}{K_w|}\right)^2 R_r^4 R_{\mathrm{eff}}^3. \tag{35}$$

For uni-modal size distributions of the type described by Eq. (6) the ratio of the lidar–radar effective radius to the normal effective radius is given by

$$R_r = \left[\frac{(\gamma+5)(\gamma+4)(\gamma+3)}{(\gamma+2)^3}\right]^{1/4}, \tag{36}$$

which varies between 1.13 for $\gamma = 9$ and 1.28 corresponding to $\gamma = 3$. Thus for uni-modal distributions there is a well-constrained relationship between reflectivity and the product of the water content and the cube of the effective radius. However it is well known that this is not the case in general if even small amounts of drizzle are present (e.g. Fox and Illingworth, 1997a). In particular, the value of R_r yielded by Eq. (34) represents a lower limit and multi-modal distributions can yield much higher values (Donovan and van Lammeren, 2001). This will be considered in more detail in Sect. 4.3 and Appendix B.

The continuous red lines in the bottom-panel of Fig. 11 show the true total reflectivity of the drizzle and cloud droplets combined, while the light-blue line shows the contribution of just the cloud droplets. It can be seen that the reflectivity predicted by the lidar results is a consistently better match to the cloud-only reflectivity. This is expected due to the lack of sensitivity of the lidar signals to the presence of optical thin drizzle. This result implies that it will be useful to compare the radar reflectivity derived from the lidar inversion results to actual observation (as will be done in Sect. 4.2). Agreement, however can only be expected in non-drizzling conditions. Cases where the observed Z_e is greater than the predicted reflectivity levels may indicate the presence of drizzle. However cases where the observed Z_e are less than the reflectivity levels predicted on the basis of the lidar inversion results via Eq. (35) are not physical and would indicate a problem with the observations or the inversion procedure (e.g. convergence to the wrong minimum) or with the radar calibration.

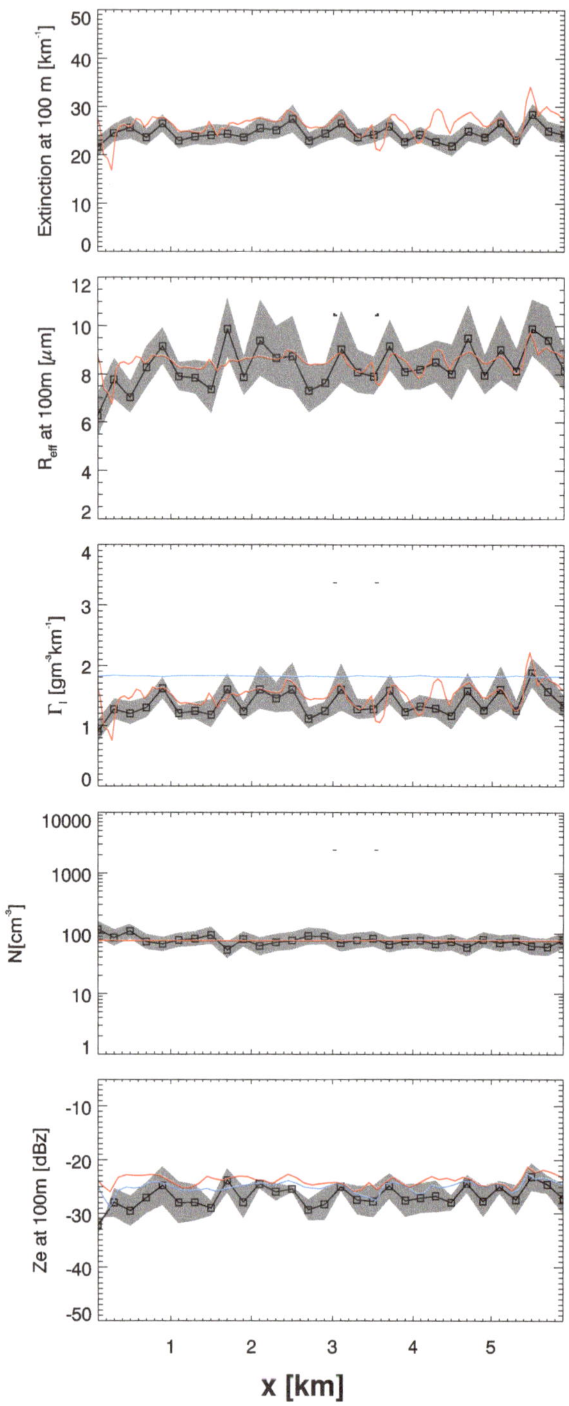

Figure 11. Results of the retrieval applied to the simulated lidar data along with the radar reflectivity simulated using the lidar results. Here the black lines show the retrieval results with the grey bands indicating the estimated 1-sigma uncertainty range. The red lines show the true values extracted directly from the LES-derived model fields. The light-blue line in the LWC panel indicates the value of the adiabatic (Γ_l) slope at cloud base. The light-blue line in the reflectivity panel indicates the true reflectivity levels if the contribution of the drizzle mode is removed.

As well as the results directly presented here, several other trials were conducted using the same scene but with the fixed cloud-number density set to lower and higher values as well as trials where the a priori values of C_r and δ^C were perturbed. It was generally found that the results were largely insensitive to errors in δ^C but errors in C_r were important. For example, it was found that a 5 % error in C_r coupled to a similar a priori error estimate couple leads to errors in R_{eff}, Γ_l and N of 10–15 %. Runs were also conducted where the assumed lidar FOV was changed from the true value. For example, if the true FOV was 1 mrad but the look-up-tables corresponding to 0.5 mrads were used to conduct the inversion then it was found that R_{eff} was overestimated by a factor of about 20–25 %, while Γ_l was overestimated by about a factor of 27–30 %, leading to an underestimation of N by close to a factor of 2.

4 Application to Cabauw observations

In this section, we describe the application of the depolarisation lidar inverse procedure to a substantial number of instances of actual observations. The inversion procedure was applied to about 150 selected periods ranging in time from a few 10s of minutes to several hours of boundary layer (BL) stratus clouds observed at the CESAR measurement site in the central Netherlands using a depolarisation lidar operating at 355 nm. In particular, cases from May 2008 (coinciding with the European Integrated project on Aerosol, Cloud, Climate, and Air Quality Interactions (EUCARI) impact campaign, www.atm.helsinki.fi/eucaari/.) as well as cases from January and July 2010 were selected. The observational data used in this study are freely available from the CESAR database (http://www.cesar-database.nl/).

The actual data record of UV-depolarisation lidar observations is much more extensive than the limited number of cases presented here; however the immediate aim here is not to conduct an exhaustive analysis of the results but to demonstrate the consistency and realism of of the depolarisation inversion results. A more extensive application and analysis is intended to be the focus of future work.

4.1 Measurements and case selection

The UV-depolarisation lidar at Cabauw is a commercial Leosphere ALS-450 lidar operating at 355 nm which has separate parallel and perpendicular channels. The system has been in operation at Cabauw since mid-2007 with breaks in the record ranging from weeks to several months. The data was acquired with a vertical resolution of 15 m and a temporal resolution of about 30 s. The depolarisation inter-channel calibration factor and the corresponding cross-talk parameters were estimated using the method described in Donovan and Apituley (2013a, b). The values of C_r and δ^C were found to be stable between instrument servicing which occurred be-

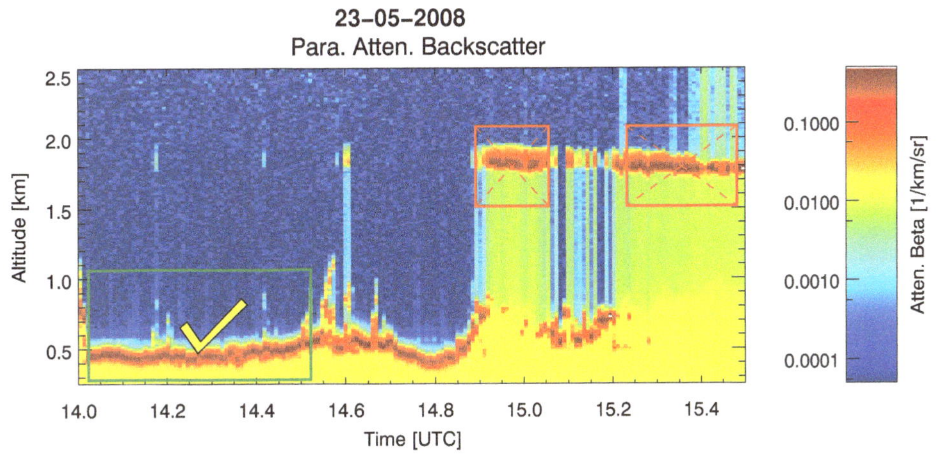

Figure 12. Illustration of the case selection criteria. Here all three of the boxed areas satisfy the conditions of being well-defined stratus water layers. However only the green outlined region appears to be connected to the surface. The data consist of measurements made using the ALS-450 system at Cabauw.

tween intervals ranging from a few months to a year. However within certain periods the cross-talk (δ^C) appeared to vary quasi-diurnally by up to 50 % (possibly linked to the temperature of the unit). The field of view of the lidar was found to be stable between servicing. The FOV of the lidar system was estimated by fitting an overlap function to lidar signals acquired during selected cloudless periods with low well-mixed BL aerosol burdens in a procedure similar in nature to those described in Guerrero-Rascado et al. (2010). The overlap model used was produced by convolving Eq. (7.72) of Measures (1984) with a Gaussian function in order to model the effects of an divergent emitted laser beam. The resulting overlap model is a function of the separation of the transmitter and receiver optical axes, the effective beam and receiver diameters as well as the effective beam divergence and receiver FOV. The separation between the emission and receiver optical axes and the beam and receiver diameters were found by physically making measurements on the device itself. The fits then yielded estimates of the effective beam divergence and the receiver FOV. As was the case with the C_r parameter, the FOV was found to be stable between instrument services and, depending on the particular time interval, the FOV was found to vary between about 0.5 and 1.5 mrad.

An example of the type of observation that was selected for analysis is presented in Fig. 12. It is our intention to focus on well-defined warm cloud layers. Further, as will be presented and discussed later in Sect. 4.5, we wish to compare our derived cloud-number density estimates to aerosol number concentration measurements made near the surface. Thus we further limit our focus to layers that appear to be physically linked to well-mixed boundary layers. In Fig. 12 all three of the boxed regions are well-defined stratus layers. However the higher altitude regions are clearly above the top of the boundary layer as indicated by the sharp gradient in lidar signal present at about 2.4 km.

As well as the lidar measurements, we also make use of the 35 GHz lidar observations at Cabauw. The cloud radar is a vertically pointing Doppler radar with a vertical resolution of 89 m and a temporal resolution of approximately 15 s. Further details of this system are given in Leijnse et al. (2010). For the periods involved in this study the radar reflectivity calibration uncertainty is thought to be in the range of 2–3 dBZ.

4.2 Examples

Sample lidar and radar data as a function of altitude and time are shown in Fig. 13 for 15 January 2011 from 16:00 to 18:00 UTC. Here a stratus layer is present with the cloud base varying between 0.75 and 0.85 km. The lidar data have a vertical resolution of 15 m and a temporal resolution of 30 s. The corresponding normalised attenuated backscatter as a function of distance from the altitude of the peak parallel return (binned to a temporal resolution of 3 min) as well as two sample inversion results are shown in Fig. 14. By comparing the lidar data shown in Fig. 14 against that shown in Fig. 13 it can be seen that the profile-to-profile variation is indeed reduced. The sample fit results indicate that the observed signal profiles largely conform to our expectations based on the look-up-table values themselves and the LES simulation-based results discussed earlier (e.g. those presented in Fig. 10); however some differences may be noted. First, the cloud base is generally not as sharply defined in the actual measurements as in the LES-based simulations. One possible reason for this is presence of drizzle, especially likely in the earlier profile which is below an area of elevated radar reflectivity. Another likely factor is the existence of small-scale variability at scales finer than the resolution of the LES simulations.

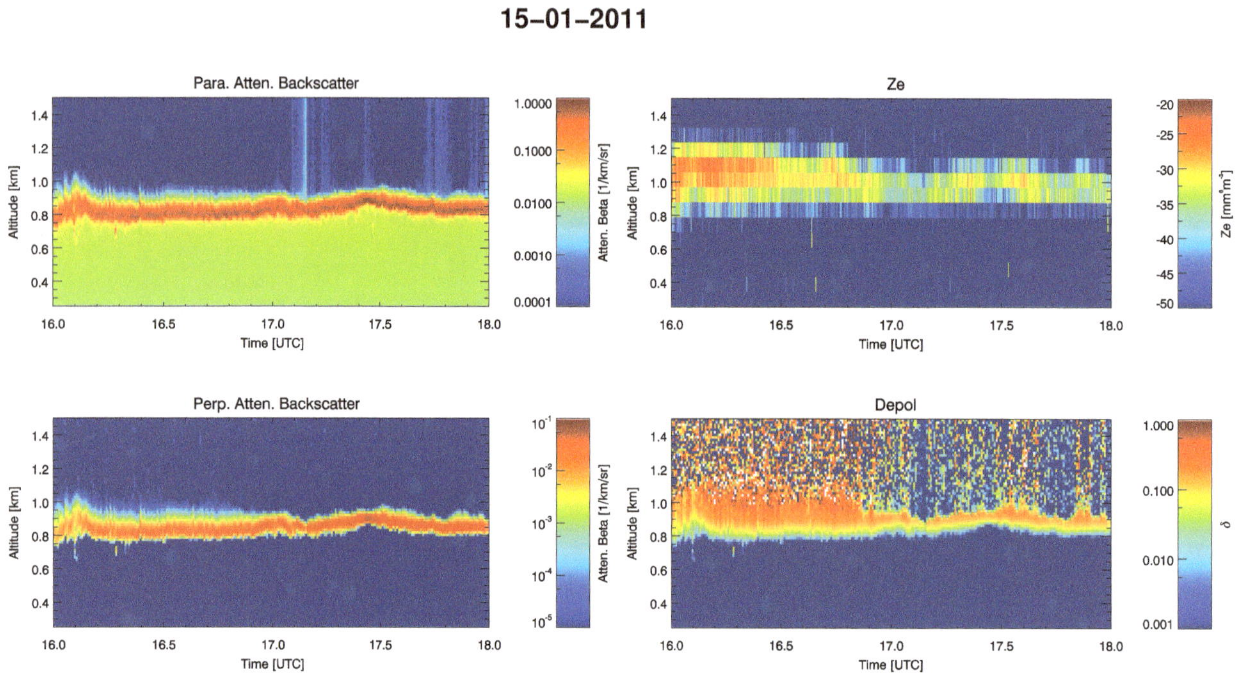

Figure 13. Observed lidar and radar signals for 15 January 2011 at Cabauw from 16:00 to 18:00 UTC.

Still another reason may be due to the presence of *not-quite* activated but still strongly growing aerosol present just below cloud base. Another notable difference between the LES simulation-based results and the observation is that the observed depolarisation ratio above 100–150 m from cloud base is often (but not always) less than expected on the basis of the look-up-table calculations and the LES-based simulations. This is presumably due to the departure of the real-cloud structure from our assumption of constant LWC slope and constant N (due to e.g. the effects of mixing at cloud top). That this observed behaviour is linked to slight non-linear effects in the lidar signal detection can also not be strictly ruled out.

In spite of these two main differences, generally very good fits for the first 100–150 m from cloud base are found. Time series of the inversion results corresponding to Fig. 14 are shown in Fig. 15. Here it can be seen that $R'_{\mathrm{eff},100}$ appears to have been fairly constant at about 4 μm and is retrieved with an estimated error of about 30 %, while the Γ_1 values are generally about 40 % of the adiabatic value. The cloud number concentration values are fairly constant with an average value of about 400 cm^{-3} and an estimated uncertainty on the order of 25 %. A comparison between the reflectivity predicted using the lidar inversion results using Eq. (35) and the observed values is shown in the middle right panel of Fig. 15. In order to conduct the comparison, the radar data were binned to the same time resolution as the lidar inversion results. To avoid the effects of partially filled radar bins near the cloud base, for each inversion time step, the altitude limits corresponding to the first radar height bin fully above the cloud base returned by the inversion procedure were found. These altitude limits were then used to average the lidar predicted Z_e. Here it can be seen that, similar to the LES-based results, e.g. bottom panel of Fig. 11, the results are generally within a few dBZ of each other with the observations higher by about 2–3 dBZ. This bias is consistent with the presence of low amounts of drizzle. Given the uncertainty in the radar calibration one can not be conclusive but the fact that the agreement between 16.9 and 17.0 UTC is in the region with the lowest reflectivities is also suggestive of drizzle being the cause of the offset. Past 17.0 UTC, the lidar results predict more reflectivity than was observed. This is not physical and points either to a problem in the lidar retrieval for this time period or, which is considered more likely in this case, that here partially filled radar bins likely could not be avoided. This is based on the fact that for this time period the cloud was likely physically thinner than 200 m which is equivalent to about 2 radar pixels in height.

Results from a second example time period corresponding to 4 January 2011 between about 18 and 19.7 h UTC are shown in Fig. 16. Here it can be seen that retrieved parameters are roughly in the same range as the results shown in the previous figure; however in general the estimated uncertainties in the retrieved quantities are more variable and generally larger. This may be linked to the fact that the lidar observations contain more profile-to-profile variability than the previous case or the fact that drizzle is more prevalent in this case. This is evident by examining the reflectivity panel along

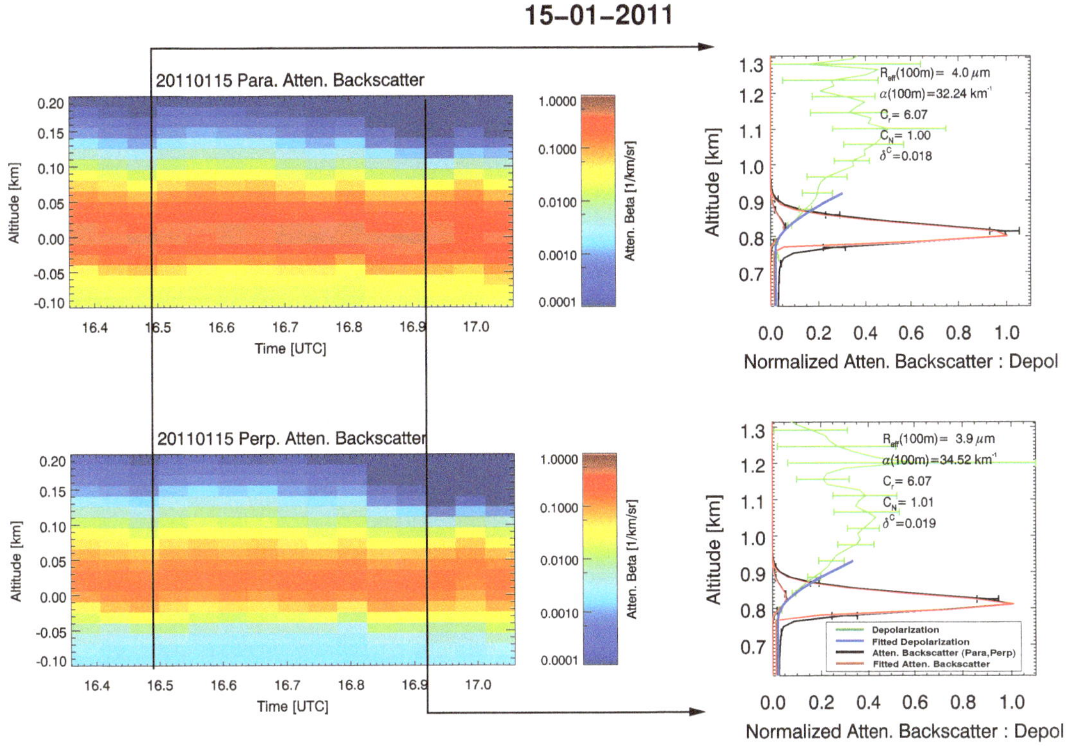

Figure 14. Normalised parallel and perpendicular attenuated backscatters as a function of altitude from the peak of the observed parallel backscatter profile (left panels) corresponding to the data shown in Fig. 13. Example fit results are shown on the right for 16.49 and 16.92 UTC.

with the panel in which the comparison between the lidar derived and observed Z_e is shown. The regions of detectable reflectivity present below the lidar-derived cloud base and the occurrence of reflectivities values above -25 dBZ are both indicative of the presence of drizzle.

By comparing the predicted and observed Z_e values for this case it can be seen that good agreement between the lidar-derived Z_e values and the actual radar observations is present past about 11.25 UTC, which is associated with cloud-base region reflectivities below about -35 dBZ. For earlier time periods the observed radar reflectivity is substantially higher than the lidar predicted values. These periods are associated with cloud-base reflectivities above -30 to -35 dBZ which are known to be associated with the presence of drizzle at cloud base (e.g. Tonttila et al., 2011; Wang and Geerts, 2003).

4.3 Comparison with cloud radar reflectivity measurements

As illustrated by the two specific example comparisons between the observed radar reflectivity and that modelled using Eq. (35) presented in the previous section, the observed reflectivity values are often apparently impacted by the presence of drizzle. This notion is explored in a more quantitative fashion in Appendix B where we use a bi-modal cloud and drizzle size distribution representation together with Eqs. (32)–(35) applied to the full 3-month set of cases. As discussed in detail within Appendix B, it was found that the lidar predicted Z_e values are largely consistent with the full set of co-located radar observations (see Fig. B2). In particular it was found that

- Instances with observed values of Z_e below those expected from the application of Eq. (35) are rare.

- Drizzle, on average, makes a substantial contribution to the observed cloud-base region reflectivity for reflectivities above about -35 dBZ.

- The relationship between the observed and predicted reflectivities are broadly quantitatively consistent with those predicted using a bi-modal size distribution model where the ratio of the drizzle mode number density to the cloud-droplet number density is on the order of 10^{-4} to 10^{-1} and values of LWC in the range of 0.05–0.1 g m^{-3}, respectively.

At this point in time, due to the lack of an independent means of assessing the drizzle contribution to the reflectivities, the comparison between the lidar predicted and observed cloud-base region reflectivities can not be taken as definitive validation of the lidar results. However it can be ro-

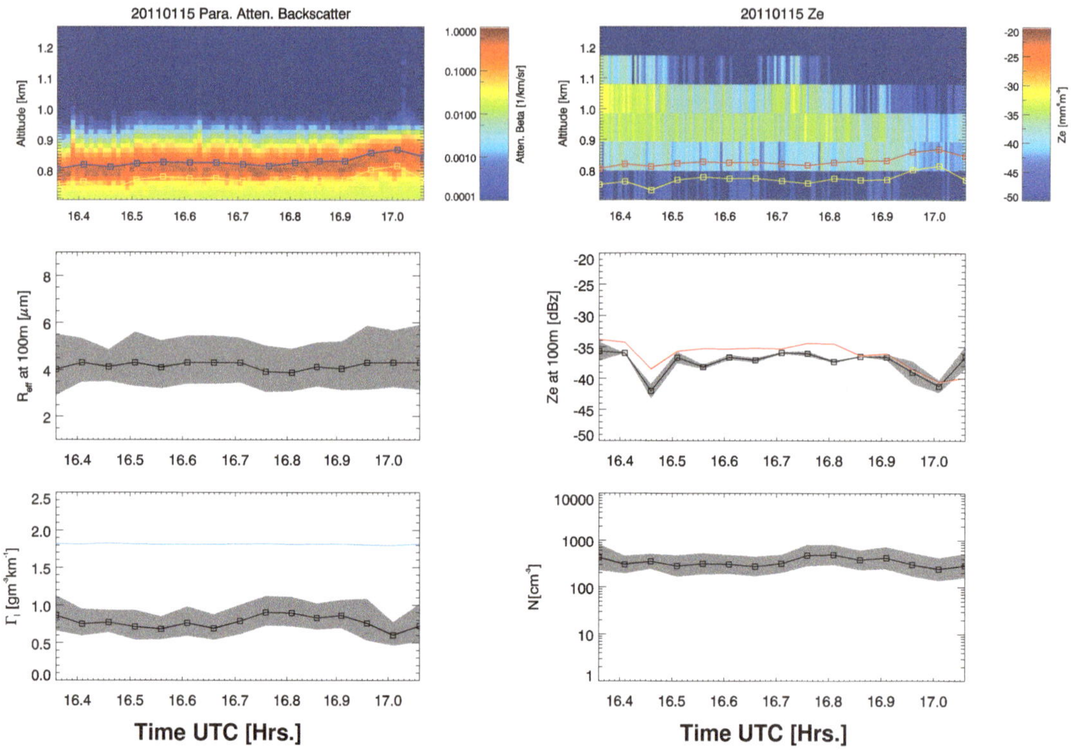

Figure 15. Retrieved time series of $R_{\mathrm{eff},100}$, Γ_l and N for 15 January 2011 from about 16.4 to 17.1 UTC. The light-blue line in the Γ_l plot indicates the adiabatic limit at cloud base. The black line in the Z_e panel shows the reflectivity predicted by Eq. (35) corresponding to the first 100 m radar bin fully above the estimated cloud base while the red line shows the corresponding actual radar observed value. The radar calibration uncertainty (not indicated) is thought to be in the range of 3 dBZ.

bustly stated that the lidar results are indeed physically consistent with the observed radar reflectivities.

4.4 LWC near cloud base

In addition to the comparison with the radar observations, an other independent evaluation criteria to judge the realism of the lidar results is the comparison of the lidar-derived Γ_l values with the corresponding adiabatic values (Γ_a). Using temperature and pressure data for Cabauw extracted from atmospheric analyses, the adiabatic liquid water mixing ratio lapse rate was calculated for the times and cloud-base altitudes of the lidar observations. A comparison between the adiabatic values and the observed values are shown in Fig. 17. Here it can be seen that the lidar observations imply a cloud-based adiabatic fraction of $0.451 \perp 0.007$. Only a few observations approach the adiabatic limit and none exceed it in a statistically significant manner. It can be noted that the sub-adiabatic fraction values seen here are within the range of previous in situ-based observations (e.g. Arabas et al., 2009; Pawlowska et al., 2006; Szczodrak et al., 2001) which were interpreted to be largely the result of entrainment at cloud base. Further, it is interesting to note that recently a new method of calculating cloud droplet number concentration in the cloud-base region was proposed by Pinsky et al. (2012). This leads to the finding that the ratio of supersaturation to the liquid water mixing ratio at the altitude of maximum supersaturation should be universal. That is, at the height of maximum supersaturation, which for stratus clouds is reached within a few 10s of metres from cloud base, the adiabatic fraction should be independent of updraught velocity or number density. Pinsky et al. (2012) predicted that this universal value should be equal to 0.44 (see Eq. 11 of Pinsky et al., 2012). The value of 0.44 compares very favourably with our finding of 0.451 ± 0.007, although more work and consideration would have to be done to properly judge the significance of this result.

4.5 Comparison with Scanning Mobility Particle Sizer aerosol measurements

At this point we feel that enough confidence in the depolarisation lidar-derived products has been accumulated so that a preliminary comparison between aerosol number concentrations and lidar-derived cloud-base number concentrations is feasible. As well as the remote-sensing equipment, Cabauw also hosts a number of in situ probes including a Scanning Mobility Particle Sizer (SMPS) instrument which measures aerosol size distributions between diameters of 10 and 470 nm. As described in Mensah et al. (2012), the SMPS

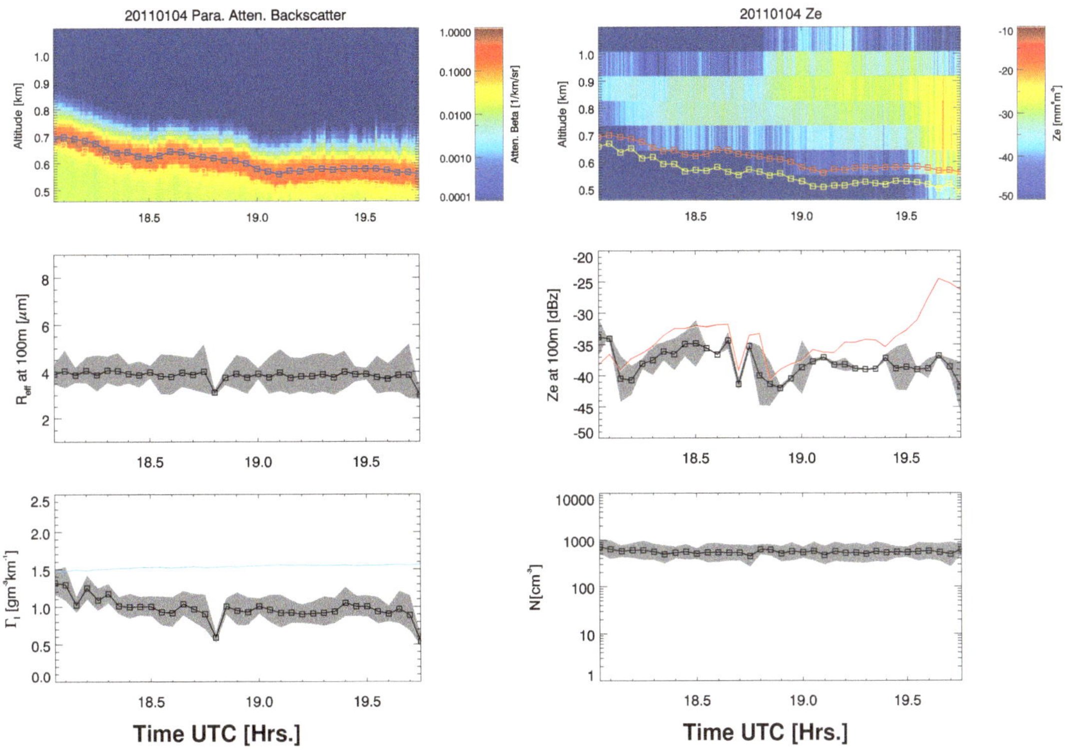

Figure 16. As Fig. 15 except for data corresponding to 4 January 2011.

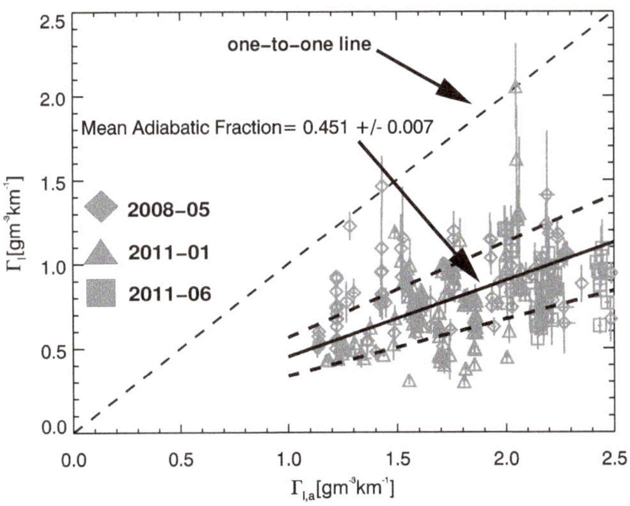

Figure 17. Adiabatic cloud-base liquid water lapse-rate values ($\Gamma_{l,a}$) and the corresponding lidar-derived values (Γ_l). The thin-dashed line represents the one-to-one line (adiabatic fraction = 1), while the thicker solid and dashed lines show the observed relationship based on the chi-square mean observed fraction and corresponding uncertainty.

instrument is housed in the basement of the Cabauw meteorological tower but the instrument is connected to a laminar flow sampling tube with an inlet at 60 m elevation so that the sampled air is expected to be more representative of the BL as a whole. Loss of some particles on the sampling tube walls does occur but this has been corrected for and, for the measurements used here, is not expected to be a significant source of uncertainty.

Previous aircraft-based studies have found correlations between aerosol number density and cloud-droplet number concentration. For example, by using number concentrations of aerosols measured with an Passive Cavity Aerosol Spectrometer Probe (PCASP) (which measures particles with diameter between 0.13 and 2 μm) and co-mounted Forward Scattering Spectrometer Probe (FSSP) cloud-droplet measurements, Gultepe and Isaac (1996) were able to demonstrate statistically significant relationships between the aerosol and cloud-droplet measurements. The observed relationship between the lidar-derived cloud-number densities N_d and the tower-based SMPS measurements is shown in Fig. 18. Here, following Pringle et al. (2009), the aerosol number concentrations shown are representative of particles with diameter greater than 50 nm. This was done to be consistent with the earlier data upon which the previous empirical relationship are based. The aerosol concentrations were also adjusted for the difference in air density between the ground and cloud base by assuming that the aerosol number density mixing ratio is conserved.

A number of empirical relationships relating aerosol number concentration to cloud-droplet number density for warm stratus clouds under different conditions were compiled by Gultepe and Isaac (1996) and Pringle et al. (2009). In Fig. 18

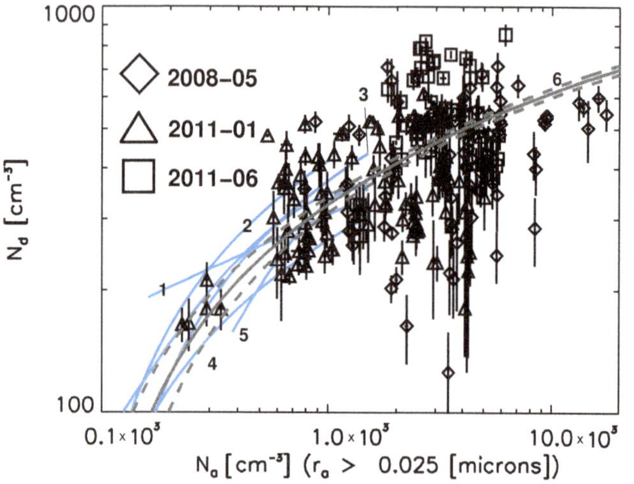

Figure 18. Lidar-derived cloud-base number density (N_d) and SMPS ground-based number density of aerosols with radii greater than 0.025 μm adjusted for density at the cloud-base altitude (N_a). The symbols follow the same conventions as Fig. B1. The light-blue lines (labelled 1–5) represent previously defined independent empirical relationships based on in situ measurements made with the aid of aircraft. Line-1 corresponds to Eq. (37), 2 corresponds to Eq. (38), 3 corresponds to Eq. (39), 4 corresponds to Eq. (40) and 5 corresponds to Eq. (41) The grey lines (6) show the fit to the points produced using Eq. (42) along with the 1-sigma error bands.

lines 1–5 represent independent relationships draw from previous aircraft based work and are listed in Gultepe and Isaac (1996). Here line 1 corresponds to

$$\log_{10}(N_\mathrm{d}) = 0.257\log_{10}(1.22N_\mathrm{a}) + 1.95, \quad (37)$$

which was originally found by Leaitch et al. (1992). Line 2 corresponds to

$$N_\mathrm{d} = -765.5 + 395.71\log_{10}(N_\mathrm{a}). \quad (38)$$

Line 3 corresponds to

$$N_\mathrm{d} = -698.4 + 356.61\log_{10}(N_\mathrm{a}) \quad (39)$$

and line 4 corresponds to

$$N_\mathrm{d} = -382.15 + 215.83\log_{10}(N_\mathrm{a}). \quad (40)$$

Lines 2–4 were found by Gultepe and Isaac (1996). Line 5 is given by

$$N_\mathrm{d} = -27.9 + 0.568N_\mathrm{a} - 2.1 \times 10^{-4}N_\mathrm{a}^2 \quad (41)$$

and was originally found by Martin et al. (1994). Line 6 represents the relationship found in this present work via chi-square fitting and is given by

$$N_\mathrm{d} = (-547 \pm 8) + (291 \pm 4)\log_{10}(N_\mathrm{a}). \quad (42)$$

The fit error bounds here were found using a bootstrap method (Press et al., 2007). All of the above relationships are roughly consistent with each other and with the findings of this present study. That they are similar, even though the aerosol chemical composition may have been different between the different studies and time periods, may be due to the fact that, so long as the aerosol is not hydrophobic, size plays a dominant role in determining whether a given aerosol particle can act as a CCN or not (Dusek et al., 2006). The fact that the results obtained using the lidar-derived cloud-base cloud-number concentrations and tower-based measurements aerosol measurements are consistent with earlier completely independent studies strongly supports the validity of the lidar inversion results and further supports the notion that under apparently well-mixed BL conditions (as assessed by evaluation of the lidar backscatter measurements) the tower-based aerosol number concentration measurements are indeed representative of the BL as a whole.

5 Conclusions

In this work a novel method for determining cloud-base properties by exploiting the signature of multiple scattering on depolarisation lidar signal was developed. The method is novel yet firmly based on older established ideas and principles. The inversion procedure has not been evaluated against direct measurements (e.g. coincident in situ measurements). However even at this arguably preliminary stage, we have a high degree of confidence in the results. This confidence is based on the following considerations:

1. There is a rather direct connection between the variables determining the relevant lidar radiative transfer (e.g. Extinction and effective cloud particle radius) and the cloud physical parameters of interest (e.g. liquid water content and cloud droplet number concentration).

2. Application of the method to LES generated clouds shows that, within reasonable limits, the method is robust to deviations from strict adiabatic cloud structure, the presence of drizzle and variations in cloud-base altitude.

3. Under low-reflectivity conditions, where the reflectivity contribution of the drizzle droplets can be neglected, it has been demonstrated that lidar results can be used to predict the observed radar reflectivity within the uncertainty of the radar calibration. Under general circumstances, where drizzle is present the range of bi-modal size distribution parameters required for consistency between the lidar and radar measurements are well within the range of accepted values.

4. Cloud-base LWC values were found to never exceed the adiabatic limit by an amount outside of the respective error estimates. Further, the observed average cloud-base adiabatic fraction is consistent with the range of previous observations.

5. The results obtained by comparing the polarisation lidar-derived cloud-base cloud-droplet number concentrations with tower-based aerosol number concentrations yields a relationship consistent with completely independently derived relationships based on previous in situ aircraft-based measurements.

The evaluation examples presented in this work represent a small fraction of the data available from the Cabauw site. A more extensive application of the method to the Cabauw data should be conducted. Additional, further validation work (possibly involving the use of in situ cloud measurements from, for example, the EUCARI/IMPACT (Kulmala et al., 2011) campaign) should be carried out. In this work, we have used results from a commercial UV depolarisation lidar that was not developed with this application in mind. Future developments could be imagined involving the optimisation of instrument parameters (e.g. wavelength, FOV, dynamic range) directed towards the implementation of the method developed here including the possible integration of multiple fields-of-views. Further, it should be noted that, depending mainly on the instrument FOVs, the methods described in this work may be applicable to a large body of existing lidar measurements made by depolarisation lidars operating in the UV as well as in the visible (e.g. 532 nm) wavelength ranges.

A key variable lacking in the examination of the relationship between the cloud-droplet number concentrations and the aerosol number concentrations is knowledge of the characteristics of the vertical velocities at cloud base. Such information may be difficult to reliably extract from radar Doppler observations (as indicated by the almost constant presence of drizzle at cloud base) but could be reliably supplied by Doppler lidar measurements. Future studies involving paired Doppler and depolarisation lidars are thus recommended.

The technique described in this work is specific to the case of upward looking terrestrial depolarisation lidars. The larger footprints involved and the change from viewing cloud top instead of cloud bottom means that the specific technique used in this work is not applicable to spaceborne lidars. If a technique similar to the one presented in this work were applicable to spaceborne lidars then the global information so obtained could be very valuable, so the matter is worth considering.

The characteristics of the depolarisation return from water clouds has been successfully exploited using CALIPSO lidar observations as a means to determine cloud phase (Hu et al., 2007), but it is unclear at this point if the approach used here could be usefully adapted to the case of spaceborne lidars. In order for this to occur, a suitable model of cloud-top conditions to serve the analogous role of the simple cloud-base model used in this work would have to be identified or formulated. Second, extensive simulations including, ultimately, the effects of expected noise levels, would have to be carried out to determine what, if any, cloud microphysical information may be recoverable. In particular, the larger footprints associated with space-based lidars lead to relationships between such quantities as the integrated backscatter and the integrated depolarisation ratio (Hu et al., 2001, 2007) which seem to be weakly dependent on the cloud microphysics when compared to the case with ground-based observations. Nevertheless, CALIPSO lidar observations and related simulations do suggest that microphysical information may indeed be recoverable (see Fig. A3 and the related discussion) but further dedicated work would have to be carried out to establish this.

Appendix A: Lidar MC model

The ECSIM lidar MC model is similar in principle to the MC model described by Hu et al. (2001) but, in addition, it is fully 3-D and can calculate the spectral-polarisation state of the lidar signal and employs a number of variance reduction techniques in order to increase the computational efficiency. Here a brief overview of the model is given along with a number of validation examples focusing on the depolarisation. The spectral state of the return (e.g. Doppler shifting and molecular thermal spectral broadening of the lidar signal under multiple-scattering conditions) is not relevant for this present work and its calculation is not discussed here. For more information see the description of the `lid_filter` and `lidar` program descriptions within the ECSIM Models and Algorithm Document (Donovan et al., 2010).

A1 Polarisation

In order to accurately model the behaviour of the lidar depolarisation signal under multi-scattering conditions it is not sufficient to specify the phase functions and polarisation-dependent backscatter coefficient. It is necessary to specify the full-scattering matrix. The relationship between the Stokes vectors of the scattered and incident electromagnetic fields (with respect to the scattering plane) can be written as

$$\begin{pmatrix} I \\ Q \\ U \\ V \end{pmatrix} = \begin{pmatrix} P_{11} & P_{12} & P_{13} & P_{14} \\ P_{21} & P_{22} & P_{23} & P_{24} \\ P_{31} & P_{32} & P_{33} & P_{34} \\ P_{41} & P_{42} & P_{43} & P_{44} \end{pmatrix} \begin{pmatrix} I_o \\ Q_o \\ U_o \\ V_o \end{pmatrix}. \quad \text{(A1)}$$

If the target scatterers are rotationally symmetric or randomly oriented then this relationship is reduced to

$$\begin{pmatrix} I \\ Q \\ U \\ V \end{pmatrix} = \begin{pmatrix} P_{11} & P_{12} & 0 & 0 \\ P_{12} & P_{22} & 0 & 0 \\ 0 & 0 & P_{33} & P_{34} \\ 0 & 0 & -P_{34} & P_{44} \end{pmatrix} \begin{pmatrix} I_o \\ Q_o \\ U_o \\ V_o \end{pmatrix}. \quad \text{(A2)}$$

The phase matrix is defined in terms of the scattering plane (i.e. the plane defined by the incoming and outgoing scattered photon paths). Thus the Stokes vector of the incoming radiation and the resulting vector describing the scattered radiation must be rotated with respect to the scattering plane. The Stokes vector resulting from a photon scattered through an angle (Θ) is given by

$$\boldsymbol{S} = \mathbf{L}(\pi - i_2)\mathbf{P}(\Theta)\mathbf{L}(-i_1)\boldsymbol{So}, \quad \text{(A3)}$$

where $\boldsymbol{So}$ is the incoming Stokes vector and $\boldsymbol{S}$ is the Stokes vector associated with the scattered radiation. $\mathbf{P}(\Theta)$ is the scattering phase matrix and $\mathbf{L}$ is the transformation matrix for the Stokes parameters (Liou, 2002):

$$\mathbf{L}(\chi) = \begin{pmatrix} 1 & 0 & 0 & 0 \\ 0 & \cos(2\chi) & \sin(2\chi) & 0 \\ 0 & -\sin(2\chi) & \cos(2\chi) & 0 \\ 0 & 0 & 0 & 1 \end{pmatrix}. \quad \text{(A4)}$$

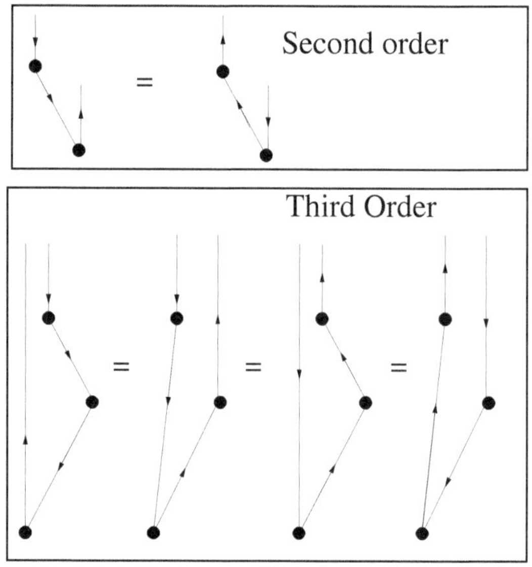

Figure A1. Schematic representation of the various types of paths involving forward and backscatter events for second- and third-order scattering.

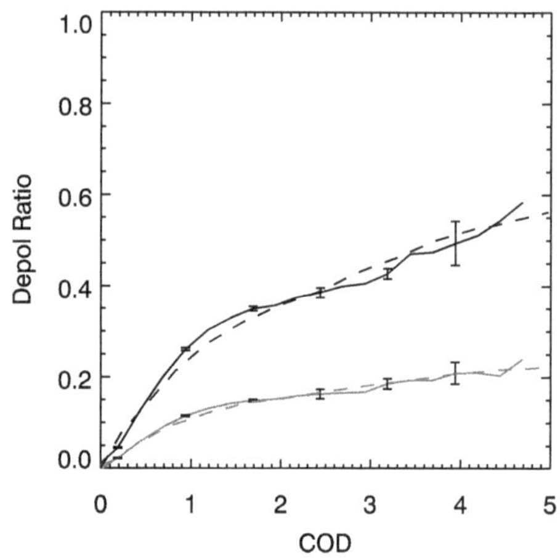

Figure A2. Linear and circular depolarisation profiles in a C1 cloud at a distance of 2 km for a FOV of 1.75 mrad and a wavelength of 700 nm as a function of cloud optical depth (COD). The solid line shows the ECSIM results while the dashed lines shown the results calculated using an approximate analytical approach (Chaikovskaya, 2008).

The angles i_1 and i_2 in Eq. (A3) are given by

$$\cos(i_1) = \frac{-\mu + \mu' \cos(\Theta)}{\pm(1-\cos(\Theta)^2)^{1/2}(1-\mu'^2)^{1/2}} \quad \text{(A5)}$$

and

$$\cos(i_2) = \frac{-\mu' + \mu' \cos(\Theta)}{\pm(1-\cos(\Theta)^2)^{1/2}(1-\mu^2)^{1/2}}, \quad \text{(A6)}$$

where μ' is the z component of the direction cosine of the incoming photon and μ is the z component of the direction

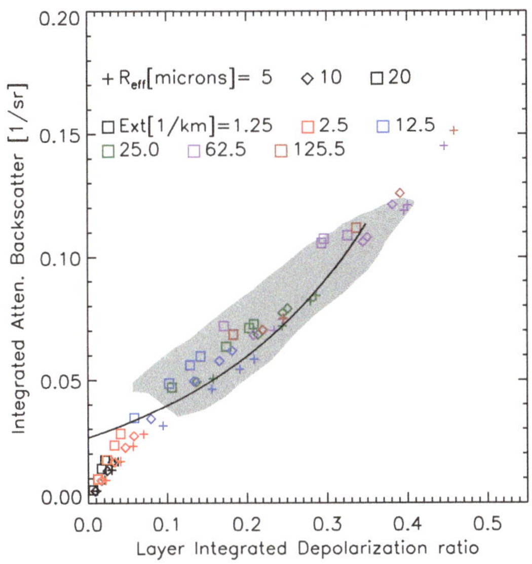

Figure A3. Layer integrated integrated backscatter vs. layer integrated depolarisation ratio. The symbols show the results of 532 nm CALIPSO ECSIM simulations for different layer effective radii and extinction coefficients as indicted in the figure legend. The grey-shaded area represents the area within the "frequency-of-occurrence = 20" level shown in Fig. 3 of Hu et al. (2007) for actual CALIPSO observations. The line denotes $\gamma = \frac{1}{2S_c}\frac{(1+\delta)^2}{(1-\delta)^2}$ ($S_c = 18$ sr) following and using the notation of Hu et al. (2007).

cosine of the scattered photon. The plus sign is to be used when $\pi < \phi - \phi' < 2\pi$ and the minus sign is to be used otherwise.

A2 MC model

A MC approach models the propagation of the laser photons in a stochastic manner. Photons are, in effect, launched and propagated within a graded extinction field. In a pure MC procedure, photons are launched from their source (in this case the laser) with an initial direction vector. The photon then travels a distance Δs before interacting with a scatterer or absorber. The probability density function for the step size (Δs) follows Beer's law:

$$\mathcal{P}(\Delta s \leq r) = 1.0 - \exp\left[-\int_0^r \alpha(r')\mathrm{d}r'\right], \tag{A7}$$

where α is the total extinction coefficient. For each photon packet Δs is determined stochastically using a suitable pseudo-random number generator according to Eq. (A7). Once an interaction has occurred the type and size of particle encountered is determined stochastically according to the contribution each particle present at that grid point makes to the total attenuation. At each event a particle may be scattered or absorbed. This too is determined stochastically according to the single-scatter albedo of the appropriate interacting particle. If it has been determined that a scattering event (rather than an absorption event) has taken place then the direction of the photon packet is changed according to its phase function. The polarisation state of the return signal is accounted by keeping track of the Stokes vector of the scattered photons (see Sect. A1). The measured linear depolarisation ratio corresponds to

$$\delta_{\mathrm{lin}} = \frac{(I-Q)}{(I+Q)}. \tag{A8}$$

A2.1 Variance reduction

In a pure MC approach photons are tracked until they are absorbed, detected or exit the simulation area. This approach is simple and accurate. However it is not very efficient as any given photon has a very small chance of being scattered back to the lidar receiver. Thus it is desirable to modify the basic MC approach to increase its computational efficiency.

In order to increase the efficiency of the calculation, the ECSIM MC model analytically calculates the amount of unscattered energy transmitted from the lidar present at each altitude and then proceeds to calculate the higher-order scattering by launching a number of appropriately weighted photon packets from each altitude bin (here 2000–5000 photons per altitude bin where are typically used). As the photon packets propagate and scatter, for each scattering event, the relative amount of signal received by the lidar is analytically calculated based on the packet weight, the optical thickness and distance between the event and receiver and the phase function of the scatterer in question together with the scattering angle back to the receiver. This contribution to the detected power is stored and in order to conserve energy the same amount of energy is then removed from the packet by appropriately reducing its weight.

In order to further increase the efficiency of the calculation, the well-know technique of forcing scattering of the photon packets to occur within a specified distance from the receiver axis (in this case 5 times the receiver field-of-view footprint) was implemented (Platt, 1981).

The aforementioned approach is exact and orders of magnitude faster than a pure MC approach. However by invoking an approximation concerning the contribution of different photon packet paths, the computational efficiency may be improved further still.

Since cloud particles are usually large compared to visible wavelengths their phase function is strongly peaked in the forward direction (Eloranta, 1998). Hence the most probable MC photon packet paths involve multiple forward scatters. However rare events, but with very large relative contributions, will happen when a backscatter occurs in the determination of the packet path so that the packet now travels towards the receiver, and thus the calculation of the direct contribution for the subsequent scattering order to the received signal involves a forward scatter. These contributions can lead to *spikes* in the modelled detected signal and adversely increase the variance (Barker et al., 2003). It is possible to eliminate the contribution of these events by only explicitly counting the contribution to the return signal in the case where the scattering angle back to the receiver is greater than 90°. By eliminating the contribution of forward scatters directly to the receiver the convergence of the MC calculation will be rapid. However this will lead to a large underestimation of the lidar signal. This underestimation may be accounted for if we consider the contribution from the various types of paths involving two or more scatterers by appealing to the reciprocity theorem (Katsev et al., 1997; Bissonnette, 2005). In Fig. A1 the various paths (limited to one large-angle backscatter) for second- and third-order scattering are presented. For second-order scattering it can be seen that only two different types of paths exist (namely forward scatter then backscatter and backscatter then forward scatter) and upon consideration, the contribution of both types of paths should be the equal. Hence by not counting the backscattering then forward-scattering events the second-order signal will be underestimated by a factor of 2. For third-order scattering we conclude that the underestimation will be a factor of 4. By similar reasoning the factors, in general, are found to be

$$F_{\mathrm{path}} = 2^{n-1}, \tag{A9}$$

where F_{path} is the underestimation factor and n is the scattering order.

A3 Validation

The ECSIM MC code has been compared to other lidar MC codes as with generally excellent agreement being found. For example, for a number of benchmark Cirrus cases the results of the ECSIM MC code was compared with the results from the MYSTIC MC code (Buras and Mayer, 2011) run in a lidar configuration. The agreement was found to be excellent (Petzold et al., 2011). MYSTIC employs a number of variance reductions strategies but, in general, they are different from the ones employed by the ECSIM model. The MYSTIC comparison did not however consider polarisation. Focusing here on the depolarisation, we present two illustrative examples, including comparisons with observations, focused on depolarisation in water clouds.

A3.1 Comparison with independent model results

Simulated linear and circular lidar depolarisation ($\delta_c = (I+V)/(I-V)$) profiles in a C1 cloud ($R_{\mathrm{eff}} = 6\,\mu$m) (Deirmendjian, 1969) at a distance of 2 km for a FOV of 1.75 mrad and a wavelength of 700 nm are shown in Fig. A2. Here both results generated by the ECSIM model and an approximate analytical approach due to Chaikovskaya (2008) are shown. The agreement between the ECSIM MC results and analytical approach is seen to be good, although differences at the 10 % level are seen near the leading edge of the cloud, and is similar to the level of agreement seen with comparisons between the analytical approach and other MC models (Chaikovskaya, 2008).

A3.2 Comparison with CALIPSO lidar observations

It has been previously noted that a robust and tight relationship between layer integrated attenuated backscatter and layer integrated multiple-scattering depolarisation ratio for water layers exists, particularly in the case of space-based lidar (Hu et al., 2007). The relationship between these two quantities for a range of different idealised homogeneous water clouds as calculated by the ECSIM MC lidar model for the CALIPSO lidar instrument parameters along with the actual approximate range observed by CALIPSO are shown in Fig. A3. It can be seen that a very good correspondence between the simulation and observations exists. We regard this result as good evidence of the accuracy of the ECSIM MC lidar results even in the case of large instrument footprints.

A3.3 Summary

Based on the comparisons with other MC codes and independent analytical calculations the ECSIM lidar MC calculations are robust for both ground-based and space-based simulations of lidar multiple scattering in water clouds. The ability of the ECSIM model to replicate the relationship between integrated depolarisation ratio and integrated backscatter as observed by the CALIPSO lidar, in particular, is regarded as a strong "stress test" of the code since it involves a much wider range of angles and higher orders of scattering compared to ground-based simulations.

Appendix B: Comparison with radar reflectivity observations

The good agreement between the observed cloud-base reflectivity values and those predicted on the basis of the lidar inversion results under apparently non-drizzling conditions is qualitatively supportive of the lidar results being accurate. In order to assess the realism of the lidar results in a more quantitative manner, we will examine the relationship between the observed cloud-base reflectivities and the values predicted on the basis of the lidar inversion results in more detail.

As a starting point, we will assume that for the cases we have selected that the lidar results are representative of the cloud properties, and that the drizzle water contents and number densities are small compared to the cloud water content and number densities so that the lidar-derived LWC is approximately equal to the total water content. This allows us to compare the lidar-derived cloud-based LWC values with those estimated using the radar reflectivity alone. A comparison between the cloud-base liquid water contents retrieved by the application of the lidar inversion procedure for the entirety of our data set and the correspond-

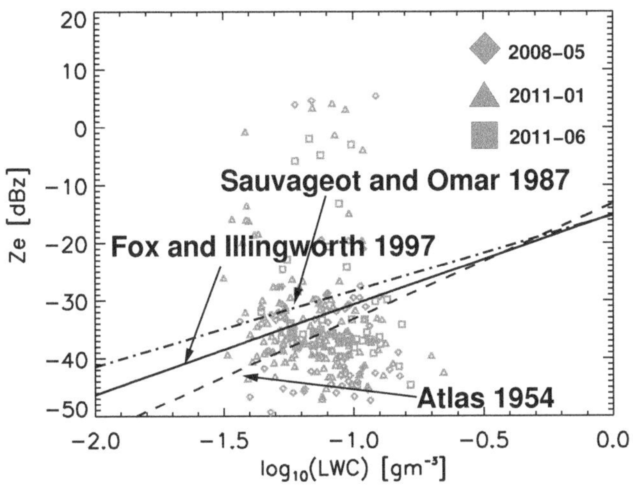

Figure B1. Relationship between observed cloud-base reflectivity and corresponding cloud-base reflectivity predicted from the depolarisation lidar inversion results. The lines correspond to three different Z_e vs. LWC relationships as labelled, for details consult the text.

ing observed reflectivities is presented in Fig. B1. Each data point corresponds to a average over a half-hour of continuous measurements. For comparison, three different relationships for "non-drizzling" clouds taken from Fox and Illingworth (1997b) are shown. As described in Fox and Illingworth (1997b) the line labelled ATLAS (1954) is described by

$$Z_e = 0.048\mathrm{LWC}^2, \tag{B1}$$

the line labelled Savageot and Omar (1987) is given by

$$Z_e = 0.03\mathrm{LWC}^{1.31}, \tag{B2}$$

and the line labelled Fox and Illingworth (1997) is described by

$$Z_e = 0.031\mathrm{LWC}^{1.56}, \tag{B3}$$

where for all three relationships Z_e is in units of $\mathrm{mm}^6\,\mathrm{m}^{-3}$ and LWC is in units of $\mathrm{g\,m}^{-3}$. Here it can be seen that if we assume that observed values of reflectivity over about -30 dBZ indicate the presence of drizzle, then the range of the data is plausibly consistent with Eqs. (B1)–(B3) but it is difficult to say anything more conclusive than this.

If we assume a model that takes into account the bi-modal structure of the cloud droplets together with the drizzle droplets, then we can carry out a more quantitative evaluation. Following Baedi et al. (2000), if we model a water cloud size distribution as the sum of a modified-gamma distribution together with an exponential distribution to describe the drizzle mode we can write

$$\frac{\mathrm{d}N(r)}{\mathrm{d}r} = \frac{N_{o,1}}{R_m}\frac{1}{(\gamma-1)!}\left(\frac{r}{R_{m,1}}\right)^{\gamma-1}\exp\left[-r/R_m\right] \tag{B4}$$

$$+ N_{o,2}\frac{1}{R_{m,2}}\exp\left[-r/R_{m,2}\right].$$

For this type of bimodal distribution, R_{eff} is given by

$$R_{\mathrm{eff}} = \frac{N_{o,1}(\gamma+2)!R_{m,1}^3 + N_{o,2}R_{m,2}^3 3!}{N_{o,1}(\gamma+1)!R_{m,1}^2 + N_{o,2}R_{m,2}^2 2!}, \tag{B5}$$

while R'_{eff} is given by

$$R'_{\mathrm{eff}} = \left[\frac{N_{o,1}(\gamma+5)!R_{m,1}^6 + N_{o,2}R_{m,2}^6 6!}{N_{o,1}(\gamma+1)!R_{m,1}^2 + N_{o,2}R_{m,2}^2 2!}\right]^{0.25}. \tag{B6}$$

If $R_{m,1} = R_{\mathrm{eff},1}/(\gamma+2)$ is specified along with the ratio $N_{o,2}/N_{o,1}$ and the cloud water content, Eqs. (B5) and (B6) can then be used together with Eq. (35) to model the relationship between R_r (Eq. 34) and Z_e. If we assume that the drizzle number densities are low enough such that $R_{\mathrm{eff}} = R_{\mathrm{eff},1}$ the observed value of cloud-base region Z_e along with the lidar-derived R_{eff} and LWC values can be used along with Eq. (35) to generate observed estimates of R_r. A scatter plot of the values of R_r derived from the lidar inversion results together with the radar observations and those generated using Eqs. (B5) and (B6) is shown in Fig. B2. Each data point corresponds to a maximum of a half-hour average of continuous measurements. The estimated variation of the mean of the observed Z_e and the corresponding variance of the mean of the lidar-based results (and accounting for the errors that are expected to be correlated in time and thus not reduced by short-term averaging e.g. the error due to the uncertainty in k) were used to generate uncertainty estimates for each point.

In order to generate the theoretical R_r relationships, $N_{o,2}/N_{o,1}$, LWC and $R_{\mathrm{eff},1}$ were fixed as indicated within the figure. $R_{m,2}$ was then varied between 1 and 80 μm. The ranges of LWC and $R_{\mathrm{eff},1}$ used were found by using the respective mean and standard deviations of the lidar based inversion results. By looking at the light-blue lines in Fig. B2 it can be seen that as $R_{\mathrm{eff},1}$ and LWC increase, the resulting R_r values shift to higher values. The behaviour of the dashed red lines illustrate the effect of changing the value of $N_{o,2}/N_{o,1}$. Here it can be seen that the lower value of $N_{o,2}/N_{o,1}$ des not change the shape of the R_v–Z_e relationship but it does limit the maximum extent of the relationship in terms of the maximum Z_e than can be reached for a given maximum value of $R_{m,2}$ (here fixed at 80 μm for all the shown curves). Larger values of $N_{o,2}/N_{o,1}$ enable larger values of Z_e to be reached; however the drizzle mode will start dominate the total LWC and R_{eff} and this leads to a slight flattening out of the R_v–Z_e curve.

In general, a good match between the observed and theoretical R_v–Z_e relationship can be made using the observed ranges of LWC and R_{eff}. If an upper limit of $R_{m,2}$ of 80 μm is imposed then the observations are consistent with $N_{o,2}/N_{o,1}$ being in roughly in the range of 0.01–0.001. A smaller/larger imposed maximum value of $R_{m,2}$ would lead to correspondingly larger/smaller $N_{o,2}/N_{o,1}$ values. There is not enough

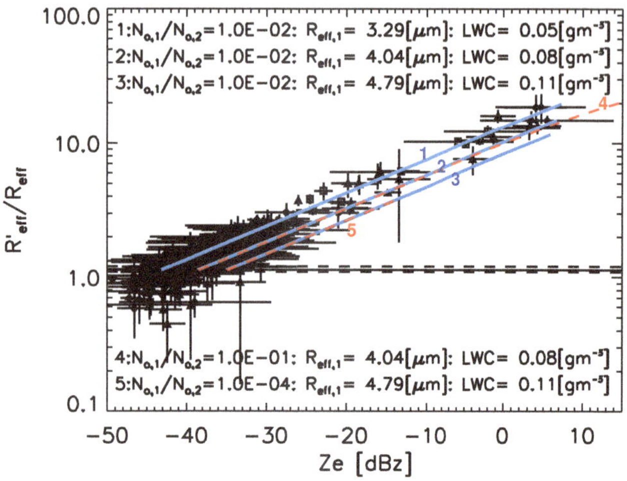

Figure B2. Values of $R'_{\mathrm{eff}}/R_{\mathrm{eff}}$ obtained by combining the lidar inversion results and the observed cloud-base reflectivities. The different symbol types correspond to different time periods as described in Fig. B1. The horizontal lines correspond to the uni-modal limit (Eq. 8) with $\gamma = 9 \pm 4$). The light-blue lines correspond to the relationships predicted using Eqs. (B5) and (B6) found by varying $R_{\mathrm{m},2}$ with the other parameters fixed to their the indicated respective values.

information to retrieve separate $N_{\mathrm{o},2}/N_{\mathrm{o},1}$ and $R_{\mathrm{m},2}$ values, but it can be concluded that the ranges of these parameters implied by the observations are consistent with the range of values expected from earlier in situ investigations of cloud-base conditions (Tonttila et al., 2011; Wang and Geerts, 2003; Liu et al., 2008). It is interesting to note that strictly uni-modal behaviour appears to be associated with only very low reflectivities (less than −35 to −40 dBZ). This "drizzle threshold" is much lower than some thresholds employed in earlier studies (e.g. the threshold for the presence of drizzle of −15 dBZ used in by Sauvageot and Omar (1987). However as discussed in Wang and Geerts (2003) and Liu et al. (2008), the idea of a threshold is a subjective one and different drizzle thresholds are based on different metrics (e.g. reflectivity, LWC, or number density fraction) appropriate to different applications. Further it can be stated the results presented here are consistent with previous studies which have indeed noted the increasing relative influence of drizzle mode droplets to the total radar reflectivity as cloud base is approached (e.g. Wang and Geerts, 2003).

Acknowledgements. The aerosol size distribution measurements used in this paper were partly funded by European Union Seventh Framework Program (FP7/2007–2013) under grant agreement no. 262254.

The LES simulations used in this work were sponsored by the Dutch National Computing Facilities Foundation (NCF) who provided the use of supercomputer facilities.

Edited by: U. Wandinger

References

Arabas, S., Pawlowska, H., and Grabowski, W. W.: Effective radius and droplet spectral width from in-situ aircraft observations in trade-wind cumuli during RICO, Geophys. Res. Letts., 36, L11803, doi:10.1029/2009gl038257, 2009.

Baedi, R., Dewit, J., Russchenberg, H., Erkelens, J., and Poiares-Baptista, J.: Estimating effective radius and liquid water content from radar and lidar based on the CLARE98 dataset, Phys. Chem. Earth, 25, 1057–1062, doi:10.1016/s1464-1909(00)00152-0, 2000.

Barker, H. W., Goldstein, R. K., and Stevens, D. E.: Monte Carlo Simulation of Solar Reflectances for Cloudy Atmospheres, J. Atmos. Sci., 60, 1881–1894, doi:10.1175/1520-0469(2003)060<1881:MCSOSR>2.0.CO;2, 2003.

Behrendt, A. and Nakamura, T.: Calculation of the calibration constant of polarization lidar and its dependency on atmospheric temperature, Opt. Express, 10, 805–817, 2002.

Bissonnette, L. R.: Lidar and multiple scattering, in: Lidar range-resolved optical remote sensing of the atmosphere, edited by: C. Weitkamp, Springer series in optical sciences, 102, ISBN 0387400753, 456 pp., 2005.

Bissonnette, L. R., Roy, G., and Roy, N.: Multiple-scattering-based lidar retrieval: method and results of cloud probings, Appl. Optics, 44, 5565–5581, 2005.

Buras, R. and Mayer, B.: Efficient unbiased variance reduction techniques for Monte Carlo simulations of radiative transfer in cloudy atmospheres: the solution, J. Quant. Spectrosc. Ra., 112, 434–447, doi:10.1016/j.jqsrt.2010.10.005, 2011.

Cao, X., Roy, G., Roy, N., and Bernier, R.: Comparison of the relationships between lidar integrated backscattered light and accumulated depolarization ratios for linear and circular polarization for water droplets, fog oil, and dust, Appl. Optics, 48, 4130–4141, 2009.

Chaikovskaya, L.: Remote sensing of clouds using linearly and circularly polarized laser beams: techniques to compute signal polarization, in: Light Scattering Reviews 3, edited by: Kokhanovsky, A. A., Springer Praxis Books, 191–228, Springer, Berlin Heidelberg, doi:10.1007/978-3-540-48546-9_6, 2008.

Chaikovskaya, L. I. and Zege, E. P.: Theory of polarized lidar sounding including multiple scattering, J. Quant. Spectrosc. Ra., 88, 21–35, doi:10.1016/j.jqsrt.2004.01.002, 2004.

Deirmendjian, D.: Electromagnetic scattering on spherical polydispersions, R (Rand Corporation), American Elsevier Pub. Co., 1969.

de Roode, S. R. and Los, A.: The effect of temperature and humidity fluctuations on the liquid water path of non-precipitating closed-cell stratocumulus clouds, Q. J. Roy. Meteor. Soc., 134, 403–416, doi:10.1002/qj.222, 2008.

Donovan, D., Voors, R., van Zadelhoff, G.-J., and Acarreta, J.-R.: ECSIM Model and Algorithms Document, KNMI Tech. Rep.: ECSIM-KNMI-TEC-MAD01-R, available at: http://www.knmi.nl/~zadelhof/file/ECSIM-KNMI-TEC-MAD01-122-R.pdf (last access: 23 September 2014), 2010.

Donovan, D. P. and Apituley, A.: Practical depolarization-ratio-based inversion procedure: lidar measurements of the Eyjafjallajkull ash cloud over the Netherlands, Appl. Optics, 52, 2394–2415, 2013a.

Donovan, D. P. and Apituley, A.: Practical depolarization-ratio-based inversion procedure: lidar measurements of the Eyjafjallajkull ash cloud over the Netherlands: erratum, Appl. Optics, 52, 4062–4062, 2013b.

Donovan, D. P. and van Lammeren, A. C. A. P.: Cloud effective particle size and water content profile retrievals using combined lidar and radar observations – 1. Theory and examples, J. Geophys. Res., 106, 27425–27448, 2001.

Dusek, U., Frank, G. P., Hildebrandt, L., Curtius, J., Schneider, J., Walter, S., Chand, D., Drewnick, F., Hings, S., Jung, D., Borrmann, S., and Andreae, M. O.: Size matters more than chemistry for cloud-nucleating ability of aerosol particles., Science, 312, 1375–1378, 2006

Eloranta, E. W.: Practical Model for the Calculation of Multiply Scattered Lidar Returns, Appl. Optics, 37, 2464–2472, 1998.

Fox, N. I. and Illingworth, A. J.: The potential of a spaceborne cloud radar for the detection of stratocumulus clouds, J. Appl. Meteorol., 36, 676–687, doi:10.1175/1520-0450-36.6.676, 1997a.

Fox, N. I. and Illingworth, A. J.: The retrieval of stratocumulus cloud properties by ground-based cloud radar, J. Appl. Meteorol., 36, 485–492, doi:10.1175/1520-0450(1997)036<0485:TROSCP>2.0.CO;2, 1997b.

Guerrero-Rascado, J. L., ao Costa, M. J., Bortoli, D., Silva, A. M., Lyamani, H., and Alados-Arboledas, L.: Infrared lidar overlap function: an experimental determination, Opt. Express, 18, 20350–20359, 2010.

Gultepe, I. and Isaac, G. A.: The relationship between cloud droplet and aerosol number concentrations for climate models, Int. J. Climatol., 16, 941–946, doi:10.1002/(SICI)1097-0088(199608)16:8<941::AID-JOC57>3.0.CO;2-O, 1996.

Heus, T., van Heerwaarden, C. C., Jonker, H. J. J., Pier Siebesma, A., Axelsen, S., van den Dries, K., Geoffroy, O., Moene, A. F., Pino, D., de Roode, S. R., and Vilà-Guerau de Arellano, J.: Formulation of the Dutch Atmospheric Large-Eddy Simulation (DALES) and overview of its applications, Geosci. Model Dev., 3, 415–444, doi:10.5194/gmd-3-415-2010, 2010.

Hogan, R. J.: Fast approximate calculation of multiply scattered lidar returns, Appl. Optics, 45, 5984–5992, 2006.

Hu, Y., Vaughan, M., Liu, Z., Lin, B., Yang, P., Flittner, D., Hunt, B., Kuehn, R., Huang, J., Wu, D., Rodier, S., Powell, K., Trepte, C., and Winker, D.: The depolarization – attenuated backscatter relation: CALIPSO lidar measurements vs. theory, Opt. Express, 15, 5327–5332, 2007.

Hu, Y.-X., Winker, D., Yang, P., Baum, B., Poole, L., and Vann, L.: Identification of cloud phase from PICASSO-CENA lidar depolarization: a multiple scattering sensitivity study, J. Quant. Spectrosc. Ra., 70, 569–579, doi:10.1016/S0022-4073(01)00030-9, 2001.

Jacobs, A., Barkmeijer, J., Siebesma, A., van der Plas, E., Wichers Schreur, B., Roozekrans, J., Holtslag, A., Steeneveld, G., Ronda, R., de Roode, S. R., Mendez Gomez, R., van den Brink, P., Haanstra, J., and Smit, L.: The Impact of Climate Change on the Critical Weather Conditions at Schiphol Airport (Impact), Dutch Knowledge for Climate Programme Office, ISBN 9789490070571, 76 pp., 2012.

Katsev, I. L., Zege, E. P., Prikhach, A. S., and Polonsky, I. N.: Efficient technique to determine backscattered light power for various atmospheric and oceanic sounding and imaging systems, J. Opt. Soc. Am. A, 14, 1338–1346, 1997.

Khairoutdinov, M. and Kogan, Y.: A new cloud physics parameterization in a large-eddy simulation model of marine stratocumulus, Mon. Weather Rev., 128, 229–243, 2000.

Kim, D., Cheong, H., Kim, Y., Volkov, S., and Lee, J.: Optical depth and multiple scattering depolarization in liquid clouds, Opt. Rev., 17, 507–512, doi:10.1007/s10043-010-0091-7, 2010.

Kulmala, M., Asmi, A., Lappalainen, H. K., Baltensperger, U., Brenguier, J.-L., Facchini, M. C., Hansson, H.-C., Hov, Ø., O'Dowd, C. D., Pöschl, U., Wiedensohler, A., Boers, R., Boucher, O., de Leeuw, G., Denier van der Gon, H. A. C., Feichter, J., Krejci, R., Laj, P., Lihavainen, H., Lohmann, U., McFiggans, G., Mentel, T., Pilinis, C., Riipinen, I., Schulz, M., Stohl, A., Swietlicki, E., Vignati, E., Alves, C., Amann, M., Ammann, M., Arabas, S., Artaxo, P., Baars, H., Beddows, D. C. S., Bergström, R., Beukes, J. P., Bilde, M., Burkhart, J. F., Canonaco, F., Clegg, S. L., Coe, H., Crumeyrolle, S., D'Anna, B., Decesari, S., Gilardoni, S., Fischer, M., Fjaeraa, A. M., Fountoukis, C., George, C., Gomes, L., Halloran, P., Hamburger, T., Harrison, R. M., Herrmann, H., Hoffmann, T., Hoose, C., Hu, M., Hyvärinen, A., Hõrrak, U., Iinuma, Y., Iversen, T., Josipovic, M., Kanakidou, M., Kiendler-Scharr, A., Kirkevåg, A., Kiss, G., Klimont, Z., Kolmonen, P., Komppula, M., Kristjánsson, J.-E., Laakso, L., Laaksonen, A., Labonnote, L., Lanz, V. A., Lehtinen, K. E. J., Rizzo, L. V., Makkonen, R., Manninen, H. E., McMeeking, G., Merikanto, J., Minikin, A., Mirme, S., Morgan, W. T., Nemitz, E., O'Donnell, D., Panwar, T. S., Pawlowska, H., Petzold, A., Pienaar, J. J., Pio, C., Plass-Duelmer, C., Prévôt, A. S. H., Pryor, S., Reddington, C. L., Roberts, G., Rosenfeld, D., Schwarz, J., Seland, Ø., Sellegri, K., Shen, X. J., Shiraiwa, M., Siebert, H., Sierau, B., Simpson, D., Sun, J. Y., Topping, D., Tunved, P., Vaattovaara, P., Vakkari, V., Veefkind, J. P., Visschedijk, A., Vuollekoski, H., Vuolo, R., Wehner, B., Wildt, J., Woodward, S., Worsnop, D. R., van Zadelhoff, G.-J., Zardini, A. A., Zhang, K., van Zyl, P. G., Kerminen, V.-M., S Carslaw, K., and Pandis, S. N.: General overview: European Integrated project on Aerosol Cloud Climate and Air Quality interactions (EUCAARI) – integrating aerosol research from nano to global scales, Atmos. Chem. Phys., 11, 13061–13143, doi:10.5194/acp-11-13061-2011, 2011.

Leaitch, W. R., Isaac, G. A., Strapp, J. W., Banic, C. M., and Wiebe, H. A.: The relationship between cloud droplet number concentrations and anthropogenic pollution: observations and climatic implications, J. Geophys. Res., 97, 2463–2474, doi:10.1029/91JD02739, 1992.

Leijnse, H., Uijlenhoet, R., van de Beek, C. Z., Overeem, A., Otto, T., Unal, C. M. H., Dufournet, Y., Russchenberg, H. W. J., Figueras i Ventura, J., Klein Baltink, H., and Holleman, I.: Precipitation measurement at CESAR, the Netherlands, J. Hydrometeorol., 11, 1322–1329, doi:10.1175/2010JHM1245.1, 2010.

Liou, K.: Preface, in: An Introduction to Atmospheric Radiation, edited by: Liou, K., vol. 84 of International Geophysics, xiii–xiv, Academic Press, doi:10.1016/S0074-6142(02)80015-8, 2002.

Liou, K.-N. and Schotland, R. M.: Multiple backscattering and depolarization from water clouds for a pulsed lidar system, J. Atmos. Sci., 28, 772–784, doi:10.1175/1520-0469(1971)028<0772:MBADFW>2.0.CO;2, 1971.

Liu, Y., Geerts, B., Miller, M., Daum, P., and McGraw, R.: Threshold radar reflectivity for drizzling clouds, Geophys. Res. Lett., 35, L03807, doi:10.1029/2007GL031201, 2008.

Lohmann, U. and Feichter, J.: Global indirect aerosol effects: a review, Atmos. Chem. Phys., 5, 715–737, doi:10.5194/acp-5-715-2005, 2005.

Lu, M.-L. and Seinfeld, J. H.: Effect of aerosol number concentration on cloud droplet dispersion: a large-eddy simulation study and implications for aerosol indirect forcing, J. Geophys. Res., 111, D02207, doi:10.1029/2005JD006419, 2006.

Malinka, A. V. and Zege, E. P.: Possibilities of warm cloud microstructure profiling with multiple-field-of-view Raman lidar, Appl. Optics, 46, 8419–8427, 2007.

Martin, G. M., Johnson, D. W., and Spice, A.: The measurement and parameterization of effective radius of droplets in warm stratocumulus clouds, J. Atmos. Sci., 51, 1823–1842, doi:10.1175/1520-0469(1994)051<1823:TMAPOE>2.0.CO;2, 1994.

McComiskey, A., Feingold, G., Shelby Frisch, A., Turner, D. D., Miller, M. A., Chiu, J. C., Min, Q., and Ogren, J. A.: An assessment of aerosol-cloud interactions in marine stratus clouds based on surface remote sensing, J. Geophys. Res., 114, D09203, doi:10.1029/2008JD011006, 2009.

Measures, R.: Laser Remote Sensing: Fundamentals and Applications, Krieger Publishing Company, ISBN 9780894646195, 510 pp., 1984.

Mensah, A. A., Holzinger, R., Otjes, R., Trimborn, A., Mentel, Th. F., ten Brink, H., Henzing, B., and Kiendler-Scharr, A.: Aerosol chemical composition at Cabauw, The Netherlands as observed in two intensive periods in May 2008 and March 2009, Atmos. Chem. Phys., 12, 4723–4742, doi:10.5194/acp-12-4723-2012, 2012.

Miles, N. L., Verlinde, J., and Clothiaux, E. E.: Cloud droplet size distributions in low-level stratiform clouds, J. Atmos. Sci., 57, 295–311, doi:10.1175/1520-0469(2000)057<0295:CDSDIL>2.0.CO;2, 2000.

Pal, S. R. and Carswell, A. I.: Polarization Properties of Lidar Backscattering from Clouds, Appl. Optics, 12, 1530–1535, 1973.

Pawlowska, H., Grabowski, W. W., and Brenguier, J.-L.: Observations of the width of cloud droplet spectra in stratocumulus, Geophys. Res. Lett., 33, L19810, doi:10.1029/2006GL026841, 2006.

Petzold, A., Esselborn, M., Weinzierl, B., Ehret, G., Ansmann, A., Müller, D., Donovan, D., van Zadelhoff, G.-J., Berthier, S., Wiegner, M., Gasteiger, J., Buras, R., Mayer, B., Lajas, D., and Wehr, T.: ICAROHS – Inter-Comparison of Aerosol Retrievals and Observational Requirements for Multi-Wavelength

HSRL Systems, Final Report, Tech. Rep. 22169/NL/CT–STSE-ICAROHS TN1, ESA, 2011.

Pinsky, M., Khain, A., Mazin, I., and Korolev, A.: Analytical estimation of droplet concentration at cloud base, J. Geophys. Res., 117, D18211, doi:10.1029/2012JD017753, 2012.

Platt, C. M. R.: Remote sounding of high clouds. III: Monte Carlo calculations of multiple-scattered lidar returns, J. Atmos. Sci., 38, 156–167, doi:10.1175/1520-0469(1981)038<0156:RSOHCI>2.0.CO;2, 1981.

Pounder, N. L., Hogan, R. J., Várnai, T., Battaglia, A., and Cahalan, R. F.: A variational method to retrieve the extinction profile in liquid clouds using multiple-field-of-view lidar, J. Appl. Meteorol. Clim., 51, 350–365, doi:10.1175/JAMC-D-10-05007.1, 2012.

Press, W. H., Teukolsky, S. A., Vetterling, W. T., and Flannery, B. P.: Numerical Recipes 3rd Edition: The Art of Scientific Computing, Cambridge University Press, New York, NY, USA, 3rd Edn., 2007.

Pringle, K. J., Carslaw, K. S., Spracklen, D. V., Mann, G. M., and Chipperfield, M. P.: The relationship between aerosol and cloud drop number concentrations in a global aerosol microphysics model, Atmos. Chem. Phys., 9, 4131–4144, doi:10.5194/acp-9-4131-2009, 2009.

Rodgers, C.: Inverse Methods for Atmospheric Sounding: Theory and Practice, Series on Atmospheric Oceanic and Planetary Physics, Volume 2, World Scientific Publishing Company Inc., Incorporated, ISBN 9789810227401, 238 pp., 2000.

Roy, G. and Cao, X.: Inversion of water cloud lidar signals based on accumulated depolarization ratio, Appl. Optics, 49, 1630–1635, 2010.

Roy, G., Bissonnette, L., Bastille, C., and Vallée, G.: Retrieval of Droplet-Size Density Distribution from Multiple-Field-of-View Cross-Polarized Lidar Signals: Theory and Experimental Validation, Appl. Optics, 38, 5202–5211, 1999.

Sassen, K.: Polarization in lidar, in: Lidar, edited by: Weitkamp, C., vol. 102 of Springer Series in Optical Sciences, 19–42, Springer, New York, doi:10.1007/0-387-25101-4_2, 2005.

Sassen, K. and Petrilla, R. L.: Lidar depolarization from multiple scattering in marine stratus clouds, Appl. Optics, 25, 1450–1459, 1986.

Sassen, K. and Zhao, H.: Lidar multiple scattering in water droplet clouds: toward an improved treatment, Opt. Rev., 2, 394–400, doi:10.1007/s10043-995-0394-2, 1995.

Sauvageot, H. and Omar, J.: Radar reflectivity of cumulus clouds, J. Atmos. Ocean. Tech., 4, 264–272, doi:10.1175/1520-0426(1987)004<0264:RROCC>2.0.CO;2, 1987.

Schmidt, J., Wandinger, U., and Malinka, A.: Dual-field-of-view Raman lidar measurements for the retrieval of cloud microphysical properties, Appl. Optics, 52, 2235–2247, 2013.

Sun, Y.-Y. and Li, Z.-P.: Depolarization of polarized light caused by high altitude clouds. 2: Depolarization of lidar induced by water clouds, Appl. Optics, 28, 3633–3638, 1989.

Szczodrak, M., Austin, P. H., and Krummel, P. B.: Variability of optical depth and effective radius in marine stratocumulus clouds, J. Atmos. Sci., 58, 2912–2926, doi:10.1175/1520-0469(2001)058<2912:VOODAE>2.0.CO;2, 2001.

Tonttila, J., O'Connor, E. J., Niemelä, S., Räisänen, P., and Järvinen, H.: Cloud base vertical velocity statistics: a comparison between an atmospheric mesoscale model and remote sensing observations, Atmos. Chem. Phys., 11, 9207–9218, doi:10.5194/acp-11-9207-2011, 2011.

Veselovskii, I., Korenskii, M., Griaznov, V., Whiteman, D. N., McGill, M., Roy, G., and Bissonnette, L.: Information content of data measured with a multiple-field-of-view lidar, Appl. Optics, 45, 6839–6848, 2006.

Voors, R., Donovan, D., Acarreta, J., Eisinger, M., Franco, R., Lajas, D., Moyano, R., Pirondini, F., Ramos, J., and Wehr, T.: ECSIM: the simulator framework for EarthCARE, Proc. SPIE, 6744, 67441Y–67441Y–11, doi:10.1117/12.737738, 2007.

Wang, J. and Geerts, B.: Identifying drizzle within marine stratus with W-band radar reflectivity, Atmos. Res., 69, 1–27, doi:10.1016/j.atmosres.2003.08.001, 2003.

4

An intercomparison study of analytical methods used for quantification of levoglucosan in ambient aerosol filter samples

K. E. Yttri[1], J. Schnelle-Kreis[2], W. Maenhaut[3,6], G. Abbaszade[2], C. Alves[4], A. Bjerke[1], N. Bonnier[14], R. Bossi[5], M. Claeys[6], C. Dye[1], M. Evtyugina[4], D. García-Gacio[7], R. Hillamo[8], A. Hoffer[9], M. Hyder[10,15], Y. Iinuma[11], J.-L. Jaffrezo[12], A. Kasper-Giebl[13], G. Kiss[9], P. L. López-Mahia[7], C. Pio[4], C. Piot[12,16], C. Ramirez-Santa-Cruz[13], J. Sciare[14], K. Teinilä[8], R. Vermeylen[6], A. Vicente[4], and R. Zimmermann[2]

[1]NILU – Norwegian Institute for Air Research, P.O. Box 100, 2027 Kjeller, Norway
[2]Helmholtz Zentrum München, Cooperation group Comprehensive Molecular Analytics, 85764 Neuherberg, Germany
[3]Department of Analytical Chemistry, Ghent University, 9000 Gent, Belgium
[4]Centre for Environmental and Marine Studies, Department of Environment, University of Aveiro, 3810-193 Aveiro, Portugal
[5]Department of Environmental Science, Aarhus University, Frederiksborgvej 399, 4000 Roskilde, Denmark
[6]Department of Pharmaceutical Sciences, University of Antwerp, Campus Drie Eiken, 2610 Antwerp, Belgium
[7]Group of Analytical Chemistry (QANAP), Institute of Environment (IUMA), Department of Analytical Chemistry, Faculty of Sciences, University of A Coruña, Campus de A Coruña, 15071 A Coruña, Spain
[8]Finnish Meteorological Institute, Atmospheric Composition Research, Aerosol Research, Erik Palménin Aukio 1, 00560, Helsinki, Finland
[9]MTA-PE Air Chemistry Research Group, P.O. Box 158, 8201Veszprém, Hungary
[10]Center for Analysis and Synthesis, Department of Chemistry, Lund University, Sweden
[11]Leibniz-Institut für Troposphärenforschung, Permoserstr. 15, 04138 Leipzig, Germany
[12]Univ. Grenoble Alpes, LGGE, 38000 Grenoble, France, CNRS, LGGE, 38000 Grenoble, France
[13]Institute of Chemical Technologies and Analytics, Vienna University of Technology, Getreidemarkt 9/164, 1060 Vienna, Austria
[14]Laboratoire des Sciences du Climat et de l'Environnement, CEA-CNRS-UVSQ, 91190 Gif-sur-Yvette, France
[15]Pakistan Space and Upper Atmosphere research commission (SUPARCO), Lahore, Pakistan
[16]Université de Savoie, LCME, 73376 Le Bourget du lac, France

Correspondence to: K. E. Yttri (key@nilu.no)

Abstract. The monosaccharide anhydrides (MAs) levoglucosan, galactosan and mannosan are products of incomplete combustion and pyrolysis of cellulose and hemicelluloses, and are found to be major constituents of biomass burning (BB) aerosol particles. Hence, ambient aerosol particle concentrations of levoglucosan are commonly used to study the influence of residential wood burning, agricultural waste burning and wildfire emissions on ambient air quality. A European-wide intercomparison on the analysis of the three monosaccharide anhydrides was conducted based on ambient aerosol quartz fiber filter samples collected at a Norwegian urban background site during winter. Thus, the samples' content of MAs is representative for BB particles originating from residential wood burning. The purpose of the intercomparison was to examine the comparability of the great diversity of analytical methods used for analysis of levoglucosan, mannosan and galactosan in ambient aerosol filter samples. Thirteen laboratories participated, of which three applied high-performance anion-exchange chromatography (HPAEC), four used high-performance liquid chromatography (HPLC) or ultra-performance liquid chromatography (UPLC) and six resorted to gas chromatography (GC). The analytical methods used were of such diversity that they should be considered as thirteen different analytical methods. All of the thirteen laboratories reported levels of levoglucosan, whereas nine reported data for mannosan and/or galactosan. Eight of the thirteen laboratories reported levels for all three isomers.

The accuracy for levoglucosan, presented as the mean percentage error (PE) for each participating laboratory, varied from −63 to 20 %; however, for 62 % of the laboratories the mean PE was within ±10 %, and for 85 % the mean PE was within ±20 %. For mannosan, the corresponding range was −60 to 69 %, but as for levoglucosan, the range was substantially smaller for a subselection of the laboratories; i.e. for 33 % of the laboratories the mean PE was within ±10 %. For galactosan, the mean PE for the participating laboratories ranged from −84 to 593 %, and as for mannosan 33 % of the laboratories reported a mean PE within ±10 %.

The variability of the various analytical methods, as defined by their minimum and maximum PE value, was typically better for levoglucosan than for mannosan and galactosan, ranging from 3.2 to 41 % for levoglucosan, from 10 to 67 % for mannosan and from 6 to 364 % for galactosan. For the levoglucosan to mannosan ratio, which may be used to assess the relative importance of softwood versus hardwood burning, the variability only ranged from 3.5 to 24 %.

To our knowledge, this is the first major intercomparison on analytical methods used to quantify monosaccharide anhydrides in ambient aerosol filter samples conducted and reported in the scientific literature. The results show that for levoglucosan the accuracy is only slightly lower than that reported for analysis of SO_4^{2-} (sulfate) on filter samples, a constituent that has been analysed by numerous laboratories for several decades, typically by ion chromatography and which is considered a fairly easy constituent to measure. Hence, the results obtained for levoglucosan with respect to accuracy are encouraging and suggest that levels of levoglucosan, and to a lesser extent mannosan and galactosan, obtained by most of the analytical methods currently used to quantify monosaccharide anhydrides in ambient aerosol filter samples, are comparable.

Finally, the various analytical methods used in the current study should be tested for other aerosol matrices and concentrations as well, the most obvious being summertime aerosol samples affected by wildfires and/or agricultural fires.

1 Introduction

The monosaccharide anhydrides (MAs) levoglucosan, galactosan and mannosan are products of the incomplete combustion and pyrolysis of cellulose and hemicelluloses, and are thus constituents of biomass burning (BB) aerosol (Hornig et al., 1985; Simoneit, 1999). The presence of MAs has been demonstrated in emissions from wood combustion (Oros and Simoneit, 1999; Fine et al., 2001; Nolte et al., 2001; Schmidl et al., 2008; Frey et al., 2009; Gonçalves et al., 2010, 2011; Orasche et al., 2012) as well as from wildfires, prescribed and agricultural fires (dos Santos et al., 2002; Gao et al., 2003; Sillanpää et al., 2005; Mazzoleni et al., 2007; Sullivan et al., 2008; Alves et al., 2010; Saarnio et al., 2010a; Vicente et al., 2012; Wang et al., 2011). Recently, it has been shown that combustion of peat (Iinuma et al., 2007; Kourtchev et al., 2011) as well as of certain types of brown coal (Fabbri et al., 2009) could be additional sources of ambient aerosol concentrations of MAs in regions where such fuels are utilised.

Levoglucosan exhibits the highest concentration of the MAs in BB aerosol; consequently, it has been applied for source apportionment of ambient particulate matter (PM). Additional information can be drawn from the PM MAs pattern, e.g. it has been used to estimate the impact from combustion of different wood types; however, no clear picture of the variability of MAs concentrations and ratios from wood combustion has been found. Some studies have reported a high variability depending on burning conditions or type of wood with filtration-based sampling (Schmidl et al., 2008; Orasche et al., 2012, 2013) or by application of aerosol mass spectrometry (Weimer et al., 2008; Elsasser et al., 2013). Other studies have shown relatively little variation in MAs emissions during wood combustion under varying conditions (Jordan and Seen, 2005; Frey et al., 2009) or with different types of wood (Fine et al., 2001, 2002, 2004). Size distribution measurements of MAs in particulate emission samples from wood combustion have shown that most MAs are associated with PM $< 1.2\,\mu m$ (Frey et al., 2009). These results are in agreement with wintertime observations showing that most levoglucosan is present in the PM_1 fraction of the ambient aerosol (Yttri et al., 2005; Agarwal et al., 2010; Krumal et al., 2010).

The atmospheric concentration of levoglucosan is highly variable as it depends not only on the source strength, but also on meteorological parameters. During wintertime in Europe residential wood burning is the major source of levoglucosan, and the observed concentration is typically $< 1\,\mu g\,m^{-3}$ (Oliveira et al., 2007; Yttri et al., 2009; Krumal et al., 2010; Schnelle-Kreis et al., 2010; Caseiro and Oliveira, 2012; Herich et al., 2014). In the evening and during night-time higher concentrations ($> 1\,\mu g\,m^{-3}$) can be observed (Elsasser et al., 2012; Saffari et al., 2013). Agricultural fires are reported to cause ambient levoglucosan concentrations well above $1\,\mu g\,m^{-3}$ (Pietrogrande et al., 2013), and concentrations $> 7\,\mu g\,m^{-3}$ have been reported for haze episodes in tropical regions (Claeys et al., 2010; He et al., 2010).

Early investigations suggested that levoglucosan is resistant to atmospheric acid-catalysed degradation (Fraser and Lakshmanan, 2000). The low vapour pressure of levoglucosan (Oja and Suuberg, 1999; Booth et al., 2011) as well as the minor impact of temperature on the observed volatility of levoglucosan in ambient PM (Ruehl et al., 2011) further supports its stability and thus its suitability as a tracer species. Recent investigations on the stability of levoglucosan in aqueous solution suggest that "oxidation of levoglucosan in atmospheric deliquescent particles is at least as fast as that of other atmospherically relevant organic compounds and levoglucosan may not be as stable as previously thought in the atmosphere, especially under high relative humidity conditions" (Hoffmann et al., 2010; Teraji and Arakaki, 2010). Further, oligomerisation of levoglucosan by Fenton chemistry in aqueous solution has been observed (Holmes and Petrucci, 2007). The decay of levoglucosan upon ageing of wood combustion aerosol has been studied in smog chamber experiments. When exposed to typical summertime OH radical concentrations, the atmospheric lifetime of levoglucosan in BB particles was found to be 0.7–2.2 days (Hennigan et al., 2010) under BB plume conditions and estimated to range up to 2 weeks for atmospheric background conditions (Slade and Knopf, 2013). The impact of the abovementioned effects on the levoglucosan content in the (true) ambient aerosol still remains to be documented; however, these are indications that levoglucosan may be depleted in the atmosphere, and that possible degradation should thus be taken into account when using levoglucosan as a tracer of BB emissions in source apportionment studies. Nonetheless, levoglucosan is a valuable tracer to investigate the impact of residential wood burning in urban (Oliveira et al., 2007; Timonen et al., 2008; Caseiro et al., 2009; Szidat et al., 2009; Bari et al., 2010; Schnelle-Kreis et al., 2010; Saffari et al., 2013), rural background (Gelencsér et al., 2007; Puxbaum et al., 2007, van Drooge and Ballesta, 2009) and remote areas (May et al., 2009; Yttri et al., 2014), especially during winter, when wood combustion is used for domestic heating.

As levoglucosan is commonly used as a tracer of emissions from biomass combustion, increasing efforts have been put into levoglucosan quantification during the last decade. Gas chromatographic methods have been established since the end of the last century, whereas liquid chromatographic methods were more recently developed and have been gaining attention. A review of existing methods was published in 2006 (Schkolnik and Rudich, 2006) and derivatisation methods for analysis of MAs using gas chromatography (GC) were covered by Hsu et al. (2007). Thus, only a brief overview on actual existing methods and developments will be given here.

Gas chromatography-mass spectrometry (GC–MS) methods for separation and quantification of MAs are widespread. Most GC methods used for analysis of MAs in environmental samples make use of solvent extraction followed by trimethylsilylation (Simoneit, 1999; Pashynska et al., 2002; Simpson et al., 2004; Medeiros and Simoneit, 2007). Reactive pyrolysis, i.e. thermally assisted alkylation of MAs, established in pyrolysis research (Fabbri et al., 2002), has also been applied to environmental samples (Blazso et al., 2003), and more recently, thermal extraction methods have been developed. Lin et al. (2007) investigated solid-phase microextraction for pre-concentration of levoglucosan from filter samples prior to GC analysis, whereas Orasche et al. (2011) developed a method for in situ derivatisation of MAs in filter samples followed by GC–MS. The most recent advance was made by Ma et al. (2010), who developed a two-dimensional GC method, which uses heart-cutting and thermal extraction of MAs from filter samples without the need of chemical derivatisation.

High-performance liquid chromatography combined with mass spectrometry (HPLC-MS) and high-performance anion-exchange chromatography (HPAEC) are the most commonly used methods for analysis of MAs in extracts based on polar solvents. Methods such as capillary electrophoresis with pulsed amperometric detection (CE-PAD) (Garcia et al., 2005), chloride attachment in liquid chromatography combined with negative ion electrospray mass spectrometry (Wan and Yu, 2007) and ion-exclusion high-performance liquid chromatography followed by spectroscopic detection (Schkolnik et al., 2005) have been reported in the scientific literature but do not seem to be widely applied to MA analysis. Gao et al. (2003) were the first to report the use of electrospray ionisation mass spectrometry (ESI-MS) for quantification of MAs, and, at present, HPLC combined with mass spectrometric (Dye and Yttri, 2005; Larsen et al., 2006; Wan and Yu, 2007; Piot et al., 2012) or aerosol charge detection (HPLC-ACD) (Dixon and Baltzell, 2006) are widespread. HPAEC coupled with pulsed amperometric detection (PAD) (Engling et al., 2006; Caseiro et al., 2007; Iinuma et al., 2009; Piazzalunga et al., 2010) or with mass spectrometric detection (HPAEC-MS) (Saarnio et al., 2010b) have been employed more recently.

The objective of the current intercomparison is to examine the comparability of the great diversity of analytical methods used for analysis of levoglucosan, mannosan and galactosan in ambient aerosol filter samples.

2 Experimental

2.1 Site description and measurement period

Aerosol filter samples were collected at the urban background site Kjeller (59°58′ N, 11°3′ E), on the outskirts of the small town of Lillestrøm (20 000 inhabitants), located 20 km east of Oslo (Norway). The samples were collected during the period 11–16 December 2008. The sampling inlets were installed approximately 4 m a.g.l.

2.2 Aerosol filter samples collection, storage, preparation and shipment

Six ambient aerosol filter samples were provided using two high-volume samplers with total suspended particulate matter (TSPM) inlets collecting aerosols on quartz fibre filters (Munktel MK 360; 150 mm in diameter; batch number 403743). The instruments were operated at a flow rate of $21.5\ m^3\ h^{-1}$. The sampling time ranged from 24 to 48 h.

After exposure, the filter samples were folded once, with the exposed side facing inwards, wrapped in aluminium foil,

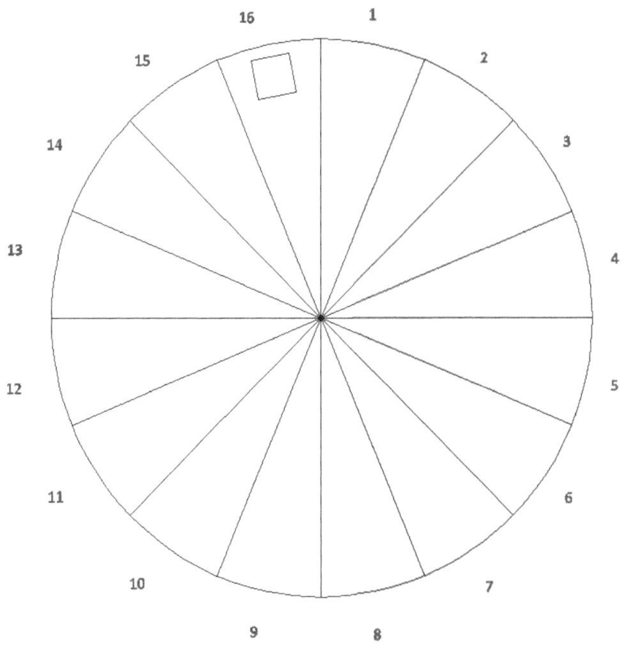

Figure 1. Schematic drawing illustrating the subdivision of one of the six filter samples subjected to the intercomparison. The square in sector 16 illustrates the $1.0\,cm^2$ punch taken from each of the 16 sectors per filter sample to perform a homogeneity test of the filter samples. Each of the sectors sent out to the participating laboratories had an exposed filter area of $8.6\,cm^2$, i.e. when accounting for the removal of the $1.0\,cm^2$ punch.

put in a zip-lock polyethylene bag and stored in a freezer at $-18\,°C$ to prevent any form of degradation of the analytes. Two field blanks were collected, which were treated in exactly the same manner regarding handling, transport and storage as the filters being exposed.

Each of the sampled filter samples was divided into 16 equally large sectors, providing a total of 96 sectors, each with an exposed area of $9.6\,cm^2$. A $1.0\,cm^2$ punch was taken from each of the 16 sectors to perform a homogeneity test of the filters subjected to the intercomparison (Sect. 2.4.1). Hence, each of the sectors sent to the participating laboratories had an exposed filter area of $8.6\,cm^2$ (Fig. 1).

Each sector was given a number, which combined with the filter sample identification gave the sector a unique identification. The sectors were then assigned to the participating laboratories following a randomising approach.

For shipment, each sector was wrapped in aluminium foil and placed in a petri slide. The six petri slides, to be received by each of the participating laboratories, were placed in a zip-lock polyethylene bag and sent in an envelope to the participants by mail. Those receiving the samples were directed to store them in a freezer at $-18\,°C$ until analysis. The deadline for reporting the results was set to be within 70 days after shipment.

2.3 The data set – background, scope and application

Fifteen laboratories located in twelve different European countries reported their interest in the current intercomparison exercise and received sectors from the filter samples for analysis. Two of the fifteen laboratories did not submit results. Hence, the results presented in the current study are based on the results of thirteen different laboratories. All thirteen laboratories reported levels for levoglucosan, whereas nine reported levels for mannosan and galactosan. Eight of thirteen laboratories reported levels for all three isomers, whereas one reported for levoglucosan and mannosan, and one for levoglucosan and galactosan.

A brief overview of the various analytical methods used by the participating laboratories in the current intercomparison is provided in Table 1, including information about the method's capability to separate the isomeric compounds levoglucosan, mannosan and galactosan, the method's limit of detection (LOD), the instrument being used for separation and detection of the analytes, the solvent(s) used for extraction and whether derivatisation of the analytes was applied. Complementary information can be found in Table B1, including extraction procedure, derivatisation agent and recovery standard. A full description of the various analytical methods can be found in Appendix A: in Sect. A1 for those using HPAEC, Sect. A2 for those using HPLC and ultra-performance liquid chromatography (UPLC) and Sect. A3 for those applying GC.

All of the major methods used for analysis of levoglucosan, mannosan and galactosan in ambient aerosol filter samples, and which have been reported in the scientific literature so far, are represented in the present intercomparison. This adds additional strength to any conclusion to be drawn from the study. In addition, the descriptions of the various methods show that none of them can be considered completely identical; i.e. they differ to a large extent with respect to crucial parameters, such as extraction procedure and derivatisation agent, to fulfil such a criterion. This, however, brings an important asset to the intercomparison, as the goal is to evaluate the comparability of the great diversity of analytical methods used for analysis of levoglucosan, mannosan and galactosan in ambient aerosol filter samples. On the contrary, this great diversity prevents comparison of the performance of different subclasses of analytical methods, e.g. GC-based versus LC-based methods.

2.4 Quality assurance

2.4.1 Homogenity of filter samples

A uniformity test was performed to ensure that the two co-located filter samplers generated samples with a homogenous distribution of the analytes.

Each of the six filter samples analysed in the intercomparison were divided into 16 equally large sectors. From each

Table 1. Overview and short description of the analytical methods used by the participating laboratories in the present intercomparison, including the method's capability for isomer separation, the method's limit of detection (LOD), the instrument used for separation and detection of the analytes, the solvent(s) used for extraction and whether derivatisation of the analytes was applied.

Lab no.	Isomer separation	LOD ($ng\,m^{-3}$)[1]	Instrument	Solvent	Derivatisation
1	Yes	0.28	HPLC–HR-TOFMS (ESI-)	tetrahydrofuran	No
2	Yes	2.2	IC–QMS (ESI-)	deionised H_2O	No
3	Yes	5.0	HPAEC–PAD	deionised H_2O	No
4	Yes	0.05	GC–MS	methanol	Yes
5	No	0.05	GC–QMS	dichloromethane/methanol	Yes
6	Yes	1.8	LC–MS (ESI+)	methanol	No
7	Yes	0.004	HPAEC–PAD	deionised H_2O	No
8	Yes	3.0	LC–MS-MS (ESI-)	deionised H_2O	No
9	Yes	0.02	GC–MS	dichloromethane/methanol	Yes
10	Yes	0.19	UPLC–MS-MS	deionised H_2O/methanol	No
11	Yes	0.02	GC–MS	dichloromethane/methanol	Yes
12	Yes	0.004	TD-GC-TOFMS	none	Yes
13	Yes	0.05	GC-QMS	dichloromethane/methanol	Yes

LOD calculated for an air volume of 2.3 $m^3\,h^{-1}$ for 24 hours and an exposed filter area of 12 cm^2. For explanation of abbreviations see Table C1.

sector one 1.0 cm^2 punch was taken and subjected to levoglucosan, mannosan and galactosan analysis using the analytical method applied by lab. 1 (See Appendix A for a detailed description of the analytical methods used by the participating laboratories), i.e. a total of 96 analyses were performed for each of the three isomers. The results from this uniformity test are shown in Table B2 and are briefly summarised as follows.

For levoglucosan, the mean relative standard deviation (RSD) for all six filter samples was 11 ± 2.7 %. The highest RSD was observed for filter sample 3 (16 %) and the lowest for filter sample 2 (9.1 %). For mannosan, the mean RSD was 19 ± 9.9 %, with the lowest RSD for filter sample 1 (10 %) and the highest for filter sample 4 (35 %). For galactosan, the mean RSD was 24 ± 15 %, with the lowest RSD for filter sample 2 (11 %) and the highest for filter sample 3 (45 %). Note that the percentage provided for the RSD also includes the uncertainty of the analytical method (which is < 5 %); hence, the filter homogeneity is in fact up to 5 % lower than that stated above. Two outliers were detected for galactosan (one in filter sample 4 and one in filter sample 5) and one for mannosan (filter sample 5), when running the Grubbs test for outliers. These outliers did not affect the results of the intercomparison, as the sectors from which they were taken were sent out to laboratories which either did not submit results at all or which did not submit results for mannosan (lab. 13), or the sector was not assigned to any of the participating laboratories. Consequently, the results from the homogeneity test demonstrate that the collected filter samples are sufficiently homogenous to be used for an intercomparison.

Finally, it can be questioned to which extent a 1.0 cm^2 punch from each sector is representative for addressing the homogeneity of the filter samples as it accounts for no more than 10 % of the total exposed filter area. However, the size of the punch used for the homogeneity test should not compromise the participation of laboratories using analytical methods with detection limits varying over a wide range.

2.4.2 Outliers – Grubbs test

The Grubbs test was used to detect outliers amongst the results submitted by the various laboratories. For levoglucosan, one outlier was detected for filter samples 1 and 2 and both were reported by the same laboratory (lab. 7). For galactosan, one outlier was detected for each of the six filter samples and all were reported by the same laboratory (lab. 9). No outliers were observed for mannosan. Values defined as outliers were excluded from calculation of the theoretical median value, but included in the measured arithmetic mean value (see Sect. 2.5 for explanation of theoretical and measured value).

2.4.3 Samples with levels below limit of detection

Two laboratories (labs. 3 and 8) reported values of galactosan below the established detection limit of their analytical method for filter samples 3–6. These values were not included in the calculated theoretical median value nor in the measured mean value.

2.5 Calculation of the analytical methods performance

The outcomes of the intercomparison are presented in Figs. 2–4 as aggregated results, whereas the aggregated data reported by the laboratories, which are used as input for Figs. 2–4, are listed in Table B3. The results are presented in terms of the percentage error (PE); i.e. the PE was calcu-

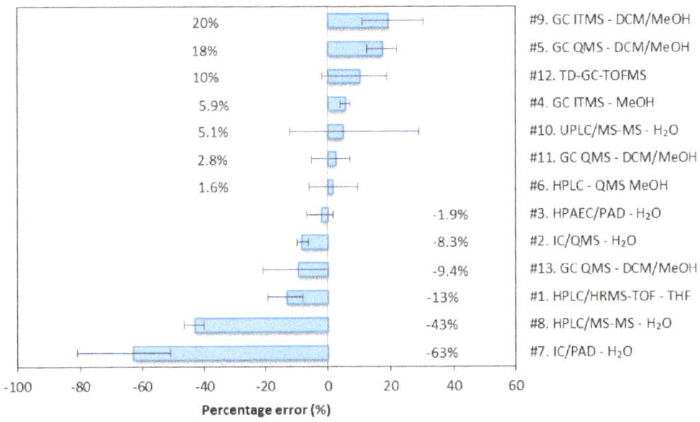

Figure 2. Calculated mean percentage error (PE) for each of the thirteen laboratories reporting levels of levoglucosan in the current intercomparison. The PE was calculated according to Eq. (1) (Sect. 2.5) for each of the six sectors received by the participating laboratories. The mean PE for each laboratory accounts for the PE calculated for all six sectors. The laboratory number, ranging from 1–13, the analytical method and solvent used for extraction are mentioned at the right of the figure.

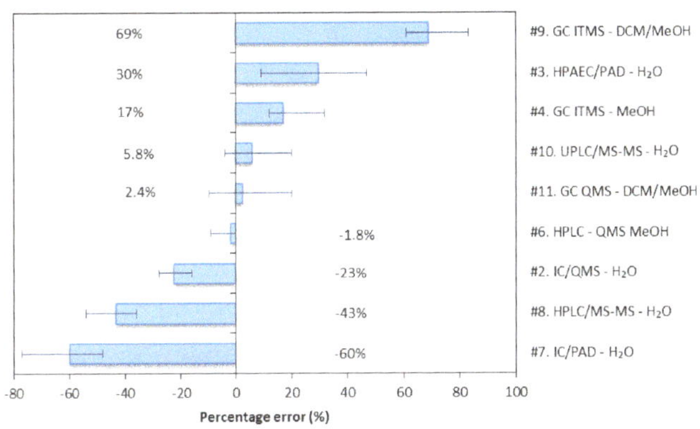

Figure 3. Calculated mean percentage error (PE) for each of the nine laboratories reporting levels of mannosan in the current intercomparison. The PE was calculated according to Eq. (1) (Sect. 2.5) for each of the six sectors received by the participating laboratories. The mean PE for each laboratory accounts for the PE calculated for all six sectors. The laboratory number, ranging from 1–13, the analytical method and solvent used for extraction are mentioned at the right of the figure.

lated for each of the participating laboratories for each of the six filter samples according to Eq. (1). The arithmetic mean PE for each laboratory, accounting for all six filters, was then subsequently calculated (Figs. 2–4).

$$\text{Percentage error (PE)} = \frac{\text{Measured} - \text{Theoretical}}{\text{Theoretical}} \times 100, \quad (1)$$

where measured is the value of the analyte, e.g. levoglucosan, reported by the actual laboratory for one of the six filter samples and which is having its accuracy tested versus

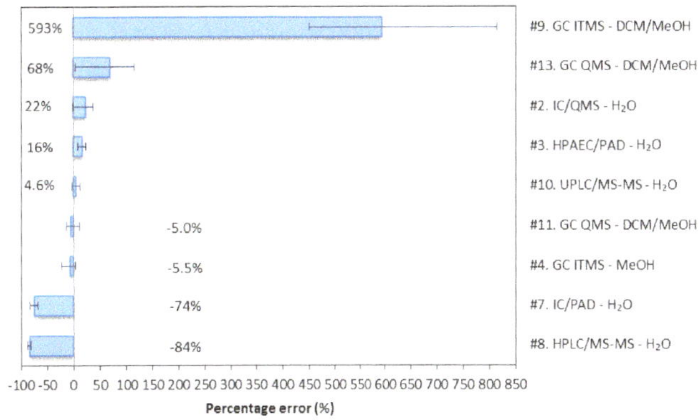

Figure 4. Calculated mean percentage error (PE) for each of the nine laboratories reporting levels of galactosan in the current intercomparison. The PE was calculated according to Eq. (1) (Sect. 2.5) for each of the six sectors received by the participating laboratories. The mean PE for each laboratory accounts for the PE calculated for all six sectors. The laboratory number, ranging from 1–13, the analytical method and solvent used for extraction are mentioned at the right in the figure.

the theoretical value. The latter value is the median value of the analyte based on the values reported by all participating laboratories (except outliers as described above). For laboratories reporting values for two or more aliquots per sector, the measured value is represented by the arithmetic mean, as this variable better reflects the variability of the method's performance. For the theoretical value we chose to use the median in order to limit the influence of deviating results not considered as outliers.

The isomer splits, which show the relative contribution of each of the three monosaccharide anhydrides to the sum of the three monosaccharide anhydrides (ΣMA), and the levoglucosan to mannosan ratio are shown in Figs. 5–8 as aggregated results, whereas the aggregated data, based on those reported by the laboratories, are listed in Table B4. The results are discussed in detail in Sect. 3.

2.6 Analytical standard compounds

A different quality of standard compounds is a factor potentially affecting the comparability of an intercomparison. Results obtained using standard compounds without a given purity are of particular concern, but also batch-to-batch inconsistency is of potential importance.

In the present study, quantification standards of levoglucosan from three different manufacturers were employed, ranging from > 98 to 99 % purity. For mannosan the range of purity was 98 % to purity not given for standards purchased from three different manufacturers. One of the laboratories (lab. 11) used the levoglucosan standard also for mannosan and galactosan. The purity of the galactosan standards was not provided for all, except for that used by lab. 9 (98 % purity) and lab. 13 (99 % purity).

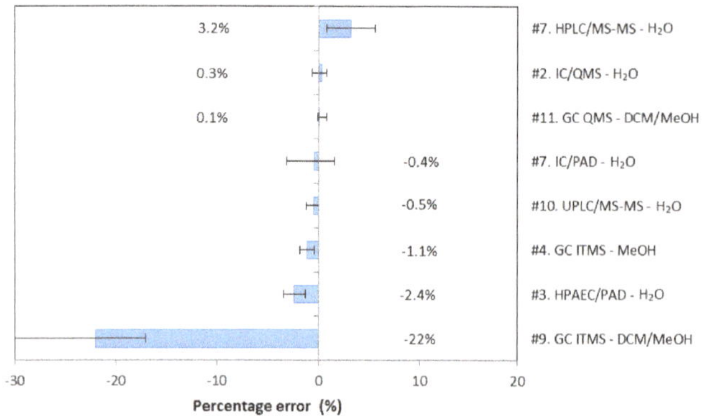

Figure 5. Calculated mean percentage error (PE) for each of the eight laboratories for which the levoglucosan to ΣMA ratio could be derived. The PE was calculated according to Eq. (1) (Sect. 2.5) for each of the six sectors received by the participating laboratories. The mean PE for each laboratory accounts for the PE calculated for all six sectors. The laboratory number, ranging from 1 to 13, the analytical method and solvent used for extraction are mentioned at the right of the figure.

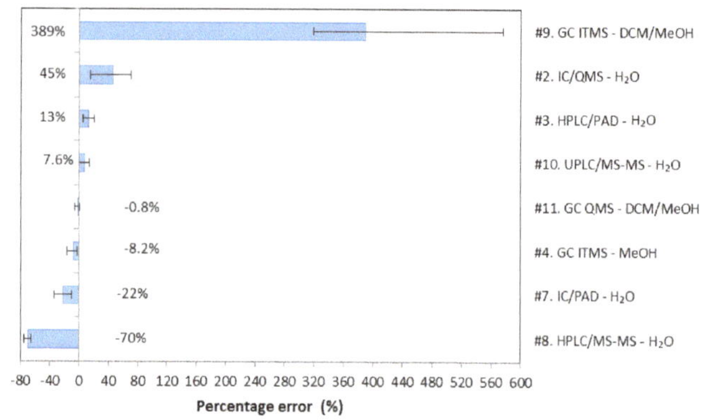

Figure 7. Calculated mean percentage error (PE) for each of the eight laboratories for which the galactosan to ΣMA ratio could be derived. The PE was calculated according to Eq. (1) (Sect. 2.5) for each of the six sectors received by the participating laboratories. The mean PE for each laboratory accounts for the PE calculated for all six sectors. The laboratory number, ranging from 1 to 13, the analytical method and solvent used for extraction are mentioned at the right of the figure.

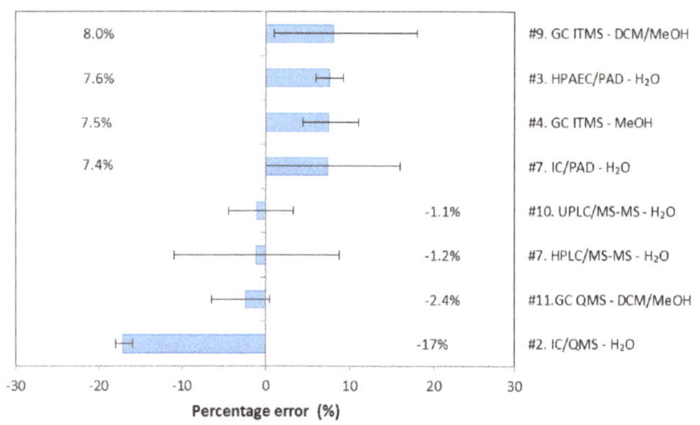

Figure 6. Calculated mean percentage error (PE) for each of the eight laboratories for which the mannosan to ΣMA ratio could be derived. The PE was calculated according to Eq. (1) (Sect. 2.5) for each of the six sectors received by the participating laboratories. The mean PE for each laboratory accounts for the PE calculated for all six sectors. The laboratory number, ranging from 1 to 13, the analytical method and solvent used for extraction are mentioned at the right of the figure.

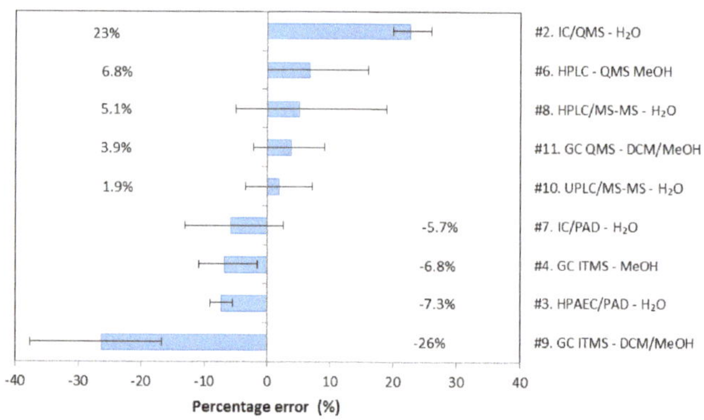

Figure 8. Calculated mean percentage error (PE) for each of the nine laboratories for which the levoglucosan to mannosan ratio could be derived. The PE was calculated according to Eq. (1) (Sect. 2.5) for each of the six sectors received by the participating laboratories. The mean PE for each laboratory accounts for the PE calculated for all six sectors. The laboratory number, ranging from 1 to 13, the analytical method and solvent used for extraction are mentioned at the right of the figure.

Nine (labs. 1, 2, 4, 5, 6, 9, 11, 12 and 13) out of the thirteen laboratories used an internal standard, but only eight (labs. 1, 2, 4, 5, 6, 9, 11 and 12) of them used it to account for potential loss of analytic compounds during the sample preparation process. $^{13}C_6$-labeled levoglucosan with a 98–99 % purity, purchased from two different manufacturers, was used by four laboratories (labs. 1, 2, 6, 12), whereas $^{2}H_7$-labeled levoglucosan (99 % purity) (lab. 9), $^{13}C_6$-labeled galactosan (purity not known) (lab. 1), $^{2}H_4$-labeled succinic acid (98 % purity) (lab. 5), *O*-L-xylanopyranoside (99 % purity) (lab. 4) and sedoheptulose (> 99 % purity) (lab. 11) were each used by one laboratory. Laboratory 13 used the internal standard (1-phenyldodecane, 99 % purity) to account for uncertainties in the injection volume.

3 Results and discussion

3.1 Levoglucosan

The median theoretical concentration of levoglucosan, calculated based on the values reported by the laboratories performing analysis of the actual compound, ranged from

552 $ng\,cm^{-2}$ (filter sample 6) to 1765 $ng\,cm^{-2}$ (filter sample 1) (Table B3). This range corresponds to an ambient concentration of levoglucosan ranging from 120 to 384 $ng\,m^{-3}$, assuming a typical low-volume sampler operating at a flow rate of 2.3 $m^3\,min^{-1}$ for 24 h, assuming an exposed filter area of 12 cm^2. This is within the range (60–900 $ng\,m^{-3}$) observed for European urban areas during wintertime sampling, as reported by Szidat et al. (2009).

For the laboratories extracting, analyzing and reporting values for two or more aliquots per sector, the repeatability was on average 4.5 ± 2.9 %. The repeatability exceeded 10 % for three of the 54 sectors, each of the three being analysed by different laboratories (labs. 1, 3 and 13).

From Fig. 2 it can be seen that the mean PE for the various analytical methods ranged from −63 to 20 %, i.e. reflecting the accuracy of the various methods. The lowest mean PE was seen for lab. 7, for which two outliers were detected. When excluding these results from the calculation, the mean PE for lab. 7 improved from −63 to −54 %, and thus also the total range including all the methods. The mean PE was within ±10 % for eight (labs. 2, 3, 4, 6, 10, 11, 12 and 13) of the thirteen laboratories, i.e. for 62 % of the laboratories, which should be considered a narrow range, and within ±20 % for eleven (labs. 1, 2, 3, 4, 5, 6, 9, 10, 11, 12, 13) of the thirteen laboratories, corresponding to 85 %. Indeed, the accuracy reported for levoglucosan analysis is only slightly lower than the results reported for analysis of SO_4^{2-} on filter samples during the latest intercomparison (intercomparison number 30) organised by EMEP (Co-operative programme for monitoring and evaluation of the long-range transmissions of air pollutants in Europe) (http://www.nilu.no/projects/ccc/intercomparison/DQO-luft-30.pdf); i.e. an accuracy ranging from −23 to 40 % for the entire data set of 64 laboratories, and from −6 to 5 % for the subselection accounting for 62 % of the laboratories. SO_4^{2-} in aerosol filter samples has been analysed by numerous laboratories for several decades, typically by ion chromatography, and should be considered as a fairly easy constituent to analyse. Hence, the comparable results obtained for levoglucosan with respect to accuracy are encouraging.

In Figs. 2–4 the error bars represent the minimum and the maximum PE observed for the six samples analysed with the respective analytical method. Hence, the range is a measure of the variability of the analytical method; i.e. the smaller the range, the better. For levoglucosan the variability ranged from 3.2 to 41 %. Two of the laboratories had a mean PE exceeding −40 %. For lab. 7 the low accuracy was accompanied by a large variability (30 %), whereas this was not the case for lab. 8 (6.5 %). For lab. 7 the combination of a low accuracy and a large variability could suggest an instrument malfunction rather than a flaw in the analytical method used. However, two of the samples reported by this laboratory were outliers, and when excluded from the calculation, the variability improved substantially (i.e. a variability of 5 %); thus, a systematic underestimation could not be excluded. It should be noted that lab. 7 later found that their instrument suffered from a technical problem that caused a random loss of signals during the sample analysis. Subsequent tests revealed that previous injections of high salt-containing samples (i.e. seawater) had contaminated the instrument. The problem was resolved by a rigorous cleaning of instrument components, including working electrode, tubing and fittings. Nevertheless, only two of the samples reported by this laboratory were considered outliers when tested; hence, the results reported from this laboratory should be included in the current data set. There were also other laboratories experiencing difficulties of various categories during the intercomparison; e.g. lab. 1 experienced possible ion suppression, mainly of levoglucosan, which was not previously observed using the actual method, causing negatively biased concentrations as well as deviating isomer splits. Furthermore, other analytic methods used in the current intercomparison could be affected by flaws as well, except that they have passed unseen. Thus, selectively removing data from the intercomparison when being aware of certain problems for one laboratory but not for others cannot easily be done without introducing additional flaws to the data set.

Despite the fact that the great diversity of the analytical methods prevents us from comparing the performance of different subclasses of analytical methods, it cannot be neglected that a certain pattern emerges from Fig. 2. Five out of the six GC-based methods participating in the intercomparison had a mean PE above the theoretical value, ranging from 2.8 to 20 %. Furthermore, four of these five GC methods were based on the approach initially developed in the USA, which subsequently has been adopted by European research laboratories, involving separation by GC, detection by mass spectrometry, extraction by methanol or dichloromethane, or a combination of the two, as well as derivatisation to improve the (gas) chromatographic behaviour of the analyte. Despite a certain variation between laboratories, the general approach outlined above is by far the most commonly used one for analyzing levoglucosan within the research community and it also has the longest record of use. Hence, refinement and experience in using this method is expected to be more extensive than for the more recently developed LC and ion chromatography (IC) methods, for which the mean PE was below the theoretical value for five of seven methods (−63 to −1.9 %) (Fig. 2). Consequently, we should not exclude the possibility that the experience associated with a more widespread analytical method could have a profound influence on the observed pattern, and that this might be an equally important factor as any differences attributed to the choice of the analytical method itself. Given that the more recently developed LC and IC methods have been taken into service on a wider scale than at the time of the current intercomparison, a follow-up intercomparison should address any potential change in the pattern seen in Fig. 2 of the current study.

3.2 Mannosan

The median theoretical concentration of mannosan, calculated based on the values reported by the laboratories performing analysis of the actual compound, ranged from $80\,ng\,cm^{-2}$ (filter sample 4) to $300\,ng\,cm^{-2}$ (filter sample 1). This range corresponds to an ambient concentration of mannosan ranging from 17 to $65\,ng\,m^{-3}$, assuming a typical low-volume sampler operating at a flow rate of $2.3\,m^3\,h^{-1}$ for 24 h, assuming an exposed filter area of $12\,cm^2$.

For the laboratories extracting, analyzing and reporting values of mannosan for two or more aliquots per sample, the repeatability was on average $5.2 \pm 3.9\,\%$. The repeatability exceeded 10 % for three of the 36 samples, and was reported by two different laboratories (labs. 8 and 9).

The mean PE, representing the accuracy, ranged from −60 to 69 % for the various analytical methods reporting values for mannosan (Fig. 3). This range is noticeably wider than that seen for levoglucosan (−63 to 20 %), and whereas 62 % of the laboratories reported a mean PE within ±10 % for levoglucosan, the corresponding percentage for mannosan was only 33 (i.e. for labs. 6, 10 and 11); 55 % of the laboratories reported a mean PE within ±23 % for mannosan (labs. 2, 4, 6, 10 and 11).

As described in Sect. 4.1, the error bars in Figs. 2–4 represent the minimum and the maximum PE observed for the actual analytical method, and thus the range is a measure of the variability of the analytical method. For mannosan the variability ranged from 9.3 to 38 %, which is a slightly smaller range than that seen for levoglucosan. For five of the nine laboratories (labs. 2, 3, 4, 8 and 11) reporting values for both levoglucosan and mannosan the variability was substantially higher, i.e. a factor of 2–6 for mannosan compared to levoglucosan. For two of the laboratories the difference was only minor (lab. 9) or non-existent (lab. 7), whereas labs. 6 and 10 had a slightly lower variability for mannosan compared to levoglucosan.

The three GC-based methods used to determine levels of mannosan had a mean PE above the theoretical value, ranging from 2.4 to 69 %, thus reflecting the general pattern seen for levoglucosan, whereas it ranged from below to above the theoretical values for the LC-based and IC-based methods.

3.3 Galactosan

The median theoretical concentration of galactosan, calculated based on the values reported by the laboratories performing analysis of the actual compound, ranged from $31\,ng\,cm^{-2}$ (filter sample 6) to $90\,ng\,cm^{-2}$ (filter sample 1). This range corresponds to an ambient concentration of galactosan ranging from 7 to $20\,ng\,m^{-3}$, assuming a typical low-volume sampler operating at a flow rate of $2.3\,m^3\,h^{-1}$ for 24 h, assuming an exposed filter area of $12\,cm^2$.

For the laboratories extracting, analyzing and reporting values for two or more aliquots per sample, the repeatability was on average $8.5 \pm 11\,\%$. The repeatability exceeded 10 % for eleven of the 34 samples, and was reported by five different laboratories (labs. 2, 3, 8, 9 and 11).

The mean PE, representing the accuracy, ranged from −84 to 593 % for the various analytical methods reporting values for galactosan. This range is noticeably wider than that seen for both levoglucosan (−63 to 20 %) and mannosan (−60 to 69 %). The very high mean PE (593 %) seen for lab. 9 is consistent with the finding that all values of galactosan reported by this laboratory were found to be outliers. When excluding these data from the calculation, the mean PE range is substantially narrowed (−84–68 %), but it is still wider than for the two other isomers. Excluding the two outliers reported by lab. 7 did not have an influence on the mean PE range including all laboratories, and it only marginally improved the mean PE for lab. 7, going from −74 to −70 %; 33 % of the laboratories (labs. 4, 10, and 11) reported a mean PE for galactosan within ±10 %. This equals the percentage found for mannosan, but it is substantially lower than that observed for levoglucosan (62 %); 55 % of the laboratories (labs. 2, 3, 4, 10, and 11) reported a mean PE within ±22 % for galactosan. It should be noted that for labs. 3 and 8, the mean PE is based on the results from only two of the six filter samples, as the value was found to be below the detection limit (BDL) for the other four.

As previously described (Sects. 4.1 and 4.2), the error bars in Figs. 2–4 represent the minimum and the maximum PE observed for the actual analytical method, and thus are a measure of the variability. For galactosan the variability ranged from 6 to 364 % when including all laboratories. For the subselection of laboratories not influenced by outliers or levels below the detection limit, the variability ranged from 16 to 112 %. For the latter subcategory of laboratories the variability for galactosan was better than that seen for mannosan for two of the laboratories (labs. 10 and 11) and poorer for the two others (labs. 2 and 4). For lab. 10, the variability was noticeably better for galactosan (16 %) also when compared to levoglucosan (41 and 31 % for levoglucosan and mannosan, respectively).

3.4 Relative contribution of levoglucosan, mannosan and galactosan to the sum of the three isomeric compounds (ΣMA)

3.4.1 Levoglucosan to ΣMA ratio

The median theoretical relative contribution of levoglucosan to ΣMA ranged from 81 to 83 % for the eight laboratories from which this ratio could be derived (Table B4); i.e. laboratories not affected by outliers of any of the three isomers or by levels below detection limit (Table B3). The levoglucosan to ΣMA ratio did not vary substantially between the laboratories included in the above-mentioned subcategory. The largest difference was observed for filter sample 1 for which lab. 3 and lab. 10 reported a value of 81 % and lab. 8 a value of 87 %. This rather small difference is to be expected given that levoglucosan is by far the major MA of the three isomeric compounds reported in the literature. Lab. 7 reported values of levoglucosan for filter samples 1 and 2, which were found to be outliers; however, the levoglucosan to ΣMA ratio for these two filter samples (80 %) did not differ substantially from that of filter samples 3–6 (82–84 %) nor from the subcategory not affected by outliers. For lab. 9 the levoglucosan to ΣMA ratio ranged from 57 to 69 %, which is substantially lower than for the other laboratories. Indeed, lab. 9 reported levels of galactosan for filter samples 1–6, which all were found to be outliers, and a markedly high mean PE for galactosan of 593 % (Fig. 4). Laboratory 9 also reported the highest mean concentration of mannosan, with a mean PE of 69 % (Fig. 3). Despite the fact that lab. 9 also reported the highest mean concentration of levoglucosan, the mean PE (20 %) was lower for this isomer than for the two others; hence, the lower levoglucosan to ΣMA ratios seen for lab. 9 can likely be attributed to an overestimation of mannosan and galactosan.

The mean PE, representing the accuracy, ranged from −22 to 3.2 % for the various laboratories for which the levoglucosan to ΣMA could be derived (Fig. 5). When excluding lab. 9, which reported outliers for filter samples 1–6 with respect to galactosan, this range is substantially reduced (−2.4 to 3.2 %). Excluding the two outliers detected for lab. 7 only had a minor effect on the mean PE for the actual laboratory, going from −0.4 to −0.6 %.

As described in Sects. 4.1–4.3, the error bars in Figs. 2–8 represent the minimum and the maximum PE observed for the actual analytical method, and thus are a measure of the method's variability. For the levoglucosan to ΣMA ratio the variability ranged from 0.9 to 13 % when including all laboratories. For the subselection of laboratories not influenced by outliers, the variability ranged from 0.9 to 4.8 %.

3.4.2 Mannosan to ΣMA ratio

The median theoretical relative contribution of mannosan to ΣMA ranged from 13 to 14 % for the laboratories from which this ratio could be derived; i.e. laboratories not affected by outliers of either of the three isomers or by levels below the detection limit. The mannosan to ΣMA ratio did not vary substantially between the laboratories included in the above-mentioned subcategory. The largest difference was observed for filter sample 1 for which lab. 2 reported a value of 11 % and labs. 1 and 3 a value of 15 %. Lab. 7 reported values of levoglucosan for filter samples 1 and 2, which were considered outliers, hence, potentially affecting the ΣMA value and the mannosan to ΣMA ratio. The mannosan to ΣMA ratio for these two filter samples (16 %) did not differ substantially from that of filter samples 3–6 (13–16 %), while they were slightly higher compared to the subcategory not affected by outliers. Lab. 9 reported levels of galactosan for filter samples 1–6 which all were considered to be outliers; however, the mannosan to ΣMA ratio (14–17 %) did not experience a similar decrease as the levoglucosan to ΣMA ratio due to the substantially elevated concentrations of galactosan; on the contrary, they were in the upper range and above that seen for the subcategory not affected by outliers; i.e. the mean PE of 69 % seen for lab. 9 with respect to mannosan seems to have counteracted this.

The mean PE, representing the accuracy, ranged from −17 to 8 % for the various laboratories for which the mannosan to ΣMA could be derived (Fig. 6). Excluding lab. 9, which reported outliers for filter samples 1–6 with respect to galactosan, and lab. 7, which reported outliers for filters 1 and 2 for levoglucosan, did not have an influence on the reported range. Excluding the two outliers detected for lab. 7 improved the mean PE for the actual laboratory from 7.4 to 3.4 %.

As previously described, the error bars in Figs. 2–8 represent the minimum and the maximum PE observed for the actual analytical method, and thus are a measure of the method's variability. For the mannosan to ΣMA ratio the variability ranged from 2 to 20 % when including all laboratories. The range did not change for the subselection of laboratories not influenced by outliers.

3.4.3 Galactosan to ΣMA ratio

The median theoretical relative contribution of galactosan to ΣMA ranged from 3.9 to 5.0 % for the laboratories from which this ratio could be derived; i.e. laboratories not affected by outliers of either of the three isomers or by levels below the detection limit. The galactosan to ΣMA ratio varied more between laboratories than seen for the two other isomers considering the above-mentioned subcategory. The largest difference was observed for filter sample 1 for which lab. 2 reported a value of 6.3 % and lab. 8 a value of 1.0 %. Lab. 7 reported values of levoglucosan for filter samples 1 and 2, which were considered outliers, hence, potentially affecting the ΣMA value and the galactosan to ΣMA ratio. The mannosan to ΣMA ratio for these two filter samples (3.5–3.6 %) was in the upper range of that seen for filter samples 3–6 (2.6–3.5 %), whereas they were slightly lower compared to the subcategory not affected by outliers. Lab. 9 reported levels of galactosan for filter samples 1–6 which all were found to be outliers. The elevated levels reported by lab. 9 with respect to galactosan also had an effect on the galactosan to ΣMA ratio, which ranged from 17 to 27 %, being substantially higher than for the subcategory not affected by outliers.

The mean PE, representing the accuracy, ranged from −70 to 389 % for the various laboratories for which the galactosan to ΣMA could be derived (Fig. 7). Excluding lab. 9, which reported outliers for filter samples 1–6 with respect to galactosan, substantially narrowed the range (−70–45 %), whereas no change was observed when excluding lab. 7, which reported outliers for filter samples 1 and 2 for levoglucosan. Excluding the two outliers detected for lab. 7 changed the mean PE for the actual laboratory from −22 to −26 %. It should be noted that for lab. 8, the mean PE (−70 %) is based on filter samples 1 and 2 only, as galactosan for filter samples 3–6 was found to be below the detection limit.

The error bars in Figs. 2–8 represent the minimum and the maximum PE observed for the actual analytical method, and thus are a measure of the method's variability. For the galactosan to ΣMA ratio the variability ranged from 6 to 257 % when including all laboratories. The range was substantially narrowed (6–56 %) when excluding lab. 9, which reported outliers for filter samples 1–6 with respect to galactosan. Excluding lab. 7, which reported outliers for filter samples 1 and 2 for levoglucosan, did not have an effect on the total range.

3.5 The levoglucosan to mannosan ratio

The levoglucosan to mannosan ratio is occasionally used to apportion the contribution of residential wood burning emissions to burning of either softwood or hardwood (Caseiro et al., 2009; Favez et al., 2010; Piazzalunga et al., 2011; Maenhaut et al., 2012), with softwood combustion giving rise to low ratios (< 4) and hardwood to high ratios (14–15) (Schmidl et al., 2008). The latter authors proposed the following equation to derive the % spruce (or softwood) burnt (relative to the total amount of softwood + hardwood burnt):

$$\%\text{spruce} = (14.8 - \text{levoglucosan}/\text{mannosan})/0.112, \qquad (2)$$

where levoglucosan / mannosan is the levoglucosan to mannosan ratio in the ambient aerosol.

This equation was derived from data that were obtained for the combustion of common hardwood (beech and oak) and softwood species (spruce and larch) in wood stoves in Austria. Besides differences in the levoglucosan to mannosan ratio between various softwood and hardwood species, the uncertainty of the analytical methods used to quantify levoglucosan and mannosan is an unknown variable potentially contributing to the overall uncertainty when performing the softwood and hardwood attribution.

The median levoglucosan to mannosan ratio did not vary much between the six filter samples collected, ranging from 5.3 to 6.2 (Table B4), suggesting a rather stable composition of the wood burnt that impacted the sampling site. When comparing the laboratories from which this ratio could be extracted, i.e. laboratories not affected by outliers for either of the two isomers or by levels below the detection limit, the largest difference observed was seen for filter sample 5 for which lab. 2 reported a levoglucosan to mannosan ratio of 6.9 and lab. 9 a value of 3.6. Lab. 7 reported values of levoglucosan for filter samples 1 and 2, which were considered outliers, thus potentially affecting the levoglucosan to mannosan ratio. The levoglucosan to mannosan ratio for these two filter samples (4.9–5.0) was in the lower range of that of filter samples 3–6 (5.1–6.5) as well as being lower compared to the subcategory not affected by outliers.

The mean PE, representing the accuracy, ranged from −26 to 23 % for the various analytical methods from which the levoglucosan to mannosan ratio could be derived (Fig. 8). In all, 78 % of the laboratories (labs. 3, 4, 6, 7, 8, 10, and 11) reported a mean PE for the levoglucosan to mannosan ratio within ±7 %, which should be considered quite a narrow range. Excluding the two outliers detected for lab. 7 changed the mean PE for the actual laboratory from −5.7 to −1.5 %, thus having no effect on the range including all laboratories.

Assuming that the levoglucosan to mannosan ratio can thus be measured to within ±7 %, one can estimate the uncertainty from the analysis on the %spruce result derived from Eq. (2). This uncertainty is ±8 % spruce points for levoglucosan to mannosan ratios close to 14.8 % (thus for %spruce values close to 0 %) and gradually decreases to ±2 % spruce points for ratios close to 0 (%spruce values close to 100 %). However, the total uncertainty is likely much larger. It should be noted that in addition to MAs, other organic aerosol species, such as syringol and guaiacol, their derivatives, and retene, could also be used to differentiate between hardwood and softwood burning (e.g. Bari et al., 2009).

The error bars in Fig. 8 represent the minimum and the maximum PE observed for the actual analytical method, and thus the range can be considered a measure of the variability. For the levoglucosan to mannosan ratio the variability ranged from 3.5 to 24 %.

4 Conclusions

In the current study we have compared the results of thirteen different analytical methods used to quantify monosaccharide anhydrides (MAs) (i.e. levoglucosan, mannosan and galactosan) in ambient aerosol filter samples. To our knowledge, this is the first major intercomparison that has been conducted and reported in the scientific literature with respect to these compounds. All major methods used for analysis of MAs in ambient aerosol filter samples, and which have been reported in the scientific literature so far, are represented in the present intercomparison.

It is shown that the accuracy for levoglucosan, presented as the mean percentage error (PE) for each participating laboratory, varied from −63 to 20 %; however, for 62 % of the laboratories the mean PE was within ±10 %, and for 85 % the mean PE was within ±20 %. The variability of the various analytical methods, as defined by their minimum and maximum PE value, ranged from 3.2 to 41 % for levoglucosan. These results show that for levoglucosan the accuracy is only slightly lower than that reported for an analysis of SO_4^{2-} on filter samples, a constituent that has been analysed by numerous laboratories for several decades, typically by ion chromatography, and which should be considered a fairly easy constituent to analyse. Hence, the results obtained for levoglucosan with respect to accuracy are encouraging and suggest that levels of levoglucosan, and to a somewhat lesser extent for mannosan and galactosan, obtained by most analytical methods currently used to quantify monosaccharide anhydrides in ambient aerosol filter samples provide comparable results.

Finally, the various analytical methods used in the current study should be tested for other aerosol matrices and concentrations as well, the most obvious being summertime aerosol samples influenced by wildfires and/or agricultural fires.

Appendix A: Detailed description of analytical methods used for quantification of levoglucosan, mannosan and galactosan in the intercomparison

A1 HPAEC–Ion chromatography (IC)

A1.1 Lab. 2

For the analysis, punches of the filters (1 cm^2) were spiked with $^{13}C_6$-levoglucosan and extracted with 5 mL deionised water under 15 min gentle rotation. 50 μL of the filtered extracts (IC Acrodisc® syringe filter 0.45 μm Supor® (PES) membrane) were used for analysis. The MAs were measured using ion chromatography (Dionex IC-3000) coupled to a quadrupole mass spectrometer (Dionex MSQ). Separations were made using a Dionex CarboPac™ PA10 guard column (2 mm i.d. × 50 mm length) and a Dionex CarboPac™ PA10 analytical column (2 mm i.d. × 250 mm length), a 2 mm ASRS-300 suppressor, a CR-ATC anion trap column, and a potassium hydroxide eluent generator. Monosaccharide anhydrides were ionized using the ESI technique. The molecular ions of levoglucosan, mannosan and galactosan were monitored at m/z 161, whereas that of $^{13}C_6$-levoglucosan was monitored at m/z 167, using the selected ion mode. Levoglucosan was used for calibration.

A1.2 Lab. 3

For the analysis, five punches of the filters with an area of 0.5 cm^2 each were extracted together with 5 mL Milli-Q grade water. Extraction was performed in polypropylene test tubes, which were first agitated with a Vortex and then put into an ultrasonic bath (20 min). After centrifugation (10 min) the aqueous extract was transferred into four microcentrifuge vials with 1 mL solution each. The solutions (injection volume 20 μL) were analysed using ion chromatography with pulsed amperometric detection (HPAEC–PAD, ICS-3000 from Dionex Corp.). The separating column was a CarboPac™ MA1, the eluent concentration ranged from 0.48 M NaOH up to 0.65 M NaOH with a flow of 0.4 mL min^{-1}. Levoglucosan, mannosan and galactosan were identified by their retention time and quantified using external standards.

A1.3 Lab. 7

For the analysis, punches of the filters (1.6 cm^2) were extracted in 0.5 mL of Milli-Q grade water under ultrasonication for 30 min. The extract was filtered through a syringe filter (0.45 μm). The samples were analysed using a Dionex ICS-3000 system. The separation was carried out on a Dionex CarboPac™ MA1 column (4 × 250 mm) with a corresponding guard column (4 × 50 mm) at room temperature. The sample injection loop was 25 μL. The eluent gradient was programmed as follows: 0.52 M NaOH from 0 to 20 min, 0.52 to 0.65 M NaOH in 15 min and held constant for 15 min. The eluent flow rate was 0.4 mL min^{-1}. The waveform used for pulsed amperometric detection was the standard quadruple potential for carbohydrate analysis. Levoglucosan, mannosan and galactosan were identified by their retention time and were quantified using external standards.

A2 HPLC

A2.1 Lab. 1

For the analysis, punches (1.5 cm^2) of the filter were spiked with $^{13}C_6$-levoglucosan and $^{13}C_6$-galactosan and extracted twice with 2 mL tetrahydrofuran under ultrasonic agitation (30 min). The filtered extracts (Teflon syringe filter, 0.45 μm) were evaporated to a total volume of 1 mL in a nitrogen atmosphere. Before analysis the sample solvent elution strength was adapted to the mobile phase by adding Milli-Q water (0.8 mL). The concentrations of the MAs were determined using high-performance liquid chromatography (HPLC) (Agilent model 1100) in combination with HR-TOFMS (high-resolution time-of-flight mass spectrometry, Micromass model LCT) operated in the negative ESI mode. Levoglucosan, mannosan and galactosan were identified on the basis of retention time and mass spectra of authentic standards. Quantification was performed using isotope-labeled standards of levoglucosan and galactosan. The mass traces at m/z 161.0455 and 167.0657 were used for quantification (approximately 50 mDa peak width).

A2.2 Lab. 6

For the analysis, filters were spiked with $^{13}C_6$-levoglucosan and extracted three times with methanol in an ultrasonic bath for 30 min. The three extracts were pooled and evaporated to dryness. The sample was reconstituted with 1 mL methanol, shaken, and filtered through a nylon filter. The analysis was performed with liquid chromatography-mass spectrometry with ESI in the positive ionisation mode. Sodium acetate (2 mM) was added to the LC mobile phase to form sodium adducts of levoglucosan and mannosan. The analytes were separated by a gradient using 10 mM sodium acetate and methanol as a mobile phase. The LC column was a Benson Polymeric BP-100 Ca++ Carbohydrate Column. Analysis was performed using selected ion monitoring (SIM) of the sodium adducts. The ion monitored for levoglucosan and mannosan was m/z 185, for $^{13}C_6$-levoglucosan it was m/z 191.

A2.3 Lab. 8

For the analysis, filter punches (4.5 cm^2) were extracted with 2 mL ultrapure water by 30 min vortex agitation; 449 μL of the filtered extracts (Acrodisc® syringe filters, 0.2 μm) were analysed by liquid chromatography (Dionex DX500) electrospray ionisation – tandem mass spectrometry (Thermo Fisher Scientific LCQ Fleet). Levoglucosan, mannosan and galactosan were identified and quantified on the basis of retention time and specific multiple reaction monitoring (MRM) transition (m/z 161–113 for levoglucosan and galactosan, m/z 161–101 for mannosan). External calibrations were performed using standards of levoglucosan, mannosan, and galactosan.

A2.4 Lab. 10

For the analysis, the entire filter punch was extracted twice with 6 mL pure water in an ultrasonic bath for 45 min. Extracts were filtered (Teflon syringe filter, 0.2 μm) and 50 μL of chloroform was added to prevent bacteria activity. Extracts were kept frozen (−18 ° C) until analysis. The analyses were performed using an ultra-performance liquid chromatography (UPLC, Waters) instrument coupled with a triple quadrupole mass spectrometer (Applied Biosystems, model API3200). Separation of MAs was achieved using an Acquity UPLC HSS T3 column, a sample loop of 10 μL, a mixed water–methanol solvent, and post-column addition of methanol. Levoglucosan, mannosan and galactosan were identified and quantified on the basis of retention time and specific MRM transition (m/z 161–113 for levoglucosan and galactosan, m/z 161–113 for mannosan). External calibrations were performed using standards.

A3 GC

A3.1 Lab. 4

For the GC–MS analyses, three filter punches (1 cm^2 each) were spiked with recovery standard (methyl *O*-L-xylanopyranoside). Each punch was extracted three times with 10 mL methanol for 5 min under ultrasonic agitation. The combined extract was reduced in volume with a rotary evaporator to about 1 mL. The filtered concentrate (Teflon syringe filter, 0.45 μm) was completely dried under a stream of nitrogen; 40 μL of BSTFA (*N*,*O*-Bis(trimethylsilyl)trifluoroacetamide) + 1 % chlorotrimethylsilane (TMCS) / pyridine (2/1) was added to the dried sample, and the mixture reacted at 70 °C for 1 h. An aliquot of 1 μL was immediately analysed by GC–MS (Thermo-Finnigan, TRACE GC2000 and Polaris Q). Quantification of the MAs was performed in the scanning mode (m/z 45–650) from the total ion chromatogram using relative response factors determined by injection of authentic standards of levoglucosan, galactosan, and mannosan.

A3.2 Lab. 5

For the GC–MS analyses, three to four filter punches (0.79 cm^2 each) were spiked with internal standard. Each sample was extracted twice with an 8 mL dichloromethane / methanol mixture (4/1) for 30 min in an ultrasonic bath. The filtered extracts (Teflon syringe filter, 0.45 μm) were evaporated to dryness under a gentle flow of nitrogen; 70 μL BSTFA and 70 μL pyridine were added to derivatise the analytes. The mixtures were reacted at 70 °C for 3 h; 1 μL of the sample was injected into the GC injector (Agilent 6890N) in the splitless mode. The EI ionisation mass spectrometer with quadrupole mass analyser (Agilent 5973) was operated in the SIM mode, the acquisition frequency was 2.74 cycles s^{-1}. The ion at m/z 333 was used for the evaluation. Calibration was performed with a standard solution of levoglucosan.

A3.3 Lab. 9

In the GC–MS method, punches (0.85 or 1.0 cm^2) of the samples were spiked with an internal standard levoglucosan ($^2H_7$) and ultrasonically extracted twice with 5 mL of a mixture of dichloromethane / methanol (2/1) for 15 min under ultrasonic agitation. The filtered extracts (glass fiber filters, 0.60 mm) were concentrated in a rotary evaporator to a small volume (ca. 200 μL), and then to dryness under a nitrogen flow; 50 μL BSTFA + 1 % TMCS was added and maintained at 70 °C for 3 h. After derivatisation, the excess of derivatisation agent was evaporated by a nitrogen flow and hexane was added to obtain a volume of 0.5 mL. The concentrations were determined using PTV-GC (programmable temperature vaporisation gas chromatography) (Thermo-Finnigan, Trace GC) in combination with a MS detector (Thermo-Finnigan, Polaris-Q). A 10 μL sample was injected and p-terphenyl was used as a recovery standard. The chromatograms were acquired in the full scan mode (m/z 50–500). Compounds were quantified according to their characteristic ion: levoglucosan (m/z 204), mannosan and galactosan (m/z 217). Quantification was performed using isotope-labeled $^2H_7$-levoglucosan (m/z 220).

A3.4 Lab. 11

For the GC–MS analyses, samples were divided into two parts (2.27 cm^2 and rest of the filter). Sedoheptulose was added as an internal standard and filters were extracted by refluxing with 300 mL dichloromethane / methanol (2/1) for 24 h. The filtered extracts (glass filtration unit, 0.45 μm) were concentrated in a rotary evaporator to a volume of about 2 mL, transferred to a vial and evaporated to dryness under a nitrogen stream. The dried extracts were re-dissolved in pyridine (100 μL) and derivatized with 200 μL BSTFA + 1 % TMCS at 70 °C for 3 h. For quantification 1 μL of the reaction mixture was injected. The concentrations of MAs were determined by gas chromatography (Agilent 6890) mass spectrometry (Agilent 7873). Levoglucosan was identified on the basis of retention time and mass spectra of an authentic standard. Quantification of the three anhydrosugars was performed in the scanning mode using relative response factors determined by injection of a levoglucosan standard.

A3.5 Lab. 12

Analyses were carried out using in situ derivatisation thermal desorption gas chromatography time-of-flight mass spectrometry (IDTD-GC-TOFMS). Filter punches (0.12–0.28 cm^2) were placed into a GC liner, internal standard ($^{13}C_6$-levoglucosan) was added, and liners were closed. For derivatisation filter punches were soaked with 10 μL N-Methyl-N-(trimethylsilyl)trifluoroacetamide (MSTFA). Closed liners were placed in an oven at 80 °C for 1h. After this derivatisation step the liners were put into the cold injector. For desorption, the temperature was raised to 300 °C and kept there for 16 min. During desorption MSTFA was added to the carrier gas. Analytes were focused on a retention gap at 70 °C followed by GC separation (Agilent 6890). The TOFMS instrument (Leco Pegasus III) was operated at a data acquisition frequency of 25 scans s^{-1}. The mass range was m/z 30–500. For quantification m/z 333 for tris-trimethylsilyl-levoglucosan and m/z 338 for tris-trimethylsilyl-$^{13}C_{6}$–levoglucosan were used. External calibration was done with native levoglucosan.

A3.6 Lab. 13

For the GC–MS analyses, filter punches (1 cm^2) were extracted three times (15, 10, 5 mL) with a mixture of dichloromethane / methanol (3/1) and subjected to ultrasonic agitation (45, 30, 15 min). The filtered extracts (Teflon syringe filter, 0.45 μm) were evaporated to dryness in a nitrogen stream. The dry mass was re-dissolved in 1 mL of dichloromethane; 50 μL of each extract solution and 5 μL of a 1,4-dithioerythritol solution (0.5 $\mu g\, mL^{-1}$) were evaporated to dryness under a stream of nitrogen at 60 ° C; 30 μL pyridine and 10 μL of MSTFA + 1 % TMCS were added, and the mixture was reacted at 80 °C for 1 h. The solution was evaporated to dryness at 60 °C under a continuous flow of nitrogen. The dry mass was dissolved in 50 μL of dichloromethane containing 1-phenyldodecane (5 $\mu g\, mL^{-1}$. For quantification 2 μL were injected. The concentrations of the anhydrosugars were determined by gas chromatography (Agilent 6890) mass spectrometry (Agilent 5973). Levoglucosan was identified on the basis of retention time and mass spectrum of an authentic standard. Quantification was performed using the selective ion monitoring mode. Compounds were quantified according to their characteristic m/z 217 and 204 for levoglucosan and m/z 246 for 1-phenyldodecane.

Appendix B: Supplementary tables

Table B1. Summarized description of the analytical methods used by the participating laboratories in the present intercomparison, including extraction procedure, derivatisation agent, recovery standard, and quantification standards.

Lab number	Separation	Detection	Extraction	Derivatisation agent	Internal SD/Recovery SD	Quantification standard(s)
2	IC	QMS	H_2O, shaking	–	$^{13}C_6$-levoglucosan (98 %)	levoglucosan (99 %), mannosan (98 %), galactosan (n.g.)
3	IC	pulsed amperometric	H_2O, ultrasonic agitation	–	–	levoglucosan (99 %), mannosan (98 %), galactosan (n.g.)
7	IC	Pulsed amperometric	H_2O, ultrasonic agitation	–	–	levoglucosan (> 98 %), mannosan (> 98 %), galactosan (n.g.)
1	HPLC	HR-TOFMS	THF, ultrasonic agitation	–	$^{13}C_6$-levoglucosan (98 %), $^{13}C_6$-galactosan (98 %)	levoglucosan (99 %), mannosan (98 %), galactosan (n.g.)
6	HPLC	QMS	MeOH, ultrasonic agitation	–	$^{13}C_6$-levoglucosan (98 %)	levoglucosan (98 %), mannosan (98 %)
8	HPLC	MS–MS	H_2O, vortex	–	–	levoglucosan (99 %), mannosan (n.g.), galactosan (n.g.)
10	UPLC	MS–MS	H_2O, ultrasonic agitation	–	–	levoglucosan (99 %), mannosan (n.g.), galactosan (n.g.)
4	GC	ITMS	MeOH, ultrasonic agitation	pyridine, BSTFA +1 % TMCS	O-L-xylanopyranoside (99 %)	levoglucosan (99 %), mannosan (98 %), galactosan (n.g.)
5	GC	QMS	DCM/MeOH, ultrasonic agitation	pyridine, BSTFA	succinic acid-$^2H_4$ (98 %)	levoglucosan (99 %)
9	GC	ITMS	DCM/MeOH, ultrasonic agitation	BSTFA + 1 % TMCS	levoglucosan-$^2H_7$(n.g.)/ p-terphenyl (99 %)	levoglucosan (99 %), mannosan (n.g.), galactosan (98 %)
11	GC	QMS	DCM/MeOH, Soxhlet	pyridine, BSTFA +1 % TMCS	sedoheptulose (> 99 %)	levoglucosan (99 %)
12	GC	TOFMS	Thermal desorption	MSTFA (in situ)	$^{13}C_6$-levoglucosan (99 %)	levoglucosan (99 %)
13	GC	QMS	DCM/MeOH, ultrasonic agitation	pyridine, MSTFA +1 % TMCS	1-phenyldodecane (99 %)	levoglucosan (99 %), galactosan (99 %)

n.g.: not given; for explanation of abbreviations see Appendix C.

Table B2. Results of the homogeneity test of the six filter samples subjected to the present intercomparison. The results were obtained from one 1.0 cm2 punch taken from each of the 16 sectors per filter; i.e. a total of 96 analyses were performed for each of the three isomers. The filter samples were analyzed according to the method described in Sect. A2.1.

	Filter sample 1	Filter sample 2	Filter sample 3	Filter sample 4	Filter sample 5	Filter sample 6
Levoglucosan						
Mean ± SD (ng cm^{-2})	1518 ± 179	1403± 127	604 ± 99	440 ± 48	541 ± 51	456 ± 48
Median (ng cm^{-2})	1478	1434	631	435	537	445
Mean/Median	1.03	0.98	0.96	1.01	1.01	1.03
RSD (%)	12	9.1	16	11	9.5	10
Outlier	–	–	–	–	–	–
Mannosan						
Mean ± SD (ng cm^{-2})	446 ± 46	370 ± 46	87 ± 22	75 ± 26	103 ± 20	86 ± 9
Median (ng cm^{-2})	437	366	93	77	109	85
Mean/Median	1.02	1.01	0.93	0.98	0.95	1.01
RSD (%)	10	12	26	35	19	10
Outlier	–	–	–	–	1	–
Galactosan						
Mean ± SD (ng cm^{-2})	112 ± 15	98 ± 11	21 ± 9.5	19 ± 7.8	34 ± 5.2	26 ± 3.9
Median (ng cm^{-2})	107	96	21	16	34	26
Mean/Median	1.04	1.02	0.99	1.13	1.00	1.02
RSD (%)	13	11	45	42	15	15
Outlier				1	1	

Table B3. Calculated mean (±SD) and median concentrations as well as maximum and minimum concentrations of levoglucosan, mannosan and galactosan in the six filter samples subjected to the intercomparison, as obtained by the various methods applied. Number of values included in the calculated concentrations, outliers and values below the detection limit (BDL) are listed. Outliers and values BDL are not included in the calculated concentrations, nor in the listed maximum and minimum values.

Sample code	Isomer	Mean ± SD (ng cm^{-2})	Median	Max	Min	*n*	Outliers	BDL
Filter 1								
	Levoglucosan	1730 ± 302	1765	2190	965	12	1 (lab. 7)	
	Mannosan	275 ± 122	300	489	68	9		
	Galactosan	76 ± 40	90	124	11	8	1 (lab. 9)	
Filter 2								
	Levoglucosan	1592 ± 317	1631	2132	873	12	1 (lab. 7)	
	Mannosan	256 ± 105	263	441	65	9		
	Galactosan	68 ± 34	79	101	14	8	1 (lab. 9)	
Filter 3								
	Levoglucosan	594 ± 141	627	765	280	13		
	Mannosan	101 ± 37	99	166	47	9		
	Galactosan	39 ± 18	40	68	12	6	1 (lab. 9)	2 (labs. 3 and 8)
Filter 4								
	Levoglucosan	544 ± 123	570	678	268	13		
	Mannosan	87 ± 33	80	146	42	9		
	Galactosan	37 ± 20	33	71	11	6	1 (lab. 9)	2 (labs. 3 and 8)
Filter 5								
	Levoglucosan	566 ± 134	586	758	258	13		
	Mannosan	110 ± 45	110	188	45	9		
	Galactosan	36 ± 20	35	70	11	6	1 (lab. 9)	2 (labs. 3 and 8)
Filter 6								
	Levoglucosan	521 ± 122	552	670	270	13		
	Mannosan	99 ± 36	100	162	51	9		
	Galactosan	35 ± 20	31	68	9	6	1 (lab. 9)	2 (labs. 3 and 8)

Table B4. Calculated mean (±SD) and median ratios, as well as maximum and minimum ratios of levoglucosan (levo), mannosan (manno) and galactosan (galacto), to the sum of the three isomers (ΣMAs) including the mannosan to levoglucosan ratio in the six filter samples subjected to the intercomparison, as obtained by the various methods applied. Number of values included in the calculated values are listed. Outliers and values BDL are not included in the calculated concentrations, nor in the listed maximum and minimum values.

Sample code	Isomer	Mean ± SD	Median	Max	Min	*n*
Filter 1						
	levo/ΣMA (%)	82 ± 2.1	82	87	81	6
	manno/ΣMA (%)	14 ± 1.4	14	15	11	6
	galacto/ΣMA (%)	4.0 ± 1.7	4.2	6.3	1.0	6
	levo/manno	5.9 ± 0.9	5.8	7.2	4.5	8
Filter 2						
	levo/ΣMA (%)	82 ± 1.2	83	83	80	6
	manno/ΣMA (%)	14 ± 1.4	14	15	12	6
	galacto/ΣMA (%)	3.8 ± 1.4	3.9	5.5	1.4	6
	levo/manno	5.8 ± 0.7	5.7	7.1	4.9	8
Filter 3						
	levo/ΣMA (%)	82 ± 0.7	82	83	81	5
	manno/ΣMA (%)	13 ± 1.2	13	14	11	5
	galacto/ΣMA (%)	4.7 ± 0.9	5.0	5.8	3.5	5
	levo/manno	6.0 ± 1.0	6.0	7.5	4.5	9
Filter 4						
	levo/ΣMA (%)	82 ± 0.8	82	84	82	5
	manno/ΣMA (%)	13 ± 1.2	13	14	11	5
	galacto/ΣMA (%)	4.8 ± 0.9	4.7	6.3	3.4	5
	levo/manno	6.2 ± 1.1	6.4	7.7	4.6	9
Filter 5						
	levo/ΣMA (%)	81 ± 0.8	82	82	80	5
	manno/ΣMA (%)	14 ± 1.5	14	16	12	5
	galacto/ΣMA (%)	4.3 ± 1.0	4.0	6.1	3.4	5
	levo/manno	5.3 ± 1.0	5.7	6.9	3.6	9
Filter 6						
	levo/ΣMA (%)	81 ± 0.6	81	82	80	5
	manno/ΣMA (%)	14 ± 1.5	14	16	12	5
	galacto/ΣMA (%)	4.4 ± 1.5	4.0	6.8	2.6	5
	levo/manno	5.3 ± 0.9	5.6	6.7	3.8	9

Appendix C: Abbreviations used

ACD	Aerosol charge detection
BB	Biomass burning
BSTFA	*N*,*O*-Bis(trimethylsilyl)trifluoroacetamide
CE	Capillary electrophoresis
DCM	Dichloromethane
ESI	Electrospray ionisation
GC	Gas chromatography
HPAEC	High-performance anion-exchange chromatography
HPLC	High-performance liquid chromatography
HR-TOFMS	High-resolution time-of-flight mass spectrometry
ITMS	Ion trap mass spectrometry
MAs	Monosaccharide anhydrides
MS	Mass spectrometry
MS–MS	Tandem mass spectrometry
MSTFA	*N*-Methyl-*N*-(trimethylsilyl)trifluoroacetamide
MeOH	Methanol
PAD	Pulsed amperometric detection
PM	Particulate matter
QMS	Quadrupole mass spectrometry
SO_4^{2-}	Sulfate
THF	Tetrahydrofuran
TOFMS	Time-of-flight mass spectrometry
TMCS	Chlorotrimethylsilane
UPLC	Ultra-performance liquid chromatography

Acknowledgements. The work was supported by the Co-operative Programme for Monitoring and Evaluation of the Long-range Transmission of Air pollutants in Europe (EMEP) under UN-ECE and the European Union Seventh Framework Programme (FP7/2007-2013) under the ACTRIS project (grant agreement no. 262254).

Edited by: P. Herckes

References

Agarwal, S., Aggarwal, S. G., Okuzawa, K., and Kawamura, K.: Size distributions of dicarboxylic acids, ketoacids, α-dicarbonyls, sugars, WSOC, OC, EC and inorganic ions in atmospheric particles over Northern Japan: implication for long-range transport of Siberian BB and East Asian polluted aerosols, Atmos. Chem. Phys., 10, 5839–5858, doi:10.5194/acp-10-5839-2010, 2010.

Alves, C. A., Gonçalves, C., Evtyugina, M., Pio, C. A., Mirante, F., and Puxbaum, H.: Particulate organic compounds emitted from experimental wildland fires in a Mediterranean ecosystem, Atmos. Environ., 44, 2750–2759, 2010.

Bari, M. A., Baumbach, G., Kuch, B., and Scheffknecht, G.: Wood smoke as a source of particle-phase organic compounds in residential areas, Atmos. Environ., 43, 4722–4732, 2009.

Bari, M. A., Baumbach, G., Kuch, B., and Scheffknecht, G.: Temporal variation and impact of wood smoke pollution on a residential area in southern Germany, Atmos. Environ., 44, 3823–3832, 2010.

Blazso, M., Janitsek, S., Gelencsér, A., Artaxo, P., Graham, B., Andreae, M. O.: Study of tropical organic aerosol by thermally assisted alkylation-gas chromatography mass spectrometry, J. Anal. Appl. Pyrol., 68–69, 351–369, 2003.

Booth, A. M., Montague, W. J., Barley, M. H., Topping, D. O., McFiggans, G., Garforth, A., and Percival, C. J.: Solid state and sub-cooled liquid vapour pressures of cyclic aliphatic dicarboxylic acids, Atmos. Chem. Phys., 11, 655–665, doi:10.5194/acp-11-655-2011, 2011.

Caseiro, A. and Oliveira, C.: Variations in wood burning organic marker concentrations in the atmospheres of four European cities, J. Environ. Monitor., 14, 2261–2269, 2012.

Caseiro, A., Marr, I. L., Claeys, M., Kasper-Giebl, A., Puxbaum, H., and Pio, C. A.: Determination of saccharides in atmospheric aerosol using anion-exchange high-performance liquid chromatography and pulsed-amperometric detection, J. Chromatogr. A, 1171, 37–45, 2007.

Caseiro, A., Bauer, H., Schmidl, C., Pio, C. A., and Puxbaum, H.: Wood burning impact on PM_{10} in three Austrian regions, Atmos. Environ., 43, 2186–2195, 2009.

Claeys, M., Kourtchev, I., Pashynska, V., Vas, G., Vermeylen, R., Wang, W., Cafmeyer, J., Chi, X., Artaxo, P., Andreae, M. O., and Maenhaut, W.: Polar organic marker compounds in atmospheric aerosols during the LBA-SMOCC 2002 biomass burning experiment in Rondônia, Brazil: sources and source processes, time series, diel variations and size distributions, Atmos. Chem. Phys., 10, 9319–9331, doi:10.5194/acp-10-9319-2010, 2010.

Dixon, R. W. and Baltzell, G.: Determination of levoglucosan in atmospheric aerosols using high-performance liquid chromatography with aerosol charge detection, J. Chromatogr. A, 1109, 214–221, 2006.

dos Santos, C. Y. M., Azevedo, D. D., and Neto, F. R. D.: Selected organic compounds from biomass burning found in the atmospheric particulate matter over sugarcane plantation areas, Atmos. Environ., 36, 3009–3019, 2002.

Dye, C. and Yttri, K. E.: Determination of monosaccharide anhydrides in atmospheric aerosols by use of high-performance liquid chromatography combined with high-resolution mass spectrometry, Anal. Chem., 77, 1853–1858, 2005.

Elsasser, M., Crippa, M., Orasche, J., DeCarlo, P. F., Oster, M., Pitz, M., Cyrys, J., Gustafson, T. L., Pettersson, J. B. C., Schnelle-Kreis, J., Prévôt, A. S. H., and Zimmermann, R.: Organic molecular markers and signature from wood combustion particles in winter ambient aerosols: aerosol mass spectrometer (AMS) and high time-resolved GC-MS measurements in Augsburg, Germany, Atmos. Chem. Phys., 12, 6113–6128, doi:10.5194/acp-12-6113-2012, 2012.

Elsasser, M., Busch, C., Orasche, J., Schön, C., Hartmann, H., Schnelle-Kreis, J., and Zimmermann, R.: Dynamic changes of the aerosol composition and concentration during different burning phases of wood combustion, Energ. Fuel., 27, 4959–4968, 2013.

Engling, G., Carrico, C. M., Kreidenweis, S. M., Collett, J. L., Day, D. E., Malm, W. C., Lincoln, E., Hao, W. M., Iinuma, Y., and Herrmann, H.: Determination of levoglucosan in biomass combustion aerosol by high-performance anion-exchange chromatography with pulsed amperometric detection, Atmos. Environ., 40, S299–S311, 2006.

Fabbri, D., Chiavari, G., Prati, S., Vassura, I., and Vangelista, M.: Gas chromatography/mass spectrometric characterisation of pyrolysis/silylation products of glucose and cellulose, Rapid Commun. Mass Sp., 16, 2349–2355, 2002.

Fabbri, D., Torri, C., Simoneit, B. R. T., Marynowski, L., Rushdi, A. I., and Fabianska, M. J.: Levoglucosan and other cellulose and lignin markers in emissions from burning of Miocene lignites, Atmos. Environ., 43, 2286–2295, 2009.

Favez, O., El Haddad, I., Piot, C., Boréave, A., Abidi, E., Marchand, N., Jaffrezo, J.-L., Besombes, J.-L., Personnaz, M.-B., Sciare, J., Wortham, H., George, C., and D'Anna, B.: Inter-comparison of source apportionment models for the estimation of wood burning aerosols during wintertime in an Alpine city (Grenoble, France), Atmos. Chem. Phys., 10, 5295–5314, doi:10.5194/acp-10-5295-2010, 2010.

Fine, P. M., Cass, G. R., and Simoneit, B. R. T.: Chemical characterization of fine particle emissions from fireplace combustion of woods grown in the northeastern United States, Environ. Sci. Technol., 35, 2665–2675, 2001.

Fine, P. M., Cass, G. R., and Simoneit, B. R. T.: Chemical characterization of fine particle emissions from the fireplace combustion of woods grown in the southern United States, Environ. Sci. Technol., 36, 1442–1451, 2002.

Fine, P. M., Cass, G. R., and Simoneit, B. R. T.: Chemical characterization of fine particle emissions from the fireplace combustion of wood types grown in the Midwestern and Western United States, Environ. Eng. Sci., 21, 387–409, 2004.

Fraser, M. P. and Lakshmanan, K.: Using levoglucosan as a molecular marker for the long-range transport of biomass combustion aerosols, Environ. Sci. Technol., 34, 4560–4564, 2000.

Frey, A. K., Tissari, J., Saarnio, K. M., Timonen, H. J., Tolonen-Kivimaki, O., Aurela, M. A., Saarikoski, S. K., Makkonen, U., Hytonen, K., Jokiniemi, J., Salonen, R. O., and Hillamo, R. E. J.: Chemical composition and mass size distribution of fine particulate matter emitted by a small masonry heater, Boreal Environ. Res., 14, 255–271, 2009.

Gao, S., Hegg, D. A., Hobbs, P. V., Kirchstetter, T. W., Magi, B., I., and Sadilek, M.: Water-soluble organic components in aerosols associated with savanna fires in southern Africa: Identification, evolution, and distribution, J. Geophys. Res.-Atmos., 108, 8491, doi:10.1029/2002JD002324, 2003.

Garcia, C. D., Engling, G., Herckes, P., Collett, J. L., and Henry, C. S.: Determination of levoglucosan from smoke samples using microchip capillary electrophoresis with pulsed amperometric detection, Environ. Sci. Technol., 39, 618–623, 2005.

Gelencsér, A., May, B., Simpson, D., Sanchez-Ochoa, A., Kasper-Giebl, A., Puxbaum, H., Caseiro, A., Pio, C., and Legrand, M.: Source apportionment of PM2.5 organic aerosol over Europe: Primary/secondary, natural/anthropogenic, and fossil/biogenic origin, J. Geophys. Res.-Atmos., 112, D23S04, doi:10.1029/2006JD008094, 2007.

Gonçalves, C., Alves, C., Evtyugina, M., Mirante, F., Pio, C., Caseiro, A., Schmidl, C., Bauer, H., Carvalho, F.: Characterisation of PM_{10} emissions from wood stove combustion of common woods grown in Portugal, Atmos. Environ., 44, 4474–4480, 2010.

Gonçalves, C., Alves, C., Fernandes, A. P., Monteiro, C., Tarelho, L., Evtyugina, M., and Pio, C.: Organic compounds in $PM_{2.5}$ emitted from fireplace and woodstove combustion of typical Portuguese wood species, Atmos. Environ., 45, 4533–4545, 2011.

He, J., Zielinska, B., and Balasubramanian, R.: Composition of semi-volatile organic compounds in the urban atmosphere of Singapore: influence of biomass burning, Atmos. Chem. Phys., 10, 11401–11413, doi:10.5194/acp-10-11401-2010, 2010.

Hennigan, C. J., Sullivan, A. P., Collett, J. L., and Robinson, A. L.: Levoglucosan stability in biomass burning particles exposed to hydroxyl radicals, Geophys. Res. Lett., 37, L09806, doi:10.1029/2010GL043088, 2010.

Herich, H., Gianini, M. F. D., Piot, C., Močnik, G., Jaffrezo, J.-L., Besombes, J.-L., Prévôt, A. S. H., and Hueglin, C.: Overview of the impact of wood burning emissions on carbonaceous aerosols and PM in large parts of the Alpine region. Atmos. Environ., 89, 64–75, 2014.

Hoffmann, D., Tilgner, A., Iinuma, Y., and Herrmann, H.: Atmospheric stability of levoglucosan: A detailed laboratory and modeling study, Environ. Sci. Technol., 44, 694–699, 2010.

Holmes, B. J. and Petrucci, G. A.: Oligomerization of levoglucosan by Fenton chemistry in proxics of biomass burning aerosols, J. Atmos. Chem., 58, 151–166, 2007.

Hornig, J. F., Soderberg, R. H., Barefoot, A. C., and Galasyn, J. F.: Wood smoke analysis: vaporization losses of PAH from filters and levoglucosan as a distinctive marker for wood smoke, in: Polynuclear Aromatic Hydrocarbons: Mechanisms, Methods, and Metabolism, edited by: Cooke, M. and Dennis, A. J., Battelle Press, Columbus, OH, 561–568, 1985.

Hsu, C. L., Cheng, C. Y., Lee, C. T., and Ding, W. H.: Derivatization procedures and determination of levoglucosan and related monosaccharide anhydrides in atmospheric aerosols by gas chromatography-mass spectrometry, Talanta, 72, 199–205, 2007.

Iinuma, Y., Bruggemann, E., Gnauk, T., Muller, K., Andreae, M. O., Helas, G., Parmar, R., and Herrmann, H.: Source characterization of biomass burning particles: The combustion of selected European conifers, African hardwood, savanna grass, and German and Indonesian peat, J. Geophys. Res.-Atmos., 112, D08209, doi:10.1029/2006JD007120, 2007.

Iinuma, Y., Engling, G., Puxbaum, H., and Herrmann, H.: A highly resolved anion-exchange chromatographic method for determination of saccharidic tracers for biomass combustion and primary bio-particles in atmospheric aerosol, Atmos. Environ., 43, 1367–1371, 2009.

Jordan, T. B. and Seen, A. J.: Effect of airflow setting on the organic composition of woodheater emissions, Environ. Sci. Technol., 39, 3601–3610, 2005.

Kourtchev, I., Hellebust, S., Bell, J. M., O'Connor, I. P., Healy, R. M., Allanic, A., Healy, D., Wenger, J. C., and Sodeau, J. R.: The use of polar organic compounds to estimate the contribution of domestic solid fuel combustion and biogenic sources to ambient levels of organic carbon and PM2.5 in Cork Harbour, Ireland, Sci. Total Environ., 409, 2143–2155, 2011.

Krumal, K., Mikuska, P., Vojtesek, M., and Vecera, Z.: Seasonal variations of monosaccharide anhydrides in PM_1 and $PM_{2.5}$ aerosol in urban areas, Atmos. Environ., 44, 5148–5155, 2010.

Larsen, R. K., Schantz, M. M., and Wise, S. A.: Determination of levoglucosan in particulate matter reference materials, Aerosol Sci. Technol., 40, 781–787, 2006.

Lin, L., Lee, M. L., and Eatough, D. J.: Gas chromatographic analysis of organic marker compounds in fine particulate matter using solid-phase microextraction, J. Air Waste Manage., 57, 53–58, 2007.

Ma, Y., Hays, M. D., Geron, C. D., Walker, J. T., and Gatari Gichuru, M. J.: Technical Note: Fast two-dimensional GC-MS with thermal extraction for anhydro-sugars in fine aerosols, Atmos. Chem. Phys., 10, 4331–4341, doi:10.5194/acp-10-4331-2010, 2010.

Maenhaut, W., Vermeylen, R., Claeys, M., Vercauteren, J., Matheeussen, C., and Roekens, E.: Assessment of the contribution from wood burning to the PM_{10} aerosol in Flanders, Belgium, Sci. Total Environ., 437, 226–236, 2012.

May, B., Wagenbach, D., Hammer, S., Steier, P., Puxbaum, H., and Pio, C.: The anthropogenic influence on carbonaceous aerosol in the European background, Tellus B, 61, 464–472, 2009.

Mazzoleni, L. R., Zielinska, B., and Moosmüller, H.: Emissions of levoglucosan, methoxy phenols, and organic acids from prescribed burns, laboratory combustion of wildland fuels, and residential wood combustion, Environ. Sci. Technol., 41, 2115–2122, 2007.

Medeiros, P. M. and Simoneit, B. R. T.: Analysis of sugars in environmental samples by gas chromatography-mass spectrometry, J. Chromatogr. A, 1141, 271–278, 2007.

Nolte, C. G., Schauer, J. J., Cass, G. R., and Simoneit, B. R. T.: Highly polar organic compounds present in wood smoke and in the ambient atmosphere, Environ. Sci. Technol., 35, 1912–1919, 2001.

Oja, V. and Suuberg, E. M.: Vapor pressures and enthalpies of sublimation of D-glucose, D-xylose, cellobiose, and levoglucosan, J. Chem. Eng. Data, 44, 26–29, 1999.

Oliveira, C., Pio, C., Alves, C., Evtyugina, M., Santos, P., Goncalves, V., Nunes, T., Silvestre, A J. D., Palmgren, F.,

Wahlin, P., and Harrad, S.: Seasonal distribution of polar organic compounds in the urban atmosphere of two large cities from the North and South of Europe, Atmos. Environ., 41, 5555–5570, 2007.

Orasche, J., Schnelle-Kreis, J., Abbaszade, G., and Zimmermann, R.: Technical Note: In-situ derivatization thermal desorption GC-TOFMS for direct analysis of particle-bound non-polar and polar organic species, Atmos. Chem. Phys., 11, 8977–8993, doi:10.5194/acp-11-8977-2011, 2011.

Orasche, J., Seidel, T., Hartmann, H., Schnelle-Kreis, J., Chow, J. C., Ruppert, H., and Zimmermann, R.: Comparison of emissions from wood combustion. Part 1: Emission factors and characteristics from different small-scale residential heating appliances considering particulate matter and polycyclic aromatic hydrocarbon (PAH)-related toxicological potential of particle-bound organic species, Energ. Fuel., 26, 6695–6704, 2012.

Orasche, J., Schnelle-Kreis, J., Schoen, C., Hartmann, H., Ruppert, H., Arteaga-Salas, J. M., and Zimmermann, R.: Comparison of emissions from wood combustion. Part 2: Impact of combustion conditions on emission factors and characteristics of particle-bound organic species and polycyclic aromatic hydrocarbon (PAH)-related toxicological potential, Energ. Fuel., 27, 1482–1491, 2013.

Oros, D. R. and Simoneit, B. R. T.: Identification of molecular tracers in organic aerosols from temperate climate vegetation subjected to biomass burning, Aerosol Sci. Technol., 31, 433–445, 1999.

Pashynska, V., Vermeylen, R., Vas, G., Maenhaut W., and Claeys, M.: Development of a gas chromatographic/ion trap mass spectrometric method for the determination of levoglucosan and saccharidic compounds in atmospheric aerosols. Application to urban aerosols, J. Mass Spectrom., 37, 1249–1257, 2002.

Piazzalunga, A., Fermo, P., Bernardoni, V., Vecchi, R., Valli, G., and De Gregorio, M. A.: A simplified method for levoglucosan quantification in wintertime atmospheric particulate matter by high-performance anion-exchange chromatography coupled with pulsed amperometric detection, Int. J. Environ. Anal. Chem., 90, 934–947, 2010.

Piazzalunga, A., Belis, C., Bernardoni, V., Cazzuli, O., Fermo, P., Valli, G., and Vecchi, R.: Estimates of wood burning contribution to PM by the macro-tracer method using tailored emission factors, Atmos. Environ., 45, 6642–6649, 2011.

Pietrogrande, M. C., Bacco, D., and Rossi, M.: Chemical characterization of polar organic markers in aerosols in a local area around Bologna, Italy, Atmos. Environ., 75, 279–286, 2013.

Piot, C., Jaffrezo, J.-L., Cozic, J., Pissot, N., El Haddad, I., Marchand, N., and Besombes, J.-L.: Quantification of levoglucosan and its isomers by High Performance Liquid Chromatography – Electrospray Ionization tandem Mass Spectrometry and its applications to atmospheric and soil samples, Atmos. Meas. Tech., 5, 141–148, doi:10.5194/amt-5-141-2012, 2012.

Puxbaum, H., Caseiro, A., Sanchez-Ochoa, A., Kasper-Giebl, A., Claeys, M., Gelencsér, A., Legrand, M., Preunkert, S., and Pio, C.: Levoglucosan levels at background sites in Europe for assessing the impact of biomass combustion on the European aerosol background, J. Geophys. Res.-Atmos., 112, D23S05, doi:10.1029/2006JD008114, 2007.

Ruehl, C. R., Ham, W. A., and Kleeman, M. J.: Temperature-induced volatility of molecular markers in ambient airborne particulate matter, Atmos. Chem. Phys., 11, 67–76, doi:10.5194/acp-11-67-2011, 2011.

Saarnio, K., Aurela, M., Timonen, H., Saarikoski, S., Teinila, K., Makela, T., Sofiev, M., Koskinen, J., Aalto, P. P., Kulmala, M., Kukkonen, J., and Hillamo, R.: Chemical composition of fine particles in fresh smoke plumes from boreal wild-land fires in Europe, Sci. Total Environ., 408, 2527–2542, 2010a.

Saarnio, K., Teinila, K., Aurela, M., Timonen, H., and Hillamo, R.: High-performance anion-exchange chromatography-mass spectrometry method for determination of levoglucosan, mannosan, and galactosan in atmospheric fine particulate matter, Anal. Bioanal. Chem., 398, 2253–2264, 2010b.

Saffari, A., Daher, N., Samara, C., Voutsa, D., Kouras, A., Manoli, E., Karagkiozidou, O., Vlachokostas, C., Moussiopoulos, N., Shafer, M. M., Schauer, J. J., and Sioutas, C.: Increased biomass burning due to the economic crisis in Greece and its adverse impact on wintertime air quality in Thessaloniki, Environ. Sci. Technol., 47, 13313–13320, 2013.

Schkolnik, G. and Rudich, Y.: Detection and quantification of levoglucosan in atmospheric aerosols: A review, Anal. Bioanal. Chem., 385, 26–33, 2006.

Schkolnik, G., Falkovich, A. H., Rudich, Y., Maenhaut, W., and Artaxo, P.: New analytical method for the determination of levoglucosan, polyhydroxy compounds, and 2-methylerythritol and its application to smoke and rainwater samples, Environ. Sci. Technol., 39, 2744–2752, 2005.

Schmidl, C., Marr, L. L. Caseiro, A., Kotianova, P., Berner, A., Bauer, H., Kasper-Giebl, A., and Puxbaum, H.: Chemical characterisation of fine particle emissions from wood stove combustion of common woods growing in mid-European Alpine regions, Atmos. Environ. 42, 126–141, 2008.

Schnelle-Kreis, J., Kunde, R., Schmoeckel, G., Abbaszade, G., Gaderer, M., Diemer, J., Ott, H., and Zimmermann, R.: Anteil von Partikelemissionen aus Holzverbrennung an PM10-Feinstaubimmissionen im städtischen Umfeld am Beispiel von Augsburg – Teil 1: Emissions- und Immissionsmessungen, Gefahrst. Reinhalt. L., 5, 203–209, 2010.

Sillanpää, M., Saarikoski, S., Hillamo, R., Pennanen, A., Makkonen, U., Spolnik, Z., Van Grieken, R., Koskentalo, T. A., and Salonen, R. O.: Chemical composition, mass size distribution and source analysis of long-range transported wildfire smokes in Helsinki, Sci. Total Environ., 350, 119–135, 2005.

Simoneit, B. R. T.: A review of biomarker compounds as source indicators and tracers for air pollution, Environ. Sci. Pollut., 6, 159–169, 1999.

Simoneit, B. R. T., Schauer, J. J., Nolte, C. G., Oros, D. R., Elias, V. O., Fraser, M. P., Rogge, W. F., and Cass, G. R.: Levoglucosan, a tracer for cellulose in biomass burning and atmospheric particles, Atmos. Environ., 33, 173–182, 1999.

Simpson, C. D., Dills, R. L., Katz, B. S., and Kalman, D. A.: Determination of levoglucosan in atmospheric fine particulate matter, J. Air Waste Manage., 54, 689–694, 2004.

Slade, J. H. and Knopf, D. A.: Heterogeneous OH oxidation of biomass burning organic aerosol surrogate compounds: assessment of volatilisation products and the role of OH concentration on the reactive uptake kinetics, Phys. Chem. Chem. Phys., 15, 5898–5915, 2013.

Sullivan, A. P., Holden, A. S., Patterson, L. A., McMeeking, G. R., Kreidenweis, S. M., Malm, W. C., Hao, W. M., Wold, C. E., and Collett Jr., J. L.: A method for smoke marker measurements and its potential application for determining the contribution of biomass burning from wildfires and prescribed fires to ambient PM2.5 organic carbon, J. Geophys. Res.-Atmos., 113, D22302, doi:10.1029/2008JD010216, 2008

Szidat, S., Ruff, M., Perron, N., Wacker, L., Synal, H.-A., Hallquist, M., Shannigrahi, A. S., Yttri, K. E., Dye, C., and Simpson, D.: Fossil and non-fossil sources of organic carbon (OC) and elemental carbon (EC) in Göteborg, Sweden, Atmos. Chem. Phys., 9, 1521–1535, doi:10.5194/acp-9-1521-2009, 2009.

Teraji, T. and Arakaki, T.: Bimolecular rate constants between levoglucosan and hydroxyl radical: Effects of pH and temperature, Chem. Lett., 39, 900–901, 2010.

Timonen, H., Saarikoski, S., Tolonen-Kivimäki, O., Aurela, M., Saarnio, K., Petäjä, T., Aalto, P. P., Kulmala, M., Pakkanen, T., and Hillamo, R.: Size distributions, sources and source areas of water-soluble organic carbon in urban background air, Atmos. Chem. Phys., 8, 5635–5647, doi:10.5194/acp-8-5635-2008, 2008.

van Drooge, B. L. and Ballesta, P. P.: Seasonal and daily source apportionment of polycyclic aromatic hydrocarbon concentrations in PM10 in a semirural European area, Environ. Sci. Technol., 43, 7310–7316, 2009.

Vicente, A., Alves, C., Monteiro, C., Nunes, T., Mirante, F., Cerqueira, M., Calvo, A., Pio, C.: Organic speciation of aerosols from wildfires in central Portugal during summer 2009, Atmos. Environ., 57, 186–196, 2012.

Wan, E. C. H. and Yu, J. Z.: Analysis of sugars and sugar polyols in atmospheric aerosols by chloride attachment in liquid chromatography/negative ion electrospray mass spectrometry, Environ. Sci. Technol., 41, 2459–2466, 2007.

Wang, G., Chen, C., Li, J., Zhou, B., Xie, M., Hu, S., Kawamura, K., and Chen,Y.: Molecular composition and size distribution of sugars, sugar-alcohols and carboxylic acids in airborne particles during a severe urban haze event caused by wheat straw burning, Atmos. Environ., 45, 2473–2479, doi:10.1016/j.atmosenv.2011.02.045, 2011.

Weimer, S., Alfarra, M. R., Schreiber, D., Mohr, M., Prévôt, A. S. H., and Baltensperger, U.: Organic aerosol mass spectral signatures from wood-burning emissions: Influence of burning conditions and wood type, J. Geophys. Res.-Atmos., 113, D10304, doi:10.1029/2007JD009309, 2008.

Yttri, K. E., Dye, C., Slørdal, L. H., and Braathen, O. A.: Quantification of monosaccharide anhydrides by liquid chromatography combined with mass spectrometry: Application to aerosol samples from an urban and a suburban site influenced by small-scale wood burning, J. Air Waste Manage., 55, 1169–1177, 2005.

Yttri, K. E., Dye, C., Braathen, O.-A., Simpson, D., and Steinnes, E.: Carbonaceous aerosols in Norwegian urban areas, Atmos. Chem. Phys., 9, 2007–2020, doi:10.5194/acp-9-2007-2009, 2009.

Yttri, K. E., Lund Myhre, C., Eckhardt, S., Fiebig, M., Dye, C., Hirdman, D., Ström, J., Klimont, Z., and Stohl, A.: Quantifying black carbon from biomass burning by means of levoglucosan – a one-year time series at the Arctic observatory Zeppelin, Atmos. Chem. Phys., 14, 6427–6442, doi:10.5194/acp-14-6427-2014, 2014.

5

The Midlatitude Continental Convective Clouds Experiment (MC3E) sounding network: operations, processing and analysis

M. P. Jensen[1], T. Toto[1], D. Troyan[1], P. E. Ciesielski[2], D. Holdridge[3], J. Kyrouac[3], J. Schatz[4], Y. Zhang[5], and S. Xie[5]

[1]Brookhaven National Laboratory, Upton, NY, USA
[2]Colorado State University, Fort Collins, CO, USA
[3]Argonne National Laboratory, Argonne, IL, USA
[4]ARM Climate Research Facility, Southern Great Plains site, Billings, OK, USA
[5]Lawrence Livermore National Laboratory, Livermore, CA, USA

Correspondence to: M. P. Jensen (mjensen@bnl.gov)

Abstract. The Midlatitude Continental Convective Clouds Experiment (MC3E) took place during the spring of 2011 centered in north-central Oklahoma, USA. The main goal of this field campaign was to capture the dynamical and microphysical characteristics of precipitating convective systems in the US Central Plains. A major component of the campaign was a six-site radiosonde array designed to capture the large-scale variability of the atmospheric state with the intent of deriving model forcing data sets. Over the course of the 46-day MC3E campaign, a total of 1362 radiosondes were launched from the enhanced sonde network. This manuscript provides details on the instrumentation used as part of the sounding array, the data processing activities including quality checks and humidity bias corrections and an analysis of the impacts of bias correction and algorithm assumptions on the determination of convective levels and indices. It is found that corrections for known radiosonde humidity biases and assumptions regarding the characteristics of the surface convective parcel result in significant differences in the derived values of convective levels and indices in many soundings. In addition, the impact of including the humidity corrections and quality controls on the thermodynamic profiles that are used in the derivation of a large-scale model forcing data set are investigated. The results show a significant impact on the derived large-scale vertical velocity field illustrating the importance of addressing these humidity biases.

1 Introduction

The Midlatitude Continental Convective Clouds Experiment (MC3E; Jensen et al., 2010) was a joint field campaign between the US Department of Energy's Atmospheric Radiation Measurement (ARM) Program (Mather and Voyles, 2013; Ackerman and Stokes, 2003; Stokes and Schwartz, 1994; http://www.arm.gov) and NASA's Global Precipitation Measurement Mission's (GPM) Ground Validation Program (Hou et al., 2014; http://www.nasa.gov/mission_pages/GPM/main/). It took place from 22 April through 6 June 2011 centered at the ARM Southern Great Plains (SGP) Facility in north-central Oklahoma. The campaign utilized the largest-to-date ground-based observing infrastructure in the central United States, including recent instrumentation upgrades provided through the American Recovery and Reinvestment Act of 2009, combined with remote sensing and in situ aircraft, additional radar and in situ precipitation instrumentation and an extensive sounding array. The overarching goal of the field campaign was to construct a comprehensive data set of surface-, aircraft- and satellite-based observations targeting processes important for the parameterization of convection in large-scale models and the retrieval of precipitation by spaceborne sensors over land.

A major component of the MC3E field campaign was the deployment of an enhanced radiosonde array designed to capture the large-scale variability of the vertical profile of atmospheric state variables (pressure, temperature, humidity and winds). The radiosonde array included six sites launching radiosondes at 3–6 h sampling intervals. This network (Fig. 1, Table 1) covered an area of approximately 300 × 300 km with five perimeter sounding launch sites and one central launch location.

The large-scale sounding array facilitates the description of the variability of the atmospheric state in space over the scale of a global climate model grid box and over the diurnal cycle. A major motivation for the sounding array is to provide large-scale forcing fields, through traditional budget analyses (e.g., Yanai et al., 1973; Johnson, 1984; Johnson and Ciesielski, 2013) and constrained variational objective analyses (Zhang and Lin, 1997; Zhang et al., 2001; Xie et al., 2010; Xie et al., 2014), that can be used for single-column model (SCM) and cloud-resolving model (CRM) studies (Klein and Del Genio, 2006).

Specifics of the sounding operations during the MC3E field campaign will be presented in Sect. 2. Details of the data processing will be discussed in Sect. 3, including data quality controls and corrections for known sounding humidity biases. Section 4 will discuss the sensitivity to measurement uncertainties and algorithm assumptions of the calculation of several important indices for convective studies and the derivation of a large-scale model forcing data set.

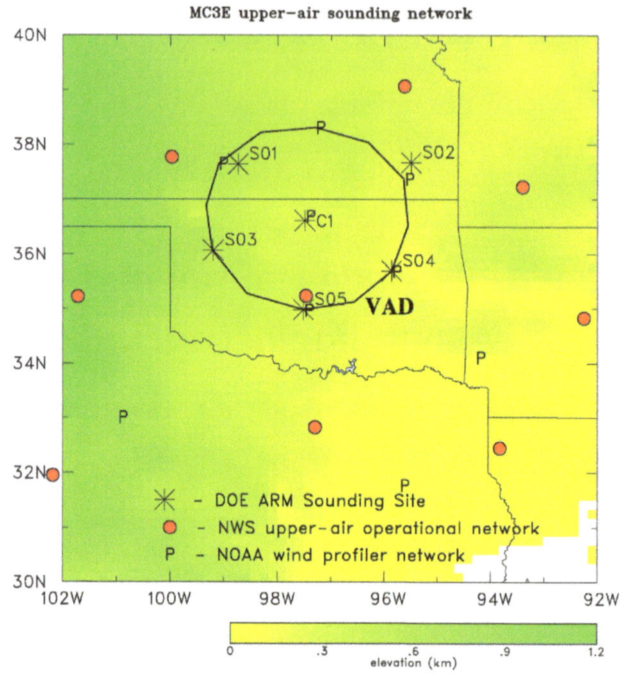

Figure 1. Map of the MC3E ground-based observing domain including the outer sounding domain: Central Facility [C1 Lamont, Oklahoma (OK) 36.695° latitude, −97.485° longitude] S1 Pratt, Kansas (KS) [37.7, −98.75]; S2 Chanute, KS [37.674, −95.488], S3 Vici, OK [36.071, −99.204]; S4 Morris, OK [35.687, −95.856]; S5 Purcell, OK [34.985, −97.522]. Also shown on this map are surrounding sites from the NWS operational sonde network and NOAA profiler network. Variational analysis domain (VAD) is indicated with a polygon. Shading indicates elevation according to the scale at bottom.

2 MC3E sounding operations

2.1 Background

The SGP ARM site was established in the spring of 1992 and consists of an array of in situ and remote sensing instrument clusters deployed over a 150 × 150 km area in north-central Oklahoma and south-central Kansas (http://www.arm.gov/sites/sgp). The site was designed to measure cloud, radiation, aerosol and atmospheric state properties over long timescales in a region that experiences a wide variety of meteorological conditions. As part of this measurement suite, the ARM program has launched radiosondes regularly from the SGP Central Facility (C1; 36.695° latitude, −97.485° longitude) since 14 July 1992. From August 2001 through the present, routine operations have consisted of four daily launches at approximately 00:30, 06:30, 11:30 and 17:30 local time. These launch times were chosen to complement the standard launch times (07:00 and 19:00 local time) by the National Weather Service. During special intensive observation periods (IOPs) the sonde launch frequency is often increased to 3-hourly at C1 and other sites in the SGP region as was done in the 1997 Summer Single Column Model IOP (Ghan et al., 2000; Xu et al., 2002) the March 2000 ARM Cloud IOP (Dong et al., 2002; Xie et al., 2005), and the 2007 Cloud and Land Surface Interaction Campaign (CLASIC, Miller et al., 2007).

2.2 Siting

During MC3E, five additional sounding locations were deployed. These sites were chosen in order to provide a sufficiently large sampling area in order to capture variability on scales larger than that of the grid size of a typical global climate model (∼ 150 km × 150 km) and to apply the constrained variational objective analysis (Zhang and Lin, 1997; Zhang et al., 2001) in order to provide large-scale forcing fields for modeling studies. Figure 1 shows a map of the MC3E enhanced sounding network including sounding stations: Central Facility at Lamont, OK (C1), Pratt, KS (S1), Chanute, KS (S2), Vici, OK (S3), Morris, OK (S4) and Purcell, OK (S5). Details of the locations of these stations are given in Table 1. Also shown on this map are surrounding sites from the National Weather Service (NWS) operational sonde network and NOAA wind profiler network. Data from the enhanced MC3E sites and surrounding networks are used to compute budgets and large-scale forcing fields over the variational analysis domain (VAD) shown by the polygon in Fig. 1.

Using a 6-day Weather Research and Forecasting (WRF) model simulation of convection over the SGP region (at 500 m horizontal resolution, 50 vertical levels and 3 h sampling intervals), we examined several different possible configurations for the sounding array by comparing the large-

Table 1. Summary of MC3E sounding array locations.

Site Designator	City, State	Latitude (deg. North)	Longitude (deg. West)	Elevation (m)
C1	Lamont, Oklahoma (OK)	36.695	97.485	315
S1	Pratt, Kansas (KS)	37.7	98.75	594
S2	Chanute, KS	37.674	95.488	297
S3	Vici, OK	36.071	99.204	628
S4	Morris, OK	35.687	95.586	217
S5	Purcell, OK	34.985	97.522	344

scale divergence calculated using a budget analysis based on array soundings compared to the high-resolution model divergence calculated over the arrays. The results showed that if the NOAA operational sounding and wind profiler networks over the central United States (shown in Fig. 1) are included in the derivation of the large-scale forcing data set, the VAD network results in relatively small biases and root mean square errors compared to other network configurations that were logistically available.

In addition to the scientific justification for the selection of the radiosonde launch locations, there were important logistical considerations. The security and safety of the operational staff was of the highest priority. For this reason, selected sites needed to be secure for both the sonde staff and equipment, and the nearby availability of a storm shelter was required in case of severe weather. An additional important consideration was the ability to leverage existing infrastructure. The three southernmost sounding locations (S3 Vici, OK; S4 Morris, OK; S5 Purcell, OK) were former ARM boundary facility sites that were historically used as sounding locations for IOPs. These sites still have the infrastructure required for regular radiosonde launches, including power, communications and housing for radiosonde equipment. The northernmost sites (S1 Pratt, KS; S2 Chanute, KS) were new sites located at local airports where space and facilities existed and were rented for use during the campaign.

2.3 Instrumentation

All sounding sites used Vaisala model RS92-SGP radiosondes attached to a 350 g helium-filled meteorological balloon. The RS92-SGP uses a capacitive wire temperature sensor with an accuracy of 0.5 °C over a range from +60 to −90 °C, a thin-film capacitor, a heated twin humidity sensor with 5 % uncertainty in relative humidity and a silicon pressure sensor with a maximum uncertainty of 1 hPa. The RS92-SGP sondes used in MC3E were manufactured in 2010–2011. The RS92-SGP data were received and recorded by Vaisala DigiCORA sounding systems at all sites. Three of these systems were of the more recent generation MW21 type (C1: Lamont, OK; S1: Pratt, KS; S2: Chanute, KS) and the remaining three sites (S3: Vici, OK; S4: Morris, OK; S5: Purcell, OK) used older MW15 systems. The older generation systems were not capable of running the latest Vaisala processing software (v3.64) with its associated corrections for humidity biases. For consistency, the v3.64 processing was turned off in the newer systems, and the humidity corrections were applied during our post-processing procedures (see Sect. 3.1). Vaisala ground check sets GC25 were used at all sites providing a pre-launch check on the radiosonde calibration and a reconditioning of the RS92-SGP relative humidity sensor. The desiccant in the GC25 ground check set was changed weekly or whenever the radiosonde relative humidity (RH) reading, while in the GC25, exceeded 1 %. Temperature, humidity, wind and pressure systems (THWAPS; Ritsche and Prell, 2011), which include Vaisala HMP233 temperature and humidity sensors, a Vaisala Model PTB201A digital barometer and an R.M. Young model 05103 wind monitor, also provided an additional calibration point at the surface at the C1, S3, S4, S5 sites. At the S1 site, Vaisala HMP233 microprocessor transmitters were used for the measurement of surface relative humidity and temperature. For the S2 site, surface meteorology observations were available from a National Weather Service Automated Surface Observing System (ASOS) station used by the airport. In order to facilitate the launch of a balloon by a single technician, a specially designed balloon launch cart was used that serves to hold the balloon in place while it is inflated and the radiosonde unit is attached.

To provide additional constraint on the radiosonde humidity observations, we used a nearby measurement of the total column water vapor amount in order to scale the integrated water vapor from the radiosonde when appropriate (see more details in Sect. 3.1). At the Central Facility, C1, observations from a two-channel (23.8 and 31.4 GHz) microwave radiometer (MWR; Morris, 2006) were used in a physical retrieval (MWRRET; Turner et al., 2007) to produce a time series (time steps of approximately 30 s) of precipitable water vapor (PWV) and liquid water path measurements. MWRs were not available for the outer sites of the radiosonde array. As an alternative, we investigated the use of PWV retrieved from GPS signal propagation delay measurements (e.g., Bevis et al., 1992) from the nearest SuomiNet (www.suominet.ucar.edu; Ware et al., 2000) GPS station. For the C1, S3 and S4 sites, the nearest GPS station was within

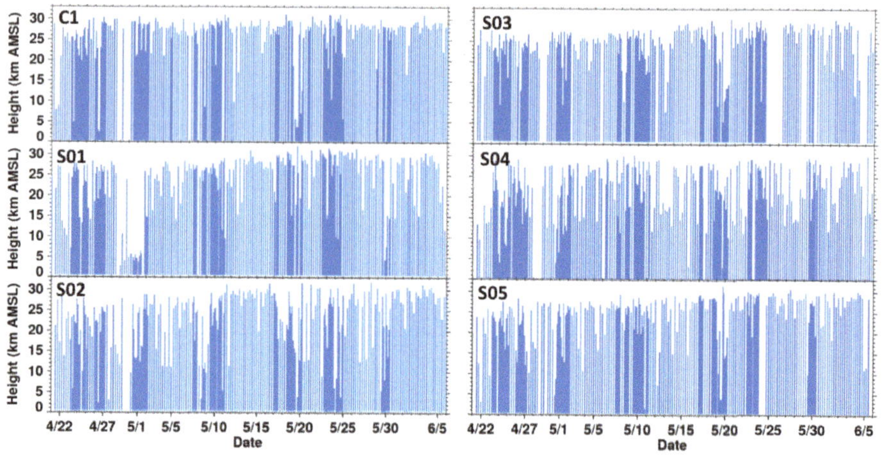

Figure 2. Sounding inventory summarizing the time of each launch and maximum height attained by each sonde for C1 (top), S1, S2, S3, S4, S5 (bottom).

1 km of the sonde launch site. For the S1, S2 and S3 sites the nearest GPS station was between 30 and 125 km from the sounding site (note that, as will be discussed in Sect. 3.1 and illustrated in Fig. 5, it was found that the GPS observations induced a dry bias in the corrected humidity profiles, and so we did not use them for scaling).

2.4 Data collection summary

Normally, under non-convective weather conditions, each of the six sites launched four soundings per day following the current schedule for routine operations at the SGP C1. On days for which convective conditions were forecast, sounding operations switched to a high-frequency launch schedule by adding launches at approximately 03:30, 09:30, 15:30, and 21:30 local time. In some cases, the high-frequency launch schedule was aborted at the discretion of the mission scientists if favorable conditions did not materialize. A launch was deemed successful if the balloon reached a pressure level of 700 hPa. If the balloon burst below this level, a second launch was attempted.

Over the course of the 6-week MC3E field campaign, a total of 1362 radiosondes were launched from the six sites. Figure 2 provides a visual inventory of the soundings, depicting information on the height attained for each launch and the frequency of sonde launches (by how close together the vertical bars are). Over the course of the campaign, a total of 8 days (24, 25, 27 April; 1, 2, 10, 11, 19, 20, 24 May) included a full high-frequency schedule of eight soundings. Seven additional days (8, 9, 17, 23, 25, 30, 31 May) included 5–7 soundings per day. Four soundings per day were launched on the remaining days of the field campaign. Note that an individual site may have fewer soundings than indicated by this list due to severe local weather or other operational considerations. During post-processing of the soundings, several data quality issues were identified and corrected when possible. A total of 14 soundings contained data quality issues (e.g., very early termination, unrealistic temperature/humidity variations) that were not correctable. In order to further summarize the success of the sounding operations, Table 2 indicates the number of soundings that sampled through a typical depth of the troposphere for each site. This table indicates that 90 % of the soundings launched during MC3E sampled at least through the depth of the troposphere. Figure 3 shows the frequency of occurrence (normalized by the total number of soundings) of the maximum altitude obtained during MC3E.

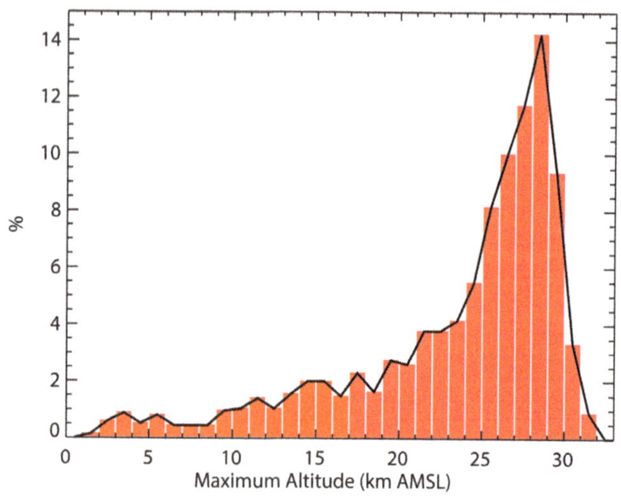

Figure 3. Frequency of occurrence (normalized by the total number of soundings) of the maximum height obtained by MC3E soundings, for all sites.

Table 2. Summary of termination height of MC3E soundings. 10th percentile tropopause height is taken from Rapid Update Cycle (RUC) analysis for period from 22 April through 6 June 2011. Early termination indicates soundings that did not sample above the 10th percentile tropopause height. The total sounding number here does not include the 14 sondes that were not correctable (see text for additional details).

Radiosonde Site	10th percentile tropopause height [hPa]	Total # soundings	# Early termination soundings (% of total)
C1 – ARM Central Facility, Lamont, OK	153.0	229	14 (6.1)
S1 – Pratt, KS	145.4	228	27 (11.8)
S2 – Chanute, KS	154.8	230	34 (14.8)
S3 – Vici, OK	143.3	209	13 (6.2)
S4 – Morris, OK	159.7	227	28 (12.3)
S5 – Purcell, OK	149.5	225	19 (8.4)
Total		1348	135 (10.0)

3 Radiosonde data processing

3.1 Automatic data quality checks

The Vaisala RS92-SGP sonde reports temperature, dew point, relative humidity, barometric pressure, wind speed and wind direction every 2 s throughout the radiosonde flight. The sounding software terminates data collection for a flight under one of the following conditions: the barometric pressure is observed to increase by more than 50 hPa (balloon has burst), several temperature measurements are missing in a row, a transmission signal is no longer received from the sounding for more than 720 s or a sensor failure is detected. The ARM program routinely includes some automated data quality checks that are based on predefined limits for the maximum, minimum and sample-to-sample change (delta) in the values of each raw variable. For the SGP locale, these limits are shown in Table 3 (Holdridge et al., 2011). One of these routine checks failed at one of more levels in 39 % of the soundings launched during MC3E. Table 4 shows the frequency of occurrence of routine data quality check failure based on the parameters defined in Table 3. Although a large percentage of sondes experienced data quality check failures at some point, the individual failures represented only a small fraction (less than 4 % total of the sonde measurements at any level). In general, failed out of bounds quality checks occurred for only a few levels in a given sounding and are dominated by missing wind observations in the upper troposphere.

3.2 Corrections to sounding humidity observations

A great deal of previous work has been devoted to defining and minimizing sources of error in Vaisala RS92-SGP humidity measurements including bias (systematic) errors, random errors and sensor time-lag errors (e.g., Milosevich et al., 2009, hereafter M09, and references therein). In particular, M09 performed a comparison of RS92-SGP radiosonde water vapor measurements with three reference instruments (cryogenic frost-point hygrometer, microwave radiometer and surface-based relative humidity probes with National Institute of Standards and Technology traceable calibrations) with well-defined accuracies. They define six separate contributions to errors in the RS92-SGP humidity measurements including (1) a mean calibration bias related to the accuracy of the Vaisala calibration model and calibration references, (2) random variability in the production of individual sondes, (3) time-lag errors caused by the slow response of the sensor at low temperatures, (4) a dry bias due to heating of the sensor by solar radiation, (5) biases related to improper application of ground check procedures and (6) precision errors due to round-off of RH to integer values.

Based on their comparative study, M09 provide an empirical mean bias correction for the RS92-SGP humidity profiles. The algorithm involves first applying a previously defined correction for the time-lag errors of the humidity sensor (Milosevich et al., 2004). Next, the empirical mean bias correction (defined in Fig. 9 and Table 1 in M09) is applied. Finally, since the M09 daytime correction was constructed for a solar zenith angle of 66°, daytime soundings require a final adjustment to account for differences from this angle. The M09 corrections are applied in the ARM SONDEADJUST value-added product (Troyan, 2012b) for the ARM SGP Central Facility site. We have extended the processing of this product to all sites in the enhanced sonde network during MC3E. It is noted that the mean bias correction was derived for clear-sky conditions and is therefore not appropriate for cloudy sky conditions. We have not taken this into account for the MC3E sounding data set and so have applied the corrections for all-sky conditions. Figure 4 shows a set of two-dimensional histograms of the difference in the dew point between the original soundings and the humidity-corrected soundings for all (panel a), daytime (panel b) and nighttime (panel c) soundings. These plots indicate that the corrections generally moisten the upper troposphere (above 8 km) and dry the lower troposphere. Upper atmosphere moistening is more pronounced during the daytime while lower troposphere drying is more prevalent during the nighttime.

Table 3. ARM defined min, max and delta data quality check limits. These limits are automatically applied to all MC3E soundings as part of the data ingest procedures. The delta check refers to the change between two adjacent data points. Data points that are outside of these limits are automatically flagged.

Variable Name	Units	Minimum	Maximum	Delta
Pressure	hPa	0.0	1100.0	10.0
Dry bulb temperature	°C	−80.0	50	10.0
Dew point temperature	°C	−110.0	50	–
Wind speed	$m\,s^{-1}$	0.0	75.0	–
Wind direction	Deg	0.0	360.0	–
Relative humidity with respect to liquid water	%	0.0	100.0	–
Eastward wind component	$m\,s^{-1}$	−100.0	100.0	–
Westward wind component	$m\,s^{-1}$	−100.0	100.0	–
Ascent rate	$m\,s^{-1}$	−10.0	20.0	5.0

Table 4. Data quality check failure rate. Summary of the percentage of observations for all sites outside of the data quality checked defined in Table 3.

Sounding Variable	% outside limits
All	3.39
Pressure	0.001
Temperature	0.001
Relative humidity/dew point temperature	1.05
Winds	2.62
Ascent rate	0.007

The SONDEADJUST value-added product (VAP) takes the sounding that has been corrected for known biases and then uses an independent retrieval of PWV to scale the radiosonde water vapor measurements (Turner et al., 2003; Cady-Pereira et al., 2008). During MC3E, there was a microwave radiometer (Turner et al., 2007; Gaustad et al., 2011) deployed at C1. We use the PWV observations from this MWR nearest the sonde launch time to scale the soundings launched from C1. Based on Global Climate Observing System (GCOS) Reference Upper-Air Network (GRUAN) RH uncertainly estimates, PWV uncertainty in RS92 sonde data is ±2 mm (Yu et al., 2015); PWV uncertainty in MWR retrievals is 0.2–0.5 mm (Turner et al., 2007; Cadeddu et al., 2013). Under some circumstances, the scaling may act to reverse the M09 corrections however, this step is justified because (1) the M09 corrections were developed under clear-sky conditions and (2) RS92 accuracy also changes for batches manufactured at different times. To help mitigate both of these correction uncertainties, scaling to a reliable independent PWV source is useful. Figure 5 shows the diurnal variation in the PWV retrieved (hourly averaged around sonde launch time) from the MWR observations and from uncorrected and corrected radiosonde observations for the C1. The mean value over all C1 soundings shows that the PWV from the corrected radiosonde observations (22.73 mm) agrees very well with the MWR observations (22.88 mm). The mean PWV bias between the MWR and corrected soundings is 0.127 mm compared to 0.396 mm between the MWR and uncorrected soundings. Clearly, the humidity corrections have improved the agreement between the soundings and MWR estimates of PWV by drying the nighttime soundings (4–5 %) and slightly moistening (2–3 %) the midday soundings.

Inherent in the notion of scaling the sonde humidity to a reliable independent measurement is the assumption that the radiosonde PWV should be the same as that observed by the independent sensor. While balloon drift and sampling scale mismatches may weaken this assumption, previous work by Turner et al. (2003) has shown that this extra constraint on radiosonde-observed water vapor profiles leads to improved modeling of clear-sky radiative fluxes. While the overall mean PWV from corrected soundings and the MWR agree extremely well, we use the ratio of the MWRRET PWV to the sonde PWV for each sounding as a constant scale factor for water vapor at all heights in the sounding in order to constrain shorter term fluctuations. Although no MWR observations are available for the extended radiosonde sites, there are GPS retrievals of PWV available from SuomiNet (http://www.suominet.ucar.edu) (Bevis et al., 1992; Ware et al., 2000) for the southern Great Plains region that were archived by the ARM External Data Center (http://www.arm.gov/xdc). Figure 5 also includes the diurnal variation of PWV retrieved from GPS observations at C1. Previous investigations have shown agreement between GPS, radiosonde and MWR observations of PWV to be within 1–2 mm (Emardson et al., 2000; Niell et al., 2001; Li et al., 2003; Garcia-Lorenz et al., 2009). Figure 5 shows that although the differences between GPS, MWR and corrected sounding PWV are within expected bounds, the GPS PWV is biased low compared to the MWR and corrected soundings. A similar dry bias in GPS PWV relative to soundings and MWR estimates of PWV was noted at several sites in a recent tropical field campaign (AMIE–ARM MJO Investigation Experiment) conducted over the Indian Ocean and west-

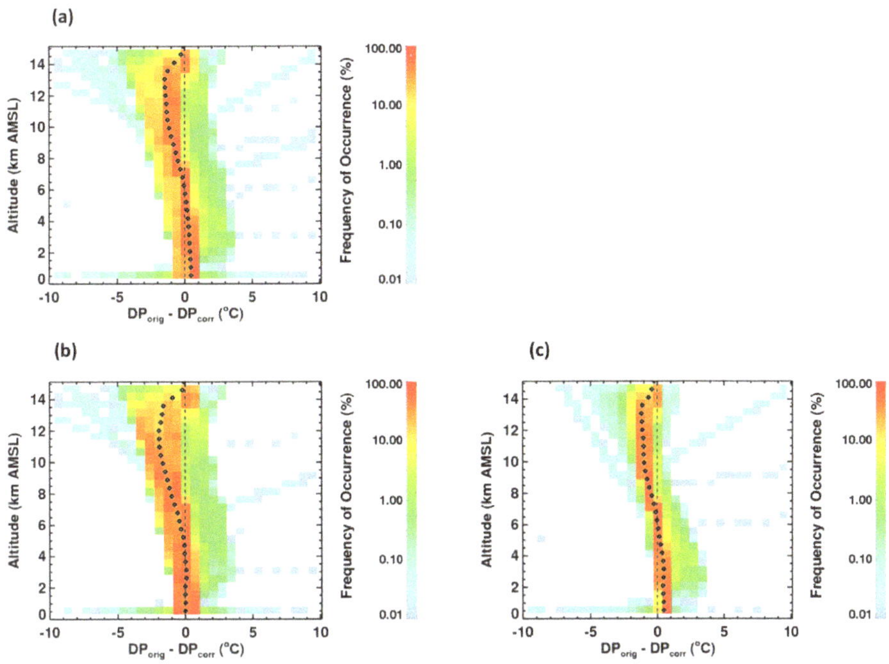

Figure 4. Two-dimensional histogram of humidity differences. Dew point (DP) difference between the original sounding and the humidity-corrected sounding for all soundings **(a)**, daytime soundings **(b)** and nighttime soundings **(c)**. The dotted black curve represents the median value at each height.

ern Pacific warm pool region (Ciesielski et al., 2014). Due to this dry bias in the GPS observations, estimates of GPS PWV were not used to scale the extended site radiosonde observations. Instead, the humidity-corrected, but un-scaled, radiosonde profiles provide our best estimate of actual moisture conditions.

3.3 Higher-order value-added data products

For C1, the corrected soundings are used as input into two higher-order products which both provide high-resolution profiles of atmospheric state for use in other derived products. The interpolated sonde (INTERPSONDE; Troyan, 2013) VAP begins with a simple linear interpolation (in time) of state variables observed by radiosondes to a 1 min time step. The water vapor profile is then scaled to the PWV retrieved from MWR observations and surface meteorological observations are attached. The main purpose of this VAP is to provide a quick estimate of atmospheric humidity in order to apply gaseous attenuation corrections to radar observations and a quick estimate of horizontal winds that can be used as a first guess in an iterative technique for dealiasing scanning radar Doppler velocity observations (Kollias et al., 2014). The INTERPSONDE VAP is further used as input to the Merged Sounding Product (MERGESONDE; Troyan, 2012a). The MERGESONDE VAP combines the state fields from INTERPSONDE with model output from the European Centre for Medium-range Weather Forecasting (ECMWF) via a weighted average dependent on the temporal distance from an actual radiosonde observation such that the output product has no contribution from the ECMWF field at times when actual sonde measurements exist. The output of the MERGESONDE VAP is a 1 min time resolution description of atmospheric state (pressure, temperature, relative humidity, winds) at 266 altitude levels in the lowest 20 km of the atmosphere. Data from the INTERPSONDE and MERGESONDE VAP are available from the ARM data archive (http://www.archive.arm.gov).

4 Analysis

4.1 Convective indices

From the MC3E soundings, a number of convective indices and other quantities important for understanding the lifecycle of convective clouds are calculated. The amount of buoyant energy available for an adiabatically rising air parcel is known as convective available potential energy (CAPE). CAPE is calculated as the vertical sum of parcel buoyancy from the level of free convection (LFC) to the equilibrium level (EL) (Moncrieff and Miller, 1976).

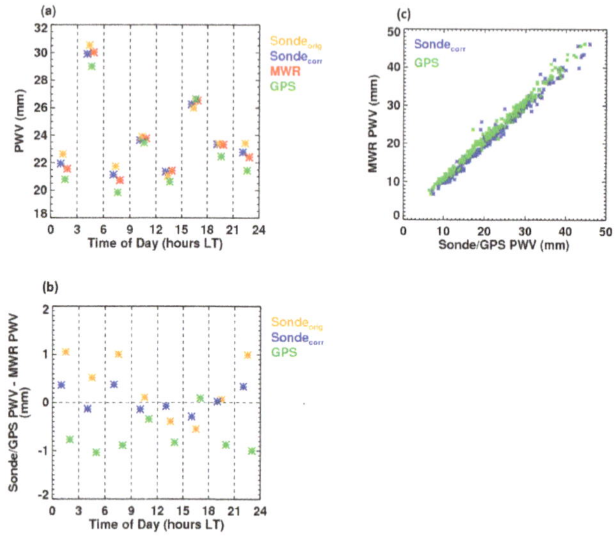

Figure 5. Diurnal variation of PWV at site C1. **(a)** Retrieved from an uncorrected sounding profile (yellow), a corrected sounding (blue), MWR observations (red) and GPS observations (green) averaged over all sondes in 3 h time windows; **(b)** absolute differences between retrieved PWV estimates; **(c)** scatterplot of retrieved PWV from corrected radiosonde profiles and GPS.

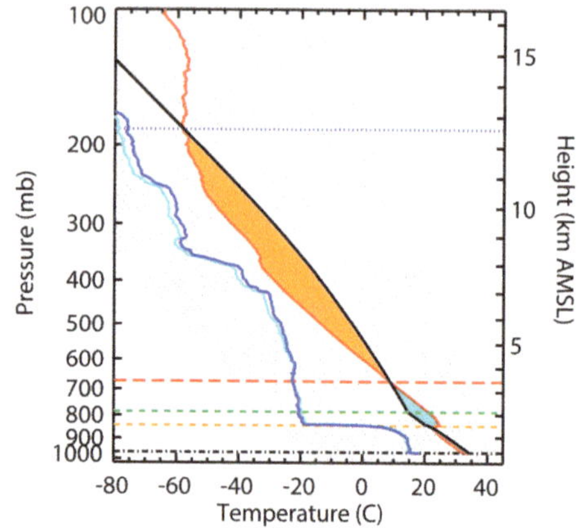

Figure 6. Graphical representation of convective indices. Central Facility radiosonde profile for 11 May 2011 02:34 UTC graphically illustrating the convective indices of CAPE (orange shading) and CIN (cyan shading) and relevant buoyancy levels including the lifting condensation level (LCL, green dashed), level of free convection (LFC, red dashed) and equilibrium level (blue dotted). The environmental virtual temperature profile is shown by the solid red line, adiabatic virtual parcel temperature is shown by the solid black line, the observed dew point temperature profile is shown with the dark blue line and the RH-corrected dew point temperature profile is shown with the cyan line.

Here we calculate CAPE using the following equation:

$$\mathrm{CAPE} = \sum_{\mathrm{LFC}}^{\mathrm{EL}} \left[\frac{(\theta_\mathrm{p} - \theta_\mathrm{v})}{\theta_\mathrm{v}} \right] g \Delta z, \tag{1}$$

where LFC is defined as the first level at which a rising parcel becomes positively buoyant, and EL is defined as the level (above the LFC) where the rising parcel changes from positive buoyancy to negative buoyancy. See Fig. 6 for a graphical representation of these levels. θ_p [K] is the virtual potential temperature of the rising parcel at height z and is determined by assuming that a rising parcel experiences dry adiabatic ascent until it reaches the lifting condensation level (LCL), beyond which the parcel rises pseudo-adiabatically (i.e., condensed water is assumed to be instantaneously removed from the rising parcel). The LCL is determined following the convention of Bolton (1980) where the temperature of the LCL is determined from the following empirical equation (Eq. 15 from Bolton, 1980):

$$T_\mathrm{LCL} = 56.0 + (1.0/(1.0/T_\mathrm{d} - 56.0) + \ln(T/T_\mathrm{d})/800), \tag{2}$$

where T_d [K] is the dew point temperature and T [K] is the dry bulb temperature. The LCL is determined by finding the height at which the temperature of a parcel rising dry-adiabatically is equal to its original T_LCL. θ_v [K] is the virtual potential temperature of the environment at height z, g (m s^{-2}) is the acceleration due to gravity and Δz is the change in height from level $i-1$ to level i (Moncrieff and Miller, 1976). The calculated CAPE value can then be interpreted as a control on the upper limit of the possible updraft velocity during convective storm initiation and growth, although this upper limit is never realized due to the impacts of entrainment, vertical pressure gradients and water loading.

The magnitude of CAPE is strongly dependent on the environmental density profile (i.e., temperature and humidity) and the choice of the level of free convection and equilibrium level through the initial parcel characteristics. The definition of initial parcel characteristics has such a profound impact on the magnitude of CAPE that different "flavors" of CAPE have been defined to specify this description. Here, we calculate CAPE by defining the initial parcel characteristics in two ways. We compute $\mathrm{CAPE_{SB}}$ using the lowest sounding observation (SBCAPE from Bunkers et al., 2002), and we compute $\mathrm{CAPE_{MU}}$ using the level of maximum virtual temperature in the lowest kilometer above ground level of the sounding (similar to the CAPE computed with the most unstable parcel in the lower atmosphere, MUCAPE, from Bukners et al., 2002). $\mathrm{CAPE_{SB}}$ is generally most relevant for situations wherein predominantly surface-based convection is of interest. $\mathrm{CAPE_{MU}}$ provides a maximum estimate of CAPE and is most often relevant in cases of elevated convection (Rochette et al., 1999).

As illustrated in Fig. 6, there is often an area of negative buoyancy below the level of free convection that is known as convective inhibition (CIN). CIN can be thought of as a measure of the work that an air parcel must do in order to reach the LFC where it will finally become positively buoyant. Mathematically, CIN can be expressed as such:

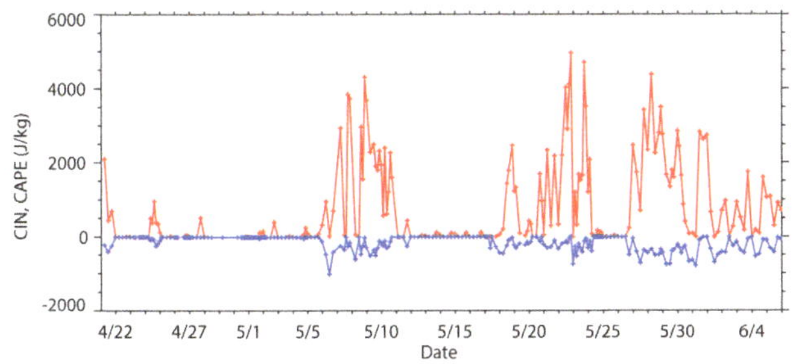

Figure 7. Time series of $CAPE_{MU}$ (red) and CIN_{MU} (blue) at the ARM C1 over the course of the MC3E field campaign, using corrected sounding humidity. Three distinct active convective periods are easily identifiable (6–11 May, 18–23 May and 26 May–6 June) with high CAPE and low CIN values.

$$\mathrm{CIN} = \sum_{\mathrm{SFC}}^{\mathrm{LFC}} \left[\frac{(\theta_{\mathrm{p}} - \theta_{\mathrm{v}})}{\theta_{\mathrm{v}}} \right] g \Delta z, \tag{3}$$

where SFC is the surface level (more details on this later) and all other variables are defined in the same manner as Eq. (1). Figure 7 shows a time series of $CAPE_{MU}$ and CIN_{MU} at the SGP C1 using uncorrected humidity soundings over the entire length of the MC3E field campaign.

4.2 Sensitivity to measurement uncertainties

The determination of the LCL, LFC, EL, CAPE and CIN are sensitive to the details of the profiles of temperature and humidity. Uncertainties in the calculation of these quantities are a function of the uncertainties and biases (e.g., the relative humidity biases discussed in Sect. 3.1) in the observed thermodynamic profile and the assumptions used in the algorithm for the calculation of each quantity. Here, we will investigate the uncertainties in the calculated CAPE/CIN due to each of these causes.

4.2.1 Differences due to humidity corrections

This subsection presents an analysis of differences found in computed convective indices based on correction of the humidity profile as defined in Sect. 3.1. For all comparisons herein, computations are made using the surface parcel defined according to the calculation of $CAPE_{MU}$.

Figure 8 shows scatterplots, for MC3E soundings at all six sites, for three convective levels: (a) LCL, (b) LFC and (c) EL, calculated with and without the humidity corrections. Each of these convective levels, important for the calculation of higher-order convective indices (e.g., CAPE and CIN), shows relatively small sensitivity to the humidity corrections.

Although 86 % of the soundings show a difference in the LCL of greater than 1 %, only 3 % show a difference greater than 10 %. During nighttime hours (00:00–12:00 UTC), when we see more drying in the boundary layer, 96 % of the soundings show a difference in LCL of greater than 1 %, while still only 3 % show a difference greater than 10 %. During the daytime hours (12:00–24:00 UTC), significantly fewer soundings (76 %) show a difference of more than 1 %, while 3.6 % show a difference greater than 10 %, similar to evening. A comparison of the mean LCL over all MC3E soundings shows the LCL with humidity corrections applied to be 39 m higher than the LCL for the original soundings with a standard deviation of the difference between the two LCL estimates of 86 m. Furthermore, the relative humidity correction generally results in higher LCLs during both day and night, with resultant LCLs 10 and 68 m higher on average, with standard deviations of 54 and 101 m, respectively.

Turning to the LFCs, 48 % of the soundings exhibit a difference of at least 1 % in the LFC, and 5 % exhibit a difference of more than 50 %. During daytime only hours (12:00–24:00 UTC) or nighttime only hours (00:00–12:00 UTC), the statistics are similar with 48.6 % of daytime and 47.4 % of nighttime soundings showing differences of at least 1 %, and 6.2 % of daytime and 4.3 % of nighttime soundings showing a difference of more than 50 %. The mean difference in LFC over all MC3E soundings is only 6 m with a standard deviation of 854 m, however the humidity correction on daytime soundings results in a positive mean difference (original less humidity-corrected) of 27 m (with a standard deviation of 989 m), while for nighttime soundings it results in a negative mean difference of 17 m (with a standard deviation of 686 m), indicating that the relative humidity correction generally results in lower LFCs during the day and higher LFCs during the night.

The ELs show differences of greater than 1 % for only 29.4 % of the soundings with 5.9 % having differences larger than 10 %. Daytime and nighttime soundings show similar statistics with 28.4 and 30.5 % exhibiting differences greater than 1 %, respectively, and 5.9 % of both showing differences of at least 10 %. The mean difference in EL over all MC3E soundings is 126 m (original less humidity-corrected), with a standard deviation of 729 m. However, during the daytime hours, the ELs of corrected soundings are 42 m (standard deviation of 744 m) lower than uncorrected, while during nighttime hours, as defined here, the difference is even greater, with corrected soundings having an EL 214 m (standard deviation of 703 m) lower on average. This follows the fact that nighttime-corrected soundings tend to have generally higher LCLs than uncorrected soundings, and more so than daytime soundings. In general, soundings that show a difference in LCL also show differences for LFC and EL.

These differences in convective levels in the atmosphere have corresponding impacts on the convective indices they are used to calculate. Figure 9 shows a scatterplot of the (a) CAPE and (b) CIN calculated without the humidity corrections versus the same quantities calculated using the corrected humidity profiles. Daytime soundings (12:00–24:00 UTC) are indicated with red markers while nighttime

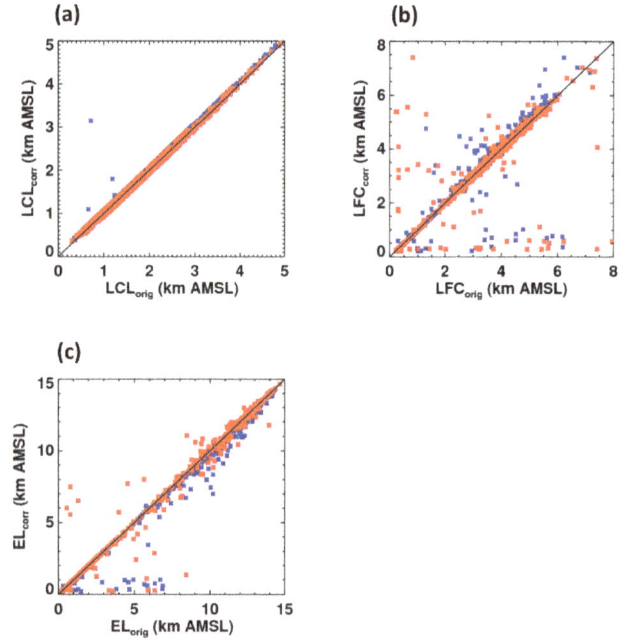

Figure 8. Scatterplots of convective levels comparing original (orig) and humidity-corrected (corr) soundings for **(a)** LCL, **(b)** LFC and **(c)** EL. Daytime soundings (12:00–24:00 UTC) are indicated with red markers while nighttime soundings are blue. Convective levels are computed using the level of maximum virtual temperature in the lowest kilometer above ground level of the sounding.

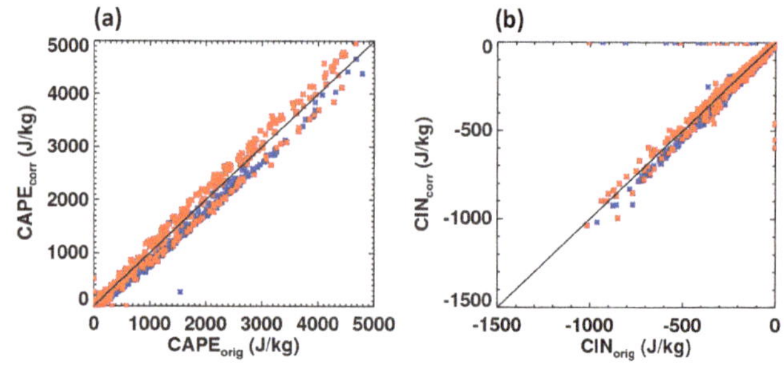

Figure 9. Scatterplots of convective indices comparing original (orig) and humidity-corrected (corr) soundings for **(a)** CAPE and **(b)** CIN. Daytime soundings (12:00–24:00 UTC) are indicated with red markers while nighttime soundings are blue.

soundings are blue. In total 42.1 % (45.2 %) of the CAPE (CIN) values agree to within 1 %. A closer look at the cases with greater differences shows that 12.2 % (7 %) of the CAPE (CIN) values disagree by more than 50 %. Because there is a correlation between the relative magnitude of the differences and the absolute value of the CAPE/CIN, we also present simple statistical measures of the absolute differences in CAPE/CIN for several different ranges, categorized by the magnitude of the CAPE/CIN value of the corrected soundings. For soundings with CAPE less than $500\,\mathrm{J\,kg^{-1}}$, the mean difference (original less humidity-corrected) in CAPE is $14\,\mathrm{J\,kg^{-1}}$ with a standard deviation of $63\,\mathrm{J\,kg^{-1}}$, while for CAPE values between 500 and $2000\,\mathrm{J\,kg^{-1}}$ the mean difference is $105\,\mathrm{J\,kg^{-1}}$ with a standard deviation of $164\,\mathrm{J\,kg^{-1}}$ and finally for CAPE values greater than $2000\,\mathrm{J\,kg^{-1}}$ the mean difference is $22\,\mathrm{J\,kg^{-1}}$ with a standard deviation of $240\,\mathrm{J\,kg^{-1}}$. For soundings with CIN values greater than $-100\,\mathrm{J\,kg^{-1}}$ the mean difference between CIN calculated from the original soundings and those with humidity corrections applied is $-9.5\,\mathrm{J\,kg^{-1}}$ with a standard deviation of $75\,\mathrm{J\,kg^{-1}}$. For soundings with CIN between -100 and $-500\,\mathrm{J\,kg^{-1}}$ the mean of the differences is $17\,\mathrm{J\,kg^{-1}}$ with a standard deviation of $40.5\,\mathrm{J\,kg^{-1}}$. Finally, soundings with CIN less than $-500\,\mathrm{J\,kg^{-1}}$ show mean differences of $39\,\mathrm{J\,kg^{-1}}$ with a standard deviation of $84\,\mathrm{J\,kg^{-1}}$. These results indicate that for many soundings there is a significant impact on the calculation of CAPE/CIN from corrections to the measured profile of humidity.

4.2.2 Differences due to surface parcel characteristics

This subsection presents an analysis of the sensitivity of convective index computation to surface parcel definition as introduced in Sect. 4.1. For all comparisons that follow, computations are made using humidity-corrected soundings for all six MC3E sites.

Figure 10 shows a comparison of the (a) LCL, (b) LFC and (c) EL for the two separate definitions of the surface parcel: (1) lowest sounding observation (SB) and (2) level of maximum virtual temperature in the lowest kilometer (MU). The scatterplots show that each of these convective levels exhibits strong sensitivity to the definition of the surface parcel characteristics. It should be noted that for most of the cases where the LCL, LFC or EL agree to within 1 %, it was found that the surface observation was also the level of maximum virtual temperature in the lowest kilometer. For the LCL, 56 % of the soundings show differences greater than 1 %, while 32 % show differences greater than 50 %. The mean difference in LCL is 510 m (maximum virtual temperature less the lowest sounding level) with a standard deviation of 906 m (i.e., when the surface parcel is defined using the lowest sounding level, the LCL tends to be lower). The LFC shows a similar pattern of differences between the surface parcel representations. In this case, 55.7 % of the soundings show a difference greater than 1 % for the two surface parcel representations while 32.9 % of the soundings show differences larger than 50 %. The mean difference in LFC is 430 m with a standard deviation of 1600 m. Similar to the LCL and LFC, the EL scatterplot also shows differences of greater than 1 % for 52.7 % of the MC3E soundings with differences larger than 50 % in 32.5 % of the soundings. The mean difference in EL is 1485 m with a standard deviation of 3992 m.

These differences in the LCL, LFC and EL translate into differences in the CAPE and CIN. Figure 11 shows scatterplots of the (a) CAPE and (b) CIN calculated using the

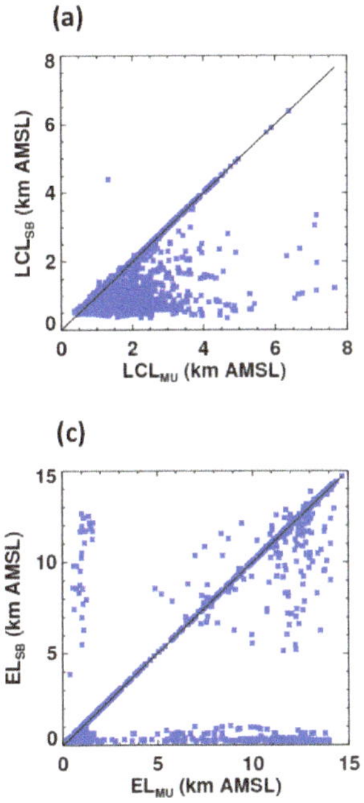

Figure 10. Scatterplots of convective levels, computed using RH-corrected soundings, comparing estimates for surface parcel characteristics defined by the level with the maximum virtual temperature in the lowest kilometer (MU) and the lowest level in the sounding (SB) for **(a)** LCL, **(b)** LFC and **(c)** EL.

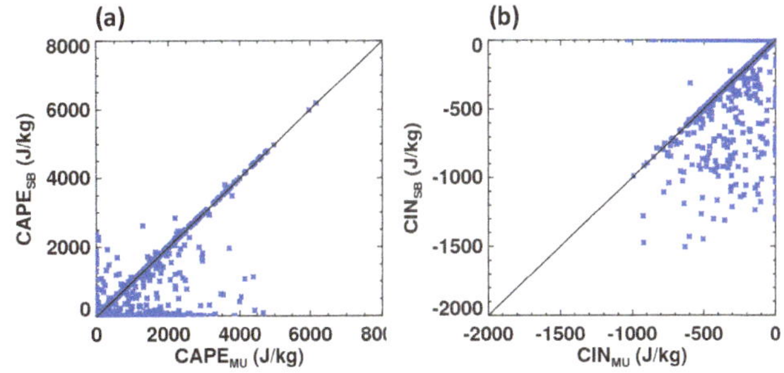

Figure 11. Scatterplots of convective indices, computed using RH-corrected soundings, comparing estimates for surface parcel characteristics defined by the level with the maximum virtual temperature in the lowest kilometer (MU) and the lowest level in the sounding (SB) for **(a)** CAPE and **(b)** CIN.

level of maximum virtual temperature in the lowest 1 km to define the surface parcel characteristics versus calculations using the lowest level observation from each sounding. For CAPE (CIN), 37.5 % (34.7 %) of the soundings show differences greater than 1 %, while 32.9 % (29.4 %) of the cases show differences greater than 50 %. That is, when there is a difference in the calculation of CAPE or CIN due to the definition of the surface parcel, it is generally a significant one. Also, recall that for many of the cases where there was no impact on the estimation of LCL, LFC or EL, the two representations of surface parcel characteristics were identical (and therefore by definition would result in no difference in any of the quantities investigated).

Again, we will consider the differences in terms of categories to account for the correlation between the relative magnitude of differences and the absolute value of the CAPE/CIN, binning here by the magnitude of the CAPE/CIN value computed using the maximum virtual temperature in the lowest 1 km. For soundings with CAPE less than $500\,\mathrm{J\,kg^{-1}}$ (using the corrected RH) the mean difference in CAPE between the two descriptions of surface parcel characteristics is $-24\,\mathrm{J\,kg^{-1}}$ with a standard deviation of $241\,\mathrm{J\,kg^{-1}}$, while for CAPE values between 500 and $2000\,\mathrm{J\,kg^{-1}}$ the mean difference is $488\,\mathrm{J\,kg^{-1}}$ with a standard deviation of $646\,\mathrm{J\,kg^{-1}}$ and finally for CAPE values greater than $2000\,\mathrm{J\,kg^{-1}}$ the mean difference is $824\,\mathrm{J\,kg^{-1}}$ with a standard deviation of $1194\,\mathrm{J\,kg^{-1}}$. For soundings with CIN values greater than $-100\,\mathrm{J\,kg^{-1}}$ the mean difference between CIN is $34\,\mathrm{J\,kg^{-1}}$ (maximum virtual temperature in lowest kilometer representation minus lowest sounding observation representation) with a standard deviation of $171\,\mathrm{J\,kg^{-1}}$. For soundings with CIN between -100 and $-500\,\mathrm{J\,kg^{-1}}$ the mean of the differences is $15.6\,\mathrm{J\,kg^{-1}}$ with a standard deviation of $284\,\mathrm{J\,kg^{-1}}$. Finally, soundings with CIN less than $-500\,\mathrm{J\,kg^{-1}}$ show mean differences of $-92\,\mathrm{J\,kg^{-1}}$ with a standard deviation of $421\,\mathrm{J\,kg^{-1}}$. These results indicate that for many soundings, there is a significant impact on the calculation of CAPE and CIN due to the representation of the surface parcel characteristics and this sensitivity is larger than that due to the sounding humidity biases found in this study.

4.2.3 Impacts on large-scale forcing

The major justification for the deployment of the sounding array during MC3E is for the derivation of a large-scale model forcing data set to drive single-column, cloud-resolving and land-surface model simulations. Towards this goal, Xie et al. (2014) use the constrained variational analysis approach of Zhang and Lin (1997) in order to derive the large-scale vertical velocity and advective tendencies of temperature and moisture over the MC3E domain. This technique uses a combination of sounding network observations, ARM extended facility observations, radar-derived surface precipitation and top-of-atmosphere radiative flux observations as constraints to adjust the sounding-observed atmospheric state variables in order to conserve column-integrated mass, moisture and static energy such that the resulting forcing data set is dynamically and thermodynamically consistent. Here, we investigate the impact of the sounding quality control and humidity corrections on the derived large-scale vertical velocity. Figure 12 shows a time series of the large-scale vertical velocity derived using the quality-controlled, humidity-corrected soundings and the difference compared to the vertical velocity derived using the raw sounding ob-

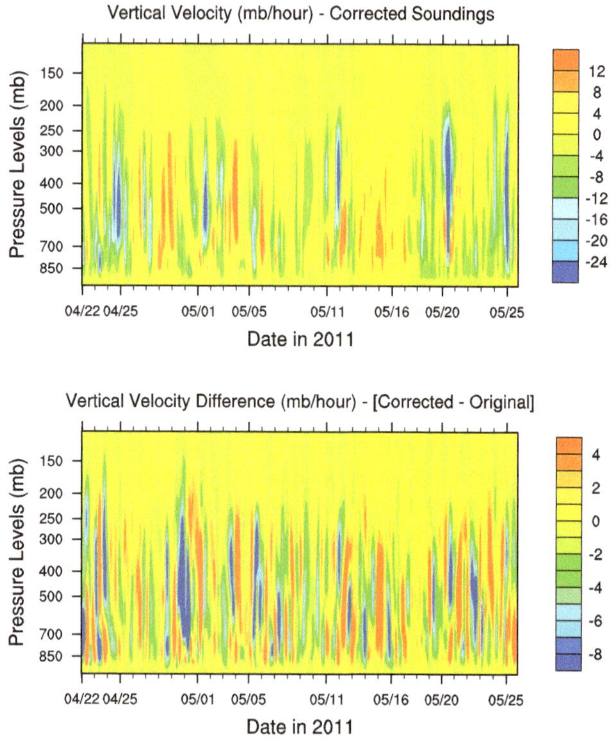

Figure 12. Comparison of large-scale forcing data set for original and corrected soundings. Derived large-scale vertical velocity using quality-controlled, humidity-corrected radiosonde profiles (top) and the difference (corrected-original) between the derived large-scale forcing using the corrected soundings and the original soundings. Please note that dates are given in month/day format.

servations. The results show differences in the derived vertical velocity ranging from +5 to $-9\,\mathrm{hPa\,h^{-1}}$. The magnitudes of these differences are significant compared to the derived vertical velocity ranging from +13 to $-25\,\mathrm{hPa\,h^{-1}}$. These differences could lead to significant differences in the simulation of convective clouds and illustrate the importance of accurate measurements of atmospheric state profiles for input to large-scale forcing estimates.

5 Summary and conclusions

An important component of the MC3E field campaign was the collection of an extensive sounding data set that included observations from six sites covering and area of 300 × 300 km.

A total of 1362 soundings were launched from this array over the course of the 6-week campaign. Routine ARM processing of the sounding data set included a set of automatic quality checks on the magnitude and variability of the observations. Known humidity biases were corrected using the empirical algorithms defined by Milosevich et al. (2009) and when available compared to independent observations of PWV. At the central ARM facility, where reliable independent measurements of PWV were available, the humidity corrections were shown to improve the agreement between the sonde and MWR PWV observations with a mean PWV difference of 0.15 mm (or 0.7 %). These corrections along with rigorous quality control of the soundings have resulted in a high-quality data set suitable for many research applications including the derivation of a model forcing data set (Xie et al., 2014).

An analysis of the impact of the corrections for known humidity biases and assumptions regarding surface parcel definitions show that both have significant impacts on the determination of convective levels and indices for many soundings. The impact of assumptions about surface parcel characteristics generally has a greater impact on these levels and indices than the humidity corrections applied in this study. Further, an investigation of the impacts of the sounding humidity corrections and quality control on the derived large-scale forcing fields indicate significant influences on the calculated large-scale vertical velocity. These differences would have important implications for model simulation of convective events.

The MC3E sounding data set (Jensen and Toto, 2014) is available from the ARM archive (www.archive.arm.gov). The raw sounding data for the Central Facility, C1, is a regular ARM data stream (sgpsondewnpn*) and can be found using the ARM archive tools. The boundary facility (S1, S2, S3, S4 and S5) humidity-corrected soundings are available from the MC3E IOP archive (http://iop.archive.arm.gov/arm-iop/2011/sgp/mc3e).

Acknowledgements. The authors wish to thank the program managers of the DOE ARM program, Wanda Ferrell, and the NASA GPM program, Ramash Karkar and Arthur Hou, for their support of the MC3E field campaign. We also wish to acknowledge the scientists and staff that participated in the deployment stage of the MC3E field campaign, particularly those who contributed to the radiosonde activities including Mike Bobbitt, Phil Cardwell, Debi Chaney, Marilyn Dowell, Mary Green, Shawn Muegge, Sandra Schovanec, Mike Shaffer, Lynda Theilen, James Cunningham, James Rowland, Robin Moore, Leah Nelson, Jim Callison, Cris Alvarez, Tony Newman, Seth Galemore, Parker Huitt, Kenneth Crick, Donald Cook, Clydie Prince, Tracy Messer, John Duncan, Cameron Summers, Raymond Cline, Bruce Sewell, Tim Ahlstrom, Daniel Harnos, Kim Reed, Joseph Wegman, Melissa Peterson, Tristan Hall, Peter Patina, Matthew Grey, Matthew Dawson, Robert Seigel, Tiffany Meyer, Elizabeth Thompson, Cort Sholten. Participation by BNL scientists has been supported by the Office of Biological and Environmental Research of the US Department of Energy as part of the Atmospheric Radiation Measurement program, the Atmospheric Systems Research Program and the Earth System Modeling Program. Paul Ciesielski acknowledges his support from the NASA PPM grant NNX13AF74G.

Edited by: J. Joiner

References

Ackerman, T. P. and Stokes, G. M.: The Atmospheric Radiation Measurement Program, Phys. Today, 56, 38–45, 2003.

Bevis, M., Businger, S., Herring, T. A., Rocken, C., Anthes, R. A., and Ware, R. H.: GPS Meteorology: Remote sensing of atmospheric water vapor using the global positioning system, J. Geophys. Res., 97, 787–801, 1992.

Bolton, D.: Computation of equivalent potential temperature, Mon. Weather Rev., 108, 1046–1053, 1980.

Bunkers, M. J., Klimowski, B. A., and Zeitler, J. W.: The importance of parcel choice and the measure of vertical wind shear in evaluating the convective environment, Preprints, 21st Conf. Severe Local Storms, San Antonio, American Meteorological Society, J117–J120, 11–16 August 2002.

Cadeddu, M. P., Liljegren, J. C., and Turner, D. D.: The Atmospheric radiation measurement (ARM) program network of microwave radiometers: instrumentation, data, and retrievals, Atmos. Meas. Tech., 6, 2359–2372, doi:10.5194/amt-6-2359-2013, 2013.

Cady-Pereira, M. W., Shephard, D. D., Turner, E. J., Mlawer, S. A., Clough, T., and Wagner, J.: Improved Daytime Column-Integrated Precipitable Water Vapor from Vaisala Radiosonde Humidity Sensors, J. Atmos. Ocean. Tech., 25, 873–883, doi:10.1175/2007JTECHA1027.1, 2008.

Ciesielski, P. E, Yu, H., Johnson, R. H., Yoneyama, K., Katsumata, M., Long, C. N., Wang, J., Loehrer, S. M., Young, K., Williams, S. F., Brown, W., Braun, J., and Van Hove, T.: Quality controlled upper-air sounding dataset for DYNAMO/CINDY/AMIE: development and corrections, J. Atmos. Ocean. Tech., 31, 741–764, 2014.

Dong, X., Mace, G. G., Minnis, P., Smith, W. L., Poellet, M., Marchand, R. T., and Rapp, A. D.: Comparison of stratus cloud properties defined from surface, GOES, and aircraft data during the March 2000 ARM Cloud IOP, J. Atmos. Sci., 59, 3265–3284, 2002.

Emardson, T. R., Johansson, J., and Elgered, G.: The systematics behavior of water vapor estimates using four years's of GPS observations, IEEE T. Geosci. Remote, 38, 324–329, 2000.

Garcia-Lorenzo, B., Castro-Almazan, J. A., Eff-Darwich, A., Munoz-Tunon, C., Pinilla-Alonso, N., Rodrigues-Espinosa, J. M., and Romero, I.: Precipitable water vapour content above the Roque de los Muchachos observatory from GPS estimations, Proc. SPIE 7475, Remote Sensing of Clouds and the Atmosphere XIV, 74751H, 29 September 2009, doi:10.1117/12.830235, 2009.

Gaustad, K. L., Turner, D. D., and McFarlane, S. A.: MWRRET Value-Added Product: The Retrieval of Liquid Water Path and Precipitable Water Vapor from Microwave Radiometer (MWR) Data Sets. DOE/SC-ARM-TR-081.2, 5 pp., 2011.

Ghan, S. J., Randall, D. A., Xu, K.-M., Cederwall, R., Cripe, D., Hack, J., Icobellis, S., Klein, S., Krueger, K., Lohmann, U., Pedretti, J., Robock, A., Rotsayn, L., Somerville, R., Stechikov, G., Sud, Y., Walker, G., Xie, S. C., Yio, J., and Zhang, M. H.: An intercomparison of single column model simulations of summertime midlatitude continental convection, J. Geophys. Res., 105, 2091–2124, 2000.

Holdridge, D. J., Prell, J. A., Ritsche, M. T., and Coulter, R.: Balloon-Borne Sounding System (SONDE) Handbook. DOE/SC-ARM-TR-029, U.S. Department of Energy, Office of Science, Office of Biological and Environmental Research, 28 pp., 2011.

Hou, A. Y., Kakar, R. K., Neeck, S., Azarbarzin, A. A., Kummerow, C. D., Kojima, M., Oki, R., Nakamura, K., and Iguchi, T.: The Global Precipitation Measurement (GPM) Mission, B. Am. Meteorol. Soc., 95, 701–722, doi:10.1175/BAMS-D-13-00164.1, 2014.

Jensen, M. P. and Toto, T.: Humidity-corrected sounding observations (SONDEADJUST). April–June 2011, 36.695° N, 97.485° W; 37.7° N, 98.75° W; 37.674° N, 95.488° W; 36.071° N, 99.204° W; 35.678° N, 95.586° W; 34.985° N, 97.522° W: Southern Great Plains Sites C1, S1, S2, S3, S4, S5. Atmospheric Radiation Measurement (ARM) Climate Research Facility Data Archive: Oak Ridge, TN, USA, doi:10.5439/1164928, 2014.

Jensen, M. P., Kollias, P., Del Genio, A. D., Giangrande, S. E., Orr, B., Petersen, W. A., Schwaller, M. R., Hou, A. Y., Rutledge, S. A., Heymsfield, G., Heymsfield, A., and Zipser, E.: Mid-Latitude Continental Convective Cloud Experiment (MC3E) Science and Implementation Plan. DOE/ARM Technical Report number DOE/SC-ARM/10-004, U.S. Department of Energy, Office of Science, Office of Biological and Environmental Research, 2010.

Johnson, R. H.: Partitioning tropical heat and moisture budgets into cumulus and mesoscale components: Implications for cumulus parameterization, Mon. Weather Rev., 112, 1590–1601, 1984.

Johnson, R. H. and Ciesielski, P. E.: Structure and properties of Madden-Julian Oscillations deduced from DYNAMO sounding arrays, J. Atmos. Sci., 70, 3157–3179, 2013.

Klein, S. A. and Del Genio, A. D.: ARM's support for GCM improvement: A white paper, U.S. Department of Energy. DOE/SC-ARM/P-06-012, U.S. Department of Energy, Office of Science, Office of Biological and Environmental Research, 2006.

Kollias, P., Jo, I., Borque, P., Tatarevic, A., Lamer, K., Bharadwaj, N., Widener, K., Johnson, K. L., and Clothiaux, E. E.: Scanning ARM cloud radars. Part II: Data quality control and processing, J. Atmos. Ocean. Tech., 31, 583–598, doi:10.1175/JTECH-D-13-00045.1, 2014.

Li, Z., Muller, J.-P., and Cross, P.: Comparison of precipitable water vapor derived from radiosonde, GPS, and Moderate-Resolution Imaging Spectroradiometer measurements, J. Geophys. Res., 108, 4651, doi:10.1029/2003JD003372, 2003.

Mather, J. H. and Voyles, J. W.: The ARM Climate Research Facility: A Review of Structure and Capabilities, B. Am. Meteorol. Soc.., 94, 377–392, 2013.

Miller, M. A.: SGP Cloud and Land Surface Interaction Campaign (CLASIC): Science and Implementation Plan. U.S. Department of Energy. DOE/SC-ARM-0703, U.S. Department of Energy, Office of Science, Office of Biological and Environmental Research, 2007.

Miloshevich, L. M., Paukkunen, A., Vömel, H., and Oltmans, S. J.: Development and validation of a time-lag correction for Vaisala radiosonde humidity measurements, J. Atmos. Ocean. Tech., 21, 1305–1327, 2004.

Miloshevich, L. M, Vomel, H., Whiteman, D. N., and Leblanc, T.: Accuracy assessment and correction of Viasala RS92 radiosonde water vapor measurements, J. Geophys. Res., 114, D11303, doi:10.1029/2008JD011565, 2009.

Moncrieff, M. W. and Miller, M. J.: The dynamics and simulation of tropical cumulonimbus and squall lines, Q. J. Roy. Meteor. Soc., 102, 373–394, 1976.

Morris, V. R.: Microwave Radiometer (MWR) handbook, DOE/SC-ARM-TR-016, U.S. Department of Energy, Office of Science, Office of Biological and Environmental Research, 20 pp., 2006.

Niell, A. E., Coster, A. J., Solheim, F. S., Mendes, V. B., Toor, P. C., Langley, R. B., and Upham, C. A.: Comparison of Measurements of Atmospheric Wet Delay by Radiosonde, Water Vapor Radiometer, GPS, and VLBI, J. Atmos. Ocean. Tech., 18, 830–850, doi:10.1175/1520-0426(2001)018<0830:COMOAW>2.0.CO;2, 2001.

Ritsche, M. T. and Prell, J.: Temperature, Humidity, Winds And Pressure System (THWAPS) handbook. DOE/ARM-TR-030, U.S. Department of Energy, Office of Science, Office of Biological and Environmental Research, 11 pp., 2011.

Rochette, S. M., Moore, J. T., and Market, P. S.: The importance of parcel choice in elevated CAPE computations, Natl. Wea. Dig., 23, 20–32, 1999.

Stokes, G. M. and Schwartz, S. E.: The Atmospheric Radiation Measurement (ARM) Program: Programmatic background and design of the cloud and radiation test bed, B. Am. Meteorol. Soc., 75, 1201–1221, 1994.

Troyan, D.: Merged Sounding Value-Added Product Technical Report. DOE/SC-ARM-TR-087, U.S. Department of Energy, Office of Science, Office of Biological and Environmental Research, 13 pp., 2012a.

Troyan, D.: Sonde Adjust Value-Added Product Technical Report. DOE/SC-ARM-TR-102. U.S. Department of Energy, Office of Science, Office of Biological and Environmental Research, 19 pp., 2012b.

Troyan, D.: Interpolated Sounding Value-Added Product Technical Report. DOE/SC-ARM/TR-124. U.S. Department of Energy, Office of Science, Office of Biological and Environmental Research, 13 pp., 2013.

Turner, D. D., Lescht, B. M., Clough, S. A., Liljigren, J. C., Revercomb, H. E., and Tobin, D. C.: Dry bias and variability in Vaisala RS80-H radiosondes: The ARM experience, J. Atmos. Ocean. Tech., 20, 117–132, doi:10.1175/15200426(2003)030<0117:DBAVIV>2.0CO;2, 2003.

Turner, D. D., Clough, S. A., Liljegren, J. C., Clothiaux, E. E., Cady-Pereira, K., and Gaustad, K. L.: Retrieving liquid water path and precipitable water vapor from Atmospheric Radiation measurement (ARM) microwave radiometers, IEEE T. Geosci. Remote, 45, 3680–3690, doi:10.1109/TGRS.2007.903703, 2007.

Ware, R. H., Fulker, D. W., Stein, S. A., Anderson, D. N., Avery, S. K., Clark, R. D., Groegemeir, K. K., Kuettner, J. P., Minster, J. B., and Sorooshian, S.: SuomiNet: A real-time national GPS network for atmospheric research and education, B. Am. Meteorol. Soc., 81, 677–694, 2000.

Xie, S., Zhang, M., Branson, M., Cederwall, R. T., Del Genio, A. D., Eitzen, Z. A., Ghan, S. J., Iacobellis, S. F., Johnson, K. L., Khairoutdinov, M., Klein, S. A., Krueger, S. K., Lin, W., Lohmann, U., Miller, M. A., Randall, D. A., Somerville, R. C. J., Sud, Y. C., Walker, G. K., Wolf, A., Wu, X., Xu, K.-M., Yio, J. J., Zhang, G., and Zhang, J.: Simulations of midlatitude frontal clouds by SCMs and CSRMs during the ARM March 2000 Cloud IOP, J. Geophys. Res., 110, D15S03, doi:10.1029/2004JD005119, 2005.

Xie, S., Hume, T., Jakob, C., Klein, S., McCoy, R., and Zhang, M.: Observed large-scale structures and diabatic heating and drying profiles during TWP-ICE, J. Climate, 23, 57–79, doi:10.1175/2009JCLI3071.1, 2010.

Xie, S., Zhang, Y., Giangrande, S. E., Jensen, M. P., McCoy, R., and Zhang, M.: Interactions between cumulus convection and its environment as revealed by the MC3E sounding array, J. Geophys. Res., 119, 11784–11808, doi:10.1002/2014JD022011, 2014.

Xu, K.-M., Cederwall, R. T., and Xie, S.: An intercomparison of cloud-resolving models with the ARM summer 1997 IOP data, Q. J. Roy. Meteorol. Soc., 128, 593–624, 2002.

Yanai, M., Esbensen, S., and Chu, J.: Determination of the bulk properties of tropical cloud clusters from large-scale heat and moisture budgets, J. Atmos. Sci., 30, 611–627, 1973.

Yu, H., Ciesielski, P., Wang, J., Kuo, H.-C., Vomel, H., and Dirksen, R.: Evaluation of humidity correction methods for Vaisala RS92 tropical sounding data, J. Atmos. Ocean. Tech., accepted, 2015.

Zhang, M. H. and Lin, J. L.: Constrained variational analysis of sounding data based on column-integrated budgets of mass, heat, moisture and momentum: Approach and application to ARM measurements, J. Atmos. Sci., 54, 1503–1524, 1997.

Zhang, M. H., Lin, J. L., Cederwall, R. T., Yiou, J. J., and Xie, S. C.: Objective analysis of ARM IOP data: Method and sensitivity, Mon. Weather Rev., 129, 295–311, 2001.

6

Tropospheric ozone and ozone profiles retrieved from GOME-2 and their validation

G. M. Miles[1], R. Siddans[1], B. J. Kerridge[1], B. G. Latter[1], and N. A. D. Richards[2]

[1]Remote Sensing Group, STFC Rutherford Appleton Laboratory, Harwell Oxford, UK
[2]School of Earth and Environment, University of Leeds, Leeds, UK

Correspondence to: G. M. Miles (georgina.miles@stfc.ac.uk)

Abstract. This paper describes and assesses the performance of the RAL (Rutherford Appleton Laboratory) ozone profile retrieval scheme for the Global Ozone Monitoring Experiment 2 (GOME-2) with a focus on tropospheric ozone. Developments to the scheme since its application to GOME-1 measurements are outlined. These include the approaches developed to account sufficiently for UV radiometric degradation in the Hartley band and for inadequacies in knowledge of instrumental parameters in the Huggins bands to achieve the high-precision spectral fit required to extract information on tropospheric ozone.

The assessment includes a validation against ozonesondes (sondes) sampled worldwide over 2 years (2007–2008). Standard deviations of the ensemble with respect to the sondes are considerably lower for the retrieved profiles than for the a priori, with the exception of the lowest subcolumn. Once retrieval vertical smoothing (averaging kernels) has been applied to the sonde profiles there is a retrieval bias of 6 % (1.5 DU) in the lower troposphere, with smaller biases in the subcolumns above. The bias in the troposphere varies with latitude. The retrieval underestimates lower tropospheric ozone in the Southern Hemisphere (SH) (15–20 % or $\sim$ 1–3 DU) and overestimates it in the Northern Hemisphere (NH) (10 % or 2 DU).

The ability of the retrieval to reflect the geographical distribution of lower tropospheric ozone, globally (rather than just ozonesonde launch sites) is demonstrated by comparison with the chemistry transport model TOMCAT. For a monthly mean of cloud-cleared GOME-2 pixels, a correlation of 0.66 is found between the retrieval and TOMCAT sampled accordingly, with a bias of 0.7 Dobson Units. GOME-2 estimates higher concentrations in NH pollution centres but lower ozone in the Southern Ocean and South Pacific, which is consistent with the comparison to ozonesondes.

1 Introduction

Ozone is an important atmospheric trace gas, absorbing ultraviolet (UV) radiation from the sun that would otherwise damage the cells of living organisms at the Earth's surface. In the stratosphere, where approximately 90 % of ozone is found, the vertical distribution determines heating rates and thereby also dynamics. The vertical distribution of stratospheric ozone is determined by the Chapman cycle (Chapman, 1930), and catalytic cycles involving nitrogen, hydrogen and halogen radicals. In the troposphere, ozone is produced though complex reaction pathways involving nitrogen oxides (NO_x) and volatile organic compounds (VOCs). Ozone is also introduced by exchange from the stratosphere, particularly at mid-latitudes. As a secondary pollutant from anthropogenic and biomass burning sources, it is an environmental hazard particularly in urban environments because it is a lung irritant. High levels of ozone have been linked to increased mortality/excess deaths when associated with localised heat wave events (Gryparis et al., 2004). Tropospheric ozone can be damaging to agriculture by increasing the failure rate of crops (Holloway et al., 2012). For these reasons, it is vitally important to monitor ozone in the troposphere as well as the stratosphere, but in situ surface observations and ozonesondes are sparse and heavily favour the Northern Hemisphere.

Tropospheric ozone is also a greenhouse gas. The uncertainty in estimates of radiative forcing from tropospheric ozone is as large as that associated with the non-well mixed greenhouse gases (IPCC, 2013) and as such good knowledge of the atmospheric concentration of tropospheric ozone is required. This uncertainty remains in part due to the reliance on atmospheric models and their spread, in addition to uncertainty about pre-industrial ozone amount. Estimates do not currently incorporate any information from satellites (IPCC, 2013). An accurate, contemporary distribution of tropospheric ozone from satellites would help to verify chemistry transport models (CTMs) and coupled chemistry-climate models (CCMs), and hence their estimates of radiative forcing and the forward projections by CCMs. The MetOp series and its successor MetOp-SG/Sentinel 5 have the potential to monitor tropospheric as well as stratospheric ozone in the decades to come.

The total atmospheric column of ozone has been measured historically via UV nadir-viewing sensors (e.g. BUV, SBUV, TOMS, SBUV-2, GOME, SCIAMACHY, OMI and GOME-2), with accuracies typically between 0.5 and 2 % (Klenk et al., 1982; Loyola et al., 2011; van Roozendael et al., 2012, and references therein). Ozone profiles have also been produced from UV nadir-sounders (e.g. Bhartia et al., 1996), however, retrieving tropospheric ozone presents a significant challenge, because $\sim$90 % of atmospheric ozone resides in the stratosphere above. Tropospheric columns have been derived by subtracting an estimate of the stratospheric component from the measured total column, using knowledge of the tropopause height and making assumptions about the ozone profile shape (e.g. Fishman and Larsen, 1987; Schoeberl et al., 2007; Ziemke et al., 2011). Tropospheric columns have also been derived in the tropics by differencing total columns in cloud-free pixels from those in nearby pixels with thick/high convective cloud (Valks et al., 2014). However, as suitable occurrences are sparse, only monthly averages are useful. Direct retrieval of tropospheric information from temperature-dependent spectral structure in the Huggins bands (320–345 nm) was first proposed by Chance et al. (1997) and has been exploited by several schemes (Munro et al., 1998; van der A et al., 2002; Liu et al., 2005, 2010; Cai et al., 2012), applied to the Global Ozone Monitoring Experiment (GOME) class of instruments.

Infrared nadir-viewing spectrometers offer complementary vertical sensitivity to tropospheric ozone, as demonstrated by the Tropospheric Emission Spectrometer (TES) (Nassar et al., 2008) and the Infrared Atmospheric Sounding Instrument (IASI) (Boynard et al., 2009).

Here, we describe and assess the performance of the RAL (Rutherford Appleton Laboratory) ozone profile retrieval scheme applied to GOME-2 measurements, with a particular focus on the troposphere. This scheme has been developed directly from that presented by Munro et al. (1998), which was the first to demonstrate retrieval of tropospheric ozone from space. Substantial improvements have been made to that algorithm and GOME-2, which was launched on MetOp-A in 2006, also improves in certain respects upon its predecessor. The RAL ozone profile optimal estimation (OE) retrieval scheme was selected for the ESA Climate Change Initiative (CCI) (Plumber, 2009) after independent comparison to the GOME-2 operational ozone profile scheme (Keppens et al., 2014). It was selected principally because of the demonstrated sensitivity to tropospheric ozone and persistently higher number of degrees of freedom for signal (DFS).

In Sect. 2 of this paper, the GOME-2 instrument will be briefly introduced, before the RAL ozone profile scheme and the principal improvements since Munro et al. (1998) are described. In Sect. 3, an error assessment is described. Section 4 presents a validation of the ozone profile scheme against global ozonesondes and a comparison to tropospheric ozone distributions from a chemistry transport model. A summary is presented in Sect. 5.

2 RAL ozone profile retrieval algorithm

2.1 GOME-2 instrument

GOME-2 is a UV/vis spectrometer with four bands that cover the 240–790 nm interval contiguously with a spectral sampling of 0.11–0.22 nm and spectral resolution of 0.24–0.53 nm that was launched in 2006 aboard ESA's MetOp-A platform (Callies et al., 2000). MetOp has a local equator crossing time of 09:30. Of principal use for the retrieval of atmospheric ozone are Bands 1 (240–315 nm) and 2 (310–403 nm), which incorporate the long-wave side of the Hartley band (200–310 nm) and the Huggins (320–360 nm) bands. The Band 1a (240–307, 240–285 nm after 10 December 2008) pixel size is 640 (across-track) × 40 km (along-track). The nominal Band 1b (307 or 285 nm) and Band 2 pixel size is 80 × 40 km (cf. 320 × 40 km for GOME). In addition to Earthshine spectrum, GOME-2 also measures a direct sun spectrum once a day. A full description of the GOME-2 instrument is given within Callies et al. (2000). GOME-2A L1b data is provided by EUMETSAT (EUMETSAT, 2006).

2.2 Retrieval algorithm

The RAL ozone profile retrieval scheme is an optimal estimation (OE) algorithm (Rodgers, 1976, 2000) which uses prior information to constrain ill-posed problems such as profile retrievals from nadir-viewing satellite instruments. OE also provides an estimate of the errors associated with retrieved parameters.

The RAL algorithm is a three-step sequential retrieval, first performing a fit to the sun-normalised radiance spectrum in Band 1 (using wavelengths between 266–307 nm) to utilise information in the long-wave tail of the Hartley band. Band 1b spectra are averaged onto Band 1a spatial pixels to improve their signal-to-noise ratio. Ozone absorption and Rayleigh scattering coefficient both decrease strongly with wavelength across this interval, yielding information predominantly on the mid-to-upper stratospheric ozone profile. In addition to the ozone profile, the retrieved parameters are a wavelength-independent Lambertian effective surface albedo, detector dark (leakage) current (in raw signal units) and a wavelength mis-registration parameter for the Earthshine spectrum with respect to the direct-sun spectrum. Rotational Raman scattering is also accounted for by retrieving a scaling factor for the theoretically calculated spectrum of in-filling by the (singly scattered) Ring effect (as modelled via the approach of Joiner et al., 1995).

The second step is to retrieve an effective surface albedo at 336 nm in Band 2. This step is important because the effective albedo retrieved from the longest wavelengths (< 307nm) in Band 1, is not appropriate in the Band 2 fit (using wavelengths from 323–335 nm) due to the differing fields of view (FoV). The retrieved ozone profile and its associated error covariance matrix from the Band 1 fit and the retrieved 336 nm effective albedo contribute to the prior information for the third and final fit in the Huggins bands (323–335 nm).

The fit in Band 1 is a direct fit of the sun-normalised radiance, r, defined as

$$r = \frac{I}{I_0}\pi, \tag{1}$$

where I is the measured Earthshine radiance and I_0 the direct-sun irradiance measurement. As such, accurate (< 1 %) radiometric calibration is required. GOME-2, as with GOME-1 and SCIAMACHY, has experienced degradation of the UV photometric throughput during its lifetime, the effects of which are greater for the shorter wavelengths (Lang et al., 2009; Lacan and Lang, 2011; Cai et al., 2012). To produce self-consistent global ozone distributions over the mission lifetime, it has been necessary to implement an empirical degradation correction to the Band 1 measurements, as outlined below in Sect. 2.3.1.

In order to obtain accurate information on tropospheric ozone, a high fitting precision in the Huggins Bands is required, < 0.1 % rms. In order to achieve this, the Band 2 retrieval fits the differential wavelength structure arising from temperature-dependent vibration–rotational structure in ozone absorption, using the logarithm of the sun-normalised radiance, with a fourth-order polynomial in wavelength subtracted in order to remove coarse-scale artefacts in the spectrum[1] and reveal the fine-scale ozone differential spectral structure. This method of fitting differential spectral structure is somewhat analogous to the DOAS approach (Platt, 1994) and is robust against instrumental effects (including some aspects of the degradation). The stringent fitting precision requirement necessitates good knowledge of the instrument's slit function, which varies across Band 2. This is achieved by an off-line fit to each direct-sun spectrum, to retrieve a scaling factor to apply to slit function key data from pre-flight characterisation (Siddans, 2003). This is done on a daily basis because the slit functions are observed to change with time (seasonally and over shorter time periods) in association with thermal cycling of the instrument focal plane. This process is discussed further in Sect. 2.3.3.

The state vector for the Band 2 retrieval step is composed of a wavelength mis-registration of the sun-normalised radiance spectrum with respect to the ozone absorption cross-section spectrum *in vacuo*, a wavelength shift between the Earthshine radiance and direct-sun irradiance spectra, the ozone profile, Ring effect scaling factor, vertical column NO_2, BrO and formaldehyde. Other species that absorb in this spectral region (such as SO_2) are modelled in the fit (based on a climatological profile shape) but not retrieved.

The retrieved ozone profile is represented in the state vector as the logarithm of the volume mixing ratio on a fixed pressure grid: surface pressure, 450, 170, 100, 50, 30, 20, 10, 5, 3, 2, 1, 0.5, 0.3, 0.17, 0.1, 0.05, 0.03, 0.017, 0.01 hPa. The forward model performs radiative transfer calculations on a finer pressure grid (approximately 2 km throughout profile), and uses the assumption that the log of ozone concentration varies linearly with log pressure between the retrieval levels. The pressure levels are herein for convenience expressed as a pressure–altitude coordinate, where an approximate equivalent altitude is assigned to a pressure profile based on the relation

$$Z^* = 16\left(3.0 - \log_{10}(p)\right), \tag{2}$$

where Z^* is in kilometres and p in hPa. This predicts approximate equivalent altitudes of the pressure grid of 0, 6, 12, 18 km then every 4 km up to 80 km. These values are usually within 2 km of the geometric altitudes calculated for hydrostatic balance. Altitudes expressed herein are Z^* altitudes. The forward model grid is finer in order to accurately model atmospheric radiative transfer. There are typically 5–6° of freedom for signal (Rodgers, 2000) for the combined Hartley–Huggins bands retrieval. This is almost independent of latitude and season. The retrieval grid over-samples the profiles in terms of the information content of typical GOME-2 measurements so the retrieval is further constrained using a priori correlations (see below).

The ozone a priori profile used is that of the McPeters et al. (2007) climatology derived in part from ozone sondes, which varies by month and latitude. The diagonal elements of the a priori error covariance matrix ($\mathbf{S}_a$) are set to the larger of the climatological % standard deviation and the following values: 0–12 km (100 %), 16 km (30 %), 20–50 km (10 %), 56 km (50 %) and 60–80 km (100 %). In practice, it is these fixed percentage values that apply in the troposphere, except at very high latitudes where the climatological standard deviation is greater. A 6 km Gaussian correlation length is imposed to specify the off-diagonal elements of the a priori covariance for the initial Band 1 step. The retrieved profile and

[1] Artefacts due for example to imperfect radiometric calibration, etalon formed from contamination of optical surfaces not in common for direct-sun and Earthshine measurements or un-modelled spectral features in UV surface sun-normalised radiance.

error covariance matrix from the Band 1 step are used as the a priori profile and to define the diagonal elements of the covariance matrix for the Band 2 steps. An 8 km Gaussian correlation length is then applied to further stabilise the Band 2 ozone retrieval in the region of the upper troposphere and lower stratosphere (UTLS).

To achieve photometric signal-to-noise adequate to retrieve tropospheric ozone information, it is necessary to average Band 2 spectra from eight adjacent GOME-2 ground pixels[2]. Averaging eight Band 2 pixels (two across-track and four along-track) to create a composite pixel of 160 × 160 km reduces photometric noise by a factor of approximately $1/\sqrt{8}$. For radiative transfer, the scheme uses a version of the GOMETRAN++ (Rozanov et al., 1997) but with a number of processing speed improvements (which do not degrade numerical accuracy). A polarisation correction based on scalar/vector LIDORT look-up tables is also implemented, as provided by BIRA (C. Lerot, personal communication, 2012). The retrieval scheme uses ECMWF Interim Re-analysis meteorological products for temperature and pressure profiles obtained from the ECMWF data server. The solar reference spectrum is that provided by Chance and Kurucz (2010). The ozone absorption cross-sections are those derived by Brion et al. (1993, 1998); Daumont et al. (1992); Malicet et al. (1995).

Although cloud may be modelled according to information from GOME-2 measurements in the O_2 A-Band (760 nm) or collocated vis/IR imagery from AVHRR/3 on MetOp, for the purposes of this exercise, cloud radiative transfer is not modelled explicitly, and instead an effective Lambertian surface albedo is co-retrieved. With this approach it is expected that the presence of cloud will lead to a negative bias in retrieved ozone, at altitudes below the cloud top, from where there is limited information.

2.2.1 Optimal estimation

The retrieval uses the standard optimal estimation algebra for the non-linear problem (Rodgers, 2000), used widely for deriving atmospheric properties from satellite measurements. An estimate of the state vector is obtained by combining measurement and prior information in accordance with their respective error covariance matrices. In the case of ozone profile retrieval from nadir UV spectral measurements such as those of GOME-2, the prior constrains what is otherwise an ill-posed problem. The solution is obtained by minimising a cost function, χ^2:

$$\chi^2 = (\boldsymbol{y} - F(\boldsymbol{x}))^T \mathbf{S}_y^{-1} (\boldsymbol{y} - F(\boldsymbol{x})) + (\boldsymbol{x}_a - \boldsymbol{x})^T \mathbf{S}_a^{-1} (\boldsymbol{x}_a - \boldsymbol{x}), \tag{3}$$

[2]This pixel averaging is not necessary to achieve adequate precision on the total column ozone retrieved from the same spectral region.

where $\boldsymbol{y}$ is the measurement vector, $\boldsymbol{x}$ and $\boldsymbol{x}_a$ are the state vector (or expected solution) and a priori vector, F is the forward model and $\mathbf{S}_y$ and $\mathbf{S}_a$ the error covariance matrices for the measurement and prior, respectively. The Levenberg–Marquardt method is used to minimise the cost function (summarised in Press et al., 1995), and the state vector is iteratively updated as follows:

$$\boldsymbol{x}_{i+1} = \boldsymbol{x}_i + \left(\mathbf{K}_i^T \mathbf{S}_y^{-1} \mathbf{K}_i + \mathbf{S}_a^{-1} + \gamma I\right)^{-1} \mathbf{K}_i^T \mathbf{S}_y^{-1} (\boldsymbol{y} - F(\boldsymbol{x}_i) + \mathbf{K}_i (\boldsymbol{x}_i - \boldsymbol{x}_a)), \tag{4}$$

where γ is the step size, depending upon which the iteration tends towards either Newtonian iteration or steepest descent (Rodgers, 2000). $\mathbf{K}$ is the weighting function at iteration i, defined as

$$\mathbf{K}_i = \frac{\partial F(\boldsymbol{x}_i)}{\partial x_i}. \tag{5}$$

The sensitivity of the retrieval to perturbations in the measurements is characterised by the gain matrix $\mathbf{G}$, of dimensions m by n, where m is the number of measurements (in the sun-normalised radiance spectrum) and n the number of retrieval levels. This is defined as follows (Rodgers, 2000):

$$\mathbf{G} = \left(\mathbf{K}^T \mathbf{S}_y^{-1} \mathbf{K} + \mathbf{S}_a^{-1}\right)^{-1} \mathbf{K}^T \mathbf{S}_y^{-1}. \tag{6}$$

The sensitivity of the retrieval to perturbations in the true state is given by the (n by n) averaging kernel matrix $\mathbf{A}$ (also herein referred to as AK):

$$\mathbf{A} = \mathbf{GK}. \tag{7}$$

Errors on the solution are characterised by the covariance matrix:

$$\mathbf{S}_x = \left(\mathbf{S}_a^{-1} + \mathbf{K}^T \mathbf{S}_y^{-1} \mathbf{K}\right)^{-1}. \tag{8}$$

The square-roots of the diagonals of this matrix are referred to here as the estimated standard deviations (ESDs) of the retrieval, and are assumed to provide a reliable measure of the error applicable to each level of the retrieved profile. The extent to which this is true is investigated in Sect. 4.1, below. $\mathbf{S}_x$ includes errors arising from measurement noise (as characterised by $\mathbf{S}_y$) and smoothing error deriving from the prior constraint (as characterised by $\mathbf{S}_a$), However, it should be noted that the covariance matrix applies to the profile as represented on the retrieval grid. It does not include smoothing errors related to finer-scale structures than the retrieval grid, and it is not formally possible to estimate errors at finer scales directly from it (as discussed in von Clarmann, 2014). $\mathbf{S}_x$ should provide a reasonable characterisation of the difference between retrieved profiles and true profiles which have been interpolated onto the retrieval grid, after having been smoothed to a commensurate resolution (see also Calisesi et al., 2005). Application of averaging kernels to the true profile allows the most appropriate means to compare with the retrieved profile. In that case smoothing errors (including effects on finer scales) are accounted for and differences between retrieval and smoothed "true" profiles should ideally be characterised by the retrieval noise covariance:

$$\mathbf{S}_n = \mathbf{G}^T \mathbf{S}_y \mathbf{G}. \tag{9}$$

2.3 Improvements to ozone profile retrieval scheme for GOME-2

GOME-2 measurements are subject to measurement errors from a variety of sources, which must be characterised on a pixel-by-pixel basis for accurate retrievals using optimal estimation. As an estimate of the photometric and dark current noise was not supplied with the Level 1b data acquired by GOME-2 before 2013, we use a model to estimate the measurement noise, based on calibration key data derived for the GOME-2 error study (Kerridge et al., 2002) now updated with calibration key data for the MetOp-A GOME-2 instrument, and similar to the model used by Nowlan et al. (2011). The noise model is described in detail in Miles et al. (2012).

2.3.1 Correction for degradation to GOME-2 UV throughput

The MetOp-A GOME-2 instrument (and instruments of its class) is subject to throughput degradation over time that is more acute at the shorter UV wavelengths (Lang et al., 2009; Lacan and Lang, 2011; Cai et al., 2012). To accommodate this, a low-order polynomial fit in wavelength and time has been derived empirically from the ratio between a climatological (in this case the same as the a priori) modelled UV sun-normalised radiance (with its associated solar viewing geometry) and the observed sun-normalised radiance spectrum. This is similar to the approach by van der A (2001) for ozone column retrieval. A detector dark current, or leakage current, in raw signal units, which is assumed constant for all detector pixels in Band 1, has been jointly fit with the low-order polynomial in order to separate the wavelength/time polynomial from this instrumental parameter, since the dark current is co-retrieved with the ozone profile and other parameters from individual Band 1 (Hartley band) measurements. A separate polynomial correction has been derived for each of the West, Nadir and East Band 1 scan positions, sampling only cloud-free data within 30° of the equator 1 day per week throughout the mission. The empirical degradation correction employed in Band 1 has resulted in a relatively stable stratospheric ozone distribution from that band. A degradation correction has not been applied in the Band 2 (Huggins bands) step and so the retrieval is still sensitive to trends in the total column ozone) although the use of differential structure greatly reduces sensitivity of the Huggins bands retrieval step to UV radiometry. The more subtle effect on ozone retrieval of differential UV degradation (for the irradiance and irradiance) in Band 2 will be a topic of future work.

2.3.2 Systematic residual from spectral fit to the Huggins bands

A systematic residual spectral signature remains from the Huggins band fit that is of the order of 0.2 % amplitude (of sun-normalised radiance). This signature has a characteristic spectral structure, which is quite persistent not only with sun–Earth viewing geometry and time. Although its origins in the solar spectral irradiance, atmosphere/surface (polarised) radiative transfer and/or instrument response have yet to be firmly established, the persistence of the spectral residual is amenable to the co-retrieval of a scaling factor, which enables an rms fit precision of $< 0.1\,\%$ to then be achieved in the Huggins bands, commensurate with the estimated photometric noise level. In practice, the leading six principal components of the systematic residual spectral signature have been determined (considering fit residuals from observations on selected days spanning the missing to date) and scaling factors for each of these included in the retrieval state vector. Variations of the retrieved scaling factors with both time and space, give some physical insights into their origin and an opportunity for future development.

Although these principal components of the systematic residual signature should not be spectrally correlated to ozone, some correlation is found between the retrieved scaling factors and tropospheric ozone under conditions that are particularly challenging, such as at high latitudes in the Northern Hemisphere spring, below high columns of stratospheric ozone and where temperature is close to isothermal over a broad layer near the tropopause.

Some quality control of the retrieved product is necessary under these circumstances, where if the line-of-sight zenith angle component of the total column ozone in step 1 (Band 1) is greater than 500 DU, the retrieved tropospheric column is unreliable and the pixel should not be used. These conditions usually coincide with an extensive near-isothermal tropopause. Since the information on the ozone profile below the stratospheric peak is principally derived from the temperature-dependent ozone spectral structure, such conditions are particularly unfavourable for high-precision retrievals in this region.

2.3.3 Retrieval of slit function width

In order to achieve the fit precision in the Huggins bands needed to retrieve tropospheric ozone, accurate knowledge of the spectral response function (or slit function) of individual detector pixels is required. The slit functions for the GOME-2 instrument were characterised prior to launch from laboratory measurements (Siddans et al., 2006), but it became apparent while in orbit that they had changed and continue to change (Cai et al., 2012). Failure to adequately char-

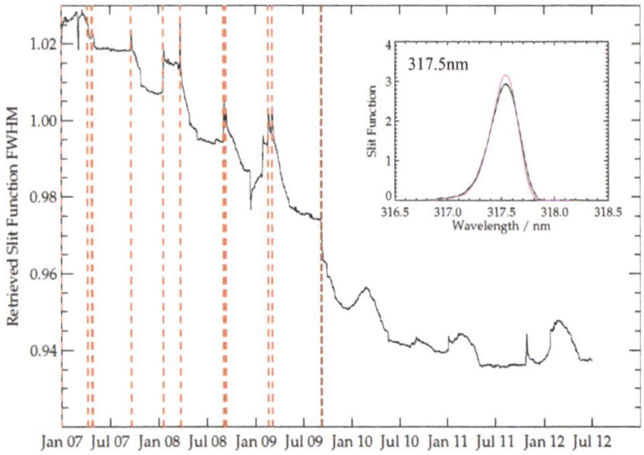

Figure 1. Variation with time of retrieved scaling factor for nominal FWHM of slit function (black solid). Red dashed lines indicate discontinuities associated with various in-orbit operations, including the second (and last) throughput test in September 2009. The inset panel shows an example of how the effective shape of the measured slit function is modified for the pixel centred at 317.5 nm, where the black line indicates start of operations (January 2007) and the pink line is the shape in January 2013.

acterise the changing slit function leads to a spurious trend (with respect to ozonesondes) in the retrieved ozone; particularly in the troposphere. To account for this, an offline slit function OE retrieval has been added to the fit of daily direct-sun measurement to the high-resolution solar reference spectrum (Chance and Kurucz, 2010) which is used to refine wavelength registration (Sect. 2.2). In addition to the series of wavelength polynomial coefficients for radiometric gain, radiometric offset and a wavelength shift/squeeze, the state vector has been extended to incorporate a single scaling factor for the full width half maxima (FWHM) of all the slit functions in the Band 2 wavelength interval from 320–340 nm. This encompasses the wavelength range needed for ozone retrieval and makes an allowance for edge effects from Legendre polynomials. The retrieved FWHM scaling factor is shown in Fig. 1 from January 2007 to July 2012. Also shown is an example of how a slit function for a single detector pixel is modified by this parameter, demonstrating the effective narrowing of the slit functions with time in this spectral region. The overall change in FWHM is in good agreement with that suggested by others (e.g. Cai et al., 2012).

3 Error analysis and retrieval characterisation

An extensive simulation study of errors pertaining to ozone profile retrieval by the RAL scheme from the GOME-1 UV spectrometer was reported by Siddans (2003). This was based on retrieval simulations for a set of standard geophysical scenarios which had been defined for the GOME-2 Error Study (Kerridge et al., 2002), which had presented a detailed error budget, based on information available at that time. The retrievals for the GOME-2 instrument in flight is found to behave broadly as predicted.

3.1 Retrieval characterisation and error analysis

Estimation of the averaging kernel for the three-step process needs to account for the fact that off-diagonals of the ozone a priori covariance used in step 3 are different to the solution error covariance output from step 1. This is done considering the sensitivity of the step 3 retrieval to changes in the a priori used in step 3, which in turn is related to the true profile by averaging kernel for step 1, as well as the sensitivity of step 3 to the measurements used in that step:

$$\mathbf{A} = \mathbf{A}_3 + (\mathbf{I} - \mathbf{A}_3)\,\mathbf{A}_1. \qquad (10)$$

$\mathbf{A}_1$ and $\mathbf{A}_3$ are the averaging kernel matrices for step 1 and 3, considered in isolation, applying Eq. (9) to the matrices used in the respective steps. In this equation, the quantity $(\mathbf{I} - \mathbf{A}_3)$ is the a priori gain matrix. In practice the equation is slightly complicated by the fact that the full state vector is not identical in the two bands (other state parameters are retrieved). The equation can be extended to include mapping of the sensitivity of the band 3 retrieval to the surface albedo retrieved in step 2, but this has negligible impact on the averaging kernel for the ozone profile.

The retrieval precision, or estimated standard deviation (ESD), as given by the square roots of diagonals of the solution error covariance matrix is generally in the few percent range in the stratosphere increasing to a few tens of percent in the lowest retrieval levels.

$\mathbf{S}_c$, the estimated covariance on subcolumn amounts, is given by

$$\mathbf{S}_c = \mathrm{M}^T \mathbf{S}_x \mathbf{M}, \qquad (11)$$

where $\mathbf{S}_x$ is the solution covariance (from the final, third retrieval step) in volume mixing ratio (VMR) units and $\mathbf{M}$ is n by $n-1$ matrix with elements which transform the mixing ratios on levels to subcolumn amount between levels. $\mathbf{M}$ has elements which are all zero except $\mathbf{M}_{(i,i)}$ and $\mathbf{M}_{(i+1,i)}$ (for $i = 1, n-1$) which have the necessary weights to perform the integration of the subcolumn, making the same assumption as the FM for the variation of ozone with pressure between the retrieval levels. The ESD on the subcolumn amounts is given by the square root diagonal elements of this matrix. Estimated retrieval noise errors can be similarly derived, applying Eq. (11) to $\mathbf{S}_n$.

An example is presented in Fig. 2 for a mid-latitude profile in Northern Hemisphere summer. In this case, the ESD on retrieval levels and layer subcolumns is typically much smaller than the a priori error throughout the profile. The retrieval noise error is around a factor of 2 smaller than the ESD. Figure 3c shows an example of how the ESD varies for a typical orbit cross-section and is also given as a ratio with the prior uncertainty in Fig. 3d. In general, at all altitudes and latitudes a reduction compared to the prior uncertainty is observed. An indication of ESD in the presence of cloud is given later in Sect. 4.

3.2 Averaging kernels

Figure 2 also shows example averaging kernels for a mid-latitude ozone profile. The AKs for retrieval levels at the surface and in the mid-troposphere show pronounced peaks in the troposphere, while for higher levels the AKs become smoother. The AKs for retrieval levels in the troposphere have tails which extend much higher, indicating an apparent sensitivity of retrieved tropospheric ozone to true perturbations occurring in the stratosphere and mesosphere. However, variability in ozone number density at the altitudes where these tails are large is in practice very small, and therefore so is its influence on the tropospheric ozone retrieval. The influence of realistic variations in stratospheric ozone on retrieved tropospheric ozone is therefore usually small, and retrieval in the troposphere generally reflects realistic tropospheric variability (as evident in the comparison to model fields shown in Sect. 4.3). Where stratospheric perturbations are unusually large these can cause spurious tropospheric signals. However, this sensitivity is described by the AKs provided along with the retrievals, so this effect can be properly taken into account when using the data. Figure 3a shows a retrieved ozone orbit cross-section, the improvement of retrieval error as compared to prior error and the combined surface and 450 hPa AKs. The largest reduction upon prior uncertainty in the example given here is found in the UTLS and lower stratospheric region (6–20 km) at mid-to-high latitudes, where it is reduced in places to less than 20 % of the prior error. In the tropics, the largest reduction is found in the mid-troposphere. The smallest reduction is found near the surface at high southern latitudes, which in the case of this orbit cross-section coincides with the southern ocean off the south coast of Australia, consistent with the averaging kernels for the lowest levels shown in Fig. 3b. It is apparent from this that there is some sensitivity to the lowest 3 km of the atmosphere, although the dominant contribution is from around 500 hPa. Most significantly, this AK has very little contribution from above 10 km and in most circumstances is quite independent of stratospheric ozone. The behaviour of AKs is critical to inter-comparisons with ozonesondes, for validation, and with model distributions, as discussed in the following section.

4 Validation and model inter-comparison

In this section the performance of the retrieval algorithm as applied to real measurements will be validated against ozonesondes and inter-compared with the global distribution predicted by a chemistry transport model.

4.1 GOME-2 ozonesonde comparison

The period of interest considered here is 2007 (start of mission operations) through 2008. This is principally because some of the characteristics of the instrument changed in September 2009 as a result of an instrument throughput test and it is more straightforward to interpret the results from GOME-2 before that event. The WOUDC/NDACC (www.woudc.org and www.ndsc.ncep.noaa.gov) and SHADOZ (Thompson et al., 2003) ozonesonde databases are used for this analysis, adopting collocation criteria of < 200 km and < 2 h, with cloud screening (effective cloud fraction of < 0.2 and a cloud top pressure of > 700 hPa) unless otherwise stated. All biases are evaluated with respect to the sonde (retrieval minus ozonesonde).

Ozonesonde measurements are known to differ in accuracy with sensor type, time, altitude and launch site. They are currently the focus of effort by the global ozonesonde community to homogenise the quality of the products (SI^2N, 2012). Spurious sondes have been eliminated in this analysis by testing whether each 4 km subcolumn for each sonde site is outside 4σ of the monthly mean for that site/subcolumn. This eliminates most aberrant sondes whilst retaining characteristic natural variability at the sonde location. Only sonde profiles that extend above 20 km are considered.

4.1.1 Subcolumns and application of averaging kernels

Sonde comparisons are performed in terms of the vertically integrated subcolumn between retrieval levels. Sondes are directly integrated using

$$C_i = D \int_{pi}^{pi+1} x(p) \cdot \mathrm{d}p, \tag{12}$$

where C_i is the subcolumn amount between vertical retrieval grid levels i and $i+1$, p is pressure, x is ozone mixing ratio, and D is a constant such that the resulting subcolumns are in Dobson units. GOME-2 subcolumns are first interpolated onto the forward model grid in a manner consistent with that used in the retrieval (see Sect. 2.2).

Direct comparisons are made between the retrieved and sonde derived subcolumns; however, it is also important to account for differences caused by retrieval smoothing using the averaging kernels. These are applied to ozonesonde profiles as described in Deeter et al. (2007), and we apply their Eq. (6) to estimate the volume mixing ratio (VMR) profile expected from the retrieval:

$$\hat{\boldsymbol{x}} = \mathbf{A}^{\mathrm{S}} \left(\boldsymbol{x}^{\mathrm{S}} - \boldsymbol{x}_{\mathrm{a}}^{\mathrm{S}} \right) + \boldsymbol{x}_{\mathrm{a}}, \tag{13}$$

where $\hat{\boldsymbol{x}}$ is the expected simulated retrieval, $\boldsymbol{x}_{\mathrm{a}}$ the a priori profile, $\boldsymbol{x}^{\mathrm{S}}$ and $\boldsymbol{x}_{\mathrm{a}}^{\mathrm{S}}$ are the sonde profile and the a priori profile, defined on the vertical grid at which the sonde profile is provided (indicated by superscript S).

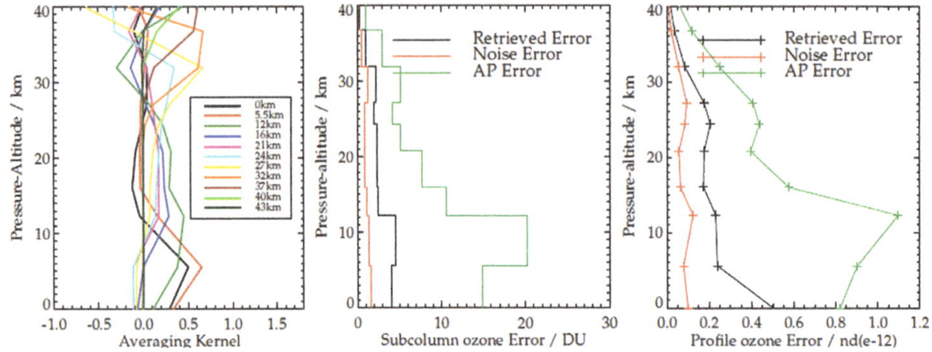

Figure 2. The left panel shows averaging kernels derived in number density units on levels for a nadir pixel at 45° N on 25 August 2008. The averaging kernels themselves are unitless but the magnitude and shape of the off-diagonal elements are very different when evaluated in either VMR or number density. The centre panel shows the associated ESD, noise-only and a priori subcolumn errors, and the panel on the right the errors for the profile.

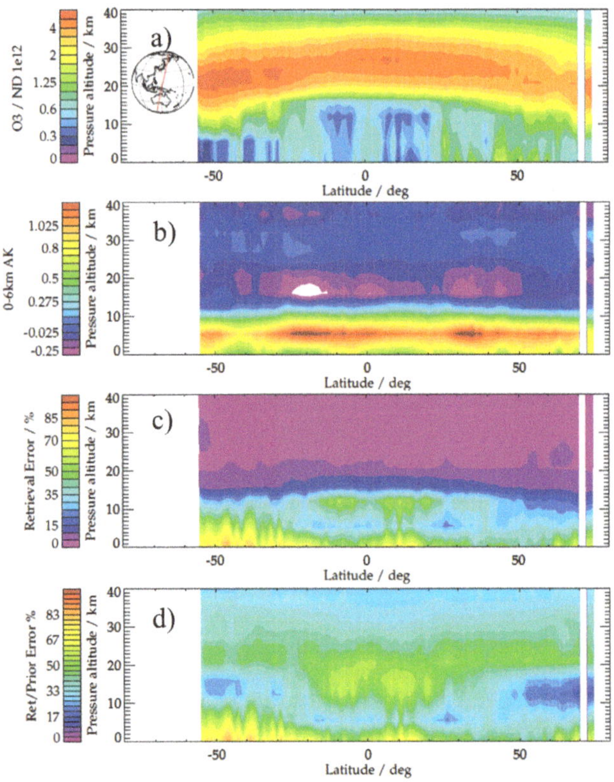

Figure 3. **(a)** An ozone cross-section on 25 August 2008 retrieved from the Band 2 (final) step for the nadir pixel. The orbit track is also indicated. **(b)** The combined surface and 450 hPa (circa 0 and 6 km) averaging kernels. **(c)** Relative retrieval error. **(d)** The associated ratio of retrieved to a priori error.

Each row of $\mathbf{A}^{\mathrm{S}}$ characterises the expected perturbation to the retrieval at a given level to perturbations in the supposed true profile, which is expressed on the relatively finely spaced sonde grid. Retrieval output files contain the mixing ratio averaging kernel $\mathbf{A}$ (square matrix), given directly by Eq. (10), whose rows describe the effect of perturbations to the true profile on the retrieval grid. The transformation of $\mathbf{A}$ to $\mathbf{A}^{\mathrm{S}}$ (which must account for the different thicknesses of the layers concerned) is achieved by first forming the layer thickness normalised averaging kernel $\mathbf{A}^{\mathrm{N}}$ using

$$\mathbf{A}^{\mathrm{N}}_{ij} = \mathbf{A}_{ij}\frac{1}{\Delta p_j}, \quad (14)$$

where Δp_j is the effective pressure thickness associated with retrieval level j:

$$\Delta p_j = \frac{1}{2}\left(p_j + p_{j+1}\right). \quad (15)$$

Here index i refers to rows of the kernel (retrieval levels) while j refers to columns (levels in the true profile). The rows of $\mathbf{A}^{\mathrm{N}}$ are then linearly interpolated to the vertical grid of the ozonesonde measurement. This is then scaled to give $\mathbf{A}^{\mathrm{S}}$ using

$$\mathbf{A}^{\mathrm{S}}_{ij} = \mathbf{A}^{\mathrm{N}}_{ij}\Delta p^{\mathrm{S}}_j, \quad (16)$$

where Δp^{S}_j is the effective thickness of sonde grid index j. Applying Eq. (12) will provide estimated mixing ratios on the retrieval grid with vertical smoothing consistent with the satellite vertical sensitivity. These are then integrated to give subcolumn amounts, in the same way as the retrieved mixing ratios (i.e. by first interpolating to the forward model grid in the appropriate manner).

4.1.2 Results

We first consider statistics from an ensemble of all ozonesondes at all sites, and then provide examples in separate latitude bands. Figure 4 shows the bias, standard deviation and correlation coefficient for a priori and retrieved ozone profiles calculated with respect to individual ozonesondes for the full ensemble. The bias is the ensemble average difference between each GOME-2 retrieved profile and the corresponding

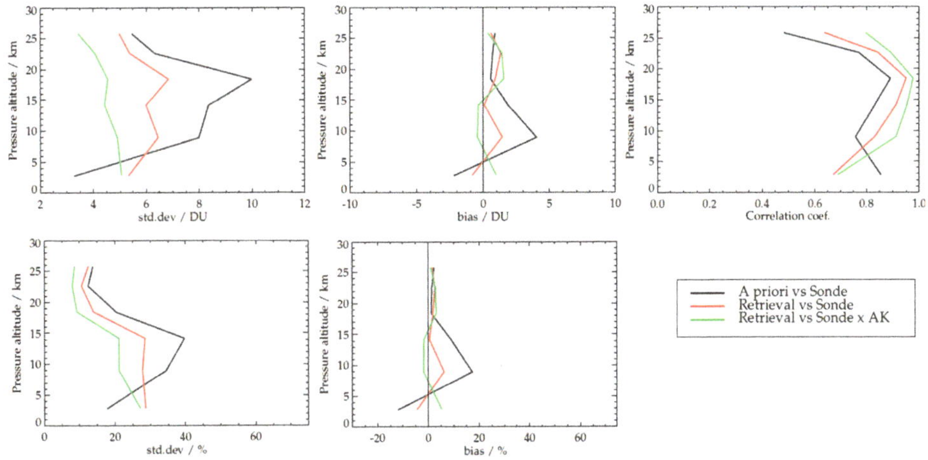

Figure 4. Statistical comparison of RAL GOME-2 ozone profiles with ozonesondes sampled worldwide for 2007–2008. Collocation criteria are given in the text. The standard deviations (left) and biases (centre) in GOME-2 minus ozonesonde values are in absolute (DU) units and as % of sonde value in the top and bottom rows, respectively. The top right panel shows the correlation coefficient. Points denote the mid-point of each subcolumn. In each case, results are shown for the a priori vs. sonde and for the retrieval vs. sonde with and without application of AKs to the ozonesonde profiles. Statistics have been derived from percentage difference calculated with respect to each individual ozonesonde.

sonde profile. The fractional bias (also shown) is the bias divided by the mean sonde amount in that layer.

The ensemble standard deviation of differences between GOME-2 retrievals and corresponding sonde profiles is an independent estimate of the (random) error on an individual retrieved profile with respect to the ozonesonde (i.e. ground-truth). The bias, fractional bias and standard deviation are also computed for the a priori profiles. When AKs are applied to the sonde profiles, the retrieval is seen to add substantial information to the a priori, except for the lowest subcolumn. This is also the case for the correlation coefficient and is due to atmospheric variability in this lowest layer as sampled by the sondes being generally smaller than the ESD. It is therefore important to note that ozonesondes only partially sample the global variability (as shown in Sect. 5) The retrieval bias with respect to sondes is rather small once AKs are applied ($\sim 6\,\%$ in the lowest layer and $< 5\,\%$ in higher layers), and substantially lower than that of the a priori. Figure 5 shows the geographical distribution ozonesondes and the number of collocated profiles used in this comparison.

Figure 6 shows histograms of the subcolumn error ratio, ER_c, defined as the difference between retrieved (C_i^{GOME2}) and sonde (C_i^{sonde}) subcolumns, normalised by the ESD_c for the subcolumn:

$$\mathrm{ER}_c = \frac{C_{i,k}^{\mathrm{GOME2}} - C_{i,k}^{\mathrm{sonde}}}{\mathrm{ESD}_{c,i,k}}, \tag{17}$$

where ESD_c denotes the estimated retrieval error for the subcolumn rather than a retrieval level. Subscripts denote individual layers (i) for each collocation (k). The analysis is performed both with and without averaging kernels applied to the sonde profile. When averaging kernels are not applied, the standard deviation of the histograms (after accounting for mean bias) is typically only slightly larger than 1, confirming that the reported ESDs provide a good estimate of the retrieval random error. When averaging kernels are applied to the sondes, standard deviations are reduced, however not as much as might be expected considering that noise-only errors are around a factor of 2 smaller than ESDs (see Fig. 2). Figure 7 shows the a priori and retrieval biases for subcolumns in Dobson units for different latitude bands as well as for the global average. Sonde agreement varies with latitude for a number of reasons, not least because of the changing vertical gradients and amount of ozone present. For the 450–170 hPa layer, the bias is seen to vary from +3 DU in the 30° S–30° N band to −3 DU in the 30–60° S, 60–90° S and 60–90° N bands. The bias exceeds +5 DU in the 60–90° S band for the 50–30 hPa and 30–20 hPa layers, which is due to both the limited vertical sensitivity and to the retrieval be-

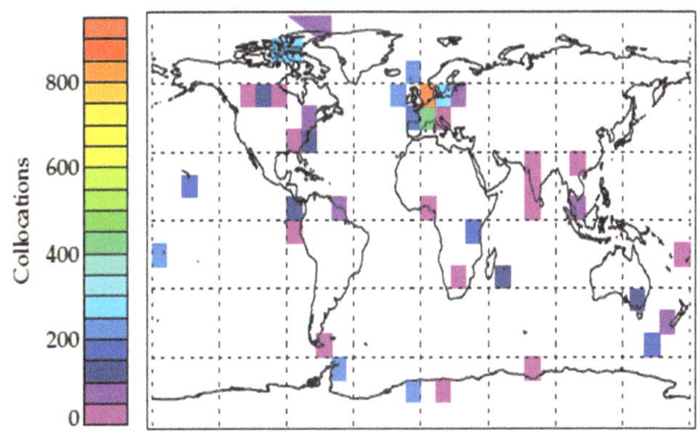

Figure 5. Geographical distribution of ozonesondes and numbers of GOME-2 profile collocations from 2007–2008 in 10 degree bins.

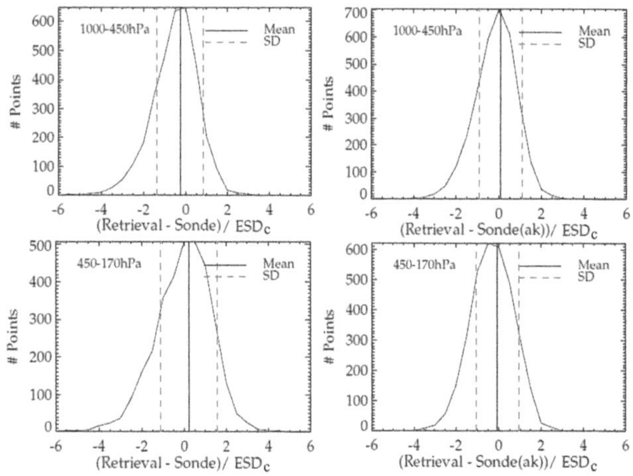

Figure 6. Histograms of differences between retrieved and sonde layer amounts relative to the estimated standard deviation (ESD) on each layer, for the lower-most and second subcolumns (top and bottom), with and without averaging kernels applied (right and left). The mean/standard deviation values are as follows: −0.184/1.12 (top left), 0.2/1.09 (top right), 0.158/1.41 (bottom left), 0.135/1.12 (bottom right).

ing influenced by an a priori profile (and its associated covariance) which is very unrepresentative of ozone hole conditions occurring in the Antarctic spring stratosphere. There is seen to be a small persistent positive bias (+2–3 DU) in the stratosphere (< 100 hPa) in all other latitude bands.

4.2 Retrieval performance in the presence of cloud

Retrievals of tropospheric ozone are affected by the presence of cloud. Extensive, thick cloud prevents photons penetrating to lower layers. As discussed in Sect. 2.2, the fitting of a surface albedo in Band 1 (270–308 nm) and in Band 2 (335 nm) partially accommodates cloud sun-normalised radiance and above-cloud scattering, so the remaining impact of cloud is obscuration of the ozone column beneath, as demonstrated in Fig. 8. Cloud information (effective cloud fraction and cloud top pressure) provided in the GOME-2 L1 data for each ground pixel from the FRESCO scheme (Fournier et al., 2004) is provided with the RAL height-resolved ozone product, so as to allow filtering by users.

4.3 Comparison to the global chemical transport model TOMCAT

Whereas ozonesondes can provide a near-truth in situ at fixed locations, they cannot necessarily indicate how well a satellite product captures the regional or global spatial distribution of ozone. Once validated quantitatively with ozonesondes, spatial agreement with CTMs can be indicative of this. These are driven by realistic atmospheric circulation (e.g. ECMWF re-analysed winds) and emission inventories, but employ differing schemes for chemistry, surface deposition, boundary layer mixing, convection and other vertical transport processes. Intercomparison of satellite data with a CTM can nonetheless be informative to evaluate both. Here we present a comparison of GOME-2 lower-tropospheric ozone with the TOMCAT CTM. We focus our comparison on the lowest layer, which is the most challenging for ozone retrieval from satellite.

4.3.1 TOMCAT chemistry transport model

A full description of the TOMCAT CTM is given elsewhere (Arnold et al., 2005; Chipperfield, 2006, and summarised in Richards et al., 2013), but it is briefly outlined here. TOMCAT is a 3-D chemical transport model which is optimised to reproduce the composition of the global troposphere. The version used here has a horizontal resolution of approximately $2.8° \times 2.8°$ and has been driven by ECMWF ERA-Interim temperature, winds and humidity (Dee et al., 2011). It operates on 31 hybrid sigma-pressure levels and the chemistry scheme and emission inventories used in this study are detailed in Richards et al. (2013). The model was spun-up for 6 months and then global O_3 fields were output four times a day at 00:00, 06:00, 12:00 and 18:00 UT. Model fields were interpolated in time and space to the satellite sampling (MetOp has an overpass time of 09:30 LT) for 2008. Lower-tropospheric ozone retrieved from GOME-2 by the RAL scheme has previously been shown to have excellent agreement with TOMCAT, in particular for the NH summer Mediterranean region (Richards et al., 2013).

4.3.2 Model comparison

Figure 9 compares GOME-2 with TOMCAT for the lowest retrieved subcolumn in August 2008. The GOME-2 data have been cloud-screened, based on cloud height and fraction from FRESCO in the L1b data, and GOME-2 AKs have been applied to the model. Geographical structure in the monthly mean distribution is seen to be represented quite consistently by GOME-2 and the model. In particular, there is seen to be agreement in locations of high ozone concentration over the Mediterranean region and Southeast China, which are typically found at this time of year, although peak values observed there by GOME-2 are higher than predicted by TOMCAT.

Consistency between GOME-2 and TOMCAT geographical distributions is indicated quantitatively by the standard deviation (4 DU) and correlation coefficient (0.66) for the August 2008 ensemble in Fig. 9. The global mean bias between GOME-2 and TOMCAT (~ 0.8 DU) for August 2008 is comparable to that between GOME-2 and ozonesondes in this layer (~ 1 DU) for the 2 years 2007–2008. Furthermore, the latitudinal dependence of the GOME-2 minus TOMCAT difference in Fig. 9 also mirrors that of the GOME-2 minus ozonesonde bias in Fig. 6 – being positive at northern mid/high-latitudes and negative at southern mid-latitudes.

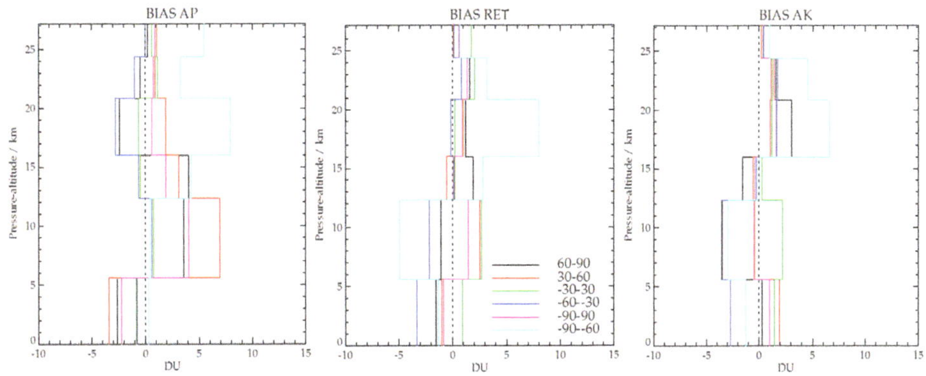

Figure 7. Bias with respect to ozonesondes as a function of latitude and pressure for subcolumns in Dobson units for the a priori (left) and retrieved (centre) profiles and for retrieved profiles with GOME-2 AKs applied to the sonde profiles (right). The pink lines indicate the averages over all latitude bands, for comparison to the black and green lines in the left hand panel of Fig. 4, which depict the same a priori and retrieval biases as % differences from the ozonesondes.

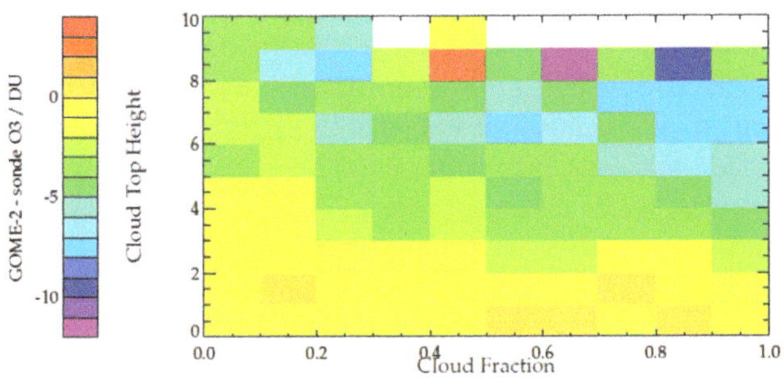

Figure 8. The lowest subcolumn ozone (surface to 450 hPa) differenced from ozonesonde subcolumn, without AKs applied and without any cloud clearing. In the presence of high/thick cloud where fewer photons can penetrate, there is less sensitivity to the lowermost ozone subcolumn.

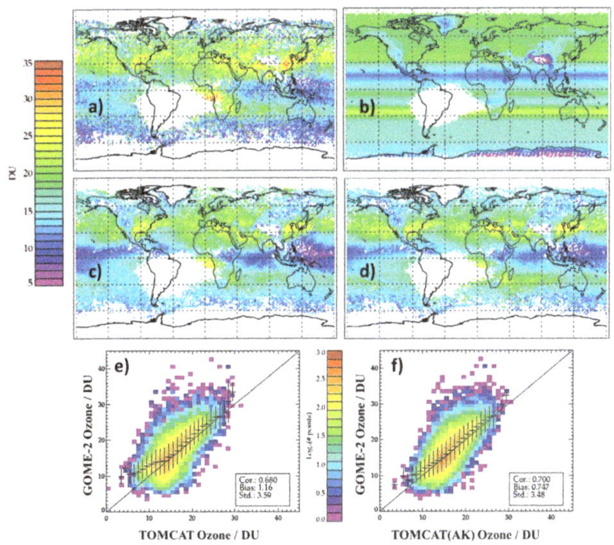

Figure 9. **(a)** GOME-2 surface to 450 hPa layer ozone gridded (1.125) monthly mean for September 2008. Pixels have been strictly cloud-cleared such that only pixels with a cloud fraction of < 0.2 and cloud top pressure of > 700 hPa remain; **(b)** a priori for GOME-2 retrieval (all pixels); **(c)** TOMCAT model with satellite sampling, **(d)** TOMCAT model with GOME-2 averaging kernels also applied; **(e)** correlation of **(a)** and **(c)** with associated bias and standard deviation; **(f)** correlation of **(a)** and **(d)**. The vertical and horizontal black lines in panels **(e)** and **(f)** indicate the respective standard deviations of those data sampled along the TOMCAT axis, and the numbers of points in $\log_{10}$ units are indicated by the colour bar.

4.3.3 Model time series comparison

Figure 10 shows monthly mean averages for the GOME-2 retrieval and its a priori, and the TOMCAT model (with GOME-2 spatial sampling) in four regions. These are the NH remote Pacific, the USA, the Mediterranean and eastern China. The remote Pacific in particular is not well sampled by ozonesondes. In the four regions selected, there is good agreement between GOME-2 and TOMCAT in the shape of the seasonal cycle in lower-tropospheric ozone. This is particularly the case for the USA and eastern China, where a double peak in the seasonal cycles is seen by both the model and the retrieval, but not the a priori. In the case of eastern China, a higher correlation is found between the model and the a priori than with GOME-2, and the former both predict lower absolute amounts of ozone in the annual cycle. However, the a priori does not capture the seasonal cycle that is present in both the model and the satellite record. In the Mediterranean, the summer peak is found to occur at a similar time in the retrieval and model but several months earlier in the prior.

5 Summary

The RAL ozone profile retrieval algorithm for nadir-viewing satellite UV spectrometers has been developed to have sensitivity to tropospheric as well as stratospheric ozone. This has been achieved by a three-step retrieval approach in which high fit precision (< 0.1 % RMS) is required in the third step to extract tropospheric information from the temperature-

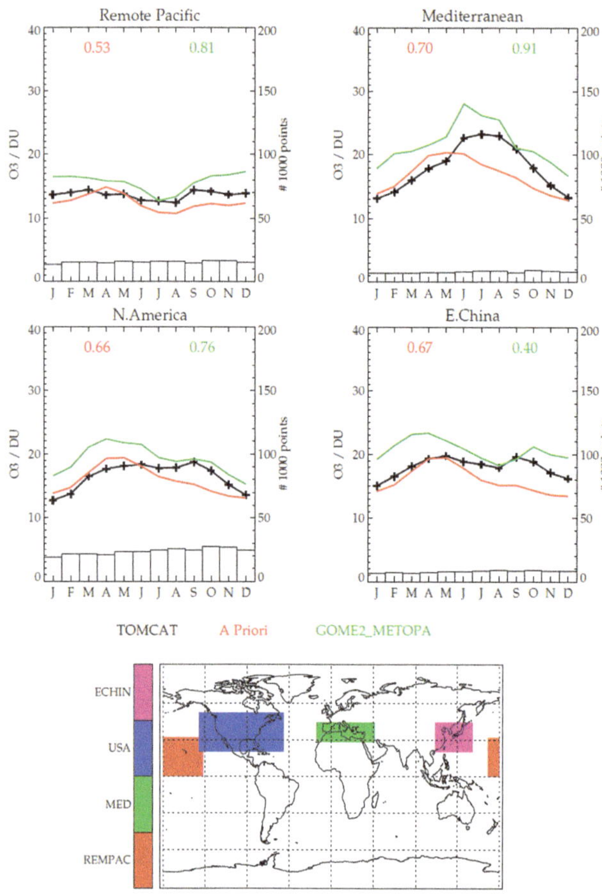

Figure 10. Time series comparison of surface to 450 hPa ozone for four regions of TOMCAT (black), GOME-2 (green) and the GOME-2 retrieval a priori/climatology in 2008. Monthly correlation coefficient of TOMCAT and the a priori (red) and GOME-2 (green) are also given for each region. In all cases GOME-2 averaging kernels have been applied to TOMCAT. Bars and second axis indicate number of measurements in each month for each region.

dependent Huggins bands (323–335 nm). The bias with respect to ozonesondes sampled worldwide over 2 years is of the order of 6 % ($\sim$ 1 DU) in the surface to 450 hPa layer and $<$ 5 % in the subcolumns above. The bias in part reflects the extent to which uncertainties in knowledge of the GOME-2 absolute UV (Hartley band) radiometry and (Huggins bands) slit function shape can be accommodated. The bias varies systematically with latitude/solar zenith angle. It is typically less than $\pm$3 DU, except in the tropical UTLS region where there is a positive bias of up to 5 DU, due to smearing of the sharp change in ozone vertical gradient near the tropopause. This corresponds to a bias of less than $\pm$20 % in the troposphere and $+$10 % in the tropical UTLS. As expected, the retrieval shows a negative bias in the troposphere in the presence of high or pervasive cloud because, for this validation exercise, cloud parameters have not been co-retrieved or explicitly modelled; their effects on UV sun-normalised radiance have been accommodated only through retrieval of an effective Lambertian albedo (and no ghost column has been added).

The GOME-2 retrieval and the CTM TOMCAT show agreement in the August 2008 monthly mean global distribution of lower tropospheric ozone and specifically in the location of high ozone concentrations over the Mediterranean and over Southeast China. Concentrations in the surface–450 hPa layer retrieved from GOME-2 are persistently higher at northern mid/high latitudes and lower at southern mid-latitudes than predicted by TOMCAT, a pattern which is consistent with the GOME-2 ozonesonde bias for 2007–2008.

Significant developments to the GOME-2 retrieval scheme are now planned. These include: (a) updating to and evaluating performance with the latest ozone spectroscopy (e.g. Serdyuchenko et al., 2014), as this has been identified as potentially important by e.g. Liu et al. (2007) and others; (b) improved modelling of the slit function shape and related changes with time, which is expected to impact upon tropospheric ozone in particular; (c) improved handling of radiometric degradation occurring in both the Hartley and Huggins UV bands over the mission lifetime, for a more accurate multi-year time series and (d) addition of the visible Chappuis bands as a fourth retrieval step, to increase ozone sensitivity in the lower troposphere over land which cannot at present be achieved with UV measurements alone.

Acknowledgements. This work is funded by the Natural Environment Research Council (NERC) through the National Centre for Earth Observation (NCEO), and in part also by ESA Climate Change Initiative and EUMETSAT Visiting Scientist Position in O3MSAF). We would like to thank T. von Clarmann and an anonymous referee for their constructive reviews of this paper.

Edited by: J. Tamminen

References

Arnold, S. R., Chipperfield, M. P., and Blitz, M. A.: A three dimensional model study of the effect of new temperature dependent quantum yields for acetone photolysis, J. Geophys.Res.-Atmos., 110, D22305, doi:10.1029/2005jd005998, 2005.

Boynard, A., Clerbaux, C., Coheur, P.-F., Hurtmans, D., Turquety, S., George, M., Hadji-Lazaro, J., Keim, C., and Meyer-Arnek, J.: Measurements of total and tropospheric ozone from IASI: comparison with correlative satellite, ground-based and ozonesonde observations, Atmos. Chem. Phys., 9, 6255–6271, doi:10.5194/acp-9-6255-2009, 2009.

Bhartia, P. K., McPeters, R. D., Mateer, C. L., Flynn, L. E., and Wellemeyer, C.: Algorithm for the estimation of vertical ozone profiles from the backscattered ultraviolet technique, J. Geophys. Res., 101, 18793–18806, doi:10.1029/96JD01165, 1996.

Brion, J., Chakir, A., Daumont, D., and Malicet, J.: High-resolution laboratory absorption cross section of O_3: Temperature effect, Chem. Phys. Lett., 213, 610–512, 1993.

Brion, J., Chakir, A., Charbonnier, J., Daumont, D., Parisse, C., and Malicet, J.: Absorption spectra measurements for the ozone molecule in the 350–830 nm region, J. Atmos. Chem., 30, 291–299, 1998.

Cai, Z., Liu, Y., Liu, X., Chance, K., Nowlan, C. R., Lang, R., Munro, R., and Suleiman, R.: Characterization and correction of Global Ozone Monitoring Experiment 2 ultraviolet measurements and application to ozone profile retrievals, J. Geophys. Res., 117, D07305, doi:10.1029/2011JD017096, 2012.

Calisesi, Y., Soebijanta, V. T., and van Oss, R.: Regridding of remote soundings: Formulation and application to ozone profile comparison, J. Geophys. Res., 110, D23306, doi:10.1029/2005JD006122, 2005.

Callies, J., Corpaccioli, E., Eisinger, M., Hahne, A., and Lefebvre, A.: GOME-2 – Metop's Second-Generation, Sensor for Operational Ozone Monitoring. ESA Bulletin number 102, May 2000, 28–36, 2000.

Chance, K. and Kurucz, R. L.: An improved high-resolution solar reference spectrum for Earth's atmosphere measurements in the ultraviolet, visible, and near infrared, J. Quant. Spectrosc. Ra., 111, 1289–1295, 2010.

Chance, K. V., Burrows, J. P., Perner, D., and Schneider, W.: Satellite measurements of atmospheric ozone profiles, including tropospheric ozone, from ultraviolet/visible measurements in the nadir geometry: a potential method to retrieve tropospheric ozone, J. Quant. Spectrosc. Ra., 57, 467–476, 1997.

Chapman, S.: A theory of upper atmosphere ozone, Mem. R. Metrol. Soc., 3, 103–125, 1930.

Chipperfield, M. P.: New version of the TOMCAT/SLIMCAT offline chemical transport model: Intercomparison of stratospheric tracer experiments, Q. J. R. Meteorol. Soc., 132, 1179–1203, doi:10.1256/qj.05.51, 2006.

Daumont, M., Brion, J., Charbonnier, J., and Malicet, J.: Ozone UV spectroscopy I: Absorption cross-sections at room temperature, J. Atmos. Chem., 15, 145–155, 1992.

Dee, D. P., Uppala, S. M., Simmons, A. J., Berrisford, P., Poli, P., Kobayashi, S., Andrae, U., Balmaseda, M. A., Balsamo, G., Bauer, P., Bechtold, P., Beljaars, A. C. M., van de Berg, L., Bidlot, J., Bormann, N., Delsol, C., Dragani, R., Fuentes, M., Geer, A. J., Haimberger, L., Healy, S. B., Hersbach, H., Hólm, E. V., Isaksen, L., Kållberg, P., Köhler, M., Matricardi, M., McNally, A. P., Monge-Sanz, B. M., Morcrette, J.-J., Park, B.-K., Peubey, C., de Rosnay, P., Tavolato, C., Thépaut, J.-N., and Vitart, F.: The ERA-Interim reanalysis: configuration and performance of the data assimilation system, Q. J. Roy. Meteorol. Soc., 137, 553–597, doi:10.1002/qj.828, 2011.

Deeter, M. N., Edwards, D. P., Gille, J. C., and Drummond, J. R.: Sensitivity of MOPITT observations to carbon monoxide in the lower troposphere, J. Geophys. Res., 112, D24306, doi:10.1029/2007JD008929, 2007.

EUMETSAT (European Organisation for the Exploitation of Meteorological Satellites): GOME-2 Level 1 Product Generation Specification Report, EPS.SYS.SPE.990011, Darmstadt, Germany, 2006.

Fishman, J. and Larsen, J. C.: Distribution of total ozone and stratospheric ozone in the tropics: Implications for the distribution of tropospheric ozone, J. Geophys. Res., 92, 6627–6634, doi:10.1029/JD092iD06p06627, 1987.

Fournier, N., Stammes, P., Acarreta, J. R., and van Geffen, J.: FRESCO Cloud Algorithm for GOME-2, Final Report, ESTEC contract number 17332/03/NL/GS, 2004.

Gryparis, A., Forsberg, B., Katsouyanni, K., Analitis, A., Touloumi, G., Schwartz, J., Samoli, E., Medina, S., Anderson, H. R., Niciu, E. M., Wichmann, H. E., Kriz, B., Kosnik, M., Skorkovsky, J., Vonk, J. M., and Dortbudak, Z.: Acute effects of ozone on mortality from the "Air pollution and health: A European approach" project, Am. J. Resp. Crit. Care, 170, 1080–1087, doi:10.1164/rccm.200403-333OC, 2004.

Hollaway, M. J., Arnold, S. R., Challinor, A. J., and Emberson, L. D.: Intercontinental trans-boundary contributions to ozone-induced crop yield losses in the Northern Hemisphere, Biogeosciences, 9, 271–292, doi:10.5194/bg-9-271-2012, 2012.

IPCC: Climate Change 2013: The Physical Science Basis. Contribution of Working Group I to the Fifth Assessment Report of the Intergovernmental Panel on Climate Change, edited by: Stocker, T. F., Qin, D., Plattner, G.-K., Tignor, M., Allen, S. K., Boschung, J., Nauels, A., Xia, Y., Bex, V., and Midgley, P. M., Cambridge University Press, Cambridge, United Kingdom and New York, NY, USA, 1535 pp., doi:10.1017/CBO9781107415324, 2013.

Joiner, J., Barthia, P. K., Cebula, R. P., Hilsenrath, E., McPeters, R. D., and Park, H.: Rotational Raman scattering (Ring effect) in satellite backscatter ultraviolet measurements, Appl. Optics, 34, 4513–4525, 1995.

Keppens, A., Lambert, J.-C., Granville, J., Miles, G., Siddans, R., van Peet, J. C. A., van der A, R. J., Hubert, D., Verhoelst, T., Delcloo, A., Godin-Beekmann, S., Kivi, R., Stübi, R., and Zehner, C.: Round-robin evaluation of nadir ozone profile retrievals: methodology and application to MetOp-A GOME-2, Atmos. Meas. Tech. Discuss., 7, 11481–11546, doi:10.5194/amtd-7-11481-2014, 2014.

Kerridge, B. J. K., Siddans, R., Latter, B. L., Burrows, J. P., Weber, M., De Beek, R., Aben, I., and Hartman, W.: GOME-2 Error Assessment Study, Final Report EUMETSAT Contract No EUM/CO/01/901/DK, 2002.

Klenk, K. F., Bhartia, P. K., Fleig, A. J., Kaveeshwar, V. G., McPeters, R. D., and Smith, P. M.: Total ozone determination from the Backscattered Ultraviolet (BUV) experiment, J. Appl. Meteorol., 21, 1672–1684, 1982.

Lacan, A. and Lang, R.: Investigation on GOME-2 throughput degradation, Final report, EUM/LEO/REP/09/0732 Issue 1.1, 16 July 2011.

Lang, R., Munro, R., Livschitz, Y., Dyer, R., and Lacan, A.: GOME-2 FM3 Long-Term In-Orbit Degradation – Basic Signatures After 2nd Throughput Test, EUMETSAT Technical report, EUM.OPS-EPS.DOC.09.0464, 2009.

Liu, X., Chance, K., Sioris, C. E., Spurr, R. J. D., Kurosu, T. P., Martin, R. V., and Newchurch, M. J.: Ozone profile and tropospheric ozone retrievals from Global Ozone Monitoring Experiment: Algorithm description and validation, J. Geophys. Res., 110, D20307, doi:10.1029/2005JD006240, 2005.

Liu, X., Chance, K., Sioris, C. E., and Kurosu, T. P.: Impact of using different ozone cross sections on ozone profile retrievals from Global Ozone Monitoring Experiment (GOME) ultraviolet measurements, Atmos. Chem. Phys., 7, 3571–3578, doi:10.5194/acp-7-3571-2007, 2007.

Liu, X., Bhartia, P. K., Chance, K., Spurr, R. J. D., and Kurosu, T. P.: Ozone profile retrievals from the Ozone Monitoring Instrument,

Atmos. Chem. Phys., 10, 2521–2537, doi:10.5194/acp-10-2521-2010, 2010.

Loyola, D. G., Koukouli, M. E., Valks, P., Balis, D. S., Hao, N., Van Roozendael, M., Spurr, R. J. D., Zimmer, W., Kiemle, S., Lerot, C., and Lambert, J.-C.: The GOME-2 total column ozone product: Retrieval algorithm and ground-based validation, J. Geophys. Res., 116, D07302, doi:10.1029/2010JD014675, 2011.

Malicet, C., Daumont, D., Charbonnier, J., Parisse, C., Chakir, A., and Brion, J.: Ozone UV spectroscopy, II. Absorption cross-sections and temperature dependence, J. Atmos. Chem., 21, 263–273, 1995.

McPeters, R. D., Lebow, G. J., and Logan, J. A.: Ozone climatological profiles for satellite retrieval algorithms, J. Geophys. Res., 112, D05308, doi:10.1029/2005JD006823, 2007.

Miles, G. M., Siddans, R. S., Tuinder, O., and Latter, B. L.: Analysis of GOME-2 level-1b data quality and degradation effects on operational ozone profile retrievals, EUMETSAT O3MSAF Visiting Scientist Final Report, 2012.

Munro, R., Siddans, R., Reburn, W. J., and Kerridge, B. J.: Direct measurement of tropospheric ozone distributions from space, Nature, 392, 168–171, doi:10.1038/32392, 1998.

Nassar, R., Logan, J. A., Worden, H. M., Megretskaia, I. A., Bowman, K. W., Osterman, G. B., Thompson, A. M., Tarasick, D. W., Austin, S., Claude, H., Dubey, M. K., Hocking, W. K., Johnson, B. J., Joseph, E., Merrill, J., Morris, G. A., Newchurch, M., Oltmans, S. J., Posny, F., Schmidlin, F. J., Vomel, H., Whiteman, D. N., and Witte, J. C.: Validation of Tropospheric Emission Spectrometer (TES) nadir ozone profiles using ozonesonde measurements, J. Geophys. Res.-Atmos., 113, D15S17, doi:10.1029/2007jd008819, 2008.

Nowlan, C. R., Liu, X., Chance, K., Cai, Z., Kurosu, T. P., Lee, C., and Martin, R. V.: Retrievals of sulfur dioxide from the Global Ozone Monitoring Experiment 2 (GOME-2) using an optimal estimation approach: Algorithm and initial validation, J. Geophys. Res., 116, D18301, doi:10.1029/2011JD015808, 2011.

Platt, U.: Differential optical absorption spectroscopy (DOAS), Chem. Anal. Series, 127, 27–83, 1994.

Plumber, S.: The ESA Climate Change Initiative – Description, ESA, Technical Note: EOP-SEP/TN/0030-09/SP, 2009.

Press, W. H., Teukolksy, S., Vetterling, W. T., and Flannery, B.: Numerical Recipes: the art of scientific computing, 2nd Edn., Cambridge University Press, 1995.

Richards, N. A. D., Arnold, S. R., Chipperfield, M. P., Miles, G., Rap, A., Siddans, R., Monks, S. A., and Hollaway, M. J.: The Mediterranean summertime ozone maximum: global emission sensitivities and radiative impacts, Atmos. Chem. Phys., 13, 2331–2345, doi:10.5194/acp-13-2331-2013, 2013.

Rodgers, C. D.: Retrieval of atmospheric temperature and composition from remote measurements of thermal radiation, Rev. Geophys.-Space Phys., 14, 609–624, 1976.

Rodgers, C. D.: Inverse Methods for Atmospheric Sounding: Theory and Practice, World Sci., Hackensack, N.J., 2000.

Rozanov, V. V., Diebel, D., Spurr, R. J. D., and Burrows, J. P.: GOMETRAN: A radiative transfer model for the satellite project GOME – the plane-parallel version, J. Geophys. Res., 102, 16683–16695, 1997.

Schoeberl, M. R., Ziemke, J. R., Bojkov, B., et al.: A trajectory-based estimate of the tropospheric ozone column using the residual method, J. Geophys. Res., 112, D24S49, doi:10.1029/2007JD008773, 2007.

Siddans, R.: Height Resolved Ozone Retrievals from Global Ozone Monitoring Experiment, PhD Thesis, University of Reading, 2003.

Siddans, R., Kerridge, B. J., Latter, B. G., Smeets, J., Otter, G., and Slijkhuis, S.: Analysis of GOME-2 slit function measurements: Final report, EUM/CO/04/1298/RM, Eur. Organ. for the Exploit. of Meteorol. Satell., Darmstadt, Germany, 2006.

SI^2N (SPARC/IO3C/IGACO-O3/NDACC): Initiative, 2011, Activity on Past changes in the Vertical Distribution of Ozone, available at: http://igaco-o3.fmi.fi/VDO/index.html (last access: 20 January 2015), 2012.

Serdyuchenko, A., Gorshelev, V., Weber, M., Chehade, W., and Burrows, J. P.: High spectral resolution ozone absorption cross-sections – Part 2: Temperature dependence, Atmos. Meas. Tech., 7, 625–636, doi:10.5194/amt-7-625-2014, 2014.

Thompson, A. M., Witte, J. C., McPeters, R. D., Oltmans, S. J., Schmidlin, F. J., Logan, J. A., Fujiwara, M., Kirchhoff, V. W. J. H., Posny, F., Coetzee, G. J. R., Hoegger, B., Kawakami, S., Ogawa, T., Johnson, B. J., Vomel, H., and Labow, G. J.: Southern Hemisphere Additional Ozonesondes (SHADOZ) 1998–2000 tropical ozone climatology: 1. Comparison with Total Ozone Mapping Spectrometer (TOMS) and ground-based measurements, J. Geophys. Res., 108, 8238, doi:10.1029/2001JD000967, 2003.

van der A, R.: Recalibration of GOME spectra for the purpose of ozone profile retrieval, KNMI Technical Note RT-236, 2001.

van der A, R. J., van Oss, R. F., Piters, A. J. M., Fortuin, J. P. F., Meijer, Y. J., and Kelder, H. M.: Ozone profile retrieval from recalibrated Global Ozone Monitoring Experiment data, J. Geophys. Res., 107, ACH2.1–ACH2.10, doi:10.1029/2001JD000696, 2002.

van Roozendael, M., Spurr, R., Loyola, D., Lerot, C., Balis, D., Lambert, J.-C., Zimmer, W., van Gent, J., van Geffen, J., Koukouli, M., Granville, J., Doicu, A., Otto, S., Fayt, C., and Zehner, C.: Sixteen years of GOME/ERS2 total ozone data: the new direct-fitting GOME Data Processor (GDP) Version 5 – Algorithm Description, J. Geophys. Res., 117, D03305, doi:10.1029/2011JD016471, 2012.

Valks, P., Hao, N., Gimeno Garcia, S., Loyola, D., Dameris, M., Jöckel, P., and Delcloo, A.: Tropical tropospheric ozone column retrieval for GOME-2, Atmos. Meas. Tech., 7, 2513–2530, doi:10.5194/amt-7-2513-2014, 2014.

von Clarmann, T.: Smoothing error pitfalls, Atmos. Meas. Tech., 7, 3023–3034, doi:10.5194/amt-7-3023-2014, 2014.

Ziemke, J. R., Chandra, S., Labow, G. J., Bhartia, P. K., Froidevaux, L., and Witte, J. C.: A global climatology of tropospheric and stratospheric ozone derived from Aura OMI and MLS measurements, Atmos. Chem. Phys., 11, 9237–9251, doi:10.5194/acp-11-9237-2011, 2011.

7

Use of neural networks in ground-based aerosol retrievals from multi-angle spectropolarimetric observations

A. Di Noia[1], O. P. Hasekamp[1], G. van Harten[2], J. H. H. Rietjens[1], J. M. Smit[1], F. Snik[2], J. S. Henzing[3], J. de Boer[2], C. U. Keller[2], and H. Volten[4]

[1]SRON Netherlands Institute for Space Research, Sorbonnelaan 2, 3584 CA Utrecht, the Netherlands
[2]Leiden Observatory, Leiden University, Niels Bohrweg 2, 2333 CA Leiden, the Netherlands
[3]Netherlands Organisation for Applied Research (TNO), Princetonlaan 6, 3584 CB Utrecht, the Netherlands
[4]National Institute for Public Health and the Environment (RIVM), Antonie van Leeuwenhoeklaan 9, 3721 MA Bilthoven, the Netherlands

Correspondence to: A. Di Noia (a.di.noia@sron.nl)

Abstract. In this paper, the use of a neural network algorithm for the retrieval of the aerosol properties from ground-based spectropolarimetric measurements is discussed. The neural network is able to retrieve the aerosol properties with an accuracy that is almost comparable to that of an iterative retrieval. By using the outcome of the neural network as first guess in the iterative retrieval scheme, the accuracy of the retrieved fine- and coarse-mode optical thickness is further improved, while for the other parameters the improvement is small or absent. The resulting scheme (neural network + iterative retrieval) is compared to the original one (look-up table + iterative retrieval) on a set of simulated ground-based measurements, and on a small set of real observations carried out by an accurate ground-based spectropolarimeter. The results show that the use of a neural-network-based first guess leads to an increase in the number of converging retrievals, and possibly to more accurate estimates of the aerosol effective radius and complex refractive index.

1 Introduction

Atmospheric aerosols influence Earth's radiative balance and climate in multiple ways. First, through their scattering and absorption properties, they modify the amount of solar radiation reaching the Earth's surface and the lowest atmospheric layers (direct effect, Yu et al., 2006, and references therein). Second, acting as cloud condensation nuclei, they influence the cloud formation processes, thereby affecting cloud microphysical properties and precipitations (indirect effect; Lohmann and Feichter, 2005, and references therein). Third, highly absorbing aerosols have a warming effect on the atmosphere, which can cause the evaporation of cloud particles and hence lead to a reduction in the cloud cover (semi-direct effect; Hansen et al., 1997; Kaufman and Koren, 2006).

The net effect of the atmospheric aerosols on climate is highly uncertain, both in magnitude and sign (Boucher et al., 2013). This uncertainty is partly due to the uncertainties in the parametrization of aerosol processes in models (Liu et al., 2012; Rosenfeld et al., 2013), but is also due to the intrinsic limitations in our current knowledge of the global distribution of the most important aerosol microphysical and optical properties (Mishchenko et al., 2004, 2005).

Further motivation for studying atmospheric aerosols is provided by their impact on human health (Anderson et al., 2012) and aviation safety (Alexander, 2013).

The sensitivity of polarized radiance measurements to the aerosol size distribution, shape and refractive index (Hansen and Travis, 1974) provides, in principle, a means for retrieving these aerosol properties from remote measurements either from space (Mishchenko and Travis, 1997; Hasekamp and Landgraf, 2007) or from the ground (Boesche et al., 2006).

A limited number of techniques for retrieving the aerosol properties using intensity and polarization of the scattered sunlight are reported in the literature. Some of these are based on look-up tables (LUTs), in that they select the aerosol model that best fits the observed total and polarized intensities from a set of pre-calculated values (e.g. Deuzé et al., 2000, 2001). Other methods consist of variational retrievals, where full radiative transfer calculations are used to minimize a nonlinear cost function with respect to a set of aerosol microphysical and optical properties. In particular, Dubovik et al. (2011) describe a variational retrieval scheme where satellite intensity and polarization measurements are combined in order to retrieve a set of aerosol (total aerosol load, aerosol size distribution in a number of size bins, fraction of spherical particles, spectrally dependent complex refractive index and aerosol layer height) and surface reflection parameters, relying on the assumption that all the aerosol size bins have the same refractive index, and imposing a number of smoothness constraints on the size distribution and on the spatial and temporal variability of the aerosol properties. An alternative approach is described in Hasekamp et al. (2011), where a different set of aerosol parameters is retrieved using a scheme based on Phillips–Tikhonov regularization, assuming a bimodal log-normal aerosol size distribution and allowing for a fine and a coarse mode characterized by different refractive indices, which are assumed to be wavelength-independent. Again, a nonlinear cost function – depending on the discrepancy between observed and simulated intensities and degrees of linear polarization (DLPs) at multiple wavelengths and viewing angles – is iteratively minimized, starting from a first guess provided by a LUT. Further details are given in Sect. 2.

The choice of a LUT-based first guess is also common to other variational aerosol retrieval schemes (e.g. Waquet et al., 2009; Knobelspiesse et al., 2011). The accuracy of a LUT retrieval depends on the amount of entries (aerosol models, other atmospheric parameters, viewing angles, etc.) stored in the table. However, enhancing the accuracy of a LUT retrieval by increasing the LUT dimension is an expensive task, in terms of memory requirements (need for bigger arrays to be stored) and computation time (more models to be tried in order to select the one that best fits the measurements).

A possible alternative to a LUT for generating a first guess in a variational aerosol retrieval scheme is represented by a neural network (NN). Neural networks are powerful tools for the mapping of nonlinear relationships, and have been successfully applied as stand-alone retrieval algorithms to several remote sensing problems, including aerosol-related applications (Han et al., 2006; Niang et al., 2006; Radosavljevic et al., 2010; Picchiani et al., 2011). Nevertheless, the output of a NN retrieval scheme is also suitable for use as the initial step of an iterative retrieval method, giving rise to a "neuro-variational" retrieval scheme. A number of neuro-variational retrieval schemes, also applied to aerosol retrievals, are already reported in the literature. Jamet et al. (2005) and Brajard et al. (2006, 2012) used a set of NNs to build a forward model that simulates satellite-measured reflectances given a set of atmospheric and oceanic parameters, and is then inverted with respect to such parameters using an iterative method. Diouf et al. (2013) further refined this scheme by using an unsupervised neural network to generate a first guess for the iterative retrieval.

In our experiments we have investigated a different type of neuro-variational retrieval algorithm. Specifically, we have trained a multilayer perceptron (MLP) neural network (Werbos, 1974) to retrieve a set of aerosol parameters from simulated intensities and DLPs at a number of wavelengths and angles, and we have used the set of aerosol parameters retrieved by the NN as first guess for the subsequent iterative scheme using a full-physics forward model and an inversion method based on Phillips–Tikhonov regularization. We will show that the use of a NN-based first guess instead of the LUT-based one can substantially improve the performance of the retrieval algorithm, with respect to the number of retrievals that reach an acceptable minimum of the cost function and with respect to the accuracy of the retrieved parameters. The resulting retrieval scheme is also less demanding in terms of memory allocation, and is faster than the scheme that uses a LUT-based first guess.

The paper is organized as follows. Section 2 gives an overview of the original retrieval algorithm, including details on the LUT. Section 3 describes the actual setup of the experiments performed in this paper. Section 4 describes the design of the NN that is then used as a replacement for the LUT. In Sect. 5 the algorithm version using a NN-based first guess is compared to that using the LUT-based first guess on a data set of simulated observations, and then on a number of measurements performed by a ground-based spectropolarimeter based on the spectral modulation technique (van Harten et al., 2014). Section 6 is dedicated to the conclusions and to the discussion of future work.

2 Variational aerosol retrieval algorithm

2.1 General setup

The retrieval scheme used in this work is a modified version of the algorithm discussed by Hasekamp et al. (2011) for PARASOL satellite data. The main modifications are the following: (i) the forward model is adapted to the simulation of ground-based measurements instead of satellite measurements, (ii) a different method is used in the choice of the regularization parameter at each iteration of the inversion (a heuristic procedure, described in the following subsection, is used instead of the L-curve method), and (iii) the state vector is defined in a different manner. In particular, with respect to the latter point, the retrieved aerosol quantities are the effective radius, the complex refractive index, and the aerosol column amount for the fine and the coarse mode.

2.2 Inversion method

According to the Phillips–Tikhonov regularization method, the retrieval cost function is defined as

$$J(\boldsymbol{x}) = [\boldsymbol{y} - \boldsymbol{F}(\boldsymbol{x})]^T \mathbf{S}_\epsilon^{-1} [\boldsymbol{y} - \boldsymbol{F}(\boldsymbol{x})] + \gamma(\boldsymbol{x} - \boldsymbol{x}_\mathrm{a})^T \mathbf{H}(\boldsymbol{x} - \boldsymbol{x}_\mathrm{a}), \quad (1)$$

where $\boldsymbol{x}$ is the state vector (i.e. the set of quantities to be retrieved), $\boldsymbol{y}$ is the vector of measurements, $\boldsymbol{F}$ is the vector-valued forward model function that maps the state vector into the measurement vector, $\boldsymbol{x}_\mathrm{a}$ is an a priori for the state vector, $\mathbf{S}_\epsilon$ is the error covariance matrix, $\mathbf{H}$ is the regularization matrix and γ is the regularization parameter. Since $\boldsymbol{F}$ is nonlinear with respect to $\boldsymbol{x}$, the minimization of J is carried out iteratively, using the Gauss–Newton method. Starting from a first guess $\boldsymbol{x}_0$ for the state vector, a sequence of solutions $\{\boldsymbol{x}_i\}$ is generated by linearizing $\boldsymbol{F}$ around $\boldsymbol{x}_i$ at each step i, and then finding the next solution $\boldsymbol{x}_{i+1}$ as the one that minimizes J for the linearized version of $\boldsymbol{F}$. The process is stopped when the goodness of fit no longer changes significantly or when a maximum number of iterations is reached. The linear approximation of $\boldsymbol{F}$ around $\boldsymbol{x}_i$ can be written as

$$\boldsymbol{F}(\boldsymbol{x}) \approx \boldsymbol{F}(\boldsymbol{x}_i) + \mathbf{K}_i(\boldsymbol{x} - \boldsymbol{x}_i), \quad (2)$$

where $\mathbf{K}_i = \partial \boldsymbol{F}/\partial \boldsymbol{x}$ evaluated for $\boldsymbol{x} = \boldsymbol{x}_i$. Depending on the adopted forward model, $\mathbf{K}_i$ can be computed either analytically or by finite differencing. By replacing $\boldsymbol{F}$ with Eq. (2) in Eq. (1), differentiating J with respect to $\boldsymbol{x}$ and setting the result to zero, the new solution is found as (Rodgers, 2000)

$$\boldsymbol{x}_{i+1} = \boldsymbol{x}_\mathrm{a} + (\mathbf{K}_i^T \mathbf{S}_\epsilon^{-1} \mathbf{K}_i + \gamma_i \mathbf{H})^{-1} \mathbf{K}_i^T \mathbf{S}_\epsilon^{-1} [\boldsymbol{y} - \boldsymbol{F}(\boldsymbol{x}_i) + \mathbf{K}_i(\boldsymbol{x}_i - \boldsymbol{x}_\mathrm{a})]. \quad (3)$$

As indicated by the subscript i, the regularization parameter γ is also updated during the iterative process. While in Hasekamp et al. (2011) this is done using the L-curve method (Hansen and O'Leary, 1993), in this version of the algorithm the choice of γ_i is made heuristically by trying different values for γ_i and selecting the one that produces the solution with the best fit between the measurements and the simulations. The goodness of fit is measured using the parameter

$$\chi^2(\gamma_i) = \frac{1}{m} \sum_{j=1}^{m} \frac{[y_j - F_j(\boldsymbol{x}_{i+1}(\gamma_i))]^2}{\sigma_j^2}, \quad (4)$$

where m is the dimension of $\boldsymbol{y}$, y_j and F_j are the jth components of $\boldsymbol{y}$ and $\boldsymbol{F}$ respectively, and σ_j^2 is the jth element on the principal diagonal of $\mathbf{S}_\epsilon$. In Eq. (4), the dependence of χ^2 and $\boldsymbol{x}_{i+1}$ on the chosen γ_i has been made explicit. According to this formalism, the chosen γ_i is thus the one that minimizes χ^2.

The regularization matrix $\mathbf{H}$ adopted in our retrieval scheme is diagonal, and the elements on its principal diagonal can be written as

$$h_{ii} = \frac{1}{(x_i^\mathrm{max} - x_i^\mathrm{min})^2} \quad i = 1, \ldots, n, \quad (5)$$

where n is the dimension of the state vector and x_i^max and x_i^min are respectively the maximum and minimum allowed values for the elements of the state vector. In the Phillips–Tikhonov inversion scheme the matrix $\mathbf{H}$ determines the relative weight between the different parameters in the side constraint, whereas γ determines the overall contribution of the side constraint, for which a proper value is found using the procedure described above.

In the original version of the algorithm, the first guess $\boldsymbol{x}_0$ is generated by a LUT. The LUT stores the coefficients of the azimuthal Fourier expansion of the Stokes vectors simulated for 576 predefined aerosol models (i.e. combinations of the aerosol parameters) at a number of wavelengths and viewing angles, stratified by solar zenith angle, surface pressure and surface albedo. The reflectances and the DLPs are computed by recombining the aforementioned coefficients for each relative azimuth angle. For all the predefined aerosol models in the LUT, a retrieval of the fine- and coarse-mode aerosol optical thickness (AOT) is performed using the LUT as forward model, and the aerosol model that provides the best fit to the measurements is selected as first guess.

The values of the aerosol parameters and of the other radiative transfer model parameters used to generate the LUT are listed in Table 1.

2.3 Forward model

The forward model provides a simulation of the instrument measurements corresponding to a certain atmospheric state (combination of aerosol and other atmospheric parameters) and a certain observation geometry. The forward model consists of two main parts. The first part relates the microphysical aerosol properties (optical thickness, single scattering albedo, phase matrix) to their optical properties in different layers of a model atmosphere. For this part of the forward model, pre-calculated optical properties as a function of the size parameters and of the real and imaginary part of the refractive index (Dubovik et al., 2006) are employed. The optical properties are pre-calculated for spheres and spheroids with an axis ratio distribution as proposed by Dubovik et al. (2006), and the aerosols are described by a mixture of spheres and spheroids. The optical properties for a given refractive index are obtained by spline interpolation from the tabulated values, and subsequently their integration over the aerosol size distribution is performed. The second part of the forward model solves the vector radiative transfer equation for given optical properties in different altitude layers in the model atmosphere. For this, the model of Hasekamp and Landgraf (2002, 2005), based on

Table 1. LUT entries for each radiative transfer model parameter. The model parameters are solar zenith angle (SZA), viewing zenith angle (VZA), surface pressure (p_s), surface albedo (a_s), aerosol effective radius (r_{eff}), effective variance (v_{eff}), refractive index real part (m_r) and imaginary part (m_i), fraction of spherical particles (n_s) and aerosol optical thickness (AOT). The superscripts f and c on the symbols of the aerosol parameters indicate the fine and the coarse mode respectively.

Parameter	Values
Wavelengths (nm)	440, 675, 870
SZA	15° to 75° in steps of 5°
VZA	0° to 65° in steps of 5°
p_s (hPa)	700, 1013
a_s(440, 675 nm)	0.02, 0.04, 0.06, 0.10, 0.15
a_s(870 nm)	0.10, 0.15, 0.20, 0.30, 0.45
r^f_{eff} (μm)	0.05, 0.1, 0.2, 0.3
v^f_{eff} (μm^2)	0.19
m^f_r	1.37, 1.45, 1.52, 1.6
m^f_i	0.001, 0.02, 0.1
n^f_s	1.0
τ^f_{aer}	0.025, 0.25, 0.5, 0.8, 1.0, 1.5, 3.0
r^c_{eff} (μm)	1.0, 2.5
v^c_{eff} (μm^2)	0.19
m^c_r	1.38, 1.44, 1.5
m^c_i	0.001, 0.05
n^c_s	0.1
τ^c_{aer}	0.025, 0.15, 0.5, 0.8, 1.0, 3.0

the Gauss–Seidel iterative method, is used. For the ground-based, upward-looking geometry, the analytical calculation of the derivatives of the measurement vector with respect to the state vector has not been implemented yet. For this reason, the derivatives are currently calculated by means of finite differencing.

3 Experimental setup

For this study, a spectropolarimeter designed for multi-angle ground-based observations based on the spectral modulation technique (Snik et al., 2009) is used. The key idea behind this technique is that the intensity of the incoming light (e.g. sunlight scattered in the atmosphere) becomes spectrally modulated after passing through a set of optical components (viz., an achromatic quarter-wave retarder, an athermal multiple-order retarder and a polarizing beam splitter). The corresponding modulation amplitude is proportional to the DLP, and the modulation phase is related to the angle of linear polarization (ALP). The DLP and the ALP of the incoming radiation can be derived from the spectrally modulated intensity through a demodulation algorithm as prescribed by Snik et al. (2009). This measurement concept eliminates the need for moving parts in the polarization measurement device. Multi-angle measurements are obtained by sequentially pointing the instrument described above toward multiple zenith directions. A complete scan of the principal plane is obtained within 5–10 min.

The family of instruments constructed according to the concept described above has been given the name Spectropolarimeter EXperiment (SPEX). The ground-based spectropolarimeter used in this work will be hereinafter referred to as groundSPEX. A more detailed description of the instrument is given in van Harten et al. (2014).

4 Design of the neural network retrieval scheme

In order to train the NN, a set of scenarios simulating groundSPEX observations has been created. Multispectral and multi-angular reflectances and DLP measured by a groundSPEX-like instrument have been simulated for different combinations of aerosol parameters, solar zenith angles and values of the surface pressure and surface albedo. In the generation of this synthetic data set, cloud-free conditions have been assumed, and it has been hypothesized that the instrument performs observations in the principal plane. The principal plane is indeed the most favourable one for the retrieval of the aerosol properties, as it allows for sampling of the aerosol scattering matrix in the widest range of scattering angles. Since groundSPEX is currently unable to look less than 6° away from the Sun because of stray light, for simplicity of modelling it has been assumed that the groundSPEX-like instrument only scans the part of the principal plane with relative azimuth angle $\widetilde{\varphi} = 180°$ (i.e. the instrument only looks away from the Sun). With these settings, more than 10^6 simulations have been produced. A fraction of the simulated data has been used to train a NN for the retrieval of the aerosol properties; another fraction has been used to test the NN and to compare the retrieval scheme using the NN-based first guess to that using the LUT-based first guess.

It has been assumed that the measurements are performed at six viewing zenith angles (VZAs) (0, 6, 20, 32, 46 and 60°) and three wavelengths (440, 675 and 870 nm) that correspond to the wavelengths at which groundSPEX has been radiometrically calibrated. This choice allowed for a straightforward application of the method to real measurements. Finally, it has been assumed that the instrument carries out its observations on a surface covered by green vegetation. A surface albedo of 0.05 has been assumed at 440 and 675 nm, whereas the 870 nm albedo has been varied between 0.3 and 0.5 in the simulations.

Approximately 8×10^5 of the simulations described above have been used to train a NN to approximate the inverse relationship between the set of aerosol parameters $\boldsymbol{x}$ and the set of spectropolarimetric measurements $\boldsymbol{y}$. Furthermore, about 1.65×10^5 independent simulations have been used to monitor the generalization ability of the NN during the training phase (validation set) so as to make sure that no overtraining was taking place, and an equal amount of simulations have been used to validate the NN retrievals after the training phase was terminated (test set).

Prior to their presentation to the NN, the multispectral and multi-angular reflectances and DLPs have been preprocessed using a linear principal component analysis (PCA), with the twofold goal of compressing and denoising the measurements. A method for processing multi-angular spectropolarimetric data using PCA prior to the presentation to a NN has been previously reported by Tsekeri et al. (2011), who separately applied the PCA to noise-free reflectances and DLPs, and selected the number of retained principal components (PCs) based on the percentage of explained variance. In this work, instead, an approach similar to that devised by Aires et al. (2002) for the processing of satellite infrared radiances was used. Specifically, the PCA was carried out on a set of simulated noise-free measurements in order to derive the transformation matrix to be applied to the measurement data. Then, a set of noisy simulated measurements was compressed and decompressed using the previously derived matrix, the reconstruction error was monitored as a function of the number of retained PCs versus the corresponding noise-free radiances, and the number of retained PCs corresponding to the minimum reconstruction root-mean-square error (RMSE) was finally chosen.

In mathematical terms, let $\boldsymbol{y}$ be a generic vector of noise-free measurements (e.g. reflectance or DLP at a number of wavelengths and angles), let $\widetilde{\boldsymbol{y}}$ be a noisy version of $\boldsymbol{y}$, and let $\mathbf{P}$ be the matrix obtained applying the PCA to $\boldsymbol{y}$. The columns of such a matrix contain an orthonormal basis of eigenvectors of $\mathbf{S}_y$, the covariance matrix of $\boldsymbol{y}$. If the truncation of $\mathbf{P}$ obtained by retaining the eigenvectors associated to the n largest eigenvalues of $\mathbf{S}_y$ is denoted by $\mathbf{P}_n$, it is possible to write the reconstructed vector of noisy measurements as $\widetilde{\boldsymbol{y}}_n^r = \mathbf{P}_n^T \mathbf{P}_n \widetilde{\boldsymbol{y}}$. Then, the error $E_n = \|\widetilde{\boldsymbol{y}}_n^r - \boldsymbol{y}\|$ over the whole data set is monitored as a function of n, and the n that minimizes E_n is chosen.

This pre-processing scheme, separately applied to the reflectances and the DLPs, finally led to 7 PCs being retained for the reflectance and 10 PCs for the DLP.

In Fig. 1, the reconstruction RMSE is plotted as a function of the number of retained PCs for the radiances and the DLPs. As can be seen from the figure, the RMSE for the reflectance initially has a steep decrease for an increasing number of PCs until a number of 6–7 PCs (the latter being the optimal number according to the results) is reached. Then the RMSE starts to slowly increase again. A similar behaviour is observed for the DLP, with the difference that the RMSE decreases until 10 PCs are used. The increase in the reconstruction RMSE for a number of PCs higher than the optimal one indicates that the least significant PCs (11 for the reflectance, 8 for the DLP) contribute more noise than signal in the reconstruction of the observed quantities, and should therefore be discarded. Since the signal reconstructed with all the PCs (in our case, 18) is the same as the original noisy signal, it can be argued that the signal reconstructed with 7 (for the reflectance) or 10 (for the DLP) PCs is on average closer to the noise-free one than is the original noisy signal, and thus that the PCA has a beneficial denoising effect on the data. Similar results, and also similar plots, have been shown by other authors who have applied a similar type of processing to infrared hyperspectral data (see, for example, Antonelli et al., 2004).

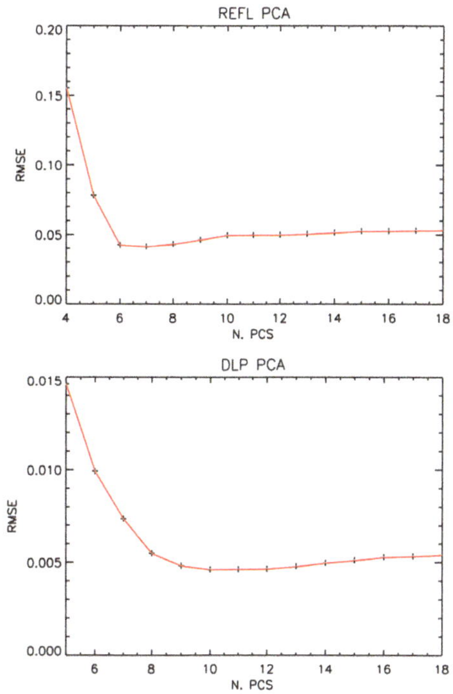

Figure 1. Reconstruction error versus number of retained principal components for the reflectances (above) and the degrees of linear polarization (below).

An important point is that the applicability of our method to real data relies on the assumptions that (i) the sample of simulated data used for the computation of the covariance matrix for the PCA is representative enough of the natural variability of the measured quantities and (ii) a realistic noise model is used when the denoising performance of the algorithm is evaluated. As will be seen in Sect. 5.2, relying on these assumptions did not seem to significantly impair the performance of our retrieval method on real data.

The settings adopted in the generation of the training data set for the NN, as well as the NN input and output variables, are summarized in Table 2, and the range assumed for each of the parameters to be retrieved is given in Table 3, together with the mean and the standard deviation of each parameter. The solar zenith angle (SZA) and the surface pressure have been used as additional input variables for the NN, in order to account for the dependence of the simulated reflectances and DLPs on these parameters. The NN has been trained using an online error backpropagation

Table 2. Simulation settings and list of the quantities used as input and output variables in the design of the aerosol retrieval neural network.

Settings	Wavelengths for reflectance and DLP: 440, 675, 870 nm Viewing zenith angles: 0, 6, 20, 32, 46, 60° Relative azimuth angle: 180° Assumed aerosol size distribution: bimodal log-normal Assumed effective variance: 0.2 μm^2 for both modes Assumed fraction of spherical particles: 100 % for fine mode, 10 % for coarse mode Assumed aerosol vertical distribution: Gaussian with peak at 1 km and FWHM* = 2 km Assumed surface type: green grass Assumed surface albedo at 440 and 675 nm: 0.05 Log-reflectance standard error: 0.026 at 440 nm, 0.046 at 675 nm, 0.076 at 870 nm DLP standard error: 0.004 at 440 nm, 0.006 at 675 and 870 nm
Inputs	Reflectance (7 principal components) DLP (10 principal components) Solar zenith angle Surface pressure
Outputs	Fine-mode effective radius ($r_{\mathrm{eff}}^{\mathrm{f}}$) Fine-mode refr. index real part ($m_{\mathrm{r}}^{\mathrm{f}}$) Fine-mode refr. index imaginary part ($m_{\mathrm{i}}^{\mathrm{f}}$) Fine-mode AOT at 550 nm ($\tau_{\mathrm{aer}}^{\mathrm{f}}$) Coarse-mode effective radius ($r_{\mathrm{eff}}^{\mathrm{c}}$) Coarse-mode refr. index real part (m_{r}^{c}) Coarse-mode refr. index imaginary part ($m_{\mathrm{i}}^{\mathrm{c}}$) Coarse-mode AOT at 550 nm ($\tau_{\mathrm{aer}}^{\mathrm{c}}$) Surface albedo at 870 nm (a_{s})

* Full width at half maximum

algorithm (Rumelhart et al., 1986) accelerated by means of a learning rate annealing technique (Bös and Amari, 1999) that consists in reducing the learning rate of the NN while the training proceeds. The choice of the online backpropagation learning instead of more sophisticated algorithms like scaling conjugate gradient (Møller, 1993) or Levenberg–Marquardt (Hagan and Menhaj, 1994) – which are reported to be, in principle, faster and more stable than the online backpropagation – is motivated by the fact that these latter algorithms require that the entire training set be presented to the NN before each update of the network weights. This can cause slow training when the training set is very large (Wilson and Martinez, 2000), as is the case in this work. The training process has been regularized by adding Gaussian noise to the simulated measurements during the training phase. This technique prevents the NN from yielding unreasonable responses when stimulated by noisy inputs (Holmström and Koistinen, 1992; Bishop, 1995b), and is therefore of critical importance for the applicability of a NN to real-world data.

As pointed out in Table 2, the NN designed following the method described above has an input layer of 19 neurons (7 reflectance PCs and 10 DLP PCs, plus SZA and surface pressure) and an output layer of 9 neurons (8 aerosol parameters plus the surface albedo at 870 nm). The number of hidden layers and neurons was selected by comparing a number of different architectures and selecting the architecture that gave the lowest RMSE on the validation set. The chosen architecture consists of 3 hidden layers with 40 neurons per layer. The hidden neurons have a hyperbolic tangent activation function (the reader may refer to Appendix A for the meaning of the term "activation function"), whereas the activation function of the output neurons is linear.

At a first glance, the result found for the number of hidden layers may seem surprising, given that NNs with more than 2 hidden layers are rarely reported in the literature, at least in remote sensing applications (in other fields of application, "deeper" NNs are less uncommon; see, for example, LeCun et al., 1989, for an interesting application of a MLP with three hidden layers). The most often reported reason why simpler NN architectures are usually chosen is perhaps that more complex architectures (i.e. NNs with more hidden layers and neurons) are more prone to overfitting, i.e. a situation where the training errors are small but poor performance is obtained for cases that are not included in the training set. This is, in principle, true, but it is important to keep in mind that the relationship between complexity and generalization also depends on the complexity of the function to be approximated and on the number of training data (the more training data, the lower the risk of overfitting for a given architecture). The fact that the NN described in this paper has been trained with a very large number of data, together with the fact that the backpropagation algorithm seems to have a bias towards

Table 3. Minimum, maximum, mean and standard deviation of the aerosol parameters used to generate the synthetic data set for this study. Please refer to Table 2 for the meaning of the symbols in the first column.

Parameter	Min	Max	Mean	SD
$r_{\text{eff}}^{\text{f}}$ (μm)	0.020	0.620	0.320	0.175
m_{r}^{f}	1.330	1.590	1.460	0.076
m_{i}^{f}	0.001	0.410	0.080	0.128
$\tau_{\text{aer}}^{\text{f}}$	0.000	3.000	0.928	0.922
$r_{\text{eff}}^{\text{c}}$ (μm)	0.800	4.000	2.400	0.980
m_{r}^{c}	1.330	1.590	1.460	0.076
m_{i}^{c}	0.001	0.410	0.080	0.128
$\tau_{\text{aer}}^{\text{c}}$	0.000	2.000	0.598	0.587
a_{s}	0.350	0.500	0.425	0.048

smooth functions even when the NN has many parameters (Lawrence et al., 1997; Caruana et al., 2001), makes the results found for the number of hidden layers more plausible than it may be intuitively expected.

Table 4 summarizes a number of error statistics used to evaluate the performance of the NN on the training, validation and test set. The RMSE, the mean absolute error (MAE), the mean bias and the Pearson correlation coefficient ρ between retrieved and actual aerosol parameters are reported for each aerosol parameter. The effective radius and the complex refractive index for each mode are provided only if the retrieved AOT of the considered mode is larger than 0.05. It can be seen that no significant differences are present in the performance of the NN on the three sets. This indicates that the NN can be reliably applied to data that are not included in the training set.

5 Use of the neural network retrievals as first guess in an iterative scheme

A further step in this study has been the use of the aerosol parameters retrieved by the NN as first guess for the variational retrieval scheme described in Sect. 2, as a replacement for the LUT. In principle, there are at least two reasons why it might be worthwhile applying a variational retrieval after the NN instead of using the NN as a stand-alone method. First, since the NN is trained using a cost function that encompasses a wide range of physical situations, its estimate might not be the optimal one for individual cases, and the application of a variational retrieval might refine the NN output by defining a case-specific cost function. Second, although some methods exist for characterizing the uncertainty associated with the output of a NN (see, for example, Dybowski and Roberts, 2001, for a general review, and Ristovski et al., 2012, for an application to aerosol retrievals), it is still not straightforward to associate the most widespread diagnostics (averaging kernels, full covariance matrix of the retrieval, degrees of freedom for signal) to NN retrievals, whereas such diagnostics can be provided using a variational retrieval scheme. Furthermore, the variational approach provides a means for checking the credibility of each individual retrieval using the χ^2 as a diagnostic.

The drawback of the proposed approach is that the advantage of a superior speed offered by a NN retrieval is lost because of the need for multiple calls to a radiative transfer model. This drawback might be overcome, in principle, by training a second NN to approximate the radiative transfer model (Schiller, 2007), but the accuracy requirements for such a network would be stringent and difficult to meet for a forward model with many independent input parameters.

In order to assess the impact of using the NN as first guess in the variational retrieval algorithm described in Sect. 2, the scheme with a NN-based first guess (hereinafter named NN + PT, where PT stands for Phillips–Tikhonov) has been compared to the scheme with a LUT-based first guess (hereinafter named LUT + PT) on two test data sets, one consisting of simulated data and one consisting of real observations. The two data sets, as well as the results of the comparisons, are described in the following subsections.

5.1 Application to simulated data

The synthetic data set chosen for the comparisons consists of 5000 combinations of the aerosol parameters, randomly drawn from the data set used to test the NN after the training process. The only reason why only a subset of the test data set has been used to compare the two retrieval algorithms, instead of the whole data set, is that running both the retrieval schemes on 1.65×10^5 data would have required much more time, while randomly selecting a few thousands cases is expected to be enough in order to draw reliable conclusions about the performance of the two retrieval schemes.

The combinations of aerosol parameters to be retrieved have been used as input for new radiative transfer simulations, aimed at creating a "pseudo-reality" in which the retrieval algorithm has to operate. In order to obtain error statistics that are as indicative as possible of those that can be obtained in an operational scenario, the values of some of the non-retrieved aerosol parameters (effective variance, fraction of spherical particles of the coarse mode, aerosol layer height) have been randomly perturbed with respect to those assumed in the training of the NN and in the retrieval algorithm by replacing the assumed values with values drawn from a uniform distribution (between 0.1 and 0.3 for the effective variance of both modes, between 0.03 and 0.06 for the surface albedo at 440 and 675 nm, between 0 and 1 for the fraction of spherical particles of the coarse mode).

The results of the comparison between the two retrieval schemes are summarized in Table 5. The variational retrievals have been assumed to have converged if the final value of the goodness of fit (χ^2 as defined in Eq. 4) was less than 2 within a maximum of 20 iterations. Again, the effec-

Table 4. Root-mean-square error (RMSE), mean absolute error (MAE), mean bias and Pearson correlation coefficient (ρ) between the retrieved and the actual aerosol parameters, computed on the training, test and validation set. Please refer to Table 2 for the meaning of the symbols in the first column.

	Training set				Validation set				Test set			
Variable	RMSE	MAE	Bias	ρ	RMSE	MAE	Bias	ρ	RMSE	MAE	Bias	ρ
r_{eff}^{f}	0.061	0.038	−0.012	0.940	0.061	0.038	−0.012	0.939	0.061	0.038	−0.012	0.939
m_{r}^{f}	0.046	0.035	0.011	0.806	0.046	0.034	0.011	0.806	0.046	0.035	0.011	0.804
m_{i}^{f}	0.048	0.020	0.004	0.930	0.048	0.020	0.004	0.930	0.048	0.020	0.004	0.931
τ_{aer}^{f}	0.166	0.105	−0.004	0.984	0.167	0.105	−0.004	0.983	0.168	0.106	−0.005	0.983
r_{eff}^{c}	0.752	0.569	−0.130	0.654	0.751	0.570	−0.126	0.654	0.754	0.571	−0.130	0.651
m_{r}^{c}	0.062	0.050	−0.001	0.569	0.062	0.050	−0.001	0.566	0.062	0.050	−0.001	0.568
m_{i}^{c}	0.085	0.038	−0.020	0.772	0.085	0.038	−0.020	0.768	0.086	0.038	−0.021	0.770
τ_{aer}^{c}	0.166	0.106	0.011	0.960	0.165	0.106	0.010	0.960	0.166	0.106	0.011	0.959
a_{s}	0.036	0.029	−0.005	0.665	0.036	0.029	−0.005	0.664	0.036	0.029	−0.005	0.663

tive radius and the complex refractive index for each mode are included in the statistics only if the retrieved AOT of the considered mode is larger than 0.05. It is clear from Table 5 that the fraction of converging retrievals for the NN + PT algorithm was considerably higher than that achieved by the LUT + PT scheme (slightly higher than 55 % for the latter, slightly lower than 95 % for the former). The second and third panel of Table 5 summarize a number of error statistics computed for the two retrieval schemes on the benchmark data set. In particular, the second panel of Table 5 reports RMSE, MAE, bias and Pearson correlation coefficient for the estimates produced by the LUT alone and at the end of the LUT + PT scheme, taking only the converging LUT + PT retrievals into account. The same is done in the third panel for the NN and the NN + PT retrievals. A slight difference between the LUT and the NN first guess is that the LUT does not fit the surface albedo at 870 nm, for which a prior of 0.4 is always assumed. However, this difference does not compromise the comparison between the two methods, because the iterative retrieval is in most cases able to recover from a wrong initial value for this parameter, as shown in Table 5.

A number of elements are noteworthy in the observed results.

1. By comparing the RMSE or the MAE of the LUT + PT to that of the NN + PT scheme (upper and lower right panels in Table 5), it is possible to see that the NN + PT scheme seems to outperform the LUT + PT scheme in the estimation of all the aerosol parameters. This is especially true for the fine-mode effective radius and refractive index (real and imaginary part), and for the imaginary part of the coarse-mode refractive index. For these variables, the decrease in the RMSE is also accompanied by a substantial rise in the correlation coefficient. The improvement in the fine-mode AOT, instead, is not very big. This is probably because the relationship between the measurements and this parameter is closer to a linear one, so that the iterative method is less dependent on the first guess for this particular variable.

2. The differences seem to a large extent attributable to the fact that the quality of the first guess produced by the NN is much better than that of the first guess produced by the LUT. This applies to all the retrieved quantities. It can be seen that the NN retrievals alone already seem to perform better than the full LUT + PT retrievals.

3. For both algorithms, the coarse-mode effective radius and real part of refractive index seem the most difficult parameters to retrieve. Even though an improvement is noticed by replacing the LUT with a NN, the errors are quite large and the correlation coefficients are below 0.50 for the effective radius and even below 0.45 for the real part of the refractive index. This behaviour is partly explained by the uncertainty in the fraction of spherical coarse-mode particles, which is assumed to be constant in the retrieval algorithm but is allowed to vary in the pseudo-reality. Furthermore, the absence of viewing angles near the forward scattering direction reduces the sensitivity to coarse-mode aerosols, because the region of enhanced intensity around such a direction – which is a distinctive feature of large particles – is not sampled.

4. By comparing the error statistics of the NN retrievals to those of the NN + PT retrievals (third panel in Table 5), no improvement is observed in many of the retrieved parameters after the application of the iterative retrieval to the NN outputs. The only parameters that improve are the fine-mode AOT and the surface albedo at 870 nm. In Fig. 2 the histograms of the absolute errors in the retrieved parameters for the NN alone and for the NN + PT schemes are shown. It is clear that the iterative retrieval generally improves the estimates of the surface albedo and of the fine- and coarse-mode AOTs with respect to the NN alone, whereas the improvement

Table 5. Root-mean-square error (RMSE), mean absolute error (MAE), mean bias and Pearson correlation coefficient (ρ) between retrieved and actual aerosol parameters. Please refer to Table 2 for the meaning of the symbols in the first column.

Method	N converging retrievals (fraction)
LUT + Phillips–Tikhonov	2756 (55.12 %)
NN + Phillips–Tikhonov	4321 (93.60 %)

	LUT only				LUT + Phillips–Tikhonov			
Variable	RMSE	MAE	Bias	ρ	RMSE	MAE	Bias	ρ
r_{eff}^{f}	0.147	0.109	−0.088	0.676	0.109	0.066	−0.032	0.752
m_{r}^{f}	0.134	0.113	0.094	0.158	0.085	0.062	0.028	0.530
m_{i}^{f}	0.097	0.045	−0.023	0.391	0.075	0.034	−0.012	0.687
τ_{aer}^{f}	0.405	0.272	−0.112	0.897	0.290	0.161	−0.034	0.946
r_{eff}^{c}	1.289	1.032	−0.599	0.142	1.430	1.024	−0.422	0.192
m_{r}^{c}	0.090	0.073	−0.012	0.100	0.095	0.074	0.002	0.320
m_{i}^{c}	0.118	0.065	−0.026	0.168	0.109	0.051	−0.029	0.444
τ_{aer}^{c}	0.559	0.333	0.019	0.581	0.356	0.202	0.017	0.784
a_{s}	0.054	0.042	−0.025	0.000	0.044	0.022	−0.001	0.727

	NN only				NN + Phillips–Tikhonov			
Variable	RMSE	MAE	Bias	ρ	RMSE	MAE	Bias	ρ
r_{eff}^{f}	0.077	0.048	0.000	0.895	0.078	0.044	0.008	0.896
m_{r}^{f}	0.052	0.038	0.010	0.742	0.059	0.041	0.010	0.707
m_{i}^{f}	0.060	0.023	−0.003	0.883	0.063	0.025	−0.007	0.869
τ_{aer}^{f}	0.226	0.144	0.027	0.971	0.210	0.117	0.026	0.975
r_{eff}^{c}	0.882	0.688	−0.218	0.505	0.948	0.723	−0.206	0.474
m_{r}^{c}	0.066	0.054	0.001	0.506	0.079	0.061	0.005	0.448
m_{i}^{c}	0.090	0.043	−0.019	0.735	0.089	0.042	−0.014	0.733
τ_{aer}^{c}	0.231	0.146	−0.011	0.922	0.229	0.129	−0.033	0.926
a_{s}	0.045	0.036	0.004	0.477	0.033	0.016	−0.002	0.816

is more limited in the effective radius and in the complex refractive index of both modes.

These results may seem to advocate the use of the NN as a stand-alone scheme without applying the iterative method afterwards. However, such a solution would result in a less complete error analysis, and in the impossibility of checking the plausibility of individual retrievals (see discussion at the beginning of Sect. 5).

A word of caution is necessary in the interpretation of Table 5, because there are some parameters (fine- and coarse-mode effective radius, coarse-mode imaginary part of refractive index) whose mean values in the training data set are close to the maximum values stored in the LUT or much larger. As a result, the comparison between the NN and the LUT alone may be unfair towards the LUT, since the error of the LUT retrieval will necessarily be large when the true value of one of these parameters is far larger than the LUT maximum. To some extent, this unfairness may also affect the comparison between the full retrievals, because in cases like those described above it may be expected that the first guess will be far from the true value and the convergence to such value will be difficult to achieve. In order to assess to what extent this issue affects our findings, we compared the LUT + PT and the NN + PT schemes on a subset of cases where all the true values of the aerosol parameters were within the LUT limits. The number of such cases was 774. On this subset, the fraction of successful retrievals was around 59 % for the LUT + PT scheme and around 88 % for the NN + PT scheme. Therefore, also looking at this subset, the main point of this comparison – i.e. that using the NN as first guess for the iterative retrieval improves the number of successful retrievals – remains largely unaffected. On the reduced data set, the accuracy of the LUT + PT retrievals was found comparable to that of the NN + PT retrievals for the fine-mode effective radius, whereas the NN + PT scheme still performed better than the LUT + PT scheme on all the other parameters.

A comparison between the LUT + PT and the NN + PT scheme in terms of speed is reported in Fig. 3, where histograms of the number of iterations needed to converge and of the total retrieval time are shown. It can be seen that the Phillips–Tikhonov scheme takes on average two iterations less to converge when it is initialized using the NN as a first guess than it does when the LUT is used as a first guess.

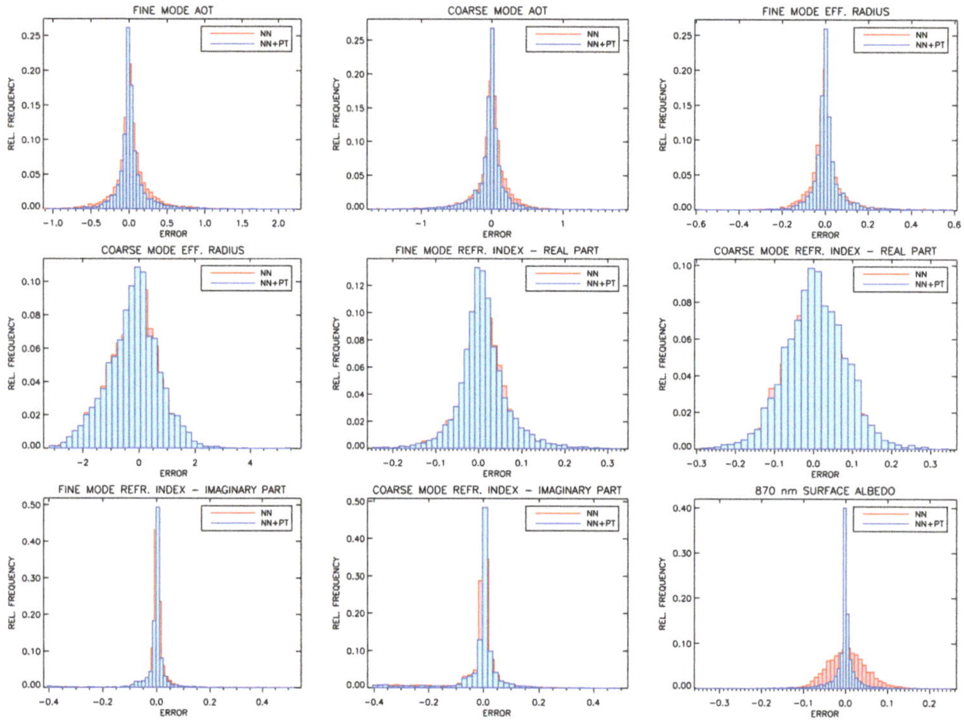

Figure 2. Histograms of the absolute errors in the retrieved parameters over the benchmark synthetic data set for the NN-only retrievals (red) and for the Phillips–Tikhonov retrievals with the NN-based first guess (blue).

The average difference in the total retrieval time is about 7 s. This difference is partly explained by the smaller number of iterations and partly by the time needed to generate the first guess. The generation of the first guess (histograms not shown) takes about 2 s when the LUT is used, and about 0.2 s when the NN is used. These 0.2 s are dominated by the radiative transfer simulation performed to evaluate the fit between the NN retrieval and the measurements, as the NN retrieval alone only takes about 4 ms. The total computation time, which is on average around 15 s for the NN + PT scheme, can be significantly reduced if the numerical computation of the Jacobians of the forward model is replaced by an analytical one. In the satellite retrieval algorithm by Hasekamp et al. (2011), where this has been done, the total retrieval time is of about 2 s, and replacing the LUT with a NN may result in a further speed-up.

5.2 Application to real measurements

The measurements used in this study have been performed using the groundSPEX instrument between 7 and 9 July 2013 and on 5 September 2013 at the Cabauw Experimental Site for Atmospheric Research (CESAR Observatory, http://www.cesar-observatory.nl), a few tens of kilometres southwest of Utrecht, in the Netherlands (Monna and Bosveld, 2013). The measurements were made under clear-sky conditions. During the July measurements, an elevated smoke layer was present over the Netherlands as a consequence of air advection from North America (KNMI, 2013). The LUT + PT and the NN + PT retrieval schemes have been applied to this set of measurements, and the retrieved aerosol parameters have been compared to those estimated by the Aerosol Robotic Network (AERONET) station located at the CESAR site.

Given the unavailability of AERONET level 2 data for the considered dates at the time of writing, level 1.5 data have been used for the comparison. The total AOTs (fine plus coarse mode) and the Ångström exponents between 440 and 675 nm estimated by groundSPEX using the LUT + PT and the NN + PT schemes have been compared to those provided by the AERONET direct sun product (Holben et al., 1998). The fine- and coarse-mode effective radii and complex refractive indices have been compared to those provided by the AERONET product based on the almucantar algorithm described by Dubovik and King (2000). As for the complex refractive index, the almucantar product provides a single value that applies to both modes and is spectrally dependent, while our algorithm provides a spectrally independent value for each mode. Therefore, prior to the comparison, the AERONET refractive indices have been averaged with respect to the wavelength, and the fine- and coarse-mode refractive indices estimated by groundSPEX have been combined in a weighted average using the AOTs of the two modes as weights.

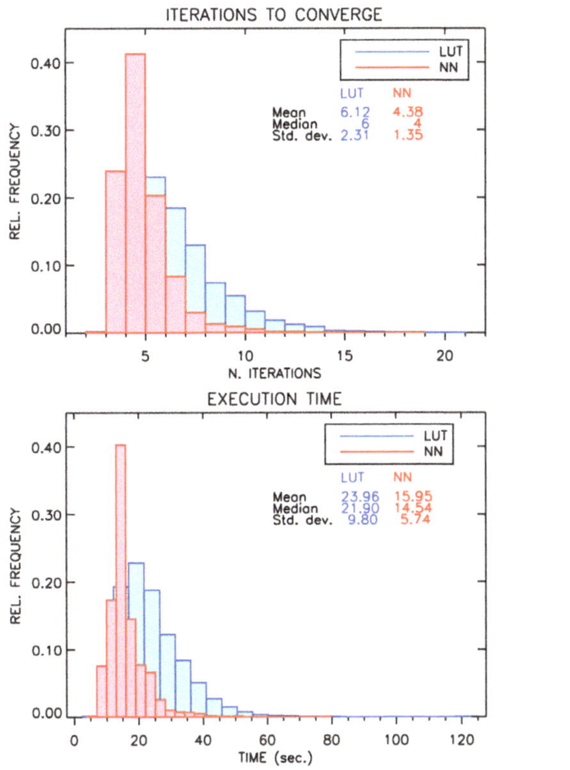

Figure 3. Histograms of the number of iterations before convergence (above) and of the total retrieval time (below) for the Phillips–Tikhonov retrievals using the LUT first guess (blue) and the NN first guess (red).

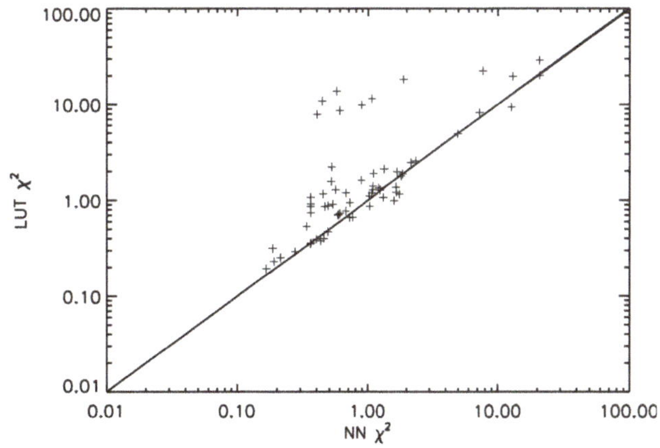

Figure 4. Plot of the final χ^2 obtained using the LUT + PT scheme (ordinate) versus that obtained using the NN + PT scheme (abscissa) for the considered groundSPEX measurements.

While AERONET total AOTs and Ångström exponents obtained from direct sun measurements are intrinsically robust reference values, AERONET almucantar retrievals of the aerosol microphysical parameters are often affected by non-negligible uncertainties (Dubovik et al., 2000). As a consequence, it is important to bear in mind that the comparison between groundSPEX and AERONET microphysical parameters should not be regarded as a validation of groundSPEX against a completely established truth. The meaning of this intercomparison is rather that of a consistency check between retrievals of a similar set of parameters made by means of two profoundly different methodologies. Because of this, cases of consistency between the retrievals reinforce our confidence in both the retrieval approaches, whereas cases of inconsistency do not necessarily imply that groundSPEX is performing worse than AERONET.

In total, 71 measurements were carried out by groundSPEX during the considered period. While this is by no means an extensive benchmark data set, it is useful in order to have at least a preliminary idea of the applicability of the NN – trained on synthetic data – to a real-world scenario. Out of the attempted retrievals, 48 of the LUT + PT retrievals satisfied the χ^2 criterion mentioned in Section 5.1, whereas the number of successful NN + PT retrievals was 59. In Fig. 4 the final χ^2 obtained by the LUT + PT scheme is plotted against that obtained by the NN + PT scheme applied to the considered set of groundSPEX measurements. Points that are close to the 1 : 1 line indicate cases where the two algorithms performed similarly. Points that are well above the 1 : 1 line indicate cases where the NN + PT scheme performed better than the LUT + PT (i.e. LUT + PT achieved a higher χ^2 than NN + PT), whereas points located well below the line represent cases where LUT + PT performed better than NN + PT. From this plot, it is straightforward to note that the number of points above the identity line is considerably larger than the number of points located below the line. This means that, besides yielding a higher number of successful retrievals, the NN + PT algorithm also has the tendency to produce final solutions that fit the measured reflectances and DLPs better than those obtained using the LUT + PT scheme.

Figures 5, 6 and 7 respectively show time series of total AOTs, fine- and coarse-mode effective radii and AOT-weighted refractive indices retrieved by the two algorithms and those estimated by AERONET during the 4 days considered in this study. Superimposed on the retrieved parameters are error bars derived by error propagation using the formalism of Rodgers (1990). According to this formalism, the retrieval error can be decomposed in a contribution due to random noise and a contribution due to the a priori error (called null space error). The null space error depends on the assumed a priori error. As explained in Sect. 2, our retrieval scheme uses the first guess produced by the LUT or by the NN as a priori. However, a method to derive the uncertainty associated with the LUT or the NN retrievals alone has not been implemented yet; therefore the a priori error is set to the reciprocal of the half width of the physical variation range assumed for each parameter. As a consequence, it is likely that the width of the error bars shown in this paper is overestimated, especially in the case of the NN retrievals. The error

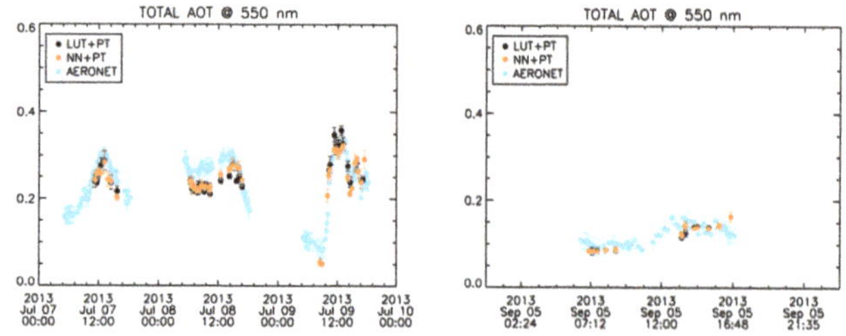

Figure 5. Time series of the total AOT retrieved by the LUT + PT scheme (black circles), the NN + PT scheme (orange circles), and the AERONET direct sun level 1.5 product (sky-blue circles): between 7 and 9 July 2013 (left) and on 5 September 2013 (right).

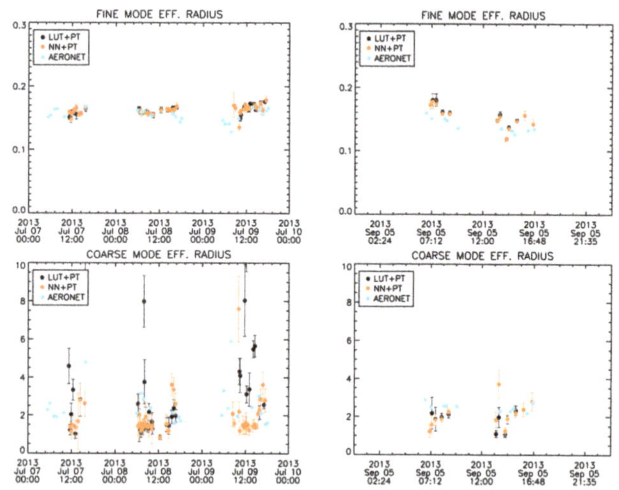

Figure 6. Time series of the fine- (above) and coarse-mode (below) effective radius retrieved by the LUT + PT scheme (black circles), the NN + PT scheme (orange circles), and the AERONET almucantar level 1.5 product (sky-blue circles): between 7 and 9 July 2013 (left) and on 5 September 2013 (right).

bars for the AERONET retrievals are the same as those used by van Harten et al. (2014).

For both algorithms, the agreement seems generally good for the AOTs (absolute deviation 95th percentile around 0.05). During the 3 days in July, the same applies to the fine-mode effective radii (absolute deviation 95th percentile around 0.02 μm) and to the real part of the AOT-weighted refractive indices (groundSPEX–AERONET differences well within the estimated AERONET error bars). On 5 September, the agreement for these parameters seems less good. It must be noted, however, that the lower AOTs that were generally observed on this latter date make the retrieval of the aerosol microphysical parameters more difficult for both groundSPEX and AERONET. The NN + PT retrievals also seem more capable than the LUT + PT retrievals of reproducing the temporal variation of the imaginary part of the refractive index, although the agreement between the values varies from day to day. This is particularly evident during 9 July (Fig. 7, lower left panel), when the

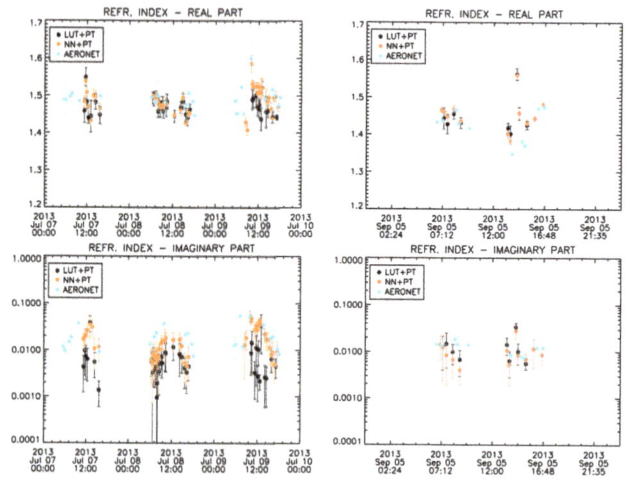

Figure 7. Time series of the real (above) and imaginary part (below) of the (AOT-weighted) refractive index retrieved by the LUT + PT scheme (black circles), the NN + PT scheme (orange circles), and the AERONET almucantar level 1.5 product (sky-blue circles): between 7 and 9 July 2013 (left) and on 5 September 2013 (right).

temporal variation of the imaginary part of the refractive index retrieved by AERONET is followed very well by the groundSPEX retrievals based on the NN + PT scheme (again, from visual inspection of Fig. 7 it is clear that the differences between groundSPEX and AERONET are well within the AERONET error bars), whereas the estimates produced using the LUT + PT scheme are more distant from the AERONET values.

A closer look at the total AOTs shown in Fig. 5 seems to indicate that a slight underestimation tendency exists in both retrieval schemes. In order to explain this fact, it might be interesting to compare the fine- and coarse-mode AOTs retrieved by groundSPEX to those provided by the AERONET product based on the spectral deconvolution algorithm (SDA) by O'Neill et al. (2003). This is done in Fig. 8. In the interpretation of this comparison, it must be kept in mind that the definition of the aerosol fine and coarse modes in our algorithm is different from that used in the SDA product, owing to differences in the assumptions made about the aerosol size distribution. Despite the fact that the groundSPEX and the SDA optical thicknesses are not rigorously comparable for the reason mentioned above, it is interesting to note that an excellent agreement exists between the groundSPEX and the SDA fine-mode AOTs, with the NN + PT that follows the diurnal cycle of the fine-mode AOT particularly well (absolute deviation 95th percentile around 0.06, but 85th percentile smaller than 0.03). Instead, the coarse-mode AOTs estimated by the two groundSPEX retrieval methods are almost always lower than those provided by the AERONET SDA product. This may suggest that the underestimation tendency in the total AOT could be, in some way, attributable to the coarse mode. However, the different definitions of the aerosol modes in

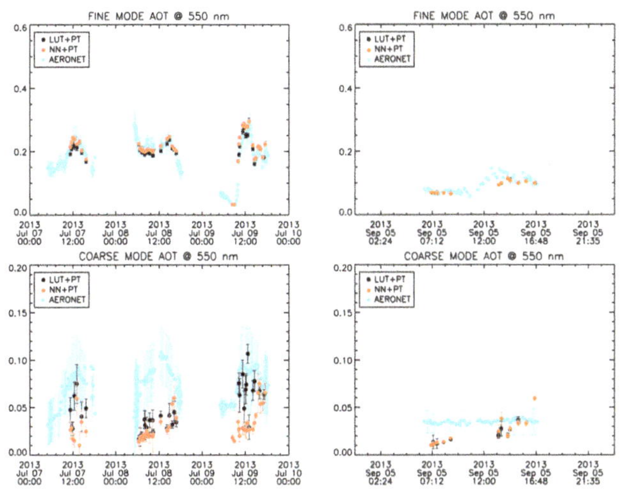

Figure 8. Time series of the fine- (above) and coarse-mode (below) AOT retrieved by the LUT + PT scheme (black circles), the NN + PT scheme (orange circles), and the AERONET SDA level 1.5 product (sky-blue circles): between 7 and 9 July 2013 (left) and on 5 September 2013 (right).

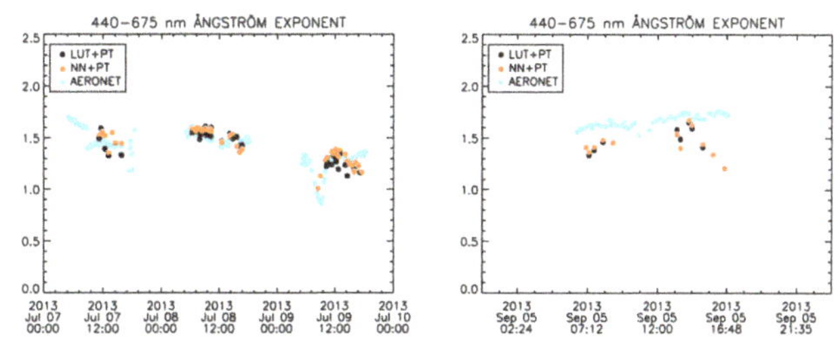

Figure 9. Time series of the 440–670 nm Ångström exponent retrieved by the LUT + PT scheme (black circles), the NN + PT scheme (orange circles), and the AERONET direct sun level 1.5 product (sky-blue circles): between 7 and 9 July 2013 (left) and on 5 September 2013 (right).

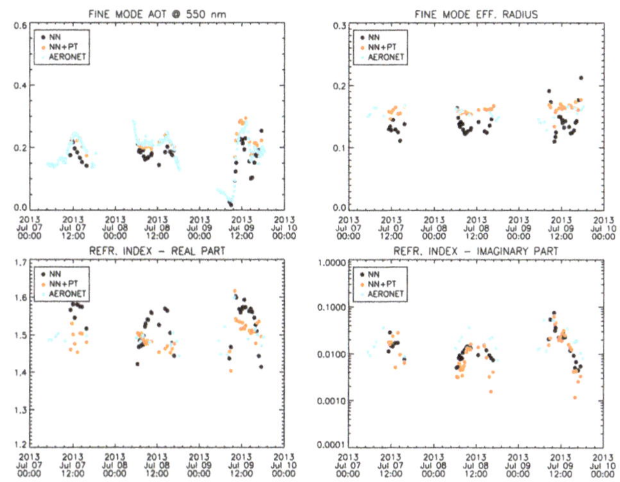

Figure 10. Fine-mode AOT (upper left) and effective radius (upper right) and AOT-weighted real (lower left) and imaginary part (lower right) of the refractive index estimated using the NN alone (black circles), the NN + PT scheme (orange circles), and reference values from AERONET (sky-blue circles).

the two algorithms, as well as the lack of a more appropriate benchmark for the groundSPEX fine- and coarse-mode AOTs, do not allow us to provide stronger evidence for this statement.

Another parameter that is also indicative of the aerosol particle size and is less dependent on the mode definition is the Ångström exponent. In Fig. 9 the Ångström exponents between 440 and 670 nm retrieved by the two groundSPEX algorithms are compared to those measured by AERONET. The agreement seems generally good during July, whereas on 5 September groundSPEX systematically estimates smaller Ångström exponents than AERONET. This would indicate that groundSPEX sees larger particles than AERONET during 5 September. However, this conclusion does not seem very consistent with the results previously shown for the coarse-mode effective radius and the coarse-mode AOTs. In view of this, no conclusive interpretation of the differences between groundSPEX and AERONET with respect to the particle size can be given at this moment in cases of low AOT.

It is interesting to compare the time series of the aerosol parameters retrieved by the NN + PT scheme to those retrieved by the NN alone. Examples of such comparisons are shown in Fig. 10 for the fine-mode AOT and effective radius, as well as for the AOT-weighted refractive index, during July 2013. It appears that the application of the iterative retrieval after the NN estimate brings the retrieved values closer to the AERONET values for the fine-mode AOT and effective radius and for the real part of the refractive index. For the imaginary part of the refractive index, the updates brought by the Phillips–Tikhonov method to the NN retrievals do not seem to improve the accuracy of the final estimates significantly. On the one hand, this might be contradicting the results shown obtained with simulated data, but on the other hand it must be kept in mind that the range of situations sampled by the real measurements during July 2013 is much narrower than that encompassed by the simulations; therefore one should not necessarily expect consistent error statistics between the two cases. In any case, the results observed so far on real data seem to suggest that there is still merit in applying a full physics retrieval after the NN inversion, even at the cost of a longer processing time.

6 Conclusions

In this paper, the use of neural networks for the retrieval of aerosol properties from ground-based multi-angle measurements of intensity and polarization has been investigated. A neural network for the retrieval of aerosol properties from ground-based multi-angular spectropolarimetric data has been developed, and has been used as a replacement for a look-up table as an initial guess for an iterative aerosol retrieval algorithm. In order to assess the impact of the neural-network-based first guess, the iterative algorithm initialized using the neural first guess and its version initialized using the LUT-based first guess have been compared on a data set consisting of simulated measurements, and on a set of real observations carried out by the groundSPEX instrument.

The synthetic data set consisted of 5000 cases generated by simulating the reflectance and the degree of linear polarization that would be measured by a groundSPEX-like instrument for a wide range of combinations of the aerosol parameters. On this data set, the replacement of the LUT-based first guess with a NN-based one has led to a remarkable increase in the fraction of successful retrievals (i.e. retrievals that achieved a goodness of fit indicator smaller than 2 within 20 iterations). While approximately 55 % of the retrievals successfully converged in the case of the LUT first-guess algorithm, almost 95 % of the retrievals converged when the NN was used instead of the LUT as initial guess. Furthermore, the converging retrievals based on the NN first guess exhibited smaller errors than those based on the LUT first guess in the coarse-mode effective radius and refractive index. The application of the iterative algorithm after the NN retrieval improved the estimates of the fine- and coarse-mode AOT with respect to the NN-only retrieval. Instead, no significant improvements were noticed in the other aerosol parameters, for which the quality of the NN output alone was found to be comparable to that of the final retrieval.

Also, when the two algorithms were applied to real measurements carried out by the groundSPEX instrument at the CESAR atmospheric observatory, a higher number of successful retrievals was observed after replacing the LUT first guess with the NN first guess. The observed improvement in the number of converging retrievals was not as large as that observed with simulated data. This might be due to the fact that the groundSPEX measurements were very concentrated in time, with small intrinsic variations in the aerosol properties, whereas the synthetic data set samples a much broader range of situations, including cases where the first guess produced by the LUT might be much worse than that produced by the NN.

The comparison between the groundSPEX aerosol retrievals and the AERONET retrievals showed a good agreement in the total AOT (absolute differences most often below 0.05) and in the AOT-weighted refractive index. In particular, replacing the LUT with the NN led to a remarkable improvement in the retrieval of the imaginary part of the refractive index. A good agreement (absolute errors often below $0.02\,\mu$m) was found in the fine-mode effective radius during the 3 days in July 2013, whereas the agreement was worse on 5 September, probably because the smaller fine-mode AOT observed on that day made the retrieval of the aerosol microphysical properties more challenging for both groundSPEX and AERONET. The coarse-mode effective radius seems, in all the cases, the most difficult parameter to retrieve, and this seems to confirm the results obtained with simulated data. A negative bias in the total AOT was observed in the groundSPEX retrievals, independent of the first guess. Qualitative comparisons against the SDA fine- and coarse-mode AOTs seem to suggest that this bias could be attributed to the coarse-mode AOT. However, a rigorous quantification of the bias using the SDA product is not possible, because the definitions of fine and coarse mode made in this product are different from those used by the groundSPEX algorithm. Also, the analysis of the retrieved Ångström exponents, which are indicative of the particle size but are independent of the mode definitions, did not lead to conclusive results, because it seemed to suggest that groundSPEX overestimates the particle sizes with respect to AERONET, which is contrary to our previous findings.

A follow-up of this work will be the extension of the method to the case of downlooking observations, such as satellite or aircraft measurements. The main difficulty of such an approach lies in training a NN for an instrument whose set of viewing angles changes from pixel to pixel, as is the case for satellite instruments. In fact, the variability of the observed geometry makes the definition of a feature set to train a NN problematic. This is not a problem for ground-based observations, where the set of viewing geometries adopted by the instrument can always be programmed so as to satisfy the needs of the retrieval algorithm. Solutions aimed at overcoming this difficulty with satellite measurements are under study.

Appendix A: Neural networks: general concepts and terminology

This section summarizes some aspects of the NN theory and language that may be useful for the reader of this paper. For a more complete treatment, the reader can refer to Bishop (1995a) or Haykin (1999).

Given the large variety of neural network models, a general definition of a neural network is difficult to give. In very broad terms, an artificial neural network can be seen as a system or model that is able to change its internal parameters through an interaction with an external environment in order to perform a given task or to accomplish a prescribed behaviour, in a way that resembles the adaptation of the synaptic connections in a biologic neural system. The process of adapting the internal parameters of a NN is what is called "learning", while the role of the environment is played by the so-called "training set". The word "network" emphasizes that a NN is usually made of simple computational units interconnected between each other. Such computational units are called "neurons". A neuron usually applies very simple computations to its input (e.g. multiplying the input by a vector of weights, computing the distance between the input and another vector, computing a simple nonlinear function, etc., depending on the particular type of NN).

Learning in a NN can take place in several ways. In "supervised" learning, the NN is presented with a set of input-output pairs, and the NN parameters are changed based on a measure of the discrepancy between the NN response to an input and the desired output corresponding to that input. In "unsupervised" learning, the NN is only presented with a set of input vectors, and the goal is to differentiate the NN response based on the region of the input space the input vector belongs to, thereby obtaining a clustering of the input space that can be used, e.g. for classification of the input vectors. In "reinforcement" learning, the goal is to train the NN to perform a sequence of actions in response to a given input. Differently from what happens in the supervised learning, the parameters of the NN are not changed based on the correct output for a given input, but based on a "critic", which is a score used to evaluate the quality of the NN response to the considered input.

Since the NN described in this paper is a supervised one, it is worthwhile providing the reader with further information about supervised learning. As said above, in supervised learning the discrepancy between the NN response to an input vector and the desired response for that vector is used to adjust the parameters of the NN. In practice, a cost function Q, based on the aforementioned discrepancy, is minimized with respect to the vector $\boldsymbol{w}$ of the NN parameters. Since Q is, in all the practical cases, a nonlinear function, its minimization is carried out iteratively, and leads to an update of the NN parameters in the form

$$\boldsymbol{w}_{i+1} = \boldsymbol{w}_i + \alpha_i \boldsymbol{d}_i, \tag{A1}$$

with $\alpha_i \in \mathbb{R}$ and $\boldsymbol{d}_i$ a descent direction for Q in a neighbourhood of $\boldsymbol{w}_i$. In this case, there always exists an $\bar{\alpha}_i$, such that if $\alpha_i < \bar{\alpha}_i$, then $Q(\boldsymbol{w}_{i+1}) < Q(\boldsymbol{w}_i)$. The value of this $\bar{\alpha}_i$ depends on the behaviour of Q around $\boldsymbol{w}_i$, and is not known a priori unless second-order information about Q is known. For this reason, learning algorithms that do not use second-order information may occasionally diverge if a step size $\alpha_i > \bar{\alpha}_i$ is chosen.

An important distinction, recalled in Sect. 4, is made between "online" and "offline" (or "batch") learning algorithms. In online algorithms, the cost function is evaluated (and the NN parameters updated accordingly) after the presentation of each training input. The cost function has the form

$$Q(\boldsymbol{w}_i, \boldsymbol{x}_i) = \frac{1}{2}\left|\left|\boldsymbol{y}_i(\boldsymbol{x}_i, \boldsymbol{w}_i) - \boldsymbol{t}_i\right|\right|^2 + \gamma R(\cdot), \tag{A2}$$

where $\boldsymbol{y}_i$ is the NN response to $\boldsymbol{x}_i$ with parameter vector $\boldsymbol{w}_i$, $\gamma \in \mathbb{R}$ and R is a regularization function that might depend, in principle, on the previous values assumed by the vector of parameters, or on higher order derivatives of the NN output with respect to the input or to the parameter vector. In offline algorithms, the update of the NN parameters takes place only once the entire training set has been presented to the network. The cost function has the form

$$Q(\boldsymbol{w}) = \frac{1}{2N}\sum_{i=1}^{N}\left|\left|\boldsymbol{y}_i(\boldsymbol{x}_i, \boldsymbol{w}_i) - \boldsymbol{t}_i\right|\right|^2 + \gamma R(\cdot), \tag{A3}$$

where N is the number of training data. It can be said that online learning algorithms realize a stochastic approximation of the gradient of Q, which individuates the descent direction. Offline algorithms usually have a more stable behaviour and a smoother convergence with respect to online algorithms, thanks to the better estimates of $\boldsymbol{d}$ they provide. This leads to a faster convergence provided that the training set is not so large. For large and redundant training sets, instead, online algorithms produce reliable estimates of $\boldsymbol{d}$, and may converge even faster than offline algorithms thanks to a much higher number of parameter updates per training cycle.

The type of supervised NN used in this paper is known as multilayer perceptron (MLP; Werbos, 1974). In this NN architecture, the neurons are organized in layers, with the outputs of the neurons in the ith layer being the inputs for those in the $(i+1)$th layer. The first layer of a MLP is called the "input layer", and does nothing except receive the input vector. The last layer is the "output layer", and is the layer that provides the ultimate NN output. Between the input and the output layer, one or more "hidden layers" are situated. Each neuron is characterized by a vector of weights $\boldsymbol{w}$ and a bias b, and its output y is related to its input $\boldsymbol{x}$ through the relationship

$$y = \varphi(\boldsymbol{w}^T \boldsymbol{x} + b), \tag{A4}$$

where φ is usually a C^∞ function called the "activation function" of the neuron. Hidden neurons always have nonlinear activation functions, the most popular being the hyperbolic tangent and the logistic function. Output neurons, instead, can either have a linear or a nonlinear activation function.

Several algorithms exist to train MLPs, either in online or in offline mode. The most famous online learning algorithm is the standard error backpropagation (Rumelhart et al., 1986), which is obtained by setting $\gamma = 0$ in Eq. (A2) and by setting $\boldsymbol{d}_i = -\nabla Q(\boldsymbol{w}_i, \boldsymbol{x}_i)$ and α_i to a fixed value η (called the "learning rate parameter") in Eq. (A1). By letting η decrease during the training of a MLP, the convergence of the standard backpropagation can be accelerated (Bös and Amari, 1999), and this solution has been used in this work (see Sect. 4). Among the most common offline algorithms are the batch backpropagation, obtained in the same way as the standard backpropagation but starting from Eq. (A3), and the scaled conjugate gradient (Møller, 1993) and Levenberg–Marquardt (Hagan and Menhaj, 1994) methods, which select the descent direction $\boldsymbol{d}_i$ and the step size α_i by approximating the Hessian matrix of Q at each iteration, thereby reducing the risk of generating a locally divergent parameter update. Offline algorithms tend to ensure a faster and more stable convergence than online algorithms provided that the training set is not extremely large. In the latter case, online algorithms may be preferable (Wilson and Martinez, 2000).

Acknowledgements. Antonio Di Noia was funded by the Principal Investigator Preparatory Programme (PIPP) of the Netherlands Space Office (NSO) under project PIPP 11-02.

The groundSPEX instrument was developed and built in the framework of the strategic research programme of the Dutch National Institute for Public Health and the Environment (RIVM).

The CESAR Observatory at Cabauw is operated by the Royal Netherlands Meteorological Institute (KNMI).

To ensure the high quality standards of the AERONET observations, funding is received from the European Union Seventh Framework Programme (FP7/2007-2013) under grant agreement no. 262254.

The anonymous reviewers who took part in the access review and in the discussion phase are gratefully acknowledged for their constructive and useful comments.

Edited by: A. Kokhanovsky

References

Aires, F., Rossow, W. B., Scott, N. A., and Chédin, A.: Remote sensing from the infrared atmospheric sounding interferometer instrument 1. Compression, denoising, and first-guess retrieval algorithms, J. Geophys. Res., 107, 4619, doi:10.1029/2001JD000955, 2002.

Alexander, D.: Volcanic ash in the atmosphere and risks for civil aviation: A study in European crisis management, Int. J. Disaster Risk Sci., 4, 9–19, doi:10.1007/s13753-013-0003-0, 2013.

Anderson, J. O., Thundiyil, J. G., and Stolbach, A.: Clearing the air: A review of the effects of particulate matter air pollution on human health, J. Med. Toxicol., 8, 166–175, doi:10.1007/s13181-011-0203-1, 2012.

Antonelli, P., Revercomb, H. E., Sromovsky, L. A., Smith, W. L., Knuteson, R. O., Tobin, D. C., Garcia, R. K., Howell, H. B., Huang, H.-L., and Best, F. A.: A principal component noise filter for high spectral resolution infrared measurements, J. Geophys. Res., 109, D23102, doi:10.1029/2004JD004862, 2004.

Bishop, C. M.: Neural Networks for Pattern Recognition, Oxford University Press, New York, NY, USA, 1995a.

Bishop, C. M.: Training with noise is equivalent to Tikhonov regularization, Neural Comput., 7, 108–116, doi:10.1162/neco.1995.7.1.108, 1995b.

Boesche, E., Stammes, P., Ruhtz, T., Preusker, R., and Fischer, J.: Effect of aerosol microphysical properties on polarization of skylight: sensitivity study and measurements, Appl. Optics, 45, 8790–8805, doi:10.1364/AO.45.008790, 2006.

Bös, S. and Amari, S.: Annealed online learning in multilayer neural networks, in: On-Line Learning in Neural Networks, edited by: Saad, D., Chap. 10, 209–229, Cambridge University Press, New York, NY, USA, doi:10.1017/CBO9780511569920.011, 1999.

Boucher, O., Randall, D., Artaxo, P., Bretherton, C., Feingold, G., Forster, P., Kerminen, V.-M., Kondo, Y., Liao, H., Lohmann, U., Rasch, P., Satheesh, S. K., Sherwood, S., Stevens, B., and Zhang, X. Y.: Clouds and aerosols, in: Climate Change 2013: The Physical Science Basis. Contribution of Working Group I to the Fifth Assessment Report of the Intergovernmental Panel on Climate Change, edited by: Stocker, T. F., Qin, D., Plattner, K.-F., Tignor, M., Allen, S. K., Boschung, J., Nauels, A., Xia, Y., Bex, V., and Midgley, P. M., Chap. 5, 571–657, Cambridge University Press, Cambridge, UK and New York, NY, USA, 2013.

Brajard, J., Jamet, C., Moulin, C., and Thiria, S.: Use of a neuro-variational inversion for retrieving oceanic and atmospheric constituents from satellite ocean colour sensor: Application to absorbing aerosols, Neural Networks, 19, 178–185, doi:10.1016/j.neunet.2006.01.015, 2006.

Brajard, J., Santer, R., Crépon, M., and Thiria, S.: Atmospheric correction of MERIS data for case-2 waters using a neuro-variational inversion, Remote Sens. Environ., 126, 51–61, doi:10.1016/j.rse.2012.07.004, 2012.

Caruana, R., Lawrence, S., and Giles, C. L.: Overfitting in neural nets: Backpropagation, conjugate gradient, and early stopping, in: Proceedings of Neural Information Processing Systems (NIPS) Conference 2000, edited by: Leen, T. K., Dietterich, T. G., and Tresp, V., 402–408, MIT Press, Cambridge, MA, USA, 2001.

Deuzé, J. L., Goloub, P., Herman, M., Marchand, A., Perry, G., Susana, S., and Tanré, D.: Estimate of the aerosol properties over the ocean with POLDER, J. Geophys. Res., 105, 15329–15346, doi:10.1029/2000JD900148, 2000.

Deuzé, J. L., Bréon, F., Devaux, C., Goloub, P., Herman, M. Lafrance, B., Maignan, F., Marchand, A., Nadal, F., Perry, G., and Tanré, D.: Remote sensing of aerosols over land surfaces from POLDER-ADEOS-1 polarized measurements, J. Geophys. Res., 106, 4913–4926, doi:10.1029/2000JD900364, 2001.

Diouf, D., Niang, A., Brajard, J., Crépon, M., and Thiria, S.: Retrieving aerosol characteristics and sea-surface chlorophyll from satellite ocean color multi-spectral sensors using a neural-variational method, Remote Sens. Environ., 130, 74–86, doi:10.1016/j.rse.2012.11.002, 2013.

Dubovik, O. and King, M. D.: A flexible inversion algorithm for retrieval of aerosol optical properties from Sun and sky radiance measurements, J. Geophys. Res., 105, 20673–20796, doi:10.1029/2000JD900282, 2000.

Dubovik, O., Smirnov, A., Holben, B. N., King, M. D., Kaufman, Y. J., Eck, T. F., and Slutsker, I.: Accuracy assessments of aerosol optical properties retrieved from Aerosol Robotic Network (AERONET) Sun and sky radiance measurements, J. Geophys. Res., 105, 9791–9806, doi:10.1029/2000JD900040, 2000.

Dubovik, O., Sinyuk, A., Lapyonok, T., Holben, B. N., Mishchenko, M., Yang, P., Eck, T. F., Volten, H., Muñoz, O., Veihelmann, B., van der Zande, W. J., Leon, J.-F., Sorokin, M., and Slutsker, I.: Application of spheroid models to account for aerosol particle nonsphericity in remote sensing of desert dust, J. Geophys. Res., 111, D11208, doi:10.1029/2005JD006619, 2006.

Dubovik, O., Herman, M., Holdak, A., Lapyonok, T., Tanré, D., Deuzé, J. L., Ducos, F., Sinyuk, A., and Lopatin, A.: Statistically optimized inversion algorithm for enhanced retrieval of aerosol properties from spectral multi-angle polarimetric satellite observations, Atmos. Meas. Tech., 4, 975–1018, doi:10.5194/amt-4-975-2011, 2011.

Dybowski, R. and Roberts, S. J.: Confidence intervals and prediction intervals for feedforward neural networks, in: Clinical Applications of Artificial Neural Networks, edited by: Dybowski, R. and Gant, V., Chap. 13, 298–326, Cambridge University Press, Cambridge, UK, doi:10.1017/CBO9780511543494.013, 2001.

Hagan, M. T. and Menhaj, M. B.: Training feedforward networks with the Marquardt algorithm, IEEE T. Neural Networ., 5, 989–993, doi:10.1109/72.329697, 1994.

Han, B., Vucetic, S., Braverman, A., and Obradovic, Z.: A statistical complement to deterministic algorithms for the retrieval of aerosol optical thickness from radiance data, Eng. Appl. Artif. Intel., 19, 787–795, doi:10.1016/j.engappai.2006.05.009, 2006.

Hansen, J., Sato, M., and Ruedy, R.: Radiative forcing and climate response, J. Geophys. Res., 102, 6831–6864, doi:10.1029/96JD03436, 1997.

Hansen, J. E. and Travis, L. D.: Light scattering in planetary atmospheres, Space Sci. Rev., 16, 527–610, doi:10.1007/BF00168069, 1974.

Hansen, P. C. and O'Leary, D. P.: The use of the L-curve in the regularization of discrete ill-posed problems, SIAM J. Sci. Comput., 14, 1487–1503, doi:10.1137/0914086, 1993.

Hasekamp, O. P. and Landgraf, J.: A linearized vector radiative transfer model for atmospheric trace gas retrieval, J. Quant. Spectrosc. Ra., 75, 221–238, doi:10.1016/S0022-4073(01)00247-3, 2002.

Hasekamp, O. P. and Landgraf, J.: Linearization of vector radiative transfer with respect to aerosol properties and its use in satellite remote sensing, J. Geophys. Res., 110, D04203, doi:10.1029/2004JD005260, 2005.

Hasekamp, O. P. and Landgraf, J.: Retrieval of aerosol properties over land surfaces: capabilities of multiple-viewing-angle intensity and polarization measurements, Appl. Optics, 46, 3332–3344, doi:10.1364/AO.46.003332, 2007.

Hasekamp, O. P., Litvinov, P., and Butz, A.: Aerosol properties over the ocean from PARASOL multiangle photopolarimetric measurements, J. Geophys. Res., 116, D14204, doi:10.1029/2010JD015469, 2011.

Haykin, S.: Neural Networks: A Comprehensive Foundation, Prentice Hall, Upper Saddle River, NJ, USA, 1999.

Holben, B. N., Eck, T. F., Slutsker, I., Tanré, D., Buis, J. P., Setzer, A., Vermote, E., Reagan, J. A., Kaufman, Y. J., Nakajima, T., Lavenu, F., Jankowiak, I., and Smirnov, A.: AERONET – A federated instrument network and data archive for aerosol characterization, Remote Sens. Environ., 66, 1–16, doi:10.1016/S0034-4257(98)00031-5, 1998.

Holmström, L. and Koistinen, P.: Using additive noise in backpropagation training, IEEE T. Neural Networ., 3, 24–38, doi:10.1109/72.105415, 1992.

Jamet, C., Thiria, S., Moulin, C., and Crépon, M.: Use of a neurovariational inversion for retrieving oceanic and atmospheric constituents from ocean color imagery: A feasibility study, J. Atmos. Ocean. Tech., 22, 460–475, doi:10.1175/JTECH1688.1, 2005.

Kaufman, Y. J. and Koren, I.: Smoke and pollution aerosol effect on cloud cover, Science, 313, 655–658, doi:10.1126/science.1126232, 2006.

KNMI: OMI News: More North American wildfire smoke observed over Europe, available at: http://www.knmi.nl/omi/news/archive/fullFiles/2013-07-11_smoke_canadian_fires_en.php (last access: 25 November 2014), 2013.

Knobelspiesse, K., Cairns, B., Ottaviani, M., Ferrare, R., Hair, J., Hostetler, C., Obland, M., Rogers, R., Redemann, J., Shinozuka, Y., Clarke, A., Freitag, S., Howell, S., Kapustin, V., and McNaughton, C.: Combined retrievals of boreal forest fire aerosol properties with a polarimeter and lidar, Atmos. Chem. Phys., 11, 7045–7067, doi:10.5194/acp-11-7045-2011, 2011.

Lawrence, S., Giles, C. L., and Tsoi, A. C.: Lessons in neural network training: Overfitting may be harder than expected, in: Proceedings of the Fourteenth National Conference on Artificial Intelligence, AAAI-97, 540–545, AAAI Press, Providence, RI, USA, 1997.

LeCun, Y., Boser, B., Denker, J. S., Henderson, D., Howard, R. E., Hubbard, W., and Jackel, L. D.: Backpropagation applied to handwritten ZIP code recognition, Neural Comput., 1, 541–551, doi:10.1162/neco.1989.1.4.541, 1989.

Liu, X., Easter, R. C., Ghan, S. J., Zaveri, R., Rasch, P., Shi, X., Lamarque, J.-F., Gettelman, A., Morrison, H., Vitt, F., Conley, A., Park, S., Neale, R., Hannay, C., Ekman, A. M. L., Hess, P., Mahowald, N., Collins, W., Iacono, M. J., Bretherton, C. S., Flanner, M. G., and Mitchell, D.: Toward a minimal representation of aerosols in climate models: description and evaluation in the Community Atmosphere Model CAM5, Geosci. Model Dev., 5, 709–739, doi:10.5194/gmd-5-709-2012, 2012.

Lohmann, U. and Feichter, J.: Global indirect aerosol effects: a review, Atmos. Chem. Phys., 5, 715–737, doi:10.5194/acp-5-715-2005, 2005.

Mishchenko, M. I. and Travis, L. D.: Satellite retrievals of aerosol properties over the ocean using polarization as well as intensity of reflected sunlight, J. Geophys. Res., 102, 16989–17013, doi:10.1029/96JD02425, 1997.

Mishchenko, M. I., Cairns, B., Hansen, J. E., Travis, L. D., Burg, R., Kaufman, Y. J., Martins, J. V., and Shettle, E. P.: Monitoring of aerosol forcing of climate from space: Analysis of measurement requirements, J. Quant. Spectrosc. Ra., 88, 149–161, doi:10.1016/j.jqsrt.2004.03.030, 2004.

Mishchenko, M. I., Cairns, B., Chowdhary, J., Geogdzhayev, I. V., Liu, L., and Travis, L. D.: Remote sensing of tropospheric aerosols from aircraft and satellites, J. Phys. Conf. Series, 6, 73–89, doi:10.1088/1742-6596/6/1/005, 2005.

Møller, M. F.: A scaled conjugate gradient algorithm for fast supervised learning, Neural Networks, 6, 525–533, doi:10.1016/S0893-6080(05)80056-5, 1993.

Monna, W. and Bosveld, F.: In Higher Spheres. 40 years of observations at the Cabauw Site, KNMI Publication 232, De Bilt, the Netherlands, available at: http://www.knmi.nl/knmi-library/knmipubmetnummer/knmipub232.pdf (last access: 25 November 2014), 2013.

Niang, A., Badran, F., Moulin, C., Crépon, M., and Thiria, S.: Retrieval of aerosol type and optical thickness over the Mediterranean from SeaWiFS images using an automatic neural classification method, Remote Sens. Environ., 100, 82–94, doi:10.1016/j.rse.2005.10.005, 2006.

O'Neill, N. T., Eck, T. F., Smirnov, A., Holben, B. N., and Thulasiraman, S.: Spectral discrimination of coarse and fine mode optical depth, J. Geophys. Res., 108, 4559, doi:10.1029/2002JD002975, 2003.

Picchiani, M., Chini, M., Corradini, S., Merucci, L., Sellitto, P., Del Frate, F., and Stramondo, S.: Volcanic ash detection and retrievals using MODIS data by means of neural networks, Atmos. Meas. Tech., 4, 2619–2631, doi:10.5194/amt-4-2619-2011, 2011.

Radosavljevic, V., Vucetic, S., and Obradovic, Z.: A data-mining technique for aerosol retrieval across multiple accuracy measures, IEEE Geosci. Remote Sens. Lett., 7, 411–415, doi:10.1109/LGRS.2009.2037720, 2010.

Ristovski, K., Vucetic, S., and Obradovic, Z.: Uncertainty analysis of neural-network-based aerosol retrieval, IEEE T. Geosci. Remote, 50, 409–414, doi:10.1109/TGRS.2011.2166120, 2012.

Rodgers, C. D.: Characterization and error analysis of profiles derived from remote sounding measurements, J. Geophys. Res., 95, 5587–5595, doi:10.1029/JD095iD05p05587, 1990.

Rodgers, C. D.: Inverse Methods for Atmospheric Sounding: Theory and Practice, World Scientific, London, UK, 2000.

Rosenfeld, D., Wood, R., Donner, L. J., and Sherwood, S. C.: Aerosol Cloud-Mediated Radiative Forcing: Highly Uncertain and Opposite Effects from Shallow and Deep Clouds, in: Climate Science for Serving Society, edited by: Asrar, G. R. and Hurrell, J. W., Chap. 5, 105–149, Springer, Dordrecht, the Netherlands, 10.1007/978-94-007-6692-1_5, 2013.

Rumelhart, D. E., Hinton, G. E., and Williams, R. J.: Learning representations by back-propagating errors, Nature, 323, 533–536, doi:10.1038/323533a0, 1986.

Schiller, H.: Model inversion by parameter fit using NN emulating the forward model – Evaluation of indirect measurements, Neural Networks, 20, 479–483, doi:10.1016/j.neunet.2007.04.022, 2007.

Snik, F., Karalidi, T., and Keller, C. U.: Spectral modulation for full linear polarimetry, Appl. Optics, 48, 1337–1346, doi:10.1364/AO.48.001337, 2009.

Tsekeri, A., Gross, B., Moshary, F., and Ahmed, S.: Potential retrieval of aerosol properties combining the multispectral, multiangle Research Scanning Polarimeter (RSP) measurements of the intensity and linear polarization of light, in: Proc. 91st American Meteorological Society Annual Meeting, Seattle, WA, USA, 2011.

van Harten, G., de Boer, J., Rietjens, J. H. H., Di Noia, A., Snik, F., Volten, H., Smit, J. M., Hasekamp, O. P., Henzing, J. S., and Keller, C. U.: Atmospheric aerosol characterization with a ground-based SPEX spectropolarimetric instrument, Atmos. Meas. Tech., 7, 4341–4351, doi:10.5194/amt-7-4341-2014, 2014.

Waquet, F., Cairns, B., Knobelspiesse, K., Chowdhary, J., Travis, L. D., Schmid, B., and Mishchenko, M. I.: Polarimetric remote sensing of aerosols over land, J. Geophys. Res., 114, D01206, doi:10.1029/2008JD010619, 2009.

Werbos, P. J.: Beyond Regression: New Tools for Prediction and Analysis in the Behavioral Sciences, Ph.D. thesis, Harvard University, Cambridge, MA, USA, 1974.

Wilson, D. R. and Martinez, T. R.: The inefficiency of batch training for large training sets, in: Proc. IEEE-INNS-ENNS International Joint Conference on Neural Networks (IJCNN '00), Vol. 2, 113–117, doi:10.1109/IJCNN.2000.857883, 2000.

Yu, H., Kaufman, Y. J., Chin, M., Feingold, G., Remer, L. A., Anderson, T. L., Balkanski, Y., Bellouin, N., Boucher, O., Christopher, S., DeCola, P., Kahn, R., Koch, D., Loeb, N., Reddy, M. S., Schulz, M., Takemura, T., and Zhou, M.: A review of measurement-based assessments of the aerosol direct radiative effect and forcing, Atmos. Chem. Phys., 6, 613–666, doi:10.5194/acp-6-613-2006, 2006.

Development and validation of inexpensive, automated, dynamic flux chambers

B. B. Almand-Hunter[1], J. T. Walker[2], N. P. Masson[1], L. Hafford[1], and M. P. Hannigan[1]

[1]Mechanical Engineering Department, University of Colorado, 427 UCB, Boulder, CO, 80303, USA
[2]Air Pollution Prevention and Control Division, National Risk Management Research Laboratory, US Environmental Protection Agency, E305-2, MD-63, Research Triangle Park, NC 27711, USA

Correspondence to: B. B. Almand-Hunter (berkeley.almand@colorado.edu)

Abstract. We developed and validated an automated, inexpensive, and continuous multiple-species gas-flux monitoring system that can provide data for a variety of relevant atmospheric pollutants, including O_3, CO_2, and NO_x. Validation consisted of conducting concurrent gas-phase dry-deposition experiments, using both dynamic flux chambers and an eddy-covariance system, in a grassy clearing in the Duke Forest (Chapel Hill, NC). Experiments were carried out in June and September under a variety of meteorological conditions. Ozone-deposition measurements from the two methods matched very well (4–10 % difference in mean flux rate) when the leaf-area index (LAI) inside the chambers was representative of the average LAI in the field. The dynamic flux chambers can be considered an accurate measurement system under these conditions.

1 Introduction

Deposition of pollutants – including ozone, nitrogen, and acidic compounds (SO_x, NO_y) – places environmental stress on sensitive vegetated landscapes and aquatic ecosystems (Driscoll et al., 2001; Williams and Tonnessen, 2000; Fangmeier et al., 1994). Examples of this stress include increased susceptibility to injury (DeHayes et al., 1999) and decreased growth for sensitive plant species, decreased water quality, toxicity to freshwater organisms, eutrophication, change in greenhouse emissions from soil (Fenn et al., 1998), reduction in biodiversity, and interference with a plant's uptake of other important cations, such as potassium (Fangmeier et al., 1994). These negative effects can be particularly pronounced at high altitudes, where buffering capacities can be below average (Fenn et al., 1998; Williams and Tonnessen, 2000; Benedict et al., 2013). There has been debate over whether ozone damage to vegetation is best quantified and regulated using ambient concentrations or atmospheric fluxes (Musselman et al., 2006). While the use of ambient concentrations is certainly much simpler, fluxes have more physical meaning.

Dry deposition, which is the process by which pollutants are transported from the atmosphere to the earth's surface without precipitation (Seinfeld and Pandis, 2006), is an important component of atmospheric deposition. This process is estimated to account for up to 50 % of total atmospheric deposition in the United States (EPA, 2010; Wesely and Hicks, 2000). Despite this sizable contribution to total atmospheric deposition, there is a shortage of direct measurements of dry deposition in the US. Because of this measurement shortage, improving deposition models is crucial. Additionally, understanding deposition and emission rates is an important piece of estimating atmospheric concentrations in the planetary boundary layer for climate and weather models. Efforts to improve deposition models are ongoing (Saylor et al., 2014; Zhang et al., 2003; Brook et al., 1999; Pleim et al., 2013), and models estimate flux well under some conditions, but fluxes determined by different models and observations can vary by a factor of 2 to 3 (Schwede et al., 2011; Wu et al., 2011; Flechard et al., 2011). Direct dry-deposition measurements are needed to improve and validate models for a variety of ecosystems and environmental conditions.

The main reason for the shortage of direct dry-deposition measurements is that the current measurement methods are prohibitively expensive and complex. This results in significant uncertainty in deposition loads, specifically regarding transfer ratios (the relationship between ambient concentrations and total deposition). Given the large spatiotemporal variability in air–surface exchange rates of reactive compounds, there is a need for low-cost, easily deployable systems to measure dry deposition directly. These measurement devices should be automated and remotely controlled, so that they can be deployed for extended periods of time without excessive maintenance.

Currently, the most accurate direct method for measuring atmospheric fluxes is eddy covariance (Seinfeld and Pandis, 2006; Turnipseed et al., 2009). Eddy covariance consists of taking high-speed measurements of concentration and three-dimensional wind velocity. The flux is computed from the covariance between the fluctuating components of wind velocity and concentration (Turnipseed et al., 2009). This method is the most mathematically robust and accurate way to acquire dry-deposition measurements, but it is expensive and technically difficult compared with indirect measurement methods (Baldocchi et al., 1988).

Another method for measuring flux, which is used more frequently to measure emissions than it is to measure deposition, is the flux chamber. Advantages of flux chambers over eddy covariance include reduced cost, the ability to determine spatial variability in deposition, the ability to take measurements in areas with complex topography and areas with non-uniform vegetation (eddy covariance typically requires an area of uniform vegetation that is $\geq 100\,m^2$), mobility, and the potential to be used with inexpensive sensors (Horst and Weil, 1994). The main drawback of using chambers for flux measurements is that they alter the environment in which they are placed. Static chambers, which are commonly used to measure emissions, significantly affect environmental conditions (Pape et al., 2009).

Dynamic flux chambers minimize the alteration of environmental conditions by constantly pumping ambient air into the chamber. Table 1 lists previous flux-chamber measurements of NO, NO_2, CO_2, and O_3. One type of flux chamber listed in Table 1 is the leaf-scale dynamic chamber, which is used to measure fluxes to and from individual leaves and branches (Breuninger et al., 2012, 2013; Geßler et al., 2000; Sparks et al., 2001; Altimir et al., 2002). While leaf-scale deposition measurements are important for understanding plant dynamics, they can be difficult to translate to the canopy scale and do not directly represent ecosystem-level flux.

Another type of chamber listed in Table 1 is the dynamic soil-flux chamber (Remde et al., 1993; Norman et al., 1997). A significant portion of the chambers listed did not have open tops, and the soil or vegetation in the chamber was only exposed to ambient conditions via air pumped into the chamber. These chambers, which are not normally open to the ambient environment, have significant drawbacks. They all block a fraction of incoming solar radiation, and in order to maintain ambient conditions they have to be moved frequently, which makes long-term or remote deployments difficult.

Several research groups have addressed these issues by developing chambers with lids that open and close automatically (Meixner et al., 1997; Pape et al., 2009; Kitzler et al., 2006). These automatic chambers operate in a normally open mode, with lids that close for just a few minutes per hour. Provided that the chambers are made out of highly transparent materials, so sunlight can reach the vegetation inside, the environmental conditions in the chamber remain very close to ambient (Pape et al., 2009).

While many chamber measurements have been made (Table 1), very few of these studies compare O_3 fluxes measured by chambers to measurements acquired via micrometeorological techniques. Several groups have compared chamber measurements of NO fluxes from soils to gradient measurements (Parrish et al., 1987; Stella et al., 2012). Norman et al. (1997) compared several types of static and dynamic chambers with each other and eddy correlation for measuring CO_2 fluxes in forest soils, but only two data points for eddy correlation were available for comparison, each representing one day. Li et al. (1999) compared chamber measurements of NO fluxes from agricultural soils with eddy-correlation measurements and found that the fluxes measured by the chambers were higher than the eddy-correlation measurements but followed a similar diurnal trend. Pape et al. (2009) compared an automatic, dynamic flux chamber with an eddy-covariance system at a grassland site and demonstrated good agreement for CO_2 deposition. Due to the fact that these comparison studies are limited in number, and sometimes did not yield good agreement between methods, further comparisons of flux chambers and micrometeorological methods are warranted.

Our research effort expands on this validation-based flux-chamber development through the creation of an automated, inexpensive, and continuous multiple-species gas-flux monitoring system, which can provide data for a variety of relevant atmospheric pollutants, including O_3, CO_2, and NO_x. The chambers have automatic lids, which keep the environment in the chambers close to ambient, and eliminate the need to regularly remove them from sampling plots. This project is unique because our chambers not only build on the limited chamber-validation literature, but also utilize an inexpensive design (< USD 2000 each). The chambers are equipped with inexpensive metal-oxide O_3 and NO_2 sensors, which cost between USD 10 and 100, and our ultimate goal is to obtain fluxes using these inexpensive sensors. The first step toward reaching that goal is to use data from established O_3, CO_2, and NO_x monitors to validate the dynamic flux-chamber measurements, which enables us to isolate the uncertainty related to the use of inexpensive sensors from chamber performance. We present preliminary results, comparing chamber fluxes to eddy-covariance fluxes for O_3 and CO_2, and present NO_x fluxes measured by the flux chamber.

Table 1. Summary of selected chamber measurements of NO_2, NO, O_3, and CO_2.

Reference	Gases measured	Surface	Chamber type	Validation method
Altimir et al. (2002)	CO_2, O_3	Scots pine shoots	dynamic, shoot chamber	compared w/typical O_3 flux values
Breuninger et al. (2012)	NO_2, NO, CO_2, O_3	Norway spruce	dynamic, branch chamber	none
Breuninger et al. (2013)	NO_2, NO, CO_2, O_3	Norway spruce	dynamic, branch chamber	none
Gut et al. (2002)	NO_2, NO, CO_2, O_3	soil	dynamic	model comparison (using ambient concentration)
Horváth et al. (2006)	NO, O_3	spruce and oak soil	dynamic	none
Kirkman et al. (2002)	NO_2, NO, O_3	pasture	dynamic	none
Li et al. (1999)	NO	agricultural soil	dynamic	chamber values larger than eddy covariance, but varied similarly with time
Meixner et al. (1997)	NO, NO_2, O_3	grassland and crops	dynamic, automated lid	none
Norman et al. (1997)	CO_2	forest soil	dynamic and static	compared 5 types of chambers and 2 eddy-covariance data points
Pape et al. (2009)	NO_2, NO, CO_2, O_3	grassland	dynamic, automatic lid	very good agreement w/eddy covariance for CO_2 (did not compare NO, NO_2, and O_3)
Parrish et al. (1987)	NO	grassland	dynamic	nighttime comparison with gradient method
Pilegaard (2001)	NO, NO_2, O_3	forest soil	dynamic, automated lid	none
Remde et al. (1993)	NO, NO_2, O_3	marsh soil	dynamic	none
Stella et al. (2012)	NO	agricultural soil	dynamic, automated lid	agreed with gradient method for low fluxes, but underestimated high fluxes
Unsworth et al. (1984)	O_3	soybeans	dynamic, open top	canopy resistances comparable to other field studies
Williams and Davidson (1993)	NO	grassland	dynamic	comparison of 2 chamber types
This study	O_3	grassland	dynamic, automatic lid	comparison with eddy covariance

2 Methods

2.1 Overview

We conducted gas-phase dry-deposition experiments in a grassy clearing in the Blackwood Division of Duke Forest in Orange County, North Carolina, USA (35.97° N, 79.09° W). The field is 480 m × 305 m, and the vegetation is primarily tall fescue (*Festuca arundinacea* Shreb.), which is a common C3 grass in the southeastern United States. Less-prominent vegetation includes C3 and C4 grasses, herbs, and forbs, which are present in smaller amounts (Fluxnet, 2013).

We used five pairs of acrylic-glass flux chambers to measure dry deposition of NO_x, O_3, and CO_2 to the grassland vegetation. Experiments were carried out in June and September, under a variety of meteorological conditions. We compared the chamber results with eddy-covariance measurements, which were conducted by the EPA at the same site.

2.2 Leaf-area index

The LAI in the field, as well as in chambers A and B, was measured on 11 November. LAI measurements in the open field were made by sampling at regular distances along 100 m transects ($n = 10$ locations) to the southwest and northwest of the eddy-covariance tower (prevailing fetch) with a LI-COR Model LAI-2000 plant canopy analyzer (LI-COR Biosciences, Lincoln, NE). LAI measurements within the chambers were made by inserting the leaf area meter through a port at the bottom of the chamber. Individual measurements consisted of one above-canopy measurement and five below-canopy measurements. Three replicate measurements were taken in each chamber. Measurements within the control chambers showed no difference between above- and below-canopy measurements.

The LAI in the field was between 2.8 and 3.5, the LAI in chamber A was between 2.4 and 2.9, and the LAI in chamber B was between 2.75 and 3.1. While the chamber-LAI measurements were on the low end of the field measurements, they were inside the range of LAI measurements in the field. The mean grass height in the field did not significantly change between June and November, and measured heights were 42.2 and 43.7 cm, respectively.

2.3 Eddy-covariance measurements

Above-canopy fluxes of CO_2, H_2O, O_3, sensible heat, and momentum were measured using an instrument package that consisted of an R.M. Young sonic anemometer (Model 81000V, Traverse City, MI), aspirated thermocouple (Model ASPTC, Campbell Scientific, Logan, UT), LI-COR (Lincoln, NE) Model 7500 (CO_2 and H_2O) open-path infrared gas analyzer (IRGA), and a custom fast chemiluminescence O_3 sensor, positioned at 2.5 m above the canopy. Gas-phase instruments were calibrated weekly by mass-flow-controlled dilution of compressed gas standards with clean air. Wind speed components (u, v, w), temperature, and air concentrations were sampled at a frequency of 10 Hz, and data were recorded on a single laptop, using a custom data-acquisition system. Data were reduced to 30 min and hourly averages using a custom SAS program (SAS Institute, 2003).

Eddy-covariance O_3 fluxes were measured with a custom sensor that follows the basic design of Guesten et al. (1992), which consists of a pump (Model BTC IIS miniature diaphragm pump, Parker, Hollis, NH), a reaction cell, and a photomultiplier tube (Model H9306 side-on photosensor, Hamamatsu, Middlesex, NJ) (Guesten et al., 1992). While

Guesten et al. (1992) are generally credited with developing the first of these systems for flux measurement applications, a number of variants of the original design have been developed over the following years (see Zahn et al., 2012). This measurement technique is known as "dry chemiluminescence", and the air sample passes over a disk, which is coated with Coumarin-1 dye (Bagus Consulting, Speyer, Germany). The reaction of O_3 with the dye results in luminescence, which is quantified by counting the resulting photons with a photomultiplier tube that views the reaction chamber from the side opposite the Coumarin disk. The ozone concentration is proportional to the number of photons produced. However, this is not an absolute measurement, and the disks have a limited lifetime over which the photon yield per unit O_3 decreases. Thus, to calculate the absolute flux, the fast sensor must be calibrated to a second collocated sensor. The second sensor measures absolute O_3 concentrations, at a frequency consistent with the timescale of the flux, which is 30 min to 1 h. The collocated instrument is a Model 205 dual-beam instrument (2B Technologies, Boulder, CO), which measures O_3 by UV absorption. Ozone fluxes were calculated from 10 Hz measurements, calibrated to the 2B sensor, using the ratio-offset method recommended by Muller et al. (2010).

For flux calculations, 10 Hz data were subjected to spike filtering, 2-D coordinate rotation, correction for the time delay between the chemical sensor and vertical velocity, application of Webb–Pearman–Leuning correction (CO_2, H_2O), and correction for high-frequency spectral attenuation (O_3) (Horst, 1997; Webb et al., 1980). The fast O_3 sensor has an effective time response of approximately 1.5 Hz. At lower frequencies, the cospectra of O_3 and vertical velocity match the shape of the temperature/vertical velocity cospectra well, with both following the generalized cospectral characteristics described by Kaimal and Finnigan (1994). In this case, the method described by Horst (1997) was used to correct for spectral attenuation at frequencies > 1.5 Hz (Horst and Weil, 1994). Tests for stationarity and the presence of fully developed turbulence were also applied (Foken and Wichura, 1996). More information about the eddy-covariance method is available in the Supplement.

2.4 Ancillary measurements

Ancillary measurements included net solar radiation (Rebs Q7.1 Net Radiometer, Campbell Scientific, Logan, UT), photosynthetically active radiation (Model LI190 quantum sensor, LI-COR, Inc., Lincoln, NE), soil heat flux (Model HP101 heat flux plate, Hukseflux USA, Inc., Manorville, NY), soil temperature (thermocouple, OMEGA Engineering, Stamford, CT), soil volumetric water (Model CS615 water content reflectometer, Campbell Scientific, Logan UT) leaf wetness (Model 237, Campbell Scientific, Logan UT), and rainfall (Model TE525 rain gauge, Campbell Scientific, Logan, UT). Data are recorded on a Campbell Scientific CR23X datalogger and reduced to 30 min and hourly averages.

Figure 1. The photo above shows a pair of flux chambers at the field site in the Duke Forest.

2.5 Flux-chamber description

The dynamic flux chambers, which are shown in Fig. 1, were constructed using clear, cylindrically shaped acrylic. The chambers were constructed in pairs, and each pair had an open-bottomed chamber, which measured deposition to the vegetation inside, and a "blank" chamber, which had an acrylic bottom, and enabled us to measure deposition in the absence of vegetation. The blank measurement represents trace-gas losses to the chamber walls as well as any chemical reactions in the chamber that are unaccounted for in the flux calculations.

All of the chambers have a 45.7 cm diameter and 0.48 cm wall thickness. Four pairs of chambers have a height of 83.8 cm, and the remaining pair has a height of 58.4 cm. The chambers were designed with this height distribution because many species of natural vegetation, including grassland, are taller than 58.4 cm. The shorter chambers were designed to measure fluxes over shorter vegetation, such as alpine tundra, which is present in sensitive areas like Rocky Mountain National Park. The shorter chambers are likely more accurate for vegetation below 59 cm tall, since they increase the ratio of vegetative surface area to volume.

The chambers were designed to minimize deposition of trace gases to the chamber walls, which was accomplished by placing the inlet and outlet holes in locations that limited contact between the flow path and the chamber walls. Ambient air enters the chambers through four holes, which are each 5.2 cm in diameter, and evenly spaced around the circumference of the chamber. The chamber outlet is at the

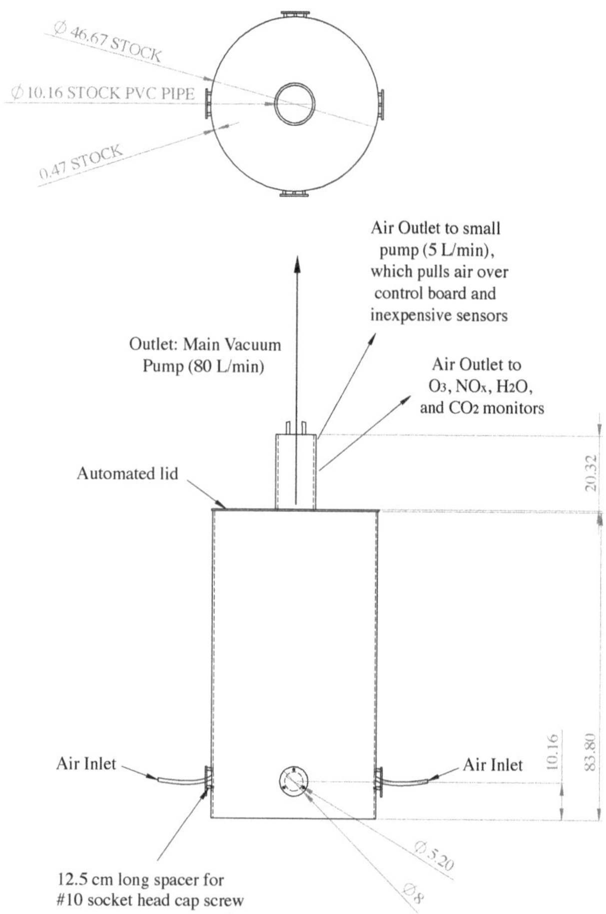

Figure 2. The plot above shows the dimensions of the chamber and the locations of the air inlets and outlet.

top of the chamber, as shown in Fig. 2. The grass outside the chamber, near the inlet holes, is removed, which prevents trace gases from depositing to external vegetation before the air stream enters the chamber.

Air is pulled through the chamber by a US General 3 CFM Two-Stage Vacuum Pump, and concentration samples were measured in one of two polytetrafluoroethylene (PTFE) tubes at the outlet. Gas-phase sampling is discussed in more detail in Sect. 2.6.

For most experiments, the pump was set to pull 80 $\mathrm{L\,min^{-1}}$ of air through the chamber. In addition to the flow induced by the vacuum pump, the 2B ozone monitor pulled approximately 1 $\mathrm{L\,min^{-1}}$; the Thermo Scientific NO_x analyzer pulled 0.1 $\mathrm{L\,min^{-1}}$; the LI-COR H_2O / CO_2 monitor pulled 0.25 $\mathrm{L\,min^{-1}}$; and the small, inexpensive pump, which pulled air over the inexpensive sensors, pulled 5 $\mathrm{L\,min^{-1}}$. Thus, the total flow rate through the chamber was 86.35 $\mathrm{L\,min^{-1}}$, which equates to a residence time of 1.5 min. Pape et al. (2009) found that other researchers have operated dynamic flux chambers with residence times ranging from 10 s to 24 min and chose to operate their dynamic flux chambers at a residence time of 40 s (Pape et al., 2009). Gillis and Miller (2000) found that changes in air-stream residence time in flux chambers caused proportional changes in mercury flux for both absorption and emission (Gillis and Miller, 2000). Aeschlimann and coworkers used a residence time of 15 s during the day and 60 s at night (Aeschlimann et al., 2005), which reflects ambient diurnal variation in friction velocity. Low residence times ensure that chambers are well mixed and minimize reactions between gases in the chamber. However, reducing residence times also reduces the difference in ambient and steady-state trace-gas concentrations in the chamber. Thus, as residence time is decreased, more precise instrumentation is required. We chose to operate our chambers with a 1.5 min residence time, because 1.5 min is sufficiently low to keep environmental conditions close to ambient yet still yield a trace-gas concentration change that is large enough to be detected by inexpensive sensors. This residence time also translates to a flow rate that can be generated with an inexpensive pump.

Another way that we reduced the cost of the chamber was by designing our own control system, using inexpensive electronic components. A customized embedded-system platform was used to automate the flux-chamber sampling system. The system is based on the low-cost M-Pod air quality monitor (Masson, 2014), with additional instrumentation for pump and actuator control. Firmware running on the common Atmel (San Jose, CA) Atmega 328 microcontroller controls both the data logging and flux-chamber sampling routine.

Each chamber runs approximately once an hour, and the main vacuum pump is off when the chamber is not sampling. Once per hour, a predefined and automated sampling schedule begins, and the vacuum pump turns on and runs with the lid open for 6.75 min. The pressure change caused by the pump can cause fluctuations in instrument readings, and this boot-up time allows the instruments to stabilize before the chamber lid closes. After the 6.75 min initialization, the chamber lid closes and remains closed for 5 min. It is important to note that the eddy-covariance measurements are fluxes averaged over a 30 min or 1 h time period, and the chamber measurements are a 5 min average, taken every 53 min.

Fluxes were calculated based on the assumption that the chamber was well mixed. A mass balance in the chamber yields the equation

$$V\frac{\mathrm{d}\mu_j(t)}{\mathrm{d}t} = Q\mu_{j,\mathrm{amb}} - Q\mu_j(t) - F_jA_\mathrm{s}, \tag{1}$$

where $\mu_j(t)$ is the mixing ratio in the chamber of gas, j, with respect to time; Q is the flow rate of air through the chamber; $\mu_{j,\mathrm{amb}}$ is the ambient mixing ratio of gas, j; t is time; A_s is the surface area of the opening at the bottom of the chamber; V is volume of the chamber; and F_j is the flux of gas, j, to the vegetation. Differentiating, $\mu_j(t)$ is found to be

Table 2. Trace gas detectors used in this study, with manufacturers' specifications.

Time period	Detection method	Manufacturer and model	Detection limit	Precision
O_3	UV light absorption	2b Technologies (Boulder, CO) 202	1.5 ppb	±1.5 ppb
NO_x	chemiluminescence	Thermo Scientific (Waltham, MA) 42S	0.4 ppb	±0.4 ppb
CO_2	non-dispersive infrared gas analyzer	LI-COR (Lincoln, NE) 7000	0 ppm	0.01 ppm
H_2O	non-dispersive infrared gas analyzer	LI-COR (Lincoln, NE) 7000	0 ppm	0.01 ppm

$$\mu_j(t) = \mu_{j,\text{amb}} - \frac{F_j A_s}{Q}(1 - e^{-\frac{Q}{V}t}). \tag{2}$$

The steady-state solution to this equation, solving for flux, is

$$F = \frac{Q}{A_s}(\mu_{j,\text{amb}} - \mu_j(\tau_{ss})), \tag{3}$$

where τ_{ss} is the time when the trace-gas concentration in the chamber reaches steady state.

2.6 Gas-phase measurements

Figure 2 shows the flow path of sample air through the chamber. Gas-phase measurements were conducted at the chamber outlet, which consisted of an 11.4 cm diameter PVC pipe. Chamber air was pulled through the outlet via the main vacuum pump. Two 4.76 mm diameter tubes were attached to the sides of the PVC pipe on one end, and instruments on the other. One tube was connected to a 2B Technologies Model 202 Ozone Monitor, Thermo Scientific Model 42S NO_x analyzer, and LI-COR 7000 H_2O/CO_2 monitor. More information about the instruments is available in Table 2.

The second tube was connected to a small vacuum pump, which moved air through the chamber control box. In addition to the control board, the box housed metal-oxide NO_x and O_3 sensors. Additional data were collected using these commercially available sensors, specifically the Sensortech (Chemlsford, UK) (formerly e2v) MICS-2611 O_3 sensor. All low-cost sensors implemented in the flux-chamber system ranged in cost from USD 10 to 100, and the O_3 sensors had a detection limit well within typical concentration changes seen in ground-flux measurements. Complex quantification schemes are necessary to quantify the sensor output properly. Such schemes incorporate correction parameters for interference effects. Inexpensive sensor technology has the potential to be incorporated into a flux-chamber system effectively, which would make widespread flux measurements a realizable objective.

2.7 Comparison of eddy-covariance and flux-chamber measurements

Theoretically, dry deposition flux (F) is proportional to the ambient concentration (C) of a trace gas at some reference height (Seinfeld and Pandis, 2006). The proportionality constant between the concentration and flux is called "deposition velocity" (v_d) (Chamberlain and Chadwick, 1953), and

$$F = -v_d C. \tag{4}$$

The deposition process has been described using a resistance analogy (Wesely and Hicks, 2000), in which species transport from the atmosphere to the surface of a material is controlled by three resistances in series.

$$v_d = \frac{1}{r_t} = \frac{1}{r_a + r_b + r_c}, \tag{5}$$

where r_t is the total resistance to deposition, r_a is the resistance to aerodynamic transport, r_b is the resistance to diffusion through the quasi-laminar boundary layer, and r_c is the resistance to uptake of a trace gas by the canopy.

This resistance analogy is based on the assumption that the atmosphere is unaltered. It is an accurate analogy for eddy-covariance measurements, but flux chambers alter the wind speed above the canopy, so the resistance analogy must be adjusted. Pape et al. (2009) proposed an alternate resistance scheme, which replaces r_a with r_{purge} and r_{mix}, which represent the purging resistance between ambient and chamber air, and mixing in the chamber, respectively. When the chamber is well mixed, r_{mix} is very small, and it can therefore be neglected in this case. r_b is replaced with a modified boundary-layer resistance, r_b^*. r_c should be modified very little by the chamber, provided the chamber does not substantially alter the environmental conditions (temperature, relative humidity) of the natural environment.

Thus, the ratio of chamber flux to ambient flux can be written as

$$\frac{F_{\text{cham}}}{F_{\text{amb}}} = \frac{r_a + r_b + r_c}{r_{\text{purge}} + r_b^* + r_c}. \tag{6}$$

In order to find $r_{\text{purge}} + r_b^*$, we conducted an experiment, placing a 15 cm dish of saturated potassium iodide solution (KI) in the bottom of a chamber that was filled with grass in the same height range as the field study. KI is an ideal ozone

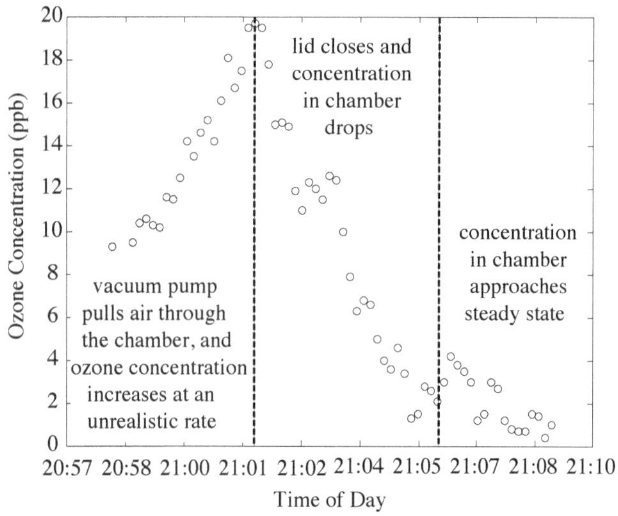

Figure 3. The plot above is an example of a run where the data could not be used to calculate a flux. The ozone concentration increases by an unreasonable amount when the chamber lid opens, which likely indicated malfunction in the 2B ozone monitor.

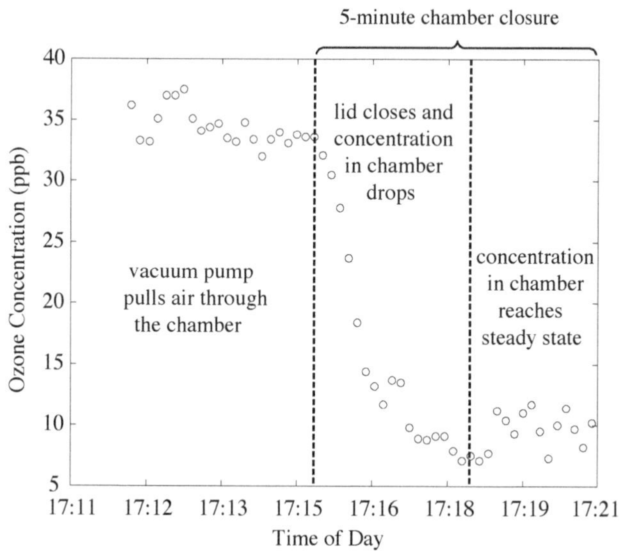

Figure 4. The plot above is an example of ozone data that can be analyzed using the steady-state mass-balance equation. The data before the lid is closed and at the end of the sample both have low noise and stay relatively constant for at least 1 min.

sink, so r_c can be approximated as zero (Galbally and Roy, 1980) in this experiment. We used the equation

$$v_d = \frac{1}{r_{purge} + r_b^* + r_c} \tag{7}$$

and the measured deposition velocity from the KI experiment to calculate $r_{purge} + r_b^*$ ($r_{purge} + r_b^* = 1/v_d = 57.5\,s\,m^{-1}$). The value of $r_{purge} + r_b^*$ describes aerodynamic and quasi-laminar boundary-layer resistance to deposition inside the chamber, so, while ozone was used to find the value, it is applicable to all gases. For future researchers, we would suggest repeating this experiment with different flow and vegetation characteristics.

We present both measured ozone fluxes and values adjusted using this resistance analogy. While this conversion factor enables chamber flux to be scaled to ambient flux, it introduces modeling assumptions and additional uncertainty to an otherwise direct measurement.

3 Results and discussion

3.1 Data processing

We collected O_3 flux data for 8 days. We used two pairs of identical tall chambers and one pair of shorter chambers. Each set of data was based on a 5-minute chamber closure, which occurred once per hour. The flux during each sampling period was assumed to be constant. Each data run was analyzed for noise and pattern, and some data sets were excluded from results.

Figure 3 is an example of a sampling period that we excluded from our results. The ozone concentration increased by an unreasonable amount when the chamber lid opened, which likely indicates malfunction in the 2B ozone monitor. Nine percent of chamber A data were excluded, 11 % of chamber B data were excluded, and 0 % of the chamber C data were excluded.

Figure 4 shows the ozone concentration in the chamber during one sampling period, as an example of ozone data that can be analyzed using the steady-state solution. The area before the decline of the ozone concentration represents the time period when the chamber lid was open. After the lid closed, the concentration began to decline and eventually reached a steady-state value. This data set met our data-quality requirements, as the data just before the lid closed and at the end of the sample both have low noise and stay relatively constant for at least 1 min. Therefore, the flux was computed using the steady-state solution (Eq. 3).

In addition to the data selection mentioned above, we also looked for short-term extreme fluctuations in the ozone time series. The first step in this process was to calculate rolling 1 min averages. Next, we found the standard deviation of the six concentration values used to calculate each 1 min average. We excluded the 1 min averages with a standard deviation greater than 3 ppb. This value was chosen because, when we looked at a histogram of the standard deviations, values greater than 3 ppb were outliers. This data-quality-check process resulted in the removal of 1.4 % of the 1 min average data.

To compute flux, we need ambient and steady-state ozone concentrations. For both of those values, we use an average over a short time window instead of a concentration at one time point to reduce uncertainty. We found the ambient ozone concentration for each cycle by calculating the mean of the last 2 min of concentration data before the chamber lid closed. We found the steady-state concentration by calculating the mean of the data between 3 and 5 min after the chamber closed. Finally, we used the ambient and steady-state concentrations we found for each data set to compute flux, using Eq. (7).

When the ambient ozone concentration is below 5 ppb, we assume that the ozone flux is zero. Ambient O_3 concentrations of 5 ppb or lower typically occur only at night, when wind speeds are low, which means that the aerodynamic resistance to deposition is high, equating to a low flux. The absolute highest flux rate that could occur, with an ambient concentration of 5 ppb, is $0.09\,\mu g\,m^{-2}\,s^{-1}$ (from Eq. 3), and a flux rate this high is very unlikely with low wind speeds. The median ozone-flux rate measured via eddy covariance, when the ambient ozone concentration was ≤ 5 ppb, during the eight-day sampling period was $0\,\mu g\,m^{-2}\,s^{-1}$, with a standard deviation of $0.05\,\mu g\,m^{-2}\,s^{-1}$.

We did not use the blank chamber data to make any adjustments to the fluxes measured by the dynamic chambers. The median difference between ambient concentration and steady-state ozone concentration was 1.9 ppb for the blank chambers. Since the uncertainty in ozone concentrations measured by the 2B ozone monitor is ± 1.5 ppb, the concentration difference is within a 95 % confidence interval for noise. Thus, correcting chamber fluxes for blank flux would only introduce more error into our measurements.

Also, the median flux measured by the blank chambers, when the open-bottom-chamber flux was nonzero, was $-0.001\,\mu g\,m^{-2}\,s^{-1}$. This value is less than 1 % of the median of the nonzero open-bottom-chamber fluxes, which was $-0.21\,\mu g\,m^{-2}\,s^{-1}$. Therefore, correcting for the blank chamber fluxes would not have a significant impact on measurements. It was encouraging that the blank fluxes were so small, since this indicated that wall losses do not have a significant impact on the flux-chamber measurements. Since wall losses were insignificant, the chamber design could be further simplified by eliminating the blank chambers.

3.2 Photochemistry in the chamber

Photochemical reactions between NO, NO_2, and O_3 can occur in the chamber and therefore must be considered in Eq. (1) (Meixner et al., 1997; Pape et al., 2009). The primary reactions of concern are

$$NO + O_3 \rightarrow NO_2 + O_2 \tag{R1}$$

and

$$NO_2 + h\nu \overset{O_2}{\rightarrow} NO + O_3, \lambda < 420\,\text{nm}. \tag{R2}$$

Pape and coworkers measured $j(NO_2)$ inside their chamber and found that the average value of $j(NO_2)$ inside the chamber was 48 % of the value outside the chamber (Pape et al., 2009). They fit a curve of $j(NO_2)$ vs. global radiation (G), and we used that curve in our calculations, since our chambers were similar in shape and material. To quantify the impact of this assumption, we calculated how increasing and decreasing $j(NO_2)$ by 25 % affects ozone flux and found that this changes ozone flux by < 1 % in all cases. The maximum flux change due to photolysis in all of our results is 1.7 %. Thus, the impact of photolysis on ozone flux was small during our study. More information about our calculation of photolysis rate can be found in the Supplement.

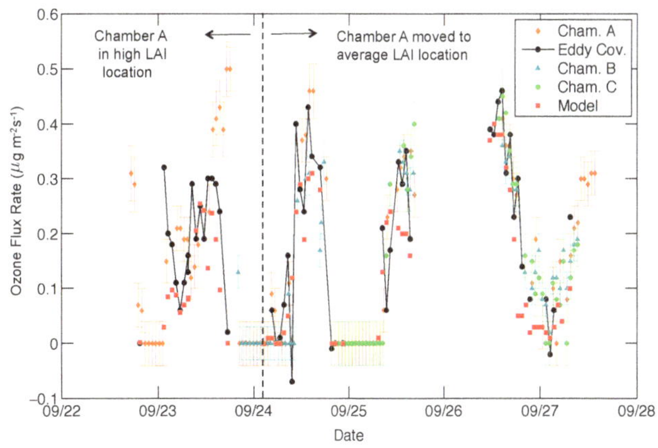

Figure 5. The plot above compares O_3 fluxes measured using eddy covariance (solid black line and black dots); surface-exchange modeling (red squares); and flux chambers A (orange diamonds), B (blue triangles), and C (green circles). The error bars represent the 95 % confidence interval. The tick marks represent midnight on the date listed.

3.3 Ozone results

We measured ozone dry deposition with flux chambers for 2 days in June, and 8 days in September. When compared with eddy-covariance measurements, flux-chamber ozone measurements were able to capture the diurnal flux trends. It is important to remember that eddy-covariance measurements are not without error. For an eddy-covariance system similar to the one used in this study, Finkelstein and Sims (2001) found that mean sampling errors for 30 min average eddy-covariance O_3 fluxes were in the range of 27–33 %.

Figure 5 shows O_3 fluxes measured via eddy covariance and flux chambers A, B and C, and also calculated using an indirect method, which combined meteorological data and surface-exchange model for the time period between 22 and 28 September. The theory used to calculate the model values is described by Wesely (1989) and Seinfeld and Pandis (2006).

The surface-exchange model results underestimated the mean eddy-covariance flux rate by 26 % between 22 and 27 September. This is a good model-to-measurement match, but it is important to remember that the models do not al-

Figure 6. Chamber A (left). Chamber B (right). The vegetation in chamber A, prior to being moved on 24 September, was not representative of the typical vegetation type or LAI at the site. As a result, flux measurements prior to the move were large when compared with measurements from other chambers and eddy covariance. The vegetation in chamber B was representative of the vegetation in the field.

ways predict flux accurately (Wu et al., 2012; Schwede et al., 2011).

Chamber A was moved from its original location in the field to a different position on 24 September. Prior to being moved, the chamber was on a plot of land with a less-prevalent vegetation type, which had a higher LAI than the dominant vegetation (see Fig. 6). After the chamber was moved to a location with more representative vegetation, the data matched the eddy-covariance results much better. Before the chamber was moved (18 and 23 September), the mean ozone-flux rate measured by eddy covariance was $-0.16\,\mu g\,m^{-2}\,s^{-1}$, and the mean chamber flux rate was $-0.23\,\mu g\,m^{-2}\,s^{-1}$, which is 48 % higher than the eddy-covariance measurement. After the move (24–27 September), the mean eddy-covariance flux rate was $-0.25\,\mu g\,m^{-2}\,s^{-1}$, and the mean flux measured by the chamber was $-0.26\,\mu g\,m^{-2}\,s^{-1}$, which is 4 % higher than the eddy-covariance measurement. This difference in measurement agreement highlights the importance of selecting a chamber placement that contains vegetation representative of the footprint of the eddy-covariance tower.

Chamber B operated from 18 to 19 September, and again from 23 to 27 September. The mean ozone flux measured by the flux chamber during this period was $-0.17\,\mu g\,m^{-2}\,s^{-1}$, which is 9 % higher than the mean eddy-covariance ozone flux during the same period ($-0.15\,\mu g\,m^{-2}\,s^{-1}$).

Chamber C, which is the shorter chamber, was operated between 18 and 19 September, and again between 24 and 27 September. The mean chamber flux measured during this period was $-0.115\,\mu g\,m^{-2}\,s^{-1}$, which was 6 % lower than the mean eddy-covariance flux during the same time period ($-0.108\,\mu g\,m^{-2}\,s^{-1}$).

In addition to the September measurements, data were collected for 4 days in June. The chambers underestimated ozone flux by 50–100 % in June, and we believe that this was because the LAI was much lower in the chambers than in the field during that time. Because we did not anticipate the spatial and temporal variability in LAI, nor its subsequent impact on flux measurements, we did not measure LAI during our June sampling period. However, we estimate, by visual inspection, that LAI in the chambers was about 50 % lower in June than in September. Further studies that measure ozone deposition with various known LAI values in the chamber could confirm the effects of changing LAI on measured flux. We will measure LAI in all future flux experiments.

There was not a systemic bias in the ozone flux data. The excellent agreement between the September flux-chamber and eddy-covariance measurements demonstrates that the flux chamber is capable of measuring ozone flux to grassland ecosystems when the LAI inside the chamber represents the average LAI in the field.

3.4 Chamber versus eddy-covariance regression analysis

Plot 7 shows a regression analysis of measured chamber flux rates and chamber flux rates that are adjusted using the chamber-to-ambient flux rate correction versus eddy covariance. While this plot is interesting, we need to be careful about placing too much emphasis on these results, since the averaging times were different for the eddy covariance (averaged over a 1 h period) and chamber measurements (5 min average). It is also important to remember that the eddy-covariance measurements have uncertainty, which Finklestein and Sims found to be 30%, on average, for half-hourly fluxes (Finkelstein and Sims, 2001).

Linear regressions were found for the measured and corrected chamber data versus eddy-covariance data. The measured data had a slope of 0.89 and an intercept of 0.03, with a coefficient of determination (R^2) of 0.64. With 95 % confidence, the slope is significantly different than 1 (p value $= 5.8 \times 10^{-24}$), and the intercept is not significantly different than 0 (p value $= 0.55$), although it is worth noting that the intercept would be significantly different than zero at a slightly higher confidence interval.

Linear regression between the corrected chamber and eddy-covariance data yielded a slope of 0.90 and an intercept of 0.002, with an R^2 of 0.77. With 95 % confidence, the slope is significantly different than 1 (p value $= 51.3 \times 10^{-33}$), and the intercept is not significantly different than zero (p value $= 0.89$). While the corrected data has a slope closer to 1 and an intercept closer to 0 than the measured data, the two lines are not significantly different in slope (p value $= 0.92$) or intercept (p value $= 0.91$). Perhaps a more important result of the correction is that it pulled the 2 most extreme chamber flux values (both $-0.5\,\mu g\,m^{-2}\,s^{-1}$), closer to agreement with the eddy-covariance values; this is reflected in the increased value of R^2.

The mean flux rate for the data in Fig. 7 is $-0.19\,\mu g\,m^{-2}\,s^{-1}$ for eddy covariance, $-0.20\,\mu g\,m^{-2}\,s^{-1}$ for the measured chamber data, and $-0.17\,\mu g\,m^{-2}\,s^{-1}$ for

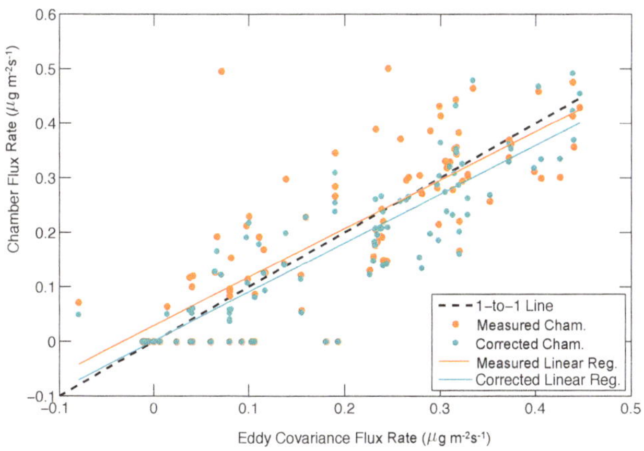

Figure 7. The plot above is a comparison of ozone fluxes measured using flux chambers with fluxes obtained via eddy covariance. The orange dots represent fluxes measured by the chambers, and the blue dots represent chamber fluxes that have been corrected using the adjusted resistance analogy. The dashed line is a 1-to-1 line. The orange line represents a linear regression of the measured chamber fluxes, and the blue line is a linear regression of the corrected chamber fluxes.

the corrected chamber data. While the mean of the corrected chamber flux rate is 9 % farther from the eddy covariance than the measured chamber flux rate, both are well within the uncertainty of the eddy-covariance measurements. Based on the reduction of extreme values and increased R^2, we believe that performing the correction improves the chamber results overall.

3.5 Seasonal flux and implications for chamber flux measurements

We completed this validation work in September, which demonstrates chamber performance in the late summer/early fall. In order to predict chamber performance in other seasons, we used seasonal meteorological data from the Duke Forest to calculate typical values of r_a, r_b, and r_c as well as the chamber-to-ambient flux correction factors (F_{cham} / F_{amb}) for winter, spring, summer, and fall. The meteorological data set includes air temperature, wind speed, friction velocity, relative humidity, global radiation, and rainfall information from 2013. We chose 1 week of representative data for each season (6–12 February, 21–27 April, 2–8 August, 2–8 November). We used a surface-exchange model, combined with the meteorological data, to calculate r_a, r_b, and r_c. The model is based on the theory in Wesely (1989) and Seinfeld and Pandis (2006). A constant value of $57.5\,s\,m^{-1}$ was used for $r_{purge} + r_b^*$ (see Sect. 2.7 for details on the calculation of $r_{purge} + r_b^*$).

As shown in Table 3, the aerodynamic (r_a) and quasi-laminar boundary-layer (r_b) resistances are similar in all seasons. The overall median canopy resistance is lowest in the summer ($110\,s\,m^{-1}$), slightly higher in the spring ($147\,s\,m^{-1}$), higher still in the fall ($263\,s\,m^{-1}$), and drastically higher in the winter ($1348\,s\,m^{-1}$). In the spring, summer, and fall, the canopy resistance makes up approximately 50 % of the total resistance to deposition, whereas in the winter it contributes 90 % of the total deposition resistance. This discrepancy in canopy resistance results in seasonal variability of the chamber-to-ambient flux correction factors. The range of correction factors (5th to 95th percentile) is much larger in the spring (0.86–2.89), summer (0.87–2.82), and fall (0.92–2.84) than in the winter (0.99–1.35).

At night, the correction factor is typically greater than 1, which has the effect of reducing the flux measured by the chamber. This accounts for the fact that the turbulence in the chamber may exceed the ambient turbulence at night. The magnitude of the nighttime correction factor is reduced in the winter as a result of the very large canopy resistance, which is so dominant that differences in aerodynamic conditions are inconsequential. During the day, the typical correction factor is close to 1 in all seasons, and sometimes below 1, which means that the turbulence in the chamber during the day typically matches the ambient conditions well but can be lower or higher than ambient at any given time.

Our validation experiments were conducted in late summer/early fall. Since the chamber-to-ambient correction factors vary the most in the spring, summer, and fall, we have demonstrated that chambers can accurately predict flux when the turbulence in the chambers differs from ambient. Due to the dominance of canopy resistance in the winter, aerodynamic differences should be negligible at that time of year. In addition to correction factors, it is important to consider overall median deposition velocities, which are greatest in the spring and summer (-0.35 and $-0.32\,cm\,s^{-1}$), slightly lower in the fall ($-0.27\,cm\,s^{-1}$), and very low in the winter ($-0.07\,cm\,s^{-1}$). Since ozone concentrations are typically higher in warm weather, fluxes are largest in the summer and spring, slightly lower in the fall, and lowest in the winter. Chamber uncertainty, as a percentage of a single flux measurement, varies from 10 to 65 % and is highest when fluxes are low. Thus, while meteorology should not impact the accuracy of chamber measurements in different seasons, uncertainty as a percentage of flux will be higher in the winter than in other seasons. It is important to remember that eddy-covariance measurements also have a higher uncertainty as a percentage of flux in the wintertime, so the method is not as strong of a benchmark as it is in the summer. In summary, we anticipate that a comparison between fluxes measured with the dynamic flux chambers and with the eddy-covariance method would yield similar agreement in all seasons except winter. In winter, both measurement methods have higher uncertainty, so a comparison might yield noisier results.

Table 3. Seasonal mean resistance, correction factor, and deposition Velocity.

		Winter	Spring	Summer	Fall
Overall	median v_d ($cm\,s^{-1}$)	0.07	0.35	0.32	0.27
	median r_a ($s\,m^{-1}$)	56	45	71	67
	median r_b ($s\,m^{-1}$)	21	20	38	30
	median r_c ($s\,m^{-1}$)	1348	147	110	263
	median F_{cham} / F_{amb}	1.01 ± 0.02	1.02 ± 0.9	1.20 ± 0.74	1.12 ± 1.0
	95th percentile F_{cham} / F_{amb}	1.35	2.89	2.82	2.84
	5th percentile F_{cham} / F_{amb}	0.99	0.86	0.87	0.92
Daytime	median v_d ($cm\,s^{-1}$)	0.11	0.56	0.63	0.38
	median r_a ($s\,m^{-1}$)	39	30	35	33
	median r_b ($s\,m^{-1}$)	15	13	21	16
	median r_c ($s\,m^{-1}$)	876	130	103	209
	median F_{cham} / F_{amb}	1.00 ± 0.02	0.94 ± 0.31	0.99 ± 0.53	0.97 ± 0.66
	95th percentile F_{cham} / F_{amb}	1.03	1.53	1.47	1.25
	5th percentile F_{cham} / F_{amb}	0.98	0.84	0.84	0.91
Nighttime	median v_d ($cm\,s^{-1}$)	0.05	0.25	0.23	0.22
	median r_a ($s\,m^{-1}$)	91	101	145	128
	median r_b ($s\,m^{-1}$)	32	44	77	56
	median r_c ($s\,m^{-1}$)	1348	245	255	282
	median F_{cham} / F_{amb}	1.04 ± 0.04	1.31 ± 1.14	1.57 ± 0.74	1.38 ± 1.1
	95th percentile F_{cham} / F_{amb}	1.44	3.65	3.24	3.1
	5th percentile F_{cham} / F_{amb}	1.0	0.97	1.06	1.0

4 Conclusions

Ozone deposition onto a grassland ecosystem was measured using dynamic flux chambers and eddy covariance. Ozone-deposition measurements from the two methods matched very well (4–10 % difference) when the LAI inside the chambers was representative of the average LAI in the field. This discrepancy is within the uncertainty of eddy covariance, and the flux chambers are considered an accurate measurement system under these conditions. There was not a bias in the chamber data, when compared with the eddy-covariance data.

When LAI inside the chambers was significantly higher or lower than the rest of the field, chamber measurements over- or underpredicted flux, respectively. A discrepancy between chamber and average LAI values can be caused by both inconsistency in vegetation density and differences in vegetation species. Eddy-covariance systems can only measure net flux to an entire fetch (> $100\,m^2$), which means that they measure a mean flux to all vegetation in the field and cannot measure flux to small patches of different vegetation types. Flux chambers are able to measure flux onto different patches of vegetation, which enables the user to understand the relative contribution of different vegetation species to total flux.

In this work, our strategy was to place every chamber on a plot of vegetation that represented the average vegetation in the field. This enabled us to confirm that the results were consistent between chambers. In the field at the Duke Forest, the minority vegetation types represent such a small fraction of the overall grassland that it is very unlikely they have a large net effect on the flux.

It would be very interesting, in future work, to intentionally place the chambers over different types of vegetation in a field and attempt to quantify what percentage of the vegetation each plot represents, and then use a weighted average of ozone fluxes onto the five types of vegetation to estimate the overall flux.

We found that the median ozone flux measured by the blank chambers, when the open-bottom-chamber flux was nonzero, was $-0.001\,\mu g\,m^{-2}\,s^{-1}$. This value is less than 1 % of the median of the nonzero open-bottom-chamber fluxes, which was $-0.21\,\mu g\,m^{-2}\,s^{-1}$. Therefore, we can conclude that we achieved the design goal of minimizing trace-gas interactions with the walls of the chamber.

CO_2 measurements were conducted for one 20 h period, and the flux chamber captured the diurnal trend in CO_2 flux. The quantity of the data was not sufficient to validate chamber performance, but the results show promise, and additional experiments will be conducted to confirm that the flux chambers can measure CO_2 deposition accurately.

Flux-chamber NO_x measurements were conducted for 4 days. Unfortunately, the eddy-covariance system for measuring NO_x was not available during this field campaign, so comparisons could not be made. However, NO_x fluxes measured by the dynamic chambers did fall in the expected range for the site. Additional experiments will be performed to confirm that the chamber NO_x-flux measurements are accurate.

The ultimate goal of our research is to operate the chambers with inexpensive sensors, and the next phase of the project is to validate performance for these sensors. Future work will also consist of measuring different species and using the chambers to measure spatial variability in dry deposition.

The Supplement related to this article is available online at doi:10.5194/amt-8-267-2015-supplement.

Acknowledgements. We are grateful for the opportunity to do this work, which was funded by the Electric Power Research Institute (EPRI). We would like to thank Corey Miller for his help building the flux chambers. We would like to thank Peter Hamlington, Nick Clements, Bill Mitchell, and Andrew Turnipseed for helpful discussions. This project would not have been possible without equipment borrowed from Christine Wiedinmyer and John Ortega, at the National Center for Atmospheric Research, as well as Joanna Gordon and Ashley Collier. Ricardo Piedrahita and Nick Masson's sensor work is the basis for the inexpensive sensor portion of this project.

Disclaimer. This document has been reviewed in accordance with US Environmental Protection Agency policy and approved for publication. The views expressed in this article are those of the author[s] and do not necessarily represent the views or policies of the US Environmental Protection Agency.

Edited by: T. F. Hanisco

References

Aeschlimann, U., Nösberger, J., Edwards, P. J., Schneider, M. K., Richter, M., and Blum, H.: Responses of net ecosystem CO_2 exchange in managed grassland to long-term CO_2 enrichment, N fertilization and plant species, Plant Cell Environ., 28, 823–833, 2005.

Altimir, N., Vesala, T., Keronen, P., Kulmala, M., and Hari, P.: Methodology for direct field measurements of ozone flux to foliage with shoot chambers, Atmos. Environ., 36, 19–29, 2002.

Baldocchi, D. D., Hincks, B. B., and Meyers, T. P.: Measuring biosphere–atmosphere exchanges of biologically related gases with micrometeorological methods, Ecology, 69, 1331–1340, 1988.

Benedict, K. B., Day, D., Schwandner, F. M., Kreidenweis, S. M., Schichtel, B., Malm, W. C., and Collett Jr., J. L.: Observations of atmospheric reactive nitrogen species in Rocky Mountain National Park and across northern Colorado, Atmos. Environ., 64, 66–76, 2013.

Breuninger, C., Oswald, R., Kesselmeier, J., and Meixner, F. X.: The dynamic chamber method: trace gas exchange fluxes (NO, NO_2, O_3) between plants and the atmosphere in the laboratory and in the field, Atmos. Meas. Tech., 5, 955–989, doi:10.5194/amt-5-955-2012, 2012.

Breuninger, C., Meixner, F. X., and Kesselmeier, J.: Field investigations of nitrogen dioxide (NO_2) exchange between plants and the atmosphere, Atmos. Chem. Phys., 13, 773–790, doi:10.5194/acp-13-773-2013, 2013.

Brook, J. R., Zhang, L., Di-Giovanni, F., and Padro, J.: Description and evaluation of a model of deposition velocities for routine estimates of air pollutant dry deposition over North America: Part I: model development, Atmos. Environ., 33, 5037–5051, 1999.

Chamberlain, A. and Chadwick, R. C.: Deposition of airborne radioiodine vapour, Nucleonics, 11, 22–25, 1953.

DeHayes, D. H., Schaberg, P. G., Hawley, G. J., and Strimbeck, G. R.: Acid rain impacts on calcium nutrition and forest health alteration of membrane-associated calcium leads to membrane destabilization and foliar injury in red spruce, BioScience, 49, 789–800, 1999.

Driscoll, C. T., Lawrence, G. B., Bulger, A. J., Butler, T. J., Cronan, C. S., Eagar, C., Lambert, K. F., Likens, G. E., Stoddard, J. L., and Weathers, K.: Acidic Deposition in the Northeastern United States: Sources and Inputs, Ecosystem Effects, and Management Strategies: The effects of acidic deposition in the northeastern United States include the acidification of soil and water, which stresses terrestrial and aquatic biota, BioScience, 51, 180–19, 2001.

EPA US: Clean Air Status and Trends Network 2010 Annual Report, Tech. rep., AMEC Environment & Infrastructure, Inc, prepared for US EPA Under Contract No. EP-W-09-028, 2010.

Fangmeier, A., Hadwiger-Fangmeier, A., Van der Eerden, L., and Jäger, H.-J.: Effects of atmospheric ammonia on vegetation – a review, Environ. Pollut., 86, 43–82, 1994.

Fenn, M. E., Poth, M. A., Aber, J. D., Baron, J. S., Bormann, B. T., Johnson, D. W., Lemly, A. D., McNulty, S. G., Ryan, D. F., and Stottlemyer, R.: Nitrogen excess in North American ecosystems: predisposing factors, ecosystem responses, and management strategies, Ecol. Appl., 8, 706–733, 1998.

Finkelstein, P. and Sims, P.: Sampling error in eddy correlation flux measurements, J. Geophys. Res., 106, 3503–3509, 2001.

Flechard, C. R., Nemitz, E., Smith, R. I., Fowler, D., Vermeulen, A. T., Bleeker, A., Erisman, J. W., Simpson, D., Zhang, L., Tang, Y. S., and Sutton, M. A.: Dry deposition of reactive nitrogen to European ecosystems: a comparison of inferential models across the NitroEurope network, Atmos. Chem. Phys., 11, 2703–2728, doi:10.5194/acp-11-2703-2011, 2011.

Fluxnet: Duke Forest Open Field, available at: http://fluxnet.ornl.gov/site/867 (last access: 14 December 2013), 2013.

Foken, T. and Wichura, B.: Tools for the quality assessment of surface-based flux measurements, Agr. Forest Meteorol., 78, 83–105, 1996.

Galbally, I. E. and Roy, C. R.: Destruction of ozone at the earth's surface, Q. J. Roy. Meteor. Soc., 106, 599–620, 1980.

Geßler, A., Rienks, M., and Rennenberg, H.: NH_3 and NO_2 fluxes between beech trees and the atmosphere–correlation with climatic and physiological parameters, New Phytol., 147, 539–560, 2000.

Gillis, A. and Miller, D. R.: Some potential errors in the measurement of mercury gas exchange at the soil surface using a dynamic flux chamber, Sci. Total Environ., 260, 181–189, 2000.

Guesten, H., Heinrich, G., Schmidt, R., and Schurath, U.: Tools for the quality assessment of surface-based flux measurements, J. Atmos. Chem., 14, 73–84, 1992.

Gut, A., Van Dijk, S., Scheibe, M., Rummel, U., Welling, M., Ammann, C., Meixner, F., Kirkman, G., Andreae, M., and Lehmann, B.: NO emission from an Amazonian rain forest soil: continuous measurements of NO flux and soil concentration, J. Geophys. Res.-Atmos., 107, LBA 24-1–LBA 24-10, doi:10.1029/2001JD000521, 2002.

Horst, T.: A simple formula for attenuation of eddy fluxes measured with first-order-response scalar sensors, Bound.-Lay. Meteorol., 82, 219–233, 1997.

Horst, T. and Weil, J.: How far is far enough? The fetch requirements for micrometeorological measurement of surface fluxes, J. Atmos. Ocean. Tech., 11, 1018–1025, 1994.

Horváth, L., Führer, E., and Lajtha, K.: Nitric oxide and nitrous oxide emission from Hungarian forest soils; linked with atmospheric N-deposition, Atmos. Environ., 40, 7786–7795, 2006.

Kaimal, J. and Finnigan, J.: Atmospheric Boundary-Layer Flows: Their Structure and Measurement, Oxford University Press, 1994.

Kirkman, G., Gut, A., Ammann, C., Gatti, L., Cordova, A., Moura, M., Andreae, M., and Meixner, F.: Surface exchange of nitric oxide, nitrogen dioxide, and ozone at a cattle pasture in Rondonia, Brazil, J. Geophys. Res.-Atmos., 107, LBA 51-1–LBA 51-17, doi:10.1029/2001JD000523, 2002.

Kitzler, B., Zechmeister-Boltenstern, S., Holtermann, C., Skiba, U., and Butterbach-Bahl, K.: Nitrogen oxides emission from two beech forests subjected to different nitrogen loads, Biogeosciences, 3, 293–310, doi:10.5194/bg-3-293-2006, 2006.

Li, Y., Aneja, V. P., Arya, S., Rickman, J., Brittig, J., Roelle, P., and Kim, D.: Nitric oxide emission from intensively managed agricultural soil in North Carolina, J. Geophys. Res.-Atmos., 104, 26115–26123, 1999.

Masson, N.: UPOD: An open-source platform for air quality monitoring, available at: http://mobilesensingtechnology.com/ (last access: 10 January 2014), 2014.

Meixner, F., Fickinger, T., Marufu, L., Serca, D., Nathaus, F., Makina, E., Mukurumbira, L., and Andreae, M.: Preliminary results on nitric oxide emission from a southern African savanna ecosystem, Nutr. Cycl. Agroecosys., 48, 123–138, 1997.

Muller, J. B. A., Percival, C. J., Gallagher, M. W., Fowler, D., Coyle, M., and Nemitz, E.: Sources of uncertainty in eddy covariance ozone flux measurements made by dry chemiluminescence fast response analysers, Atmos. Meas. Tech., 3, 163–176, doi:10.5194/amt-3-163-2010, 2010.

Musselman, R. C., Lefohn, A. S., Massman, W. J., and Heath, R. L.: A critical review and analysis of the use of exposure and flux-based ozone indices for predicting vegetation effects, Atmos. Environ., 40, 1869–1888, 2006.

Norman, J., Kucharik, C., Gower, S., Baldocchi, D., Crill, P., Rayment, M., Savage, K., and Striegl, R.: A comparison of six methods for measuring soil-surface carbon dioxide fluxes, J. Geophys. Res.-Atmos., 102, 28771–28777, 1997.

Pape, L., Ammann, C., Nyfeler-Brunner, A., Spirig, C., Hens, K., and Meixner, F. X.: An automated dynamic chamber system for surface exchange measurement of non-reactive and reactive trace gases of grassland ecosystems, Biogeosciences, 6, 405–429, doi:10.5194/bg-6-405-2009, 2009.

Parrish, D. D., Williams, E. J., Fahey, D. W., Liu, S. C., and Fehsenfeld, F. C.: Measurement of nitrogen oxide fluxes from soils: Intercomparison of enclosure and gradient measurement techniques, J. Geophys. Res.-Atmos., 92, 2165–2171, 1987.

Pleim, J. E., Bash, J. O., Walker, J. T., and Cooter, E. J.: Development and evaluation of an ammonia bidirectional flux parameterization for air quality models, J. Geophys. Res.-Atmos., 118, 3794–3806, 2013.

Pilegaard, K.: Air–soil exchange of NO, NO_2 and O_3 in forests, Water Air Soil Pollut. Focus, 1, 79–88, 2001.

Remde, A., Ludwig, J., Meixner, F. X., and Conrad, R.: A study to explain the emission of nitric oxide from a marsh soil, J. Atmos. Chem., 17, 249–275, 1993.

SAS Institute: Version 9.3 System Help, 2003.

Saylor, R. D., Wolfe, G. M., Meyers, T. P., and Hicks, B. B.: A corrected formulation of the Multilayer Model (MLM) for inferring gaseous dry deposition to vegetated surfaces, Atmos. Environ., 92, 141–145, 2014.

Schwede, D. B., Zhang, L., Vet, R., and Lear, G.: An intercomparison of the deposition models used in the CASTNET and CAPMoN networks, Atmos. Environ., 45, 1337–1346, 2011.

Seinfeld, J. H. and Pandis, S. N.: Atmospheric Chemistry and Physics, 2nd Edn., Wiley, 2006.

Sparks, J. P., Monson, R. K., Sparks, K. L., and Lerdau, M.: Leaf uptake of nitrogen dioxide (NO_2) in a tropical wet forest: implications for tropospheric chemistry, Oecologia, 127, 214–221, 2001.

Stella, P., Loubet, B., Laville, P., Lamaud, E., Cazaunau, M., Laufs, S., Bernard, F., Grosselin, B., Mascher, N., Kurtenbach, R., Mellouki, A., Kleffmann, J., and Cellier, P.: Comparison of methods for the determination of NO-O_3-NO_2 fluxes and chemical interactions over a bare soil, Atmos. Meas. Tech., 5, 1241–1257, doi:10.5194/amt-5-1241-2012, 2012.

Turnipseed, A., Burns, S., Moore, D., Hu, J., Guenther, A., and Monson, R.: Controls over ozone deposition to a high elevation subalpine forest, Agr. Forest Meteorol., 149, 1447–1459, 2009.

Unsworth, M., Heagle, A., and Heck, W.: Gas exchange in open-top field chambers – I. Measurement and analysis of atmospheric resistances to gas exchange, Atmos. Environ., 18, 373–380, 1984.

Webb, E. K., Pearman, G. I., and Leuning, R.: Correction of flux measurements for density effects due to heat and water vapour transfer, Q. J. Roy. Meteor. Soc., 106, 85–100, 1980.

Wesely, M.: Parameterization of surface resistances to gaseous dry deposition in regional-scale numerical models, Atmos. Environ., 23, 1293–1304, 1989.

Wesely, M. and Hicks, B. B.: A review of the current status of knowledge on dry deposition, Atmos. Environ., 34, 2261–2282, 2000.

Williams, E. and Davidson, E.: An intercomparison of two chamber methods for the determination of emission of nitric oxide from soil, Atmos. Environ. A-Gen., 27, 2107–2113, 1993.

Williams, M. and Tonnessen, K.: Critical loads for inorganic nitrogen deposition in the Colorado Front Range, USA, Ecol. Appl., 10, 1648–1665, 2000.

Wu, W., Zhang, G., and Kai, P.: Ammonia and methane emissions from two naturally ventilated dairy cattle buildings and the influence of climatic factors on ammonia emissions, Atmos. Environ., 61, 232–243, 2012.

Wu, Z., Wang, X., Chen, F., Turnipseed, A. A., Guenther, A. B., Niyogi, D., Charusombat, U., Xia, B., William Munger, J., and Alapaty, K.: Evaluating the calculated dry deposition velocities of reactive nitrogen oxides and ozone from two community models over a temperate deciduous forest, Atmos. Environ., 45, 2663–2674, 2011.

Zahn, A., Weppner, J., Widmann, H., Schlote-Holubek, K., Burger, B., Kühner, T., and Franke, H.: A fast and precise chemiluminescence ozone detector for eddy flux and airborne application, Atmos. Meas. Tech., 5, 363–375, doi:10.5194/amt-5-363-2012, 2012.

Zhang, L., Brook, J. R., and Vet, R.: A revised parameterization for gaseous dry deposition in air-quality models, Atmos. Chem. Phys., 3, 2067–2082, doi:10.5194/acp-3-2067-2003, 2003.

9

A semi-automated system for quantifying the oxidative potential of ambient particles in aqueous extracts using the dithiothreitol (DTT) assay: results from the Southeastern Center for Air Pollution and Epidemiology (SCAPE)

T. Fang[1], **V. Verma**[1], **H. Guo**[1], **L. E. King**[1], **E. S. Edgerton**[2], **and R. J. Weber**[1]

[1]School of Earth and Atmospheric Sciences, Georgia Institute of Technology, Atlanta, GA, USA
[2]Atmospheric Research and Analysis, Inc., Cary, NC, USA

Correspondence to: R. J. Weber (rodney.weber@eas.gatech.edu)

Abstract. A variety of methods are used to measure the capability of particulate matter (PM) to catalytically generate reactive oxygen species (ROS) in vivo, also defined as the aerosol oxidative potential. A widely used measure of aerosol oxidative potential is the dithiothreitol (DTT) assay, which monitors the depletion of DTT (a surrogate for cellular antioxidants) as catalyzed by the redox-active species in PM. However, a major constraint in the routine use of the DTT assay for integrating it with large-scale health studies is its labor-intensive and time-consuming protocol. To specifically address this concern, we have developed a semi-automated system for quantifying the oxidative potential of aerosol liquid extracts using the DTT assay. The system, capable of unattended analysis at one sample per hour, has a high analytical precision (coefficient of variation of 15 % for positive control, 4 % for ambient samples) and reasonably low limit of detection (0.31 nmol min^{-1}). Comparison of the automated approach with the manual method conducted on ambient samples yielded good agreement (slope $= 1.08 \pm 0.12$, $r^2 = 0.92$, $N = 9$). The system was utilized for the Southeastern Center for Air Pollution & Epidemiology (SCAPE) to generate an extensive data set on DTT activity of ambient particles collected from contrasting environments (urban, roadside, and rural) in the southeastern US. We find that water-soluble $PM_{2.5}$ DTT activity on a per-air-volume basis was spatially uniform and often well correlated with $PM_{2.5}$ mass ($r = 0.49$ to 0.88), suggesting regional sources contributing to the PM oxidative potential in the southeastern US. The correlation may also suggest a mechanistic explanation (oxidative stress) for observed $PM_{2.5}$ mass-health associations. The heterogeneity in the intrinsic DTT activity (per-PM-mass basis) across seasons indicates variability in the DTT activity associated with aerosols from sources that vary with season. Although developed for the DTT assay, the instrument can also be used to determine oxidative potential with other acellular assays.

1 Introduction

Epidemiological studies have associated increases in particulate matter (PM) levels with exacerbation of cardiovascular diseases (Zanobetti et al., 2014; Sun et al., 2010; Pope et al., 2004; Samet et al., 2000) and elevated incidence of respiratory disorders such as airway inflammation, bronchial muscle contraction, and asthma (Harkema et al., 2004; Schaumann et al., 2004; Aust et al., 2002; Norris et al., 1999). The mechanisms underlying these associations are not entirely clear, but reactive oxygen species (ROS) have been identified as a class of molecules that induce oxidative stress, causing cell damage (Nel, 2005; Gurgueira et al., 2002). ROS can either be adsorbed on inhaled particles or generated in vivo by target cells such as airway epithelial cells and macrophages (Li et al., 2003; Nel et al., 1998). These findings imply that aerosol oxidative potential, i.e., the ability of ambient particles to generate ROS, may be a more relevant measurement than PM mass concentration, or concentrations of specific aerosol chemical components, when attempting to link aerosols and health end points.

Both cellular (Xia et al., 2006; Kubátová et al., 2006; Bonvallot et al., 2001; Hiura et al., 1999; Antonini et al., 1998) and acellular methods (Zomer et al., 2011; Mudway et al., 2011; Jung et al., 2006; Cho et al., 2005; Venkatachari et al., 2005; Frampton et al., 1999) have been developed to measure the oxidative potential of PM. Both types of methods have advantages and disadvantages. Acellular methods require less-controlled environments and provide faster readouts of PM oxidative potential. One of the most widely used cell-free measures of particles oxidative potential is the DTT (dithiothreitol, $HSCH_2(CH(OH))_2CH_2SH$)) assay (Charrier and Anastasio, 2012; Verma et al., 2012, 2009a; Cho et al., 2005; Kumagai et al., 2002; Delfino et al., 2013). Typically, ROS in vivo are mainly produced in mitochondria and endoplasmic reticulum (ER) where molecular oxygen (O_2) are reduced to superoxide anion (O_2^-) by accepting electrons from cellular reductants, such as NADPH, or during the ER protein folding process (Alfadda and Sallam, 2012). The DTT assay simulates this electron-transfer mechanism based on the catalytic ability of redox-active species to transfer electrons from DTT to oxygen, and thus can be considered a surrogate measure of the in vivo capacity of PM to induce ROS. The rate of the reaction, commonly called DTT activity, is determined by measuring the consumption of DTT over time, which is proportional to the concentration of redox-active species in PM extracts. Researchers have identified various chemical components that may participate in the reaction, including polycyclic aromatic hydrocarbons (PAHs) (Li et al., 2003), quinones (Chung et al., 2006; Kumagai et al., 2002), transition metals (Charrier and Anastasio, 2012), water-soluble organic carbon (WSOC) (Verma et al., 2009a; Cho et al., 2005), and HUmic-LIke Substances (HULIS) (Verma et al., 2012; Lin and Yu, 2011). However, a consensus on the relative contributions of these species in the overall DTT activity of ambient particles is currently lacking.

Studies have also reported the association between DTT activity of ambient particles and cellular biomarkers such as fractional exhaled nitric oxide (FE_{NO}) – a marker of airway inflammation (Delfino et al., 2013); heme oxygenase (HO-1) – an enzyme responsive to oxidative stress (Li et al., 2003); and 3-(4,5-dimethylthiazol-2-yl)-2,5-diphenyltetrazolium bromide (MTT)-reduction activity (Steenhof et al., 2011). These studies suggest a plausible mechanistic link between DTT-assessed oxidative potential and adverse health effects of ambient particles. Routine DTT analysis as part of large-scale toxicological and health studies involving aerosol exposure is needed to further establish the health relevance of the DTT assay (and other chemical assays in general).

One major limitation in conducting the DTT assay on a large scale is its laborious and time-intensive analytical protocol, which requires precise handling practices at each step of the reaction for an accurate determination of the rate of DTT oxidation. It generally takes at least 5 h for a two-person team to analyze five samples manually, limiting the total number of samples that can be analyzed daily. Researchers have also developed alternative approaches to the traditional DTT protocol (Cho et al., 2005) for new applications. For example, a paper-based analytical device (μPAD) DTT assay was developed for personal exposure studies (Sameenoi et al., 2012b), and a microfluidic electrochemical sensor coupled with a particle-into-liquid sampler (PILS) was developed for real-time measurement of aerosol DTT activity (Koehler et al., 2014; Sameenoi et al., 2012a). We have developed a semi-automated system using programmable syringe pumps with selector valves for conducting the DTT assay on various extracts. The system is based on a simplified protocol in which the aerosol extract oxidizes DTT in a single vial. A small aliquot is withdrawn at various time intervals to determine the remaining DTT concentration and calculate the rate of DTT consumption. Particulate samples are extracted to an aqueous state and analyzed in batches using an autosampler at roughly 1 h per sample. The semi-automated system can run for 24 h unattended and can also be monitored remotely. Extraction liquids can be either deionized water or organic solvents; in the latter case the solvent is evaporated and reconstituted in deionized water. The system was validated against the manual protocol (Cho et al., 2005) using both positive controls and ambient filter samples. Detailed comparisons of the semi-automated system and traditional manual method are shown in Table S1 in the Supplement. The DTT activity of more than 500 samples collected from the southeastern United States as part of the Southeastern Center for Air Pollution and Epidemiology (SCAPE) were analyzed for this study. Here we provide a detailed description and characterization of the automated system and an overview of aerosol oxidative potential in the southeastern United States. Sources and components of $PM_{2.5}$ that produce the DTT activity are presented in Verma et al. (2014).

2 Methods

2.1 Chemicals

Dithiothreitol (DTT), Tris base, dimethyl sulfoxide (DMSO), 5,5′-dithiobis-(2-nitrobenzoic acid) (DTNB), and 9,10-phenanthraquinone (PQN) were obtained from Sigma Aldrich (St. Louis, MO, USA). Trichloroacetic acid (TCA, 10 % w/v) was obtained from LabChem Inc. (Pittsburgh, PA, USA). Monopotassium phosphate (KH_2PO_4), dipotassium phosphate (K_2HPO_4), and ethylenediaminetetraacetate (EDTA) were obtained from VWR International LLC (West Chester, PA, USA). Details of the chemical preparation and storage are provided in the Supplement.

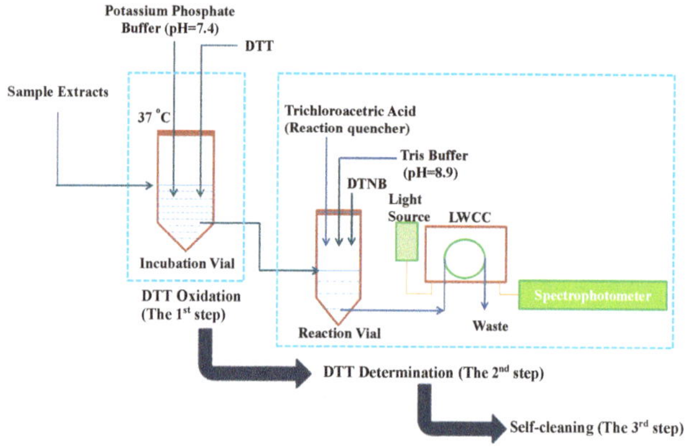

Figure 1. Automated system protocol.

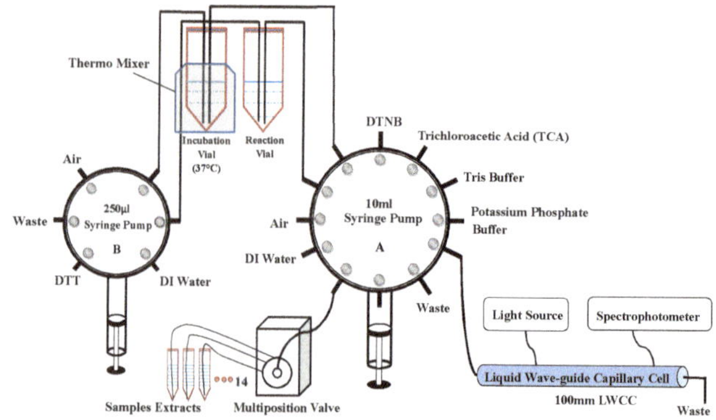

Figure 2. Automated system setup.

2.2 Automated DTT assay system development

2.2.1 Automated DTT assay system

Illustrated protocol and setup schematic of the semi-automated DTT assay system are shown in Figs. 1 and 2, respectively. There are three steps in the automated DTT method. In the first step (DTT oxidation step), 3.5 mL of the aerosol extract sample and 1 mL potassium phosphate buffer (0.5 mM) are loaded into an incubation vial (Figs. 1 and 2) (sterile polypropylene centrifuge tube, VWR International LLC, Suwanee, GA, USA) using a programmable syringe pump (Fig. 2, pump A) (Kloehn, Inc., Las Vegas, NV, USA). Then 0.5 mL of DTT (1 mM) is added to the incubation vial using another programmable syringe pump (Fig. 2, pump B). The DTT-buffer-sample mixture is incubated at 37 °C and continuously shaken at a rotational frequency of 400 rpm via a ThermoMixer (incubating accuracy: ±0.5 °C; Eppendorf North America, Inc., Hauppauge, NY, USA). DTT in the mixture is consumed over time in the incubation vial due to catalytic reaction of DTT-active components of the aerosol extract.

In the second step (DTT determination step), at a specified time (4 min) following the completion of step one, a small aliquot (100 μL) of the incubated mixture is withdrawn and transferred to another centrifuge tube (referred to as "reaction vial", wrapped in aluminum foil to prevent possible light interference), using pump B. This is mixed with 1 mL TCA (1 % w/v; the quenching agent), which had previously been added to the reaction vial by pump A. The quenched mixture with residual DTT is further mixed with 2 mL of Tris buffer (0.08 M with 4 mM EDTA) and 0.5 mL DTNB (0.2 mM) added by pump A. Reaction between the residual DTT and DTNB forms a light-absorbing product, 2-nitro-5-thiobenzoic acid (TNB), which has a high extinction coefficient of 14 150 M^{-1} cm^{-1} at 412 nm wavelength. Pump A then withdraws the final mixture from the reaction vial and pushes it through a Liquid Waveguide Capillary Cell (LWCC-M-100; World Precision Instruments, Inc., FL, USA) with an optical path length of 100 mm. The waveguide is coupled to an online spectrophotometer (Ocean Optics, Inc., Dunedin, FL, USA), which included a ultraviolet–visible (UV–Vis) light source (Ocean Optics DT-Mini-2), and a multi-wavelength light detector (USB4000 Miniature Fiber Optic Spectrometer). Absorbance intensity at 412 and 700 nm (chosen as the baseline absorbance for TNB) are recorded every 2 s using data acquisition software (SpectraSuite). The system then performs a self-cleaning using deionized water (DI water, > 18 MΩ cm^{-1}) to eliminate any residual liquid in the reaction vial, tubing, syringes, and LWCC. This second step is repeated four more times, at specific time intervals (13, 23, 30 and 41 min), generating a total of 5 data points of remaining DTT concentration with time. Following this, the automated system again performs a self-cleaning routine (third step) to ensure no carry-over in the incubation vial, tubing, syringes, and LWCC, before analyzing the next aerosol sample. A 14-port multiposition valve (VICI® Valco Instrument Co. Inc., USA) is used to consecutively select individual aerosol samples for analysis. The Kloehn control program code written for conducting the DTT assay, including hardware and software details, is included in the Supplement. The automated system is cleaned periodically (about every 15 days) by flushing at least three times with methanol, followed by six times with DI water.

2.2.2 DTT activity determination

Figure 3 shows an example of the absorbance intensity plot measured over time for a filter blank (panel a) and sample (panel b), obtained from the data acquisition software. The time intervals (4, 13, 23, 30, and 41 min) represent the incubation duration of DTT and sample in potassium phosphate buffer for each measurement. Withdrawal of the mixture containing TNB and pushing it through the LWCC (DTT determination step) causes the corresponding jumps in light absorbance at 412 nm (A_1, A_2, A_3, A_4, A_5 in Fig. 3a). After the absorbance measurement, pump A pushes DI water through the LWCC, which returns the absorbance back to the baseline value (i.e., absorbance at 412 nm equals

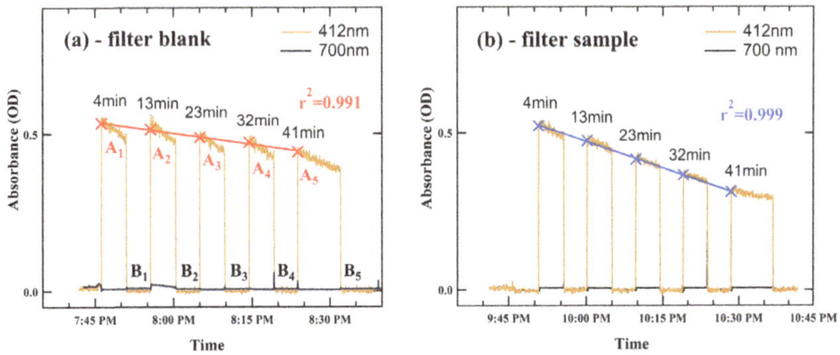

Figure 3. Example of an absorbance plot for a filter blank **(a)** and an ambient aerosol sample **(b)**.

absorbance at 700 nm; B_1, B_2, B_3, B_4, B_5 in Fig. 3a), thus generating a series of roughly square bars. A decreasing absorption intensity between successive measurements ($A_1 > A_2 > A_3 > A_4 > A_5$) for a sample reflects the DTT oxidation over time. During each specified DTT measurement time interval, absorbance decreases (angled top of square wave in Fig. 3); the average of the initial five absorbance data is taken as characteristic absorption for each interval. The rate of DTT consumption (σDTT, nmol min^{-1}) was determined from the slope and intercept of linear regression of measured absorbance versus time as follows:

$$\sigma \text{DTT} = -\sigma \text{Abs} \cdot \frac{N_0}{\text{Abs}_0}, \tag{1}$$

where σAbs is the slope of absorbance versus time; Abs_0 is the initial absorbance calculated from the intercept of the linear regression of absorbance versus time; and N_0 is the initial moles of DTT added in the reaction vial. A steeper slope corresponds to a higher rate of DTT consumption. The final DTT activity for a sample was calculated by subtracting a blank value from the sample and normalized by sample air volume or particle overall mass (or mass of a specific component) represented by the sample in the incubation vial, expressed in the units of nmol min^{-1} m^{-3} for volume-normalized DTT activity (DTTv), or nmol min^{-1} μg^{-1} for mass-normalized DTT activity (DTTm). DTTm represents the intrinsic property of particles linked to sources, while DTTv accounts for the emission rate, dilution, etc., and characterizes exposure to the aerosol. Thus,

$$\text{DTT activity} = \frac{\sigma \text{DTT}_{\text{sample}} - \sigma \text{DTT}_{\text{blank}}}{V_{\text{air}} \left(\text{or } M_{\text{particle}}\right)}, \tag{2}$$

where V_{air} and M_{particle} are the ambient air volume (m^3) and total PM mass (μg) represented by the sample in the incubation vial, respectively. For example, the total PM mass in the incubation vial for a sample with liquid concentration of 40 μg mL^{-1} would be 140 μg (3.5 mL × 40 μg mL^{-1}).

2.3 Ambient sample collection and preparation

2.3.1 Sampling

For this study, ambient $PM_{2.5}$ samples were collected over 23 h on prebaked quartz filters (Pallflex® Tissuquartz™, 8 × 10 inches) using high-volume (Hi-Vol) samplers (Thermo Anderson, flow rate normally 1.13 m^3 min^{-1}) as part of SCAPE. One sampler was fixed at a stationary site, Jefferson Street (JST), a central site representative of the Atlanta urban environment, while the other sampler (Trailer) was rotated among a roadside (RS), a near-road (Georgia Tech – GT), and a rural site (Yorkville – YRK), sampling in parallel with the fixed monitoring station (JST). Site characteristics are as follows:

1. RS site was situated within 5 m of the interstate highway (I75/85) in midtown Atlanta and was chosen for capturing immediate traffic emissions;

2. GT site was located on the rooftop of the Ford Environmental Science and Technology building at Georgia Tech, Atlanta, roughly 30 m above ground level, 840 m from the RS site, providing an intermediate location between RS and the central urban site (JST);

3. YRK site was situated in an agricultural region located approximately 70 km west of JST, representative of a rural environment.

Other sites also include an urban site in Birmingham (BHM), Alabama, and its paired rural site, Centerville (CTR), Alabama. JST, YRK, BHM, and CTR are all part of the Southeastern Aerosol Research and Characterization Study (SEARCH) network sites (Hansen et al., 2003). A map of sampling sites is provided in the Supplement. Ambient particles were collected from June 2012 to September 2013. For two periods (November 2012 and April 2013), the trailer was located at the stationary JST site for side-by-side comparisons. Table 1 provides the sampling schedule and number of filters collected at each site. In total, 503 Hi-Vol filters were collected over 15 months. Three seasons are defined based

Table 1. Sampling schedule and number of Hi-Vol filters collected at each site from June 2012 to September 2013.

Year/month season	Stationary site	Sample size	Trailer site	Sample size
2012/Jun–Jul summer	JST	31	YRK	33
2012/Jul–Aug summer	JST	37	GT	38
2012/Sep–Oct fall	JST	26	RS	29
2012/Nov	JST	13	JST	14
2012/Dec winter	JST	22	YRK	22
2013/Jan–Feb winter	JST	30	RS	31
2013/Mar winter	JST	23	GT	22
2013/Apr	JST	14	JST	14
2013/Jun–Jul summer	CTR	30	BHM	31
2013/Sep–Oct fall	GT	23	RS	20
Total number of filters				503

on the temperature profile throughout sampling period (see Supplement, Fig. S2) – summer, fall, and winter, as shown in Table 1. Collected samples were immediately wrapped in prebaked aluminum foil and stored at −18 °C until analyzed.

2.3.2 Extraction of PM filters

All filters extracted in DI water via the following method were also extracted in methanol, dried, and reconstituted in DI water for DTT analysis. Both water and methanol extracts were separated by solid-phase extraction (C-18 resin). DTT activity was analyzed with this system in order to explore the oxidative potential of chemical sub-fractions of the ambient aerosol. Thus, not including blanks and positive controls, for this project on the order of 2000 extracts were analyzed for DTT activity. For the water extraction procedure, portions of the filters (about 1/28 of the area of the Hi-Vol filter, three 1-inch-diameter punches) were punched and extracted in 15 mL of DI water in sterile polypropylene centrifuge tubes (VWR International LLC, Suwanee, GA, USA) by sonication using an Ultrasonic Cleanser (VWR International LLC, West Chester, PA, USA) for 30 min. Extracts were then filtered using PTFE 0.45 μm syringe filters (Fisherbrand™, Fisher Scientific, PA, USA) to remove insoluble components, and divided into two fractions. One fraction of at least 5 mL was analyzed for DTT activity using the automated system. The other fraction was reserved for other chemical analysis including water-soluble organic carbon and brown carbon. Here we focus only on the discussion of DTT analysis of the direct water extracts.

2.4 $PM_{2.5}$ mass concentration

$PM_{2.5}$ mass concentration were measured by a Tapered Element Oscillating Microbalance (TEOM) at the SEARCH sites – i.e., JST, YRK, BHM, and CTR by Atmospheric Research Analysis (ARA, Inc.) – and retrieved from the data archive. For the RS and GT sites, the PM mass concentrations were estimated from the sum of chemical components analyzed on the Hi-Vol filters and the summation method was calibrated with ARA TEOM at the JST site (slope $= 1.34 \pm 0.06$, intercept $= -2.46 \pm 0.59\,\mu g\,m^{-3}$, $r^2 = 0.72$, $N = 162$; see Supplement, Fig. S3). The measured components include elemental carbon (EC; Sunset Laboratory OCEC analyzer), organic mass (OC • 1.6; Turpin and Lim, 2001), water-soluble metals (measured by X-ray fluorescence method, XRF), and ammonium sulfate (assuming sulfate and ammonium are all $(NH_4)_2SO_4$ (Zhang et al., 2010), where sulfate was calculated from sulfur measured by XRF). Further description of the analytical procedure for each method will be reported in subsequent publications on the chemical data (Verma et al., 2014; Fang et al., 2015).

3 Results and discussion

3.1 Automated system performance

Performance of the automated system was assessed by conducting tests to determine instrument response, limit of detection (LOD), precision, and accuracy using both positive controls and ambient particles.

3.1.1 Automated system response to individual compound

PQN has been shown to be capable of catalyzing the oxidation of DTT (Li et al., 2009; Kumagai et al., 2002), although it is not highly water-soluble. The automated system was assessed for linearity with PQN (Fig. 4). The x axis intercept (0.11 nmol mL^{-1} in the incubation vial) represents the minimum concentration of PQN required to produce a measurable signal on the system with a pure water blank subtracted. As shown, the response of the system to PQN is highly linear with a correlation coefficient (r^2) of 0.95. At least one PQN per ambient sample batch (typically 12 samples) is used as a positive control to ensure the consistency of the system.

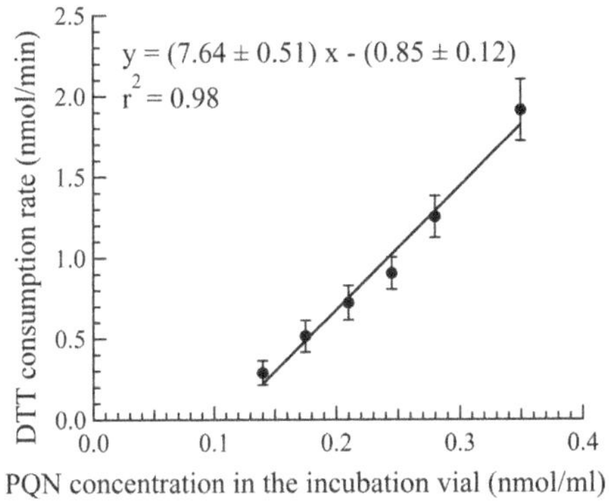

Figure 4. Blank corrected DTT consumption rate as a function of PQN (9,10-phenanthraquinone) used as a positive control. Each error bar represents the standard deviation of three independent DTT measurements on each concentration.

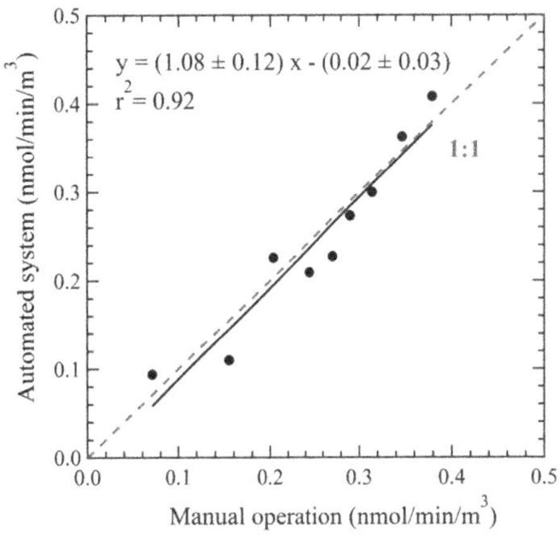

Figure 5. Comparison of the automated system with manual operation using ambient aerosol extracts ($PM_{2.5}$ samples collected from JST site, Atlanta, in December 2012) (regression analysis was done by orthogonal regression; the dotted line is 1 : 1).

3.1.2 Limit of detection

The limit of detection (LOD) of the system, defined as 3 times the standard deviation of DI water blanks ($N = 37$), is 0.31 nmol min^{-1}. Expressing the LOD in terms of the PM concentration of the sample extract ($\mu g\,mL^{-1}$) is not straightforward as it depends on many factors, including extraction efficiency and the relative fractions of DTT-active components in ambient particles. Based on the analysis of 503 Hi-Vol samples in the present study from different sites (urban, rural, and roadside), approximately 100 μg of $PM_{2.5}$ mass loading on the filter, extracted in 5 mL of DI water, was sufficient to have reliable results above LOD. The upper limit of the PM concentration is also constrained such that the DTT consumption remains less than 50 % of the initial concentration to ensure the pseudo-1st-order reaction of the DTT oxidation. For most of the PM samples collected, roughly 1/28 of the area (typical $PM_{2.5}$ mass loading = 0.2–1 mg) of a Hi-Vol filter, extracted in 15 mL of DI water, yielded a DTT consumption rate within these limits (0.8–2.6 nmol min^{-1}). In rare cases (< 10 %) of the DTT consumption exceeding 50 %, only the initial consumption data points (at least 3) were used for the rate calculation.

3.1.3 Precision

The precision of the automated system for ambient samples was assessed by separately extracting seven different equal sections (1/28 each) of the same Hi-Vol filter in 15 mL of DI water and analyzed for DTT activity. A low standard deviation of 0.081 nmol min^{-1} (coefficient of variation, CV = 4 %) indicates sufficiently high precision of the system for ambient samples.

3.1.4 Accuracy

The system was validated for accuracy by comparing the DTT activity of both positive controls and ambient PM samples obtained from the automated approach with that from the same experimental protocol performed manually (Cho et al., 2005).

Five replicates of the PQN solutions (0.21 nmol mL^{-1} in the incubation vial) were run both on the automated system and manually. The DTT consumption rate obtained from the automated system (mean ± SD of 0.77 ± 0.03 nmol min^{-1}, CV = 4.24 %) was very close to that from the manual operation (0.74 ± 0.03 nmol min^{-1}, CV = 3.97 %). As further validation, nine ambient PM samples were analyzed for DTT activity by both manual and automated approach. As shown in Fig. 5, an orthogonal fit yielded a slope (automated/manual) of 1.08 ± 0.12, intercept close to 0 (-0.02 ± 0.03 nmol $min^{-1}\,m^{-3}$), and correlation coefficient (r^2) of 0.92. Further, a paired t test shows no significant difference between the results obtained by the two methods ($t(8) < t_{critical}$, $p = 0.05$).

These tests demonstrate the robustness of the instrument as a viable alternative to the manual DTT assay, making it useful for rapid and high-throughput sample analysis for large-scale studies.

3.2 Field evaluation of the automated system

SCAPE was a coordinated multi-investigator effort to characterize ambient gas–particle mixtures in the southeastern US, to elucidate their sources, and to assess their impacts on human health. The automated system was used to measure DTT activity on the set of samples ($N = 503$) collected from multiple sites during SCAPE. Multiple DI water blanks ($N = 37$) and PQN solutions ($N = 55$), in addition to field blanks ($N = 63$) collected during the sampling, were analyzed along with the PM samples. This project provided an ideal opportunity for the field evaluation of the semi-automated DTT assay instrument. Table 2 summarizes the performance of the sys-

Table 2. Performance of the automated system as assessed by consistency of DTT consumption rate for blanks and positive controls.

Sample	Sample size	Standard deviation, nmol/min	%CV
DI blank	37	0.10	28.1
Field blank extracted in DI water	63	0.11	31.1
Positive controls (9,10-phenanthrenequinone)	55	0.19	15.0

Coefficient of variation (%CV) = SD/Avg · 100

tem as evaluated by stability of the results from both positive controls and blanks. The system remained fairly consistent throughout the analysis with reasonably small variability for both positive controls (CV = 15 %) and blanks (DI blanks: CV = 28.1 %; field blanks: CV = 26.7 %).

3.2.1 Overall method precision

The overall precision of the paired ambient sampling method utilized in this study was further assessed by comparing the DTT activity of $PM_{2.5}$ samples collected simultaneously at JST using two Hi-Vol samplers in November 2012 and April 2013 (shown in Fig. 6). The orthogonal regression yields a slope of 1.03 ± 0.05, with an intercept of 0.02 ± 0.01 nmol min^{-1} m^{-3}, and $r^2 = 0.96$. Considering the combined uncertainties from sample collection, pretreatment, and extraction, the good agreement between the two sampling systems demonstrates a high overall precision of DTT measurement for ambient samples.

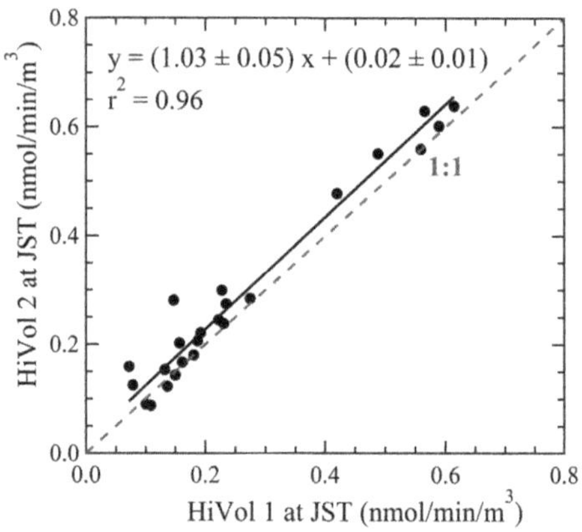

Figure 6. Overall method precision of the automated system assessed by $PM_{2.5}$ filter samples ($N = 24$) collected simultaneously using two Hi-Vol samplers deployed side-by-side at JST during November 2012 and April 2013 (regression analysis was done by orthogonal regression; the dotted line is 1 : 1).

3.2.2 DTT activity of ambient samples

The time series of both water-soluble, volume-normalized (DTTv) and water-soluble, mass-normalized (DTTm) DTT activity are shown in Fig. 7a and b, respectively. Comparing the time series of DTTv (Fig. 7a) between paired sites shows that there is generally little divergence between sites, with the exception of the JST–YRK (urban–rural) pair in winter. Some of this uniformity is due to the 23 h integrated sampling time of the Hi-Vol filters, which dampens any diurnal variability in the emission sources contributing to PM oxidative potential. Figure 8 shows the distribution of water-soluble DTTv and DTTm data of ambient $PM_{2.5}$ obtained in our study in comparison with those from other studies (Charrier and Anastasio, 2012; Verma et al., 2009b; Hu et al., 2008; Ntziachristos et al., 2007; Cho et al., 2005). Our DTTv data (Fig. 8a) are in the range reported in other studies, with the exception of RS (measurements made adjacent to a high-traffic highway), which is lower in the present study. However, comparing the DTTm levels, RS DTTm are well within the typical range of other studies, and thus our lower DTTv levels are most probably attributed to higher $PM_{2.5}$ concentrations reported in other studies sampling by roadways. Our study generally shows a broader span for urban and rural sites than reported previously, which is likely due to the much larger data set collected over a yearlong period that captured a wider range of sources and ambient conditions.

3.2.3 Seasonal and spatial variability

The spatiotemporal variability of DTTv can be used to assess the exposure to total water-soluble DTT-active species in different seasons and sites. Since DTTm is represented by DTTv divided by $PM_{2.5}$ mass concentration, which is the intrinsic property of aerosol independent of exposure levels, the spatiotemporal variability of DTTm provides information on the variation of DTT-active chemical components. Seasonal and spatial variability of aerosol oxidative potential (both DTTv and DTTm) in the southeastern US was assessed by analysis of variance (ANOVA) tests and coefficient of divergence (COD), respectively. The ANOVA tests were used to assess differences between seasons at a given site (Supplement, Tables S2 and S3), and the COD (Wilson et al., 2005) was calculated to assess spatial variability (see Table 3). The COD is calculated as follows:

$$\mathrm{COD} = \sqrt{\frac{1}{N}\sum_{i=1}^{N}\left[\frac{c_{ij}-c_{ik}}{c_{ij}+c_{ik}}\right]^2}, \tag{3}$$

where c_{ij} and c_{ik} are the 23 h averaged DTT activity (nmol min^{-1} m^{-3}) measured at site j and k, respectively, and

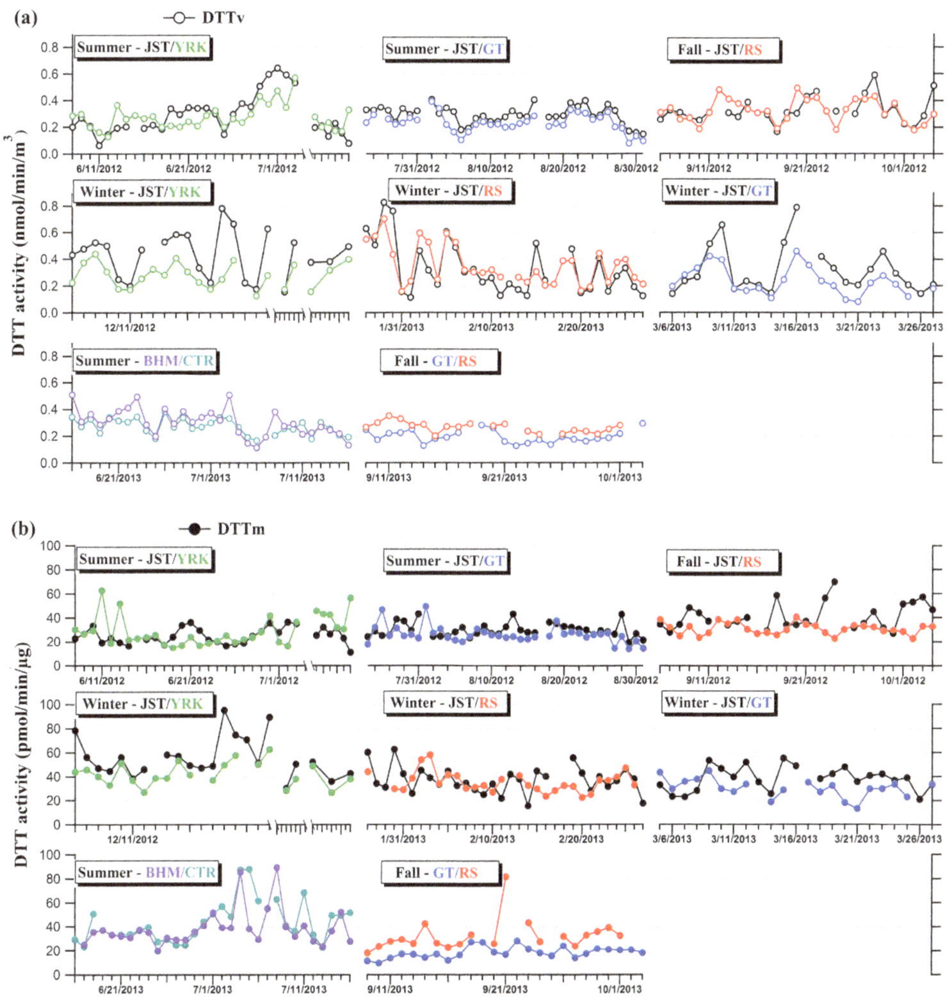

Figure 7. Water-soluble DTT activity of the ambient particles collected at paired sampling sites. The plots show **(a)** volume-normalized (DTTv) and **(b)** mass-normalized (DTTm) DTT activity at JST (urban, black) paired with YRK (rural, green), RS (roadside, red), and GT (near road, blue). Also included is the CTR (rural, cyan)–BHM (urban, purple) pair, and the GT–RS pair.

Table 3. Spatial heterogeneity of DTTv and DTTm assessed by coefficient of divergence (COD).

Season	Paired sites	CODv (DTTv)	CODm (DTTm)
Summer	JST–YRK	0.21	0.23
Summer	JST–GT	0.15	0.15
Fall	JST–RS	0.13	0.23
Winter	JST–YRK	0.25	0.15
Winter	JST–RS	0.17	0.18
Winter	JST–GT	0.23	0.22
Summer 2013	BHM–CTR	0.12	0.15
Fall 2013	GT–RS	0.18	0.31

N is the sample size. COD ranges from 0 to 1, with values close to 0 representing a homogenous distribution and those near 1 indicating heterogeneity.

Based on ANOVA tests, there was high heterogeneity across seasons for DTTv at JST ($p = 0.01$; see Supplement, Table S2), with the highest DTTv in winter (December (winter)/June (summer) = 1.51), while there was no significant seasonal variation observed at YRK, GT, or the RS site ($p > 0.01$). In comparison to DTTv, there are greater seasonal variations in DTTm; for example, average DTTm at most sites showed much higher levels in winter than summer and fall (winter/summer = 1.4, 1.2, and 2.2 for YRK, GT, and JST, respectively). ANOVA tests also validated the pronounced seasonal differences for DTTv and DTTm at JST and DTTm at other sites ($p < 0.01$; see Supplement, Tables S2 and S3). The higher seasonal differences in DTTm may suggest that the specific chemical components that contribute to the oxidative potential of particles vary between seasons and originate from different sources.

Relatively low levels of the CODs (< 0.25) (Table 3) found for both DTTv and DTTm at paired sites indicate spatial

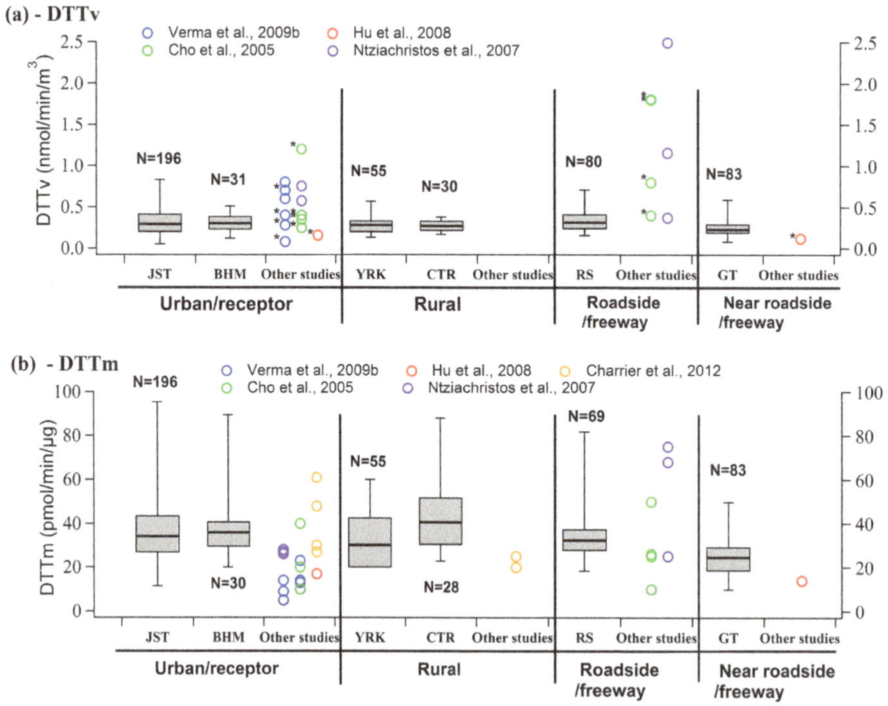

Figure 8. Distribution of water-soluble volume-normalized **(a)** and mass-normalized **(b)** DTT activity (oxidative potential) of ambient $PM_{2.5}$ in our study compared with other studies. Data from this study are expressed as median, maximal, minimal, and percentile 25 and 75. Roadside/freeway – sampling sites are located adjacent to road/highway; near roadside/freeway – sampling sites are further away but less than 1 km from road/freeway. * – numbers not presented in the paper, thus estimated from graphs.

homogeneity of water-soluble aerosol oxidative potential in the region, suggesting dominant DTT activity sources are regional in nature, such as SOA (secondary organic aerosol) or a well-mixed atmosphere of emissions from distant sources (e.g., long-range transport). Note that JST–RS and GT–RS pairs, in both fall seasons, show slightly lower CODs (more uniformity) in DTTv than in DTTm (COD for DTTv $= 0.13$ and 0.18 compared to COD for DTTm $= 0.23$ and 0.31 for JST–RS and GT–RS, respectively). This indicates that, although there may be unique local sources at the RS site, for example primary vehicle emissions and re-suspended dust, their overall contribution was not substantial compared to the regional DTT signal and so was not clearly resolved.

3.2.4 Oxidative potential and $PM_{2.5}$ mass

Figure 9 shows the correlation between DTT activity (DTTv) and $PM_{2.5}$ mass concentrations for all the sites and seasons. DTTv correlates with PM concentration in various degrees (Pearson's $r = 0.49$–0.88), but generally the correlations are high for most sites. For JST, the representative urban site, Pearson's r is particularly high ($r = 0.76$–0.82) in all seasons. Lowest DTTv–$PM_{2.5}$ mass correlations were at the roadside site, probably due to the influence of unique RS sources that contributed to mass but not significantly to DTT activity (average $PM_{2.5}$ concentrations are 8.7 and $8.2\,\mu g\,m^{-3}$ at JST compared to 10.4 and $9.5\,\mu g\,m^{-3}$ at RS in fall and winter, respectively). It is important to note that, despite the significant correlation, the slope for DTT activity versus $PM_{2.5}$ mass varies among different sites and seasons. Overall, we attribute the variation in degree of association between DTT activity and $PM_{2.5}$ mass to the varying PM chemical composition.

For example, DTT activity correlates with WSOC in summer ($R > 0.67$) but correlates better with brown carbon in winter ($R > 0.78$). Further investigation on identifying the specific sources and aerosol chemical components linked to PM oxidative potential was discussed in Verma et al. (2014).

Studies have shown a correlation between $PM_{2.5}$ mass and health end points – such as heart rate variability (HRV), blood neutrophils, and lipid changes (Tong et al., 2012) – increased oxidative DNA damage and nucleobases (Peter and Steffen, 2010), and cardiovascular mortality (Gholampour et al., 2014). The overall correlation between water-soluble DTT activity and $PM_{2.5}$ mass concentration observed in this study may help explain, at least in part, some of these associations. However the varying degree of correlation between DTT activity and $PM_{2.5}$ mass at different sites and seasons suggests that there are additional advantages of including PM oxidative potential in health studies, rather than relying on PM mass alone.

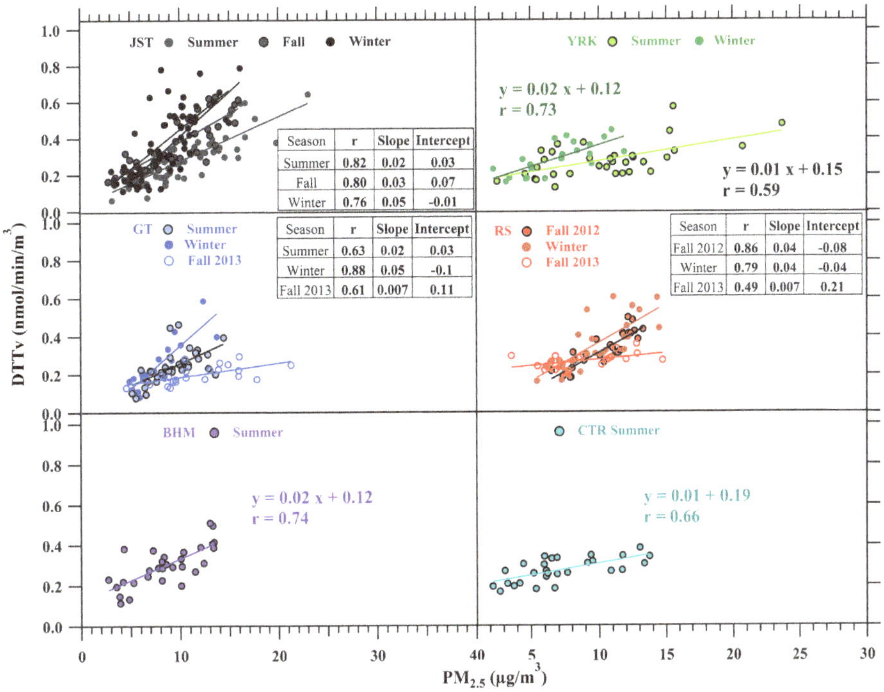

Figure 9. Correlation (Pearson's r) between water-soluble DTT activity (DTTv) and $PM_{2.5}$ mass concentration at JST (urban), YRK (rural), RS (roadside), GT (near-road), CTR (rural), and BHM (urban) sites.

4 Conclusions

An automated analytical system for quantifying the oxidative potential of aerosol liquid extracts using the DTT assay was developed. The system follows the analytical approach developed by Cho et al. (2005) and is capable of one DTT activity measurement per hour. The system response was assessed by PQN, which was used as a positive control when running a series of ambient samples. The method LOD was 0.31 nmol min^{-1}. Analytical precision based on both PQN (CV = 15 %) and ambient samples (CV = 4 %) was high. The instrument was further validated for accuracy by comparing with the manual procedure using ambient PM samples ($r^2 = 0.92$, slope = 1.08 ± 0.12). The suitability of the system for large-scale application was assessed by analyzing more than 500 filters collected in the southeastern US as part of SCAPE, the joint Emory–Georgia Tech Clean Air Research Center. The data show that water-soluble $PM_{2.5}$ DTT activity on a per-volume-of-air basis was spatially uniform and generally correlated with $PM_{2.5}$ mass ($r = 0.49$ to 0.88), indicating DTT activity in the southeast is likely, to a significant extent, related to regional sources and not dominated by a single source or a limited number of species. However, the higher seasonal heterogeneity in the intrinsic water-soluble DTT activity (per-PM-mass basis) may indicate that the dominant regional sources change with season. More in-depth analysis of the extensive data set generated with the instrument will be forthcoming. It is noted that the method can be altered to run smaller sample volumes, for situations involving samples of lower mass loadings. It has also been modified for ROS analysis using other assays (e.g., ascorbate depletion assay). The automated system presents a useful new tool for rapid and high-throughput measurement of the DTT activity of ambient particle extracts. Its application can facilitate routine use of PM oxidative potential in toxicological, panel exposures, and epidemiological studies. More work is required to identify specific compounds that are most sensitive in the DTT probe and test how they may contribute to the observed ambient DTT activity. Application of the DTT probe and other acellular and cellular assays in different regions that have a different mix of emission sources is also needed to better understand the possible links between aerosol oxidative potential and health end points. The automated DTT assay analytical instrument described here would facilitate these types of future studies. Finally, although the analytical system can provide new insight on aerosol oxidative properties from large data sets, it does not solve limitations associated with filter-based particle collection approaches, which entail artifacts due to losses of semivolatile DTT-active species during sampling and handling procedures. Adapting the DTT assay to an online system could limit sampling artifacts and provide new insights into DTT sources and atmospheric processing that more highly time-resolved data can provide.

Acknowledgements. We wish to thank the Southeastern Aerosol Research and Characterization (SEARCH) personnel for their many contributions supporting the field deployments. We also thank R. Erik Weber, Janessa Riana Rowland, and Madhusudan Kamat for assistance in lab work. This research was made possible by US EPA grant R834799. T. Fang acknowledges the support from the Oversea Study Program of Guangzhou Elite Project.

Edited by: P. Herckes

References

Alfadda, A. A. and Sallam, R. M.: Reactive oxygen species in health and disease, J. Biomed. Biotechnol., 2012, 936486, doi:10.1155/2012/936486, 2012.

Antonini, J. M., Clarke, R. W., Krishna Murthy, G. G., Sreekanthan, P., Jenkins, N., Eagar, T. W., and Brain, J. D.: Freshly generated stainless steel welding fume induces greater lung inflammation in rats as compared to aged fume, Toxicol. Lett., 98, 77–86, 1998.

Aust, A. E., Ball, J. C., Hu, A. A., Lighty, J. S., Smith, K. R., Straccia, A. M., Veranth, J. M., and Young, W. C.: Particle characteristics responsible for effects on human lung epithelial cells, Res. Rep. Health Eff. Inst., 110, 1–65, 2002.

Bonvallot, V., Baeza-Squiban, A., Baulig, A., Brulant, S., Boland, S., Muzeau, F., Barouki, R., and Marano, F.: Organic compounds from diesel exhaust particles elicit a proinflammatory response in human airway epithelial cells and induce cytochrome p450 1A1 expression, Am. J. Resp. Cell Mol., 25, 515–521, 2001.

Charrier, J. G. and Anastasio, C.: On dithiothreitol (DTT) as a measure of oxidative potential for ambient particles: evidence for the importance of soluble transition metals, Atmos. Chem. Phys., 12, 9321–9333, doi:10.5194/acp-12-9321-2012, 2012.

Cho, A. K., Sioutas, C., Miguel, A. H., Kumagai, Y., Schmitz, D. A., Singh, M., Eiguren-Fernandez, A., and Froines, J. R.: Redox activity of airborne particulate matter at different sites in the Los Angeles Basin, Environ. Res., 99, 40–47, 2005.

Chung, M. Y., Lazaro, R. A., Lim, D., Jackson, J., Lyon, J., Rendulic, D., and Hasson, A. S.: Aerosol-borne quinones and reactive oxygen species generation by particulate matter extracts, Environ. Sci. Technol., 40, 4880–4886, 2006.

Delfino, R. J., Staimer, N., Tjoa, T., Gillen, D. L., Schauer, J. J., and Shafer, M. M.: Airway inflammation and oxidative potential of air pollutant particles in a pediatric asthma panel, J. Expos. Sci. Environ. Epidemiol., 23, 466–473, 2013.

Fang, T., Verma, V., Peltier, R. E., Guo, H. Y., King, L. E., Edgerton, E. S., and Weber, R. J.: Distribution of ambient atmospheric water-soluble metals in the southeastern United States and insights into their complexation states, in preparation, 2015.

Frampton, M. W., Ghio, A. J., Samet, J. M., Carson, J. L., Carter, J. D., and Devlin, R. B.: Effects of aqueous extracts of PM_{10} filters from the Utah Valley on human airway epithelial cells, Am. J. Phys.-Lung C., 277, L960–L967, 1999.

Gholampour, A., Nabizadeh, R., Naseri, S., Yunesian, M., Taghipour, H., Rastkari, N., Nazmara, S., Faridi, S., and Mahvi, A. H.: Exposure and health impacts of outdoor particulate matter in two urban and industrialized area of Tabriz, Iran, Journal of Environmental Health Science and Engineering, 12, 27, doi:10.1186/2052-336X-12-27, 2014.

Gurgueira, S. A., Lawrence, J., Coull, B., Murthy, G. G., and Gonzalez-Flecha, B.: Rapid increases in the steady-state concentration of reactive oxygen species in the lungs and heart after particulate air pollution inhalation, Environ. Health Persp., 110, 749–755, 2002.

Hansen, D. A., Edgerton, E. S., Hartsell, B. E., Jansen, J. J., Kandasamy, N., Hidy, G. M., and Blanchard, C. L.: The Southeastern Aerosol Research and Characterization Study: Part 1-Overview, J. Air Waste Ma., 53, 1460–1471, 2003.

Harkema, J. R., Keeler, G., Wagner, J., Morishita, M., Timm, E., Hotchkiss, J., Marsik, F., Dvonch, T., Kaminski, N., and Barr, E.: Effects of concentrated ambient particles on normal and hypersecretory airways in rats, Res. Rep. Health Eff. Inst., 120, 1–68, 2004.

Hiura, T. S., Kaszubowski, M. P., Li, N., and Nel, A. E.: Chemicals in diesel exhaust particles generate reactive oxygen radicals and induce apoptosis in macrophages, J. Immunol., 163, 5582–5591, 1999.

Hu, S., Polidori, A., Arhami, M., Shafer, M. M., Schauer, J. J., Cho, A., and Sioutas, C.: Redox activity and chemical speciation of size fractioned PM in the communities of the Los Angeles-Long Beach harbor, Atmos. Chem. Phys., 8, 6439–6451, doi:10.5194/acp-8-6439-2008, 2008.

Jung, H., Guo, B., Anastasio, C., and Kennedy, I. M.: Quantitative measurements of the generation of hydroxyl radicals by soot particles in a surrogate lung fluid, Atmos. Environ., 40, 1043–1052, 2006.

Koehler, K. A., Shapiro, J., Sameenoi, Y., Henry, C., and Volckens, J.: Laboratory evaluation of a microfluidic electrochemical sensor for aerosol oxidative load, Aerosol Sci. Technol., 48, 489–497, 2014.

Kubátová, A., Dronen, L. C., Picklo, M. J., and Hawthorne, S. B.: Midpolarity and nonpolar wood smoke particulate matter fractions deplete glutathione in RAW 264.7 macrophages, Chem. Res. Toxicol., 19, 255–261, 2006.

Kumagai, Y., Koide, S., Taguchi, K., Endo, A., Nakai, Y., Yoshikawa, T., and Shimojo, N.: Oxidation of proximal protein sulfhydryls by phenanthraquinone, a component of diesel exhaust particles, Chem. Res. Toxicol., 15, 483–489, 2002.

Li, N., Sioutas, C., Cho, A. K., Schmitz, D., Misra, C., Sempf, J., Wang, M., Oberley, T., Froines, J., and Nel, A.: Ultrafine particulate pollutants induce oxidative stress and mitochondrial damage, Environ. Health Persp., 111, 455–460, 2003.

Li, Q. F., Wyatt, A., and Kamens, R. M.: Oxidant generation and toxicity enhancement of aged-diesel exhaust, Atmos. Environ., 43, 1037–1042, 2009.

Lin, P. and Yu, J. Z.: Generation of reactive oxygen species mediated by Humic-like Substances in atmospheric aerosols, Environ. Sci. Technol., 45, 10362–10368, 2011.

Mudway, I. S., Fuller, G., Green, D., Dunster, C., and Kelly, F. J.: Report: Quantifying the London specific component of PM_{10} oxidative activity, University of London, Defra, UK, 2011.

Nel, A.: Air pollution-related illness: effects of particles, Science, 308, 804–806, 2005.

Nel, A. E., Diaz-Sanchez, D., Ng, D., Hiura, T., and Saxon, A.: Enhancement of allergic inflammation by the interaction between diesel exhaust particles and the immune system, J. Allergy Clin. Immun., 102, 539–554, doi:10.1016/S0091-6749(98)70269-6, 1998.

Norris, G., YoungPong, S. N., Koenig, J. Q., Larson, T. V., Sheppard, L., and Stout, J. W.: An association between fine particles and asthma emergency department visits for children in Seattle, Environ. Health Persp., 107, 489–493, 1999.

Ntziachristos, L., Froines, J. R., Cho, A. K., and Sioutas, C.: Relationship between redox activity and chemical speciation of size-fractionated particulate matter, Particle and Fibre Toxicology, 4, 5, doi:10.1186/1743-8977-4-5, 2007.

Peter, M. and Steffen, L.: Oxidative damage to DNA and Lipids as biomarkers of exposure to air pollution, Environ. Health Persp., 118, 1126–1136, 2010.

Pope, C. A., Burnett, R. T., Thurston, G. D., Thun, M. J., Calle, E. E., Krewski, D., and Godleski, J. J.: Cardiovascular mortality and long-term exposure to particulate air pollution: epidemiological evidence of general pathophysiological pathways of disease, Circulation, 109, 71–77, 2004.

Sameenoi, Y., Koehler, K., Shapiro, J., Boonsong, K., Sun, Y., Collett, J., Volckens, J., and Henry, C. S.: Microfluidic electrochemical sensor for on-line monitoring of aerosol oxidative activity, J. Am. Chem. Soc., 134, 10562–10568, 2012a.

Sameenoi, Y., Panymeesamer, P., Supalakorn, N., Koehler, K., Chailapakul, O., Henry, C. S., and Volckens, J.: Microfluidic paper-based analytical device for aerosol oxidative activity, Environ. Sci. Technol., 47, 932–940, 2012b.

Samet, J. M., Dominici, F., Curriero, F. C., Coursac, I., and Zeger, S. L.: Fine particulate air pollution and mortality in 20 U.S. cities, 1987–1994, New Engl. J. Med., 343, 1742–1749, 2000.

Schaumann, F., Borm, P. J. A., Herbrich, A., Knoch, J., Pitz, M., Schins, R. P. F., Luettig, B., Hohlfeld, J. M., Heinrich, J., and Krug, N.: Metal-rich ambientparticles (particulate matter2.5) cause airway inflammation in healthy subjects, Am. J. Resp. Crit. Care, 170, 898–903, 2004.

Steenhof, M., Gosens, I., Strak, M., Godri, K. J., Hoek, G., Cassee, F. R., Mudway, I. S., Kelly, F. J., Harrison, R. M., Lebret, E., Brunekreef, B., Janssen, N. A. H., and Pieters, R. H. H.: In vitro toxicity of particulate matter (PM) collected at different sites in the Netherlands is associated with PM composition, size fraction and oxidative potential–the RAPTES project, Particle and Fibre Toxicology, 8, 26, doi:10.1186/1743-8977-8-26, 2011.

Sun, Q. H., Hong, X. R., and Wold, L. E.: Cardiovascular effects of ambient particulate air pollution exposure, Circulation, 121, 2755–2765, 2010.

Tong, H., Rappold, A. G., Diaz-Sanchez, D., Steck, S. E., Berntsen, J., Cascio, W. E., Devlin, R. B., and Samet, J. M.: Omega-3 fatty acid supplementation appears to attenuate particulate air pollution-induced cardiac effects and lipid changes in healthy middle-aged adults, Environ. Health Persp., 120, 952–957, 2012.

Turpin, B. J. and Lim, H.-J.: Species contributions to $PM_{2.5}$ mass concentrations: revisiting common assumptions for estimating organic mass, Aerosol Sci. Tech., 35, 602–610, 2001.

Venkatachari, P., Hopke, P. K., Grover, B. D., and Eatough, D. J.: Measurement of particle-bound reactive oxygen species in rubidoux aerosols, J. Atmos. Chem., 50, 49–58, 2005.

Verma, V., Ning, Z., Cho, A. K., Schauer, J. J., Shafer, M. M., and Sioutas, C.: Redox activity of urban quasi-ultrafine particles from primary and secondary sources, Atmos. Environ., 43, 6360–6368, 2009a.

Verma, V., Polidori, A., Schauer, J. J., Shafer, M. M., Cassee, F. R., and Sioutas, C.: Physicochemical and toxicological profiles of particulate matter in Los Angeles during the October 2007 southern California wildfires, Environ. Sci. Technol., 43, 954–960, 2009b.

Verma, V., Rico-Martinez, R., Kotra, N., King, L. E., Liu, J., Snell, T. W., and Weber, R. J.: Contribution of water-soluble and insoluble components and their hydrophobic/hydrophilic subfractions to the reactive oxygen species-generating ootential of fine ambient aerosols, Environ. Sci. Technol., 46, 11384–11392, 2012.

Verma, V., Fang, T., Guo, H., King, L., Bates, J. T., Peltier, R. E., Edgerton, E., Russell, A. G., and Weber, R. J.: Reactive oxygen species associated with water-soluble $PM_{2.5}$ in the southeastern United States: spatiotemporal trends and source apportionment, Atmos. Chem. Phys., 14, 12915–12930, doi:10.5194/acp-14-12915-2014, 2014.

Wilson, J. G., Kingham, S., Pearce, J., and Sturman, A. P.: A review of intraurban variations in particulate air pollution: Implications for epidemiological research, Atmos. Environ., 39, 6444–6462, 2005.

Xia, T., Kovochich, M., Brant, J., Hotze, M., Sempf, J., Oberley, T., Sioutas, C., Yeh, J. I., Wiesner, M. R., and Nel, A. E.: Comparison of the abilities of ambient and manufactured nanoparticles to induce cellular toxicity according to an oxidative stress paradigm, Nano Lett., 6, 1794–1807, 2006.

Zanobetti, A., Coull, B. A., Gryparis, A., Kloog, I., Sparrow, D., Vokonas, P. S., Wright, R. O., Gold, D. R., and Schwartz, J.: Associations between arrhythmia episodes and temporally and spatially resolved black carbon and particulate matter in elderly patients, Occup. Environ. Med., 71, 201–207, 2014.

Zhang, X., Hecobian, A., Zheng, M., Frank, N. H., and Weber, R. J.: Biomass burning impact on $PM_{2.5}$ over the southeastern US during 2007: integrating chemically speciated FRM filter measurements, MODIS fire counts and PMF analysis, Atmos. Chem. Phys., 10, 6839–6853, doi:10.5194/acp-10-6839-2010, 2010.

Zomer, B., Collé, L., Jedyńska, A., Pasterkamp, G., Kooter, I., and Bloemen, H.: Chemiluminescent reductive acridinium triggering (CRAT) – mechanism and applications, Anal. Bioanal. Chem., 401, 2945–2954, 2011.

10

PTR-QMS versus PTR-TOF comparison in a region with oil and natural gas extraction industry in the Uintah Basin in 2013

C. Warneke[1,2], P. Veres[1,2], S. M. Murphy[3], J. Soltis[3], R. A. Field[3], M. G. Graus[1,2,*], A. Koss[1,2], S.-M. Li[4], R. Li[1,2], B. Yuan[1,2], J. M. Roberts[2], and J. A. de Gouw[1,2]

[1]Cooperative Institute for Research in Environmental Sciences, Univ. of Colorado, Boulder, CO, USA
[2]NOAA Earth System Research Laboratory, Chemical Sciences Division, Boulder, CO, USA
[3]Department of Atmospheric Science, University of Wyoming, Laramie, WY, USA
[4]Air Quality Process Research Division, Environment Canada, Toronto, ON, Canada
[*]now at: Institute of Meteorology and Geophysics, Innsbruck University, Innsbruck, Austria

Correspondence to: C. Warneke (carsten.warneke@noaa.gov)

Abstract. Here we compare volatile organic compound (VOC) measurements using a standard proton-transfer-reaction quadrupole mass spectrometer (PTR-QMS) with a new proton-transfer-reaction time of flight mass spectrometer (PTR-TOF) during the Uintah Basin Winter Ozone Study 2013 (UBWOS2013) field experiment in an oil and gas field in the Uintah Basin, Utah. The PTR-QMS uses a quadrupole, which is a mass filter that lets one mass to charge ratio pass at a time, whereas the PTR-TOF uses a time of flight mass spectrometer, which takes full mass spectra with typical 0.1 s–1 min integrated acquisition times. The sensitivity of the PTR-QMS in units of counts per ppbv (parts per billion by volume) is about a factor of 10–35 times larger than the PTR-TOF, when only one VOC is measured. The sensitivity of the PTR-TOF is mass dependent because of the mass discrimination caused by the sampling duty cycle in the orthogonal-acceleration region of the TOF. For example, the PTR-QMS on mass 33 (methanol) is 35 times more sensitive than the PTR-TOF and for masses above 120 amu less than 10 times more. If more than 10–35 compounds are measured with PTR-QMS, the sampling time per ion decreases and the PTR-TOF has higher signals per unit measuring time for most masses. For UBWOS2013 the PTR-QMS measured 34 masses in 37 s and on that timescale the PTR-TOF is more sensitive for all masses. The high mass resolution of the TOF allows for the measurements of compounds that cannot be separately detected with the PTR-QMS, such as oxidation products from alkanes and cycloalkanes emitted by oil and gas extraction. PTR-TOF masses do not have to be preselected, allowing for identification of unanticipated compounds. The measured mixing ratios of the two instruments agreed very well ($R^2 \geq 0.92$ and within 20 %) for all compounds and masses monitored with the PTR-QMS.

1 Introduction

The measurements of volatile organic compounds (VOCs) with proton-transfer-reaction quadrupole mass spectrometry (PTR-QMS) have become a standard technique in atmospheric measurements on various platforms such as ground sites, ships and aircraft, because of its high time resolution and sensitivity. The instrument has been described and characterized in great detail over the last several years (Blake et al., 2009; de Gouw and Warneke, 2007; Hansel et al., 1995, 1999; Lindinger et al., 1998a, b; Warneke et al., 2001, 2003, 2011b) and many successful intercomparisons with other techniques such as gas chromatography mass spectrometry (GC-MS) have demonstrated the sensitivity and the selectivity of PTR-QMS (de Gouw et al., 2003; de Gouw and Warneke, 2007; Haase et al., 2012; Wisthaler et al., 2008). PTR-QMS is a chemical ionization mass spectrometry technique that detects VOCs using proton transfer reactions with H_3O^+, but has the drawback that only the nominal unit-resolution mass of one VOC at a time can be determined. To improve on the selectivity three approaches have been frequently investigated: (1) instead of H_3O^+ different reagent ions (Sulzer et al., 2012) have been used such as NO^+, O_2^+, NH_4^+ and Kr^+, (2) gas chromatographic preseparation of VOCs prior to PTR-QMS detection (Karl et al., 2001; Warneke et al., 2003) and (3) isomers are distinguished by fragmentation patterns generated with collision-induced dissociation (CID) (Müller et al., 2009; Warneke et al., 2004). In addition, other mass spectrometers have been used in place of the quadrupole such as ion traps (Mielke et al., 2008; Prazeller et al., 2003; Warneke et al., 2005a, b) and time of flight (TOF) mass spectrometers (Blake et al., 2004; Tanimoto et al., 2007).

The PTR-TOF instrument (Graus et al., 2010; Jordan et al., 2009; Müller et al., 2010) was recently developed by the University of Innsbruck and Ionicon Analytik and is commercially available. It is aimed to improve on the time response and on the selectivity of PTR-QMS by using a high-resolution TOF mass spectrometer from Tofwerk AG, which has the capability of recording mass spectra at a very high frequency (> 10 Hz) and with a high mass resolution (> 4000 m/Δ m). This mass resolution is often sufficient to distinguish between isobaric compounds.

Here we compare a commercial PTR-TOF with a standard PTR-QMS, both of which were operated side-by-side during a measurement campaign (UBWOS2013 – Uintah Basin Winter Ozone Study 2013) in an area of oil and natural gas exploration in the Uintah Basin, Utah. We evaluate the stability, sensitivity and the detection limits of both instruments and determine which instrument is more sensitive in a given measurement mode. We also describe the advantages and the additional science that can be done with PTR-TOF using the data from the UBWOS2013 campaign.

2 Experimental

2.1 UBWOS2013 field campaign

The UBWOS2013 was conducted in January and February 2013 in the Uintah Basin, Utah. This area has a very low population density, but about 10 000 active oil and natural gas wells are located within the basin. A map of the Uintah Basin study area with the oil and natural gas wells is shown in Fig. 1. This intense oil and natural gas extraction operation results in emissions of greenhouse gases, VOCs and nitrogen oxides (NO_x) (Gilman et al., 2013; Helmig et al., 2014; Howarth et al., 2011; Karion et al., 2013; Katzenstein et al., 2003; Kemball-Cook et al., 2010; Litovitz et al., 2013; Petron et al., 2012). The UBWOS2013 campaign was designed to investigate the unusual wintertime ozone production that was observed in basins with oil and natural gas exploration during strong inversions and snow-covered surfaces (Edwards et al., 2013; Helmig et al., 2014; Schnell et al., 2009).

Measurements were done at the heavily instrumented Horse Pool ground site, which is also shown in Fig. 1. During UBWOS2013, cold temperatures, snow on the ground and strong temperature inversions provided ideal conditions for wintertime ozone production and indeed ozone mixing ratios at Horse Pool exceeded 120 ppbv (parts per billion by volume) on several days. Emissions from oil and natural gas exploration in the Uintah Basin are mainly alkanes, cycloalkanes and aromatics related to the natural gas (Warneke et al., 2014), but active photochemistry produces, besides ozone, many different oxygenated VOCs (oxyVOCs). The Uintah Basin presents a unique VOC mix that is an excellent test case for demonstrating the additional capabilities of PTR-TOF, because many of the photochemically produced oxyVOCs are difficult to identify with standard PTR-QMS.

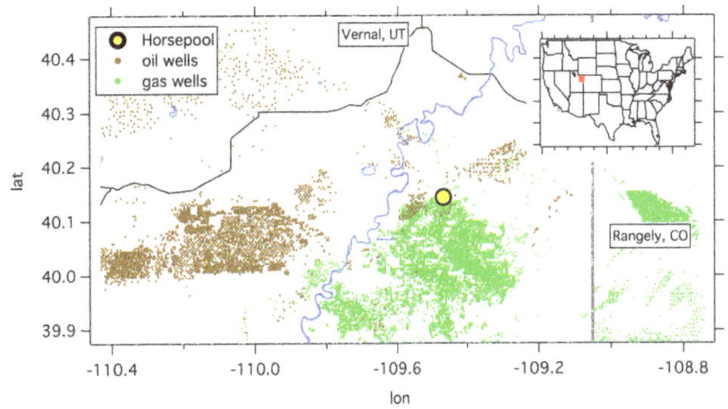

Figure 1. Map of the study area with the Horse Pool ground site in the Uintah Basin, Utah.

2.2 PTR-QMS and PTR-TOF instruments

PTR-QMS and PTR-TOF both use similar reaction chambers in which proton transfer reactions of H_3O^+ are used to ionize the VOCs of interest (Lindinger et al., 1998a). The main difference between the two instruments is the mass spectrometer, which is a quadrupole QMS420 from Pfeiffer Vacuum for the PTR-QMS and an orthogonal acceleration reflectron TOF-MS from Tofwerk AG for the PTR-TOF. Both instruments were housed in the same instrument trailer and sampled from a common inlet manifold that pulled 20 L min^{-1} through a 15 m long Teflon inlet.

The PTR-QMS (owned by the Chemical Sciences Division of NOAA) has been used extensively in various field and laboratory experiments including several ground-based, ship and airborne deployments over the past 15 years and has been characterized and described in much detail previously (de Gouw and Warneke, 2007). During this measurement campaign standard operating conditions of 2.4 mbar and 720 V resulting in an electric field/number density (E / N) of about 125 Td (Townsend) in the drift tube were used. Instrument backgrounds were determined every 3 h for about 3.5 min using a catalytic converter. Calibrations were done about every other day using three different calibration standards containing VOC mixtures in nitrogen. Using the mobile organic carbon calibration system (MOCCS; Veres et al., 2010), formaldehyde was calibrated five times and the cyclohexanes twice. The results of all the calibrations from UBWOS2013 for the PTR-QMS are shown in Fig. 2a and b. Clear differences in the gas calibration standard tanks for some compounds were observed. We have used historical calibration data to determine which of the tanks is likely to be the most accurate for each compound, and used those values in the final data analysis. Nevertheless, we define the instrument accuracy as the difference in the calibration tanks of better than 20 %. The quadrupole is a mass filter and measures only one mass at a time. During the measurements presented here the PTR-QMS was set to measure 34 masses related to VOCs with a 1 s dwell time every 37 s. A common 1 min data format was used for all instruments during UBWOS2013 and therefore either one or two 1 s dwell time measurements were used for the 1 min data.

The PTR-TOF (owned by the University of Wyoming) is a commercial instrument that was acquired from Ionicon Analytik about 1 month before the field experiment. The operating conditions were 2.2 mbar and 600 V resulting in an E / N of about 130 Td in the drift tube. Instrument backgrounds were determined similar to the PTR-QMS every 3 h for 1.5 min using a similar catalytic converter. Calibrations were done less frequently, but the same gas standards and MOCCS system were used. The calibration results from UBWOS2013 for the PTR-TOF are shown in Fig. 2c. Only three calibrations are available and the average sensitivity value was used for the entire campaign. As will be described below, the primary ion signal in the PTR-TOF was slightly unstable over the course of the campaign, but we still estimate the accuracy to be within 20 %. In the PTR-TOF, mass spectra from 1 to 500 amu with an extraction frequency of 25 kHz were pre-averaged and recorded as 10 s spectra, which were then further averaged to the 1 min UBWOS standard data format. The measured signal on m/z 21 was used to normalize to changing primary ions. A HPTDC (high-performance time-to-digital converter) was used for the TOF data acquisition. It has eight channels with 25 ps time bin width. The PTR-TOF is not run at a faster time resolution than 100 ps bin with. The number of fast channels is seven plus one trigger channel. The dead time is 10–20 ns and the event size is 4 bytes.

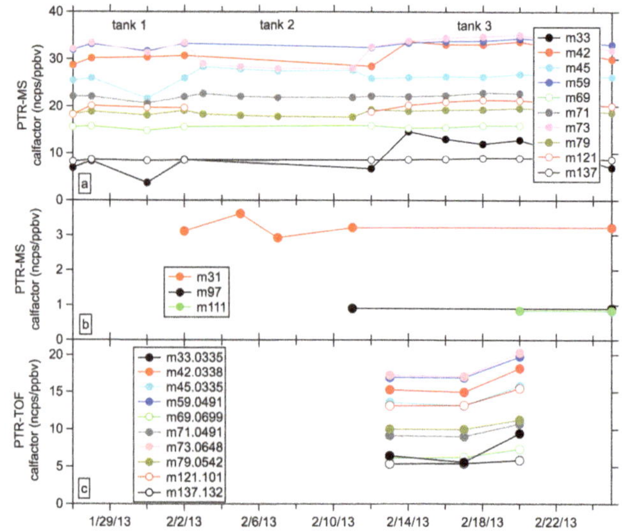

Figure 2. **(a)** Multiple VOC PTR-QMS calibration using three different calibration gas tanks during UBWOS2013: methanol (m33), acetonitrile (m42), acetaldehyde (m45), acetone (m59) isoprene (m69), methacrolein (m71), methylethylketone (m73), benzene (m79), 1,3,5-trimethylbenzene (m121) and α-pinene (m137). **(b)** PTR-QMS calibrations using the MOCCS cart for formaldehyde (m31), methylcyclohexane (m97) and dimethylcyclohexane (m111). **(c)** PTR-TOF calibration for the same compounds as in **(a)** during UBWOS2013.

3 Intercomparison

3.1 Sensitivity

The sensitivity determined during the calibration measurements of the PTR-QMS and the PTR-TOF is compared in Fig. 3. The standard way of expressing the PTR-QMS sensitivity is in units of ncps (normalized counts per second) ppbv^{-1} (de Gouw and Warneke, 2007), which normalizes the calibration signal to 10^6 H_3O^+ primary ions. Figure 3a shows that the PTR-QMS and PTR-TOF have very comparable normalized sensitivities for the individual compounds, as can be expected, because similar drift tube reaction chambers are used. The PTR-QMS sensitivity in ncps ppbv^{-1} is higher, because the PTR-QMS drift tube is 1 cm longer and 2.4 mbar are used in the drift tube at similar E / N compared to the PTR-TOF. Also, the transmission efficiency for H_3O^+ in the PTR-QMS is smaller compared to the product ions. The PTR-TOF data were analyzed with the software package described by Müller et al. (2013) and are corrected for dead time, baseline and most notably mass discrimination. The mass discrimination is caused by the sampling duty cycle in the orthogonal-acceleration region of the TOF. The duty cycle is mass dependent because different mass ions coming from the drift tube have a different velocity in the orthogonal accelerator. The smaller masses are faster and a larger fraction reaches the end of the orthogonal accelerator and is therefore lost there. The resulting mass discrimination is $[(m/z)_{max} / (m/z)]^{0.5}$. The normalization to the primary ions for both instruments and the mass discrimination are removed in Fig. 3b to compare sensitivities in actual count rates. The primary ion signal for the PTR-TOF used in Fig. 3b was 1.6×10^6 ions and for the PTR-QMS 25×10^6 ions, about a factor of 15 higher (see discussion of Fig. 6a). On average, the PTR-QMS is about a factor of 20 more sensitive than the PTR-TOF in cps ppbv^{-1}. The difference in sensitivity is dependent on the mass be-

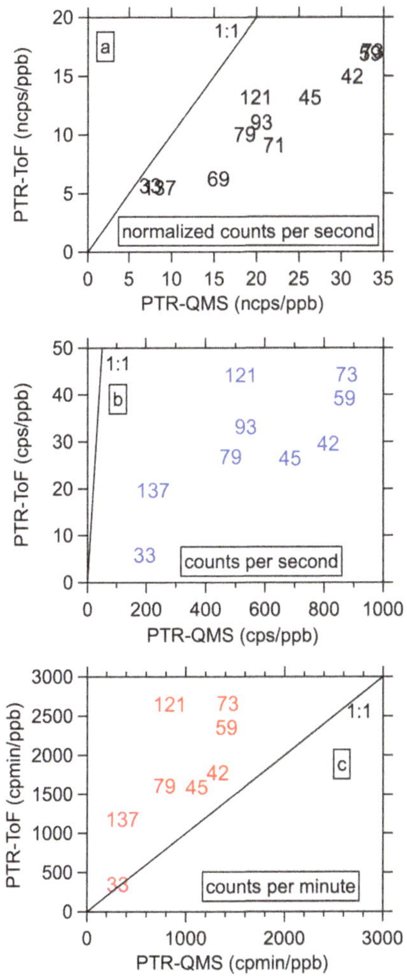

Figure 3. (a) PTR-QMS and PTR-TOF sensitivity in ncps $ppbv^{-1}$ normalized to 1×10^6 H_3O^+ ions. **(b)** Sensitivity in cps $ppbv^{-1}$ with PTR-QMS 25×10^6 and PTR-TOF 1.6×10^6 H_3O^+ ions. **(c)** Sensitivity in cpmin $ppbv^{-1}$ with PTR-QMS measuring 37 ions. The numbers in **(a)**, **(b)** and **(c)** are the nominal mass of the calibrated compounds.

cause of the mass discrimination of the PTR-TOF (Müller et al., 2013) and the transmission efficiency of the PTR-QMS (de Gouw and Warneke, 2007). The sensitivity here is expressed as cps $ppbv^{-1}$, but the PTR-QMS measures only one mass per second, while the PTR-TOF takes full mass spectra, which compensates for the lower sensitivity, if multiple masses are measured with the PTR-QMS. During UBWOS2013, the PTR-QMS measured 34 masses in 37 s duty cycles and the final data protocol required the use of 1 min data. In Fig. 3c the sensitivities from both instruments are compared in counts per minute per ppbv (cpmin $ppbv^{-1}$) and in this comparison the PTR-TOF is now the more sensitive instrument.

The ratio of the sensitivities taken from Fig. 3b of PTR-QMS / PTR-TOF is plotted in Fig. 4a and is closely related

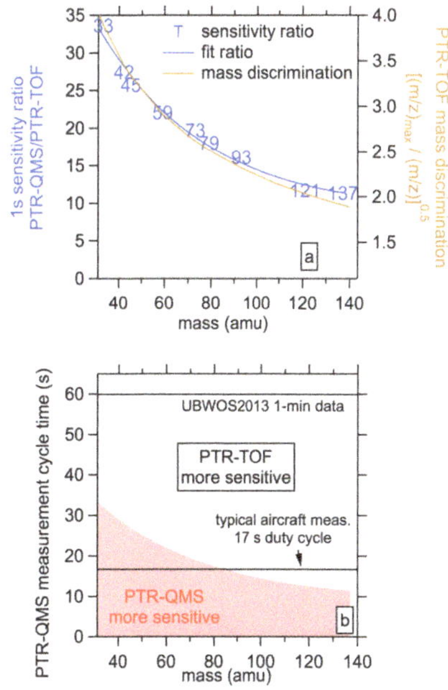

Figure 4. (a) The ratio of the 1 s sensitivities of PTR-QMS / PTR-TOF together with the PTR-TOF mass discrimination. **(b)** A mass-dependent curve showing the length of the PTR-QMS measurement cycle, where the PTR-QMS is as sensitive as the PTR-TOF (identical to the fit in **a**). In aircraft measurements a duty cycle of 17 s is typical (PTR-TOF more sensitive above mass 80) and at ground sites 1 min measurements are typical (PTR-TOF more sensitive for all masses).

to the mass discrimination of the PTR-TOF, which is also shown in Fig. 4a. This also shows that the mass discrimination of the PTR-QMS is small in comparison and the remaining mass-dependent discrimination for both instruments is very similar. This ratio then determines at what length of the PTR-QMS duty cycle, assuming 1 s dwell time per mass, the PTR-TOF becomes more sensitive, which is shown in Fig. 4b. For UBWOS2013, 1 min measurements are reported, in which case the PTR-TOF is more sensitive for all masses. During aircraft measurements such as those done during CalNex2010 (Warneke et al., 2011b), a duty cycle of about 17 s was used, in which case the PTR-QMS would be more sensitive for masses below 80 amu and the PTR-TOF above 80 amu.

3.2 Noise and detection limits

The detection limit of the instruments depends not only on the sensitivity, but also on the signal noise and the instrument background. Due to the Poisson distribution of the signal, the noise is the square root of the signal for all the product ions (if the gain on the multichannel plates (MCP) is set high enough). Primary ion signals are so large that undercounting has an influence on the noise and signal detection. This has been previously demonstrated for PTR-QMS and for in-

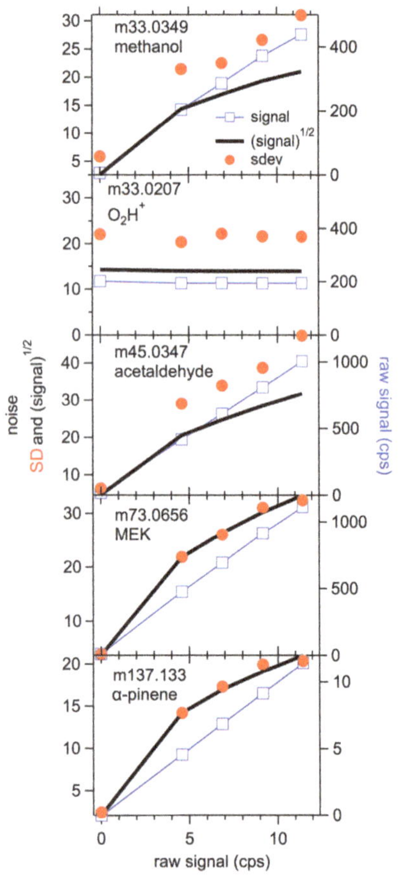

Figure 5. Signal, $(signal)^{1/2}$, and standard deviation (SD) of the signal (noise) during a typical PTR-TOF calibration measurement for some selected masses.

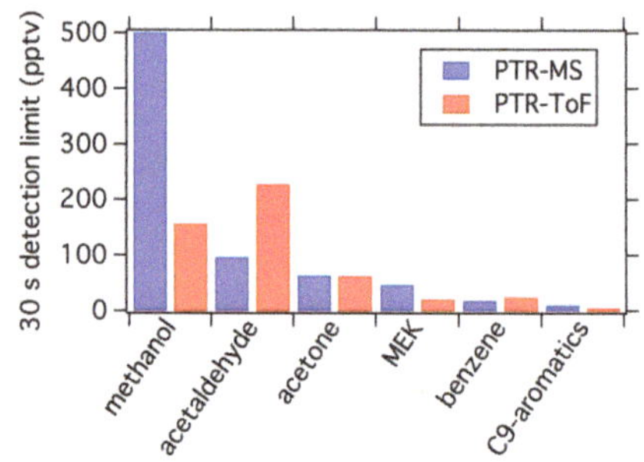

Figure 6. PTR-QMS and PTR-TOF 30 s detection limits during UBWOS2013.

struments that are operated in ion counting mode due to the dead time of the electron multiplier (de Gouw and Warneke, 2007). For a typical PTR-TOF calibration measurement, the signal, the measured noise and the square root of the signal for some selected masses are shown in Fig. 5. For most masses the noise is basically identical to the square root of the signal as can be seen in Fig. 5 for mass 73.0656 (methyl ethyl ketone) and mass 137.133 (α-pinene). There are some exceptions, which are shown by mass 33.0349 (methanol), mass 33.0207 (O_2H^+) and mass 45.0347 (acetaldehyde). The baseline is increased near peaks with very large signals, such as the primary ions (H_3O^+, O_2^+, etc.), increasing the noise on masses close to these peaks. The baseline increase can be seen in a mass spectrum in Fig. 9a.

The detection limit is calculated from the sensitivity and the noise on the background signals determined using the catalytic converter measurements. The detection limits are 3 times (S / N = 3) the standard deviation of the background measurements, where 30 s averaging times for the PTR-TOF and the 37 s measurements for the PTR-QMS were used. The results are shown for a few selected compounds in Fig. 6. The detection limits are comparable for both instruments but improve with increasing mass for the PTR-TOF, as is expected due to the increase in sensitivity with mass. Two notable exceptions in Fig. 6 are methanol and acetaldehyde. In the PTR-QMS the instrument background is comprised of mainly O_2H^+ and methanol • H^+ ions, which can be separated by mass in the PTR-TOF (Li et al., 2014). As a result, the PTR-TOF has a much smaller instrument background than the PTR-QMS and therefore also a lower detection limit. This effect can be observed for all compounds, where the background signal is comprised of more than one isobaric ion as for example acetonitrile (Dunne et al., 2012). The instrument background and therefore also the detection limit of the PTR-TOF for acetaldehyde was elevated during UBWOS2013, likely because the instrument was new and not run long enough for the background levels to drop.

3.3 Mixing ratios

PTR-TOF and PTR-QMS have recently been successfully compared (Kaser et al., 2013; Park et al., 2013). In Fig. 7 the intercomparison for acetone for UBWOS2013 is shown. In the top panel the time series for acetone on mass 59 from the PTR-QMS and on mass 59.0491 from the PTR-TOF are shown together with the respective primary ion signals. Over the whole time period the acetone measurements agree very well with a slope of 1.03 and an R^2 of 0.982.

The PTR-QMS was very stable, whereas the PTR-TOF primary ion signal was variable during the campaign. The PTR-TOF instrument was new and only briefly tested in the laboratory before being deployed during the field experiment. Therefore, various issues with the setup and the software had to be resolved during the field deployment, resulting in the large changes in primary ion signals and therefore sensitivity and detection limits. Most issues were the result of software instability, but the TOF, the MCP, and the ion source also had to be retuned during the experiment. For the measurements of the detection limit as described above, a period was chosen, where the primary ion signal was around 1.6×10^6 cps. At high MCP gain voltages, continuous operation and high signal intensities the detector may deterio-

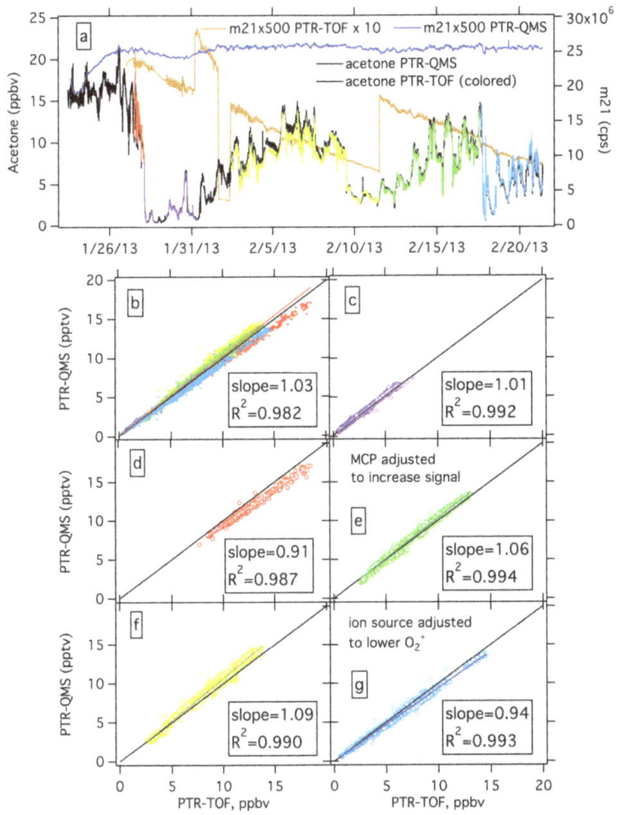

Figure 7. PTR-QMS and PTR-TOF acetone intercomparison during UBWOS2013: **(a)** time series of the PTR-TOF and PTR-QMS primary ions and acetone. The color code of the PTR-TOF acetone signal indicates when the PTR-TOF was relatively stable. **(b)** Scatter plot of PTR-QMS and PTR-TOF acetone for the whole campaign. **(c–g)** Scatter plots for separate stable periods of the PTR-TOF.

rate relatively quickly. Excessively high gain voltages may cause a more rapid deterioration, and a more frequent readjustment of the MCP gain in small steps should – all other things being equal – lead to a more constant signal. The detector deterioration due to high gain operation is a necessary trade-off to avoid an additional mass discrimination introduced by the MCP (Müller et al., 2014). Other reasons for the changes in sensitivity after software crashes and other instrument failures include the raw baseline stability, because the HPTDC was not temperature controlled, and changes in the statically set threshold levels and mass-dependent single ion peak detection at varying MCP gain voltages. All these features of the TOF operation were not monitored throughout the campaign; instead, the sensitivity of the instrument was monitored with calibrations using a calibration standard with compounds covering the mass range of interest.

The lower panels in Fig. 7 show the intercomparison separated into periods, when the PTR-TOF was relatively stable in a specific mode (except for the deterioration of the signal due to the MCP). For each individual period the comparison

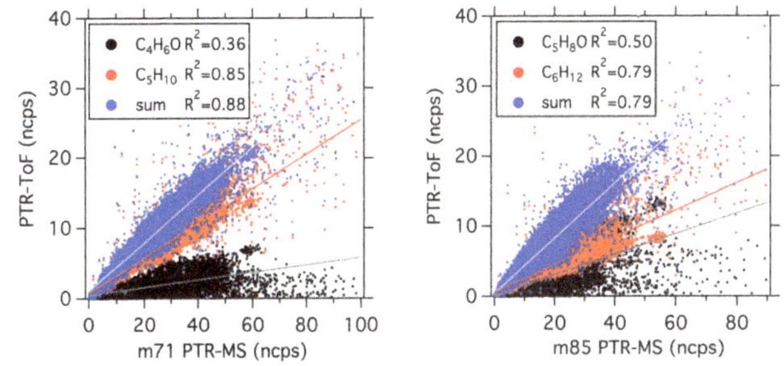

Figure 8. PTR-QMS and PTR-TOF intercomparison for mass 71 and mass 85 during UBWOS2013.

is excellent with $R^2 \geq 0.987$ and the slope varies between $\pm 9\,\%$ of unity. For the PTR-TOF only three calibrations are available and the averages from the three calibrations were used for the whole time period. This shows that without frequent calibrations an additional error of $\pm 10\,\%$ can be expected if the instrument setup is not stable during the experiment.

Many other masses routinely measured by PTR-QMS, where generally only one VOC contributes to the signal (methanol, acetaldehyde, benzene, toluene and many others), agreed well within the stated uncertainties with the PTR-TOF: $R^2 \geq 0.92$ and $\pm 20\,\%$. Benzene from the PTR-TOF and PTR-QMS agreed within 3 % and $R^2 = 0.98$. The PTR-TOF measurements of benzene were also compared with a GC-FID (gas chromatography–flame ionization detector) system; the PTR-TOF agreed with an $R^2 = 0.96$ but its results were larger by 22 %. This difference is within the stated uncertainties, but larger than observed in previous intercomparisons (Warneke et al., 2011a). Figure 8 shows the comparison of masses 71 and 85 in ncps as examples. The PTR-TOF detects two separate peaks on each of those nominal masses and identifies their atomic compositions as m71.0491 ($C_4H_6O \cdot H^+$) and m71.083 ($C_5H_{10} \cdot H^+$), and m85.0647 ($C_5H_8O \cdot H^+$) and m85.0966 ($C_6H_{12} \cdot H^+$). The PTR-QMS correlates the best with the sum of the two compounds for both masses as expected. The sensitivity of the PTR-QMS in ncps ppb^{-1} is close to a factor of 2 larger than the PTR-TOF (Fig. 3a), which is reflected in the slope in Fig. 8 as well. Other compounds that were compared are H_2S, which is discussed elsewhere (Li et al., 2014), and formaldehyde. Both of those compounds have only a slightly higher proton affinity than water and have therefore much lower sensitivities and need to be calibrated frequently (Warneke et al., 2011b). Formaldehyde calibrations for the PTR-TOF are not available, but the comparison of the signals in ncps showed an $R^2 = 0.88$.

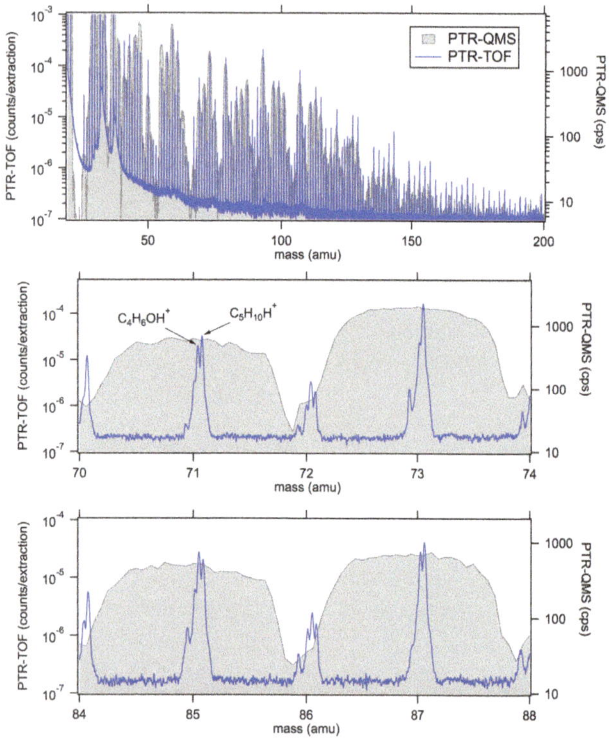

Figure 9. PTR-QMS and PTR-TOF mass scans. As an example, for m/z 71 two isobars are identified with the PTR-TOF and labeled.

4 Mass resolution and additional science possible with PTR-TOF

One of the two main differences between PTR-QMS and PTR-TOF is the achievable mass resolution: unity for the PTR-QMS and 4000–5000 for the PTR-TOF (Graus et al., 2010). Figure 9 shows mass spectra from both instruments for the full range and four selected masses. The PTR-QMS mass spectrum is an average of three cycles, where each mass (m/z 20–200) was measured for 1 s each resulting in a total of 9 min measurements. The PTR-TOF spectrum is an average of all spectra over those 9 min. Figure 9 shows that at one nominal mass, as measured by the PTR-QMS, up to four nominally isobaric ions are seen in the PTR-TOF. As an example two isobaric ions are identified with the PTR-TOF in Fig. 9 for m/z 71: $C_4H_6OH^+$ and $C_5H_{10}H^+$.

Isomeric ions have the same molecular formula and the exact same mass, whereas isobaric ions have the same nominal mass, but have a different molecular formula and a different exact mass. Therefore, the PTR-TOF can only distinguish between isobaric ions and nonisomeric ions.

The composition of each nominal mass, where the PTR-TOF detected signal in ambient air, is shown in Fig. 10. In the Uintah Basin large emissions of alkanes, cycloalkanes and aromatics from the oil and gas industry result in a rather unusual VOC composition and as a result many of the observed mass peaks are hydrocarbons. Subsequent photochemical oxidation of the alkanes produced many oxygenated compounds and VOCs with up to three oxygen atoms. In Fig. 10 the grey shaded masses are the ones that were monitored with the PTR-QMS and it can be seen that for most compounds either the hydrocarbon C_xH_y or the C_xH_yO structure dominates the signal on that nominal mass. This means that in principle with PTR-QMS the same information is obtained for the dominant peak on that mass, but the molecular identity is still not known and has to be assumed from prior knowledge of the emissions or chemistry. But in many cases multiple isobaric ions contribute to the signal at a nominal mass and the PTR-TOF provides identification of VOCs that are not separable with PTR-QMS.

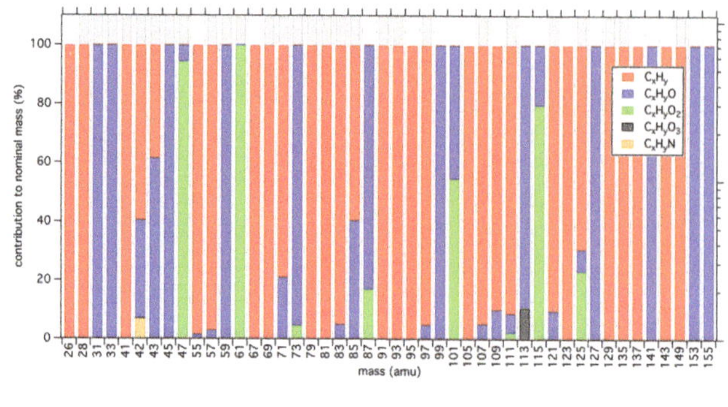

Figure 10. PTR-TOF mass contributions for all the masses that showed ambient signal during UBWOS2013. The grey bars on top indicate the masses that were monitored with the PTR-QMS.

The photo-oxidation products of some alkanes, cycloalkanes and aromatics, which are detectable with H_3O^+ ions, but have interferences on the PTR-QMS, can be quantified by PTR-TOF. The diurnal profiles for the UBWOS2013 campaign of many of those compounds are shown in Fig. 11. It can be seen that similar homologues of VOCs such as the acids, ketones or cycloketones have very similar diurnal profiles, but they are different from another. Often the signals corresponding to these compounds, especially $C_xH_yO_2$ structures, are very small and have the same nominal mass as C_xH_yO compounds. For example, mass 73 is mainly comprised of methyl ethyl ketone (C_4H_9O) and has a small contribution of methylglyoxal ($C_3H_5O_2$). Methylglyoxal is a very important compound in the chemistry leading to wintertime ozone exceedances in the Uintah Basin, because it acts as a radical precursor (Edwards et al., 2013). The detection of methylglyoxal is not possible with PTR-QMS, but feasible with PTR-TOF and has added greatly to our understanding of wintertime ozone formation.

5 Conclusions

PTR-TOF has been a significant step in the evolution and improvement of the PTR technique to monitor VOCs in the atmosphere, where the main advantages of the TOF are the

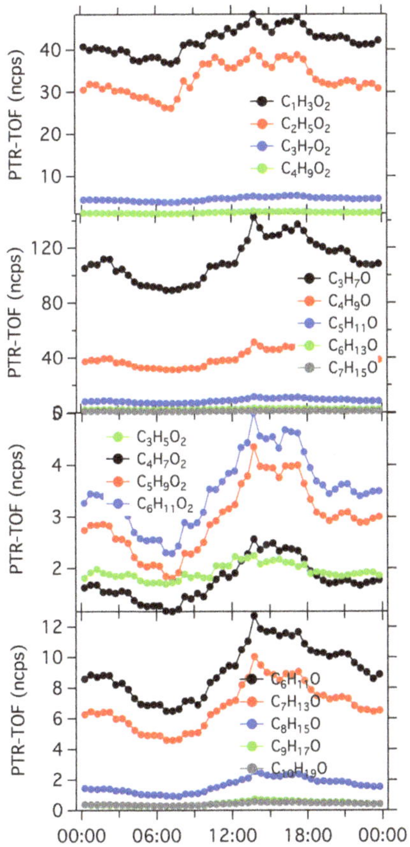

Figure 11. PTR-TOF diurnal variation of selected oxygenated VOCs during UBWOS2013. Many of those compounds cannot be unambiguously identified with PTR-QMS.

high time resolution for full mass scans and the high mass resolution of 4000–5000 compared to the unit mass resolution with PTR-QMS. The high mass resolution allows for the identification of isobaric ions. Here we compared a standard PTR-QMS with a new PTR-TOF during the UBWOS2013 field experiment in an oil and gas field in the Uintah Basin, Utah.

The setup of the measurements with the two instruments determines which instrument is more sensitive. The difference in sensitivity is dependent on the mass: the PTR-TOF is increasingly more sensitive with increasing mass, and at above 80 amu it becomes more sensitive than the PTR-QMS. In PTR-QMS, the masses that are monitored need to be chosen. If only one mass is measured with 1 s dwell time, the PTR-QMS is about a factor of 10–35 more sensitive depending on the mass. The number of masses monitored in PTR-QMS and the averaging times then determine which instrument is more sensitive: 10–35 masses for 1 s dwell times, again dependent on the mass, is the break-even point. During UBWOS2013 the PTR-QMS was set to monitor 34 masses in 37 s and data for both instruments were reported as 1 min averages. In this setup the PTR-TOF is more sensitive for all masses.

The high mass resolution of the PTR-TOF showed that many masses monitored with the PTR-QMS had contributions from only one ion, but that many others had contributions from up to four different isobaric ions. This capability gives the PTR-TOF the ability to measure small oxidation products of the main emissions from the oil and gas development (alkanes, cycloalkanes and aromatics) that play an important role in the formation of ozone. These compounds were previously not individually measurable by PTR-QMS.

Overall, the two instruments agreed for measured mixing ratios very well for all compounds where calibration gases were available and for measured count rates for all other compounds ($R^2 \geq 0.92$ and within 20 %). For masses, where more than one isobaric ion contributes to the signal, the PTR-QMS agreed with the sum of those ions observed with the PTR-TOF.

The additional analytical capabilities and the sensitivity of the current PTR-TOF version are clearly a major advance in PTR technology for VOC analysis compared to the standard PTR-QMS instruments. For measurements aboard aircraft, where PTR-QMS instruments excel by commonly monitoring only a few selected ions and thus achieve high sensitivities, PTR-TOF instruments could still be improved in sensitivity to achieve the same detection limits as PTR-QMS instruments. A promising new development for a sensitivity improvement of PTR-TOF has recently been shown by Sulzer et al. (2014), which could make PTR-TOF the ideal instrument for aircraft measurements as well.

Acknowledgements. We thank Questar Energy Products, in particular Stephanie Tomkinson, for the Horse Pool site preparation and logistical support. This work was supported in part by the Western Energy Alliance. This work was also supported by the NOAA Health of the Atmosphere Program and by the NOAA Climate Program Office – Atmospheric Composition and Climate Program. Shane Murphy was supported by the NSF grant 1215926. S.-M. Li was supported in part by Environment Canada's Clean Air Regulatory Agenda.

Edited by: E. C. Apel

References

Blake, R. S., Whyte, C., Hughes, C. O., Ellis, A. M., and Monks, P. S.: Demonstration of proton-transfer reaction time-of-flight mass spectrometry for real-time analysis of trace volatile organic compounds, Anal. Chem., 76, 3841–3845, 2004.

Blake, R. S., Monks, P. S., and Ellis, A. M.: Proton-Transfer Reaction Mass Spectrometry, Chem. Rev., 109, 861–896, 2009.

de Gouw, J. A. and Warneke, C.: Measurements of volatile organic compounds in the earths atmosphere using proton-transfer-reaction mass spectrometry, Mass Spectrom. Rev., 26, 223–257, 2007.

de Gouw, J. A., Goldan, P. D., Warneke, C., Kuster, W. C., Roberts, J. M., Marchewka, M., Bertman, S. B., Pszenny, A. A. P., and Keene, W. C.: Validation of proton transfer reaction-mass spectrometry (PTR-MS) measurements of gas-phase organic compounds in the atmosphere during the New England Air Quality Study (NEAQS) in 2002, J. Geophys. Res.-Atmos., 108, 4682, doi:10.1029/2003JD003863, 2003.

Dunne, E., Galbally, I. E., Lawson, S., and Patti, A.: Interference in the PTR-MS measurement of acetonitrile at m/z 42 in polluted urban air-A study using switchable reagent ion PTR-MS, Int. J. Mass Spectrom., 319, 40–47, 2012.

Edwards, P. M., Young, C. J., Aikin, K., deGouw, J., Dubé, W. P., Geiger, F., Gilman, J., Helmig, D., Holloway, J. S., Kercher, J., Lerner, B., Martin, R., McLaren, R., Parrish, D. D., Peischl, J., Roberts, J. M., Ryerson, T. B., Thornton, J., Warneke, C., Williams, E. J., and Brown, S. S.: Ozone photochemistry in an oil and natural gas extraction region during winter: simulations of a snow-free season in the Uintah Basin, Utah, Atmos. Chem. Phys., 13, 8955–8971, doi:10.5194/acp-13-8955-2013, 2013.

Gilman, J. B., Lerner, B. M., Kuster, W. C., and de Gouw, J. A.: Source Signature of Volatile Organic Compounds from Oil and Natural Gas Operations in Northeastern Colorado, Environ. Sci. Technol., 47, 1297–1305, 2013.

Graus, M., Mueller, M., and Hansel, A.: High Resolution PTR-TOF: Quantification and Formula Confirmation of VOC in Real Time, J. Am. Soc. Mass Spectr., 21, 1037–1044, 2010.

Haase, K. B., Keene, W. C., Pszenny, A. A. P., Mayne, H. R., Talbot, R. W., and Sive, B. C.: Calibration and intercomparison of acetic acid measurements using proton-transfer-reaction mass spectrometry (PTR-MS), Atmos. Meas. Tech., 5, 2739–2750, doi:10.5194/amt-5-2739-2012, 2012.

Hansel, A., Jordon, A., Holzinger, R., Prazeller, P., Vogel, W., and Lindinger, W.: Proton transfer reaction mass spectrometry: Online trace gas analysis at the ppb level, Int. J. Mass Spectrom., 149/150, 609–619, 1995.

Hansel, A., Jordan, A., Warneke, C., Holzinger, R., Wisthaler, A., and Lindinger, W.: Proton-transfer-reaction mass spectrometry (PTR-MS): on-line monitoring of volatile organic compounds at volume mixing ratios of a few pptv, Plasma Sources Sci. T., 8, 332–336, 1999.

Helmig, D., Thompson, C. R., Evans, J., Boylan, P., Hueber, J., and Park, J. H.: Highly Elevated Atmospheric Levels of Volatile Organic Compounds in the Uintah Basin, Utah, Environ. Sci. Technol., 48, 4707–4715, doi:10.1021/es405046r, 2014.

Howarth, R. W., Santoro, R., and Ingraffea, A.: Methane and the greenhouse-gas footprint of natural gas from shale formations, Climatic Change, 106, 679–690, 2011.

Jordan, A., Haidacher, S., Hanel, G., Hartungen, E., Maerk, L., Seehauser, H., Schottkowsky, R., Sulzer, P., and Maerk, T. D.: A high resolution and high sensitivity proton-transfer-reaction time-of-flight mass spectrometer (PTR-TOF-MS), Int. J. Mass Spectrom., 286, 122–128, 2009.

Karion, A., Sweeney, C., Petron, G., Frost, G., Hardesty, R. M., and Kofler, J.: Methane emissions estimate from airborne measurements over a western United States natural gas field, Geophys. Res. Lett., 40, 4393–4397, doi:10.1002/grl.50811, 2013.

Karl, T., Guenther, A. B., Lindinger, C., Jordan, A., Fall, R., and Lindinger, W.: Eddy covariance measurements of oxygenated volatile organic compound fluxes from crop harvesting using a redesigned proton-transfer-reaction mass spectrometer, J. Geophys. Res., 106, 24157–24167, doi:10.1029/2000JD000112, 2001.

Kaser, L., Karl, T., Schnitzhofer, R., Graus, M., Herdlinger-Blatt, I. S., DiGangi, J. P., Sive, B., Turnipseed, A., Hornbrook, R. S., Zheng, W., Flocke, F. M., Guenther, A., Keutsch, F. N., Apel, E., and Hansel, A.: Comparison of different real time VOC measurement techniques in a ponderosa pine forest, Atmos. Chem. Phys., 13, 2893–2906, doi:10.5194/acp-13-2893-2013, 2013.

Katzenstein, A. S., Doezema, L. A., Simpson, I. J., Blake, D. R., and Rowland, F. S.: Extensive regional atmospheric hydrocarbon pollution in the southwestern United States, P. Natl. Acad. Sci. USA, 100, 11975–11979, 2003.

Kemball-Cook, S., Bar-Ilan, A., Grant, J., Parker, L., Jung, J. G., Santamaria, W., Mathews, J., and Yarwood, G.: Ozone Impacts of Natural Gas Development in the Haynesville Shale, Environ. Sci. Technol., 44, 9357–9363, 2010.

Li, R., Warneke, C., Graus, M., Field, R., Geiger, F., Veres, P. R., Soltis, J., Li, S.-M., Murphy, S. M., Sweeney, C., Pétron, G., Roberts, J. M., and de Gouw, J.: Measurements of hydrogen sulfide (H2S) using PTR-MS: calibration, humidity dependence, inter-comparison and results from field studies in an oil and gas production region, Atmos. Meas. Tech., 7, 3597–3610, doi:10.5194/amt-7-3597-2014, 2014.

Lindinger, W., Hansel, A., and Jordan, A.: On-line monitoring of volatile organic compounds at pptv levels by means of proton-transfer-reaction mass spectrometry (PTR-MS) – Medical applications, food control and environmental research, Int. J. Mass Spectrom., 173, 191–241, doi:10.1016/0168-1176(97)00281-4, 1998a.

Lindinger, W., Hansel, A., and Jordan, A.: Proton-transfer-reaction mass spectrometry (PTR-MS): on-line monitoring of volatile organic compounds at pptv levels, Chem. Soc. Rev., 27, 347–354, doi:10.1039/a827347z, 1998b.

Litovitz, A., Curtright, A., Abramzon, S., Burger, N., and Samaras, C.: Estimation of regional air-quality damages from Marcellus Shale natural gas extraction in Pennsylvania, Environ. Res. Lett., 8, 014017, doi:10.1088/1748-9326/8/1/014017, 2013.

Mielke, L. H., Erickson, D. E., McLuckey, S. A., Mueller, M., Wisthaler, A., Hansel, A., and Shepson, P. B.: Development of a Proton-Transfer Reaction-Linear Ion Trap Mass Spectrometer for Quantitative Determination of Volatile Organic Compounds, Anal. Chem., 80, 8171–8177, 2008.

Müller, M., Mielke, L. H., Breitenlechner, M., McLuckey, S. A., Shepson, P. B., Wisthaler, A., and Hansel, A.: MS/MS studies for the selective detection of isomeric biogenic VOCs using a Townsend Discharge Triple Quadrupole Tandem MS and a PTR-Linear Ion Trap MS, Atmos. Meas. Tech., 2, 703–712, doi:10.5194/amt-2-703-2009, 2009.

Müller, M., Graus, M., Ruuskanen, T. M., Schnitzhofer, R., Bamberger, I., Kaser, L., Titzmann, T., Hörtnagl, L., Wohlfahrt, G., Karl, T., and Hansel, A.: First eddy covariance flux measurements by PTR-TOF, Atmos. Meas. Tech., 3, 387–395, doi:10.5194/amt-3-387-2010, 2010.

Müller, M., Mikoviny, T., Jud, W., D'Anna, B., and Wisthaler, A.: A new software tool for the analysis of high resolution PTR-TOF mass spectra, Chemometr. Intell. Lab., 127, 158–165, 2013.

Müller, M., Mikoviny, T., and Wisthaler, A.: Detector aging induced mass discrimination and non-linearity effects in PTR-ToF-MS, Int. J. Mass Spectrom., 365–366, 93–97, doi:10.1016/j.ijms.2013.12.008, 2014.

Park, J.-H., Goldstein, A. H., Timkovsky, J., Fares, S., Weber, R., Karlik, J., and Holzinger, R.: Eddy covariance emission and deposition flux measurements using proton transfer reaction – time of flight – mass spectrometry (PTR-TOF-MS): comparison with PTR-MS measured vertical gradients and fluxes, Atmos. Chem. Phys., 13, 1439–1456, doi:10.5194/acp-13-1439-2013, 2013.

Petron, G., Frost, G., Miller, B. R., Hirsch, A. I., Montzka, S. A., Karion, A., Trainer, M., Sweeney, C., Andrews, A. E., Miller, L., Kofler, J., Bar-Ilan, A., Dlugokencky, E. J., Patrick, L., Moore, C. T., Ryerson, T. B., Siso, C., Kolodzey, W., Lang, P. M., Conway, T., Novelli, P., Masarie, K., Hall, B., Guenther, D., Kitzis, D., Miller, J., Welsh, D., Wolfe, D., Neff, W., and Tans, P.: Hydrocarbon emissions characterization in the Colorado Front Range: A pilot study, J. Geophys. Res.-Atmos., 117, D04304, doi:10.1029/2011JD016360, 2012.

Prazeller, P., Palmer, P. T., Boscaini, E., Jobson, B. T., and Alexander, M.: Proton transfer reaction ion trap mass spectrometer, Rapid Commun. Mass Sp., 17, 1593–1599, 2003.

Schnell, R. C., Oltmans, S. J., Neely, R. R., Endres, M. S., Molenar, J. V., and White, A. B.: Rapid photochemical production of ozone at high concentrations in a rural site during winter, Nat. Geosci., 2, 120–122, 2009.

Sulzer, P., Edtbauer, A., Hartungen, E., Juerschik, S., Jordan, A., Hanel, G., Feil, S., Jaksch, S., Maerk, L., and Maerk, T. D.: From conventional proton-transfer-reaction mass spectrometry (PTR-MS) to universal trace gas analysis, Int. J. Mass Spectrom., 321, 66–70, 2012.

Sulzer, P., Hartungen, E., Hanel, G., Feil, S., Winkler, K., Mutschlechner, P., Haidacher, S., Schottkowsky, R., Gunsch, D., Seehauser, H., Striednig, M., Jürschik, S., Breiev, K., Lanza, M., Herbig, J., Märk, L., Märk, T. D., and Jordan, A.: A Proton Transfer Reaction-Quadrupole interface Time-Of-Flight Mass Spectrometer (PTR-QiTOF): High speed due to extreme sensitivity, Int. J. Mass Spectrom., 368, 1–5, doi:10.1016/j.ijms.2014.05.004, 2014.

Tanimoto, H., Aoki, N., Inomata, S., Hirokawa, J., and Sadanaga, Y.: Development of a PTR-TOFMS instrument for real-time measurements of volatile organic compounds in air, Int. J. Mass Sp., 263, 1–11, 2007.

Veres, P., Gilman, J. B., Roberts, J. M., Kuster, W. C., Warneke, C., Burling, I. R., and de Gouw, J.: Development and validation of a portable gas phase standard generation and calibration system for volatile organic compounds, Atmos. Meas. Tech., 3, 683–691, doi:10.5194/amt-3-683-2010, 2010.

Warneke, C., van der Veen, C., Luxembourg, S., de Gouw, J. A., and Kok, A.: Measurements of benzene and toluene in ambient air using proton-transfer-reaction mass spectrometry: calibration, humidity dependence, and field intercomparison, Int. J. Mass Spectrom., 207, 167–182, 2001.

Warneke, C., De Gouw, J. A., Kuster, W. C., Goldan, P. D., and Fall, R.: Validation of atmospheric VOC measurements by proton-transfer-reaction mass spectrometry using a gas-chromatographic preseparation method, Environ. Sci. Technol., 37, 2494–2501, 2003.

Warneke, C., Rosen, S., Lovejoy, E. R., de Gouw, J. A., and Fall, R.: Two additional advantages of proton-transfer ion trap mass spectrometry, Rapid Commun. Mass Sp., 18, 133–134, 2004.

Warneke, C., de Gouw, J. A., Lovejoy, E. R., Murphy, P. C., Kuster, W. C., and Fall, R.: Development of proton-transfer ion trap-mass spectrometry: On-line detection and identification of volatile organic compounds in air, J. Am. Soc. Mass Spectr., 16, 1316–1324, 2005a.

Warneke, C., Kato, S., De Gouw, J. A., Goldan, P. D., Kuster, W. C., Shao, M., Lovejoy, E. R., Fall, R., and Fehsenfeld, F. C.: Online volatile organic compound measurements using a newly developed proton-transfer ion-trap mass spectrometry instrument during New England Air Quality Study – Intercontinental Transport and Chemical Transformation 2004: Performance, intercomparison, and compound identification, Environ. Sci. Technol., 39, 5390–5397, 2005b.

Warneke, C., Roberts, J. M., Veres, P., Gilman, J., Kuster, W. C., Burling, I., Yokelson, R., and de Gouw, J. A.: VOC identification and inter-comparison from laboratory biomass burning using PTR-MS and PIT-MS, Int. J. Mass Spectrom., 303, 6–14, 2011a.

Warneke, C., Veres, P., Holloway, J. S., Stutz, J., Tsai, C., Alvarez, S., Rappenglueck, B., Fehsenfeld, F. C., Graus, M., Gilman, J. B., and de Gouw, J. A.: Airborne formaldehyde measurements using PTR-MS: calibration, humidity dependence, inter-comparison and initial results, Atmos. Meas. Tech., 4, 2345–2358, doi:10.5194/amt-4-2345-2011, 2011b.

Warneke, C., Geiger, F., Edwards, P. M., Dube, W., Pétron, G., Kofler, J., Zahn, A., Brown, S. S., Graus, M., Gilman, J. B., Lerner, B. M., Peischl, J., Ryerson, T. B., de Gouw, J. A., and Roberts, J. M.: Volatile organic compound emissions from the oil and natural gas industry in the Uintah Basin, Utah: oil and gas well pad emissions compared to ambient air composition, Atmos. Chem. Phys., 14, 10977–10988, doi:10.5194/acp-14-10977-2014, 2014.

Wisthaler, A., Apel, E. C., Bossmeyer, J., Hansel, A., Junkermann, W., Koppmann, R., Meier, R., Müller, K., Solomon, S. J., Steinbrecher, R., Tillmann, R., and Brauers, T.: Technical Note: Intercomparison of formaldehyde measurements at the atmosphere simulation chamber SAPHIR, Atmos. Chem. Phys., 8, 2189–2200, doi:10.5194/acp-8-2189-2008, 2008.

11

Case study on complex sporadic E layers observed by GPS radio occultations

X. Yue[1]**, W. S. Schreiner**[1]**, Z. Zeng**[1]**, Y.-H. Kuo**[1]**, and X. Xue**[2]

[1]COSMIC Program Office, University Corporation for Atmospheric Research, Boulder, CO, USA
[2]CAS Key Laboratory of Geospace Environment, Department of Geophysics & Planetary Sciences, University of Science & Technology of China, Hefei, Anhui, China

Correspondence to: X. Yue (xinanyue@ucar.edu)

Abstract. The occurrence of sporadic E (E_s) layers has been a hot scientific topic for a long time. The GNSS (global navigation satellite system)-based radio occultation (RO) has proven to be a powerful technique for detecting the global E_s layers. In this paper, we focus on some cases of complex E_s layers based on the RO data from multiple missions processed in UCAR/CDAAC (University Corporation for Atmospheric Research (UCAR) the Constellation Observing System for Meteorology, Ionosphere, and Climate (COSMIC) Data Analysis and Archive Center (CDAAC)). We first show some examples of multiple E_s layers occurred in one RO event. Based on the evaluations between colocated simultaneous RO events and between RO and lidar observations, it could be concluded that some of these do manifest the multiple E_s layer structures. We then show a case of the occurrence of E_s in a broad region during a certain time interval. The result is then validated by independent ionosondes observations. It is possible to explain these complex E_s structures using the popular wind shear theory. We could map the global E_s occurrence routinely in the near future, given that more RO data will be available. Further statistical studies will enhance our understanding of the E_s mechanism. The understanding of E_s should benefit both E_s-based long-distance communication and accurate neutral RO retrievals.

1 Introduction

The ionospheric E region is ~90–120 km above the Earth's surface. Its electrons mainly originate from soft X-ray and far ultraviolet (UV) solar radiation ionization of the neutral atmosphere. The regular E region strictly follows the photochemistry equilibrium and therefore could be represented well by a function of the solar zenith angle (Yue et al., 2006). The ionospheric E region has a relatively higher electrical conductivity and therefore plays a crucial role in the ionosphere electron dynamics. The understanding of the E region could contribute to the studies related to electrojets of high-latitude and equatorial regions, ionospheric instabilities, and ionosphere–thermosphere–magnetosphere couplings. In addition, it is a key area for long-distance radio communication. Under certain circumstances, a thin layer of enhanced ionization appears in the E region, which is called the sporadic E layer (E_s). It can be denser than either the E layer peak or even the F layer peak. E_s has been investigated extensively both theoretically and experimentally in the past century. It is generally accepted that E_s is composed of metallic ions of meteoric origin converged vertically mainly by wind shear, especially in the middle latitude region (Axford, 1963). The wind shear theory has been confirmed by many independent observations and simulations (Haldoupis, 2011; Whitehead, 1989). The occurrence of E_s is controlled by multiple factors, including tidal wind, the Earth's geomagnetic field, and meteoric deposition of metallic material in the background thermosphere. These factors result in variations of E_s occurrence with respect to local time, altitude, latitude, longitude, and season (Haldoupis, 2011). Furthermore, in the high-latitude region, the occurrence of E_s is more significantly influenced by the local electric field rather than the wind shear due to the geomagnetic field difference (Wakabayashi and Ono, 2005). Please note that the term E_s is now widely used to represent the inhomogeneity of the ionospheric E layer (e.g., layers,

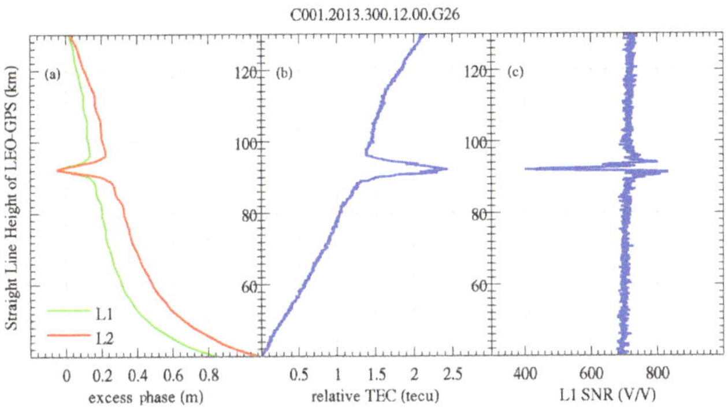

Figure 1. Straight line height variations of excess phase **(a)**, relative slant TEC **(b)**, and L1 SNR **(c)** of one selected radio occultation event observed by COSMIC FM1 on day 300 of 2013, when there was a sporadic E layer occurrence over the occultation region.

cloud, irregularity) rather than only the echo trace from the ionogram when it was discovered. To learn more about basic E_S theory and research progress, the reader can refer to these review papers and the references therein: Haldoupis (2011) and Whitehead (1989).

Traditionally, E_S is mainly observed by ground-based radars, including ionosonde and incoherent/coherent scatter radars (ISR, CSR), and occasional in situ probes carried by rockets. Recently, the GNSS (global navigation satellite system)-based radio occultation (RO) technique has shown great capability in detecting the global E_S occurrence by using high-resolution signal-to-noise ratio (SNR) data (Arras et al., 2008; Chu et al., 2014; Hocke et al., 2001; Wu et al., 2005; Yeh et al., 2012; Zeng and Sokolovskiy, 2010). GNSS RO technique has proven to be a powerful tool in monitoring climate, weather, and space weather (Anthes, 2011; Foelsche et al., 2011; Liu et al., 2010; Schreiner et al., 2007; Zhang et al., 2011; Yue et al., 2012). After the pioneer GPS/MET (Global Positioning System/Meteorology) mission, many low-Earth-orbit (LEO) satellites were launched with a RO payload, including CHAMP (Challenging Minisatellite Payload), GRACE (Gravity Recovery And Climate Experiment), SAC-C/D (Scientific Application Satellite-C/D), COSMIC (Constellation Observing System for Meteorology, Ionosphere, and Climate), C/NOFS (Communications/Navigation Outage Forecasting System), Metop-A/B (Meteorological Operational Satellite Program of Europe – A/B), and TerraSAR-X (X-band Terra SAR)/TanDEM-X (TerraSAR-X add-on for Digital Elevation Measurements). These observations have accumulated a big database with which to study the E_S. As an example, Fig. 1 shows the altitude variations of the excess phase, relative slant total electron content (TEC) along the LEO-GPS ray, and the SNR during a typical E_S event. Due to the occurrence of the E_S, both the phase and SNR show simultaneous fluctuations. The SNR has an obvious U shape, as reported by Zeng and Sokolovskiy (2010). The obvious increase in slant TEC around ~92 km implies the ionization enhancement in the E_S layer. Based on all available GPS/MET RO data, Hocke et al. (2001) did a statistical analysis on the global E_S occurrence and compared the results with the ionosonde observations. Wu et al. (2005) did a more robust statistical analysis on the global E_S using CHAMP RO data. The detailed seasonal, latitude, local time, altitude, and hemisphere variations were studied. By combining almost all the available RO data, Arras et al. (2008) showed a finer structure of E_S occurrence. Specifically, they provided direct proof of the control exerted by the Earth's geomagnetic field on E_S for the first time, and this supported the wind shear theory strongly (Haldoupis, 2011). These statistics are further confirmed by Chu et al. (2014) and Yeh et al. (2012), based on a large amount of COSMIC data. Through both simulation and real-data analysis, Zeng and Sokolovskiy (2010) revealed that the E_S cloud aligned with the wave propagation direction could cause defocusing of the GPS RO signal, accompanied by scintillation above and below the defocusing region due to the interference of direct and refracted radio waves. These effects result in specific U-shape structures in the amplitude of the GPS RO signals. This U shape will reduce and even disappear with the increase of the tilt angle of the E_S cloud with respect to the wave direction. Furthermore, RO amplitude data are also used to detect the high-altitude ionospheric irregularity through either back propagation algorithms or analysis of the amplitude variation pattern (Sokolovskiy et al., 2002; Pavelyev et al., 2012; Wickert et al., 2004).

However, some investigations have shown that the E_S could sometimes be much more complicated than a single layer. Smith and Miller (1980) reported the double-peaks E_S and even a rectangular E_S layer observed by rockets. They concluded that the complex profiles of sporadic E layers can be interpreted as an effect of unstable wind shears. Based on high-resolution ISR observations over Arecibo, Mathews et al. (2001) reported some complex E_S layers, which might be due to the modulation of polarized electric fields linked with the F region. They reproduced the observed results through 3-D numerical simulations. In addition, some investigations pointed out that the drivers responsible for the plasma instability (e.g., the Kelvin–Helmholtz instability (KHI)) could deform the E_S layer into a complex structure (Bernhardt, 2002; Cosgrove and Tsunoda, 2003). They could lift up part of the E_S layer to overlap with the original layer. Based on the SEEK-2 rocket campaign observations, Wakabayashi and Ono (2005) analyzed the double-peak E_S layer phenomena in detail. They found that the lower peak is usually more stable than the upper peak. By using simultaneous neutral wind and electric field observations, they showed that the neutral wind shear is mainly responsible for the generation of the lower layer, while the direct current (DC) electric field makes a significant contribution to the formation of the upper layer. However, the complex E_S layer can only be observed with the higher vertical resolution technique, such as ISR/CSR and rocket (Chu and Wang, 1997; Hysell et al., 2012). However, there are no corresponding routinely and globally distributed observations publicly available, especially not from the ISR and rockets; this is not cost effective. According to Haldoupis (2011), the sporadic E layer is actually not spatially or temporally sporadic either. However, it has been difficult in the past to identify the E_s layer simultaneously in a broad region due to the available sparse ground-based observations. The GNSS RO measurements can provide the quasi-vertical profile of the ionosphere globally and routinely. It offers us a potential way to study these complex E_s layer phenomena .

In this paper, we will conduct some case studies on multiple E_s events observed by COSMIC as well as other missions' atmospheric RO data, which are observed by the occultation antennas of each LEO satellite with a ~ 50 Hz resolution. Specifically, two kinds of cases will be investigated. One is multiple E_s occurring in a local region observed by one occultation. The other is multiple E_s occurring over a broad region observed by many simultaneous RO events. Multiple E_s phenomena have been touched on by some previous studies (Hocke et al., 2001; Yeh et al., 2012). Yeh et al. (2012) even conducted a statistical analysis of the multiple E_s layer distribution using ~ 3 years of COSMIC RO data. Their results show that the multiple E_s layers have a distribution that, with respect to geographic location and time, is very similar to the previous studies by Wu et al. (2005) and Arras et al. (2008). Our study has the following significance in comparison with previous studies. (1) According to Haldoupis (2011), wind shear below ~ 115 km is dominated by zonal wind, while meridional wind dominates above this altitude due to the altitude variation of the collision frequency between ions and neutrals. This will therefore result in different duration and tidal structures of the E_s layers at different altitudes. Studies on multiple E_s layers in the same location will be helpful in E_s mechanism understanding. (2) As indicated by Zeng and Sokolovskiy (2010), a nonlocal tilted E_s layer could cause the less significant fluctuation of SNR in the lower altitude of the local RO event. It increases the difficulty of identifying the multiple E_s layers by using only one RO event. In this study, we will show how to identify this phenomenon by multiple colocated RO events. (3) We will show how simultaneously observed RO events in a large broad region enable us to demonstrate that the E_s is not sporadic spatially. (4) The bending angle and therefore the refractivity/temperature in the lower atmosphere RO retrievals are influenced significantly by small-scale ionospheric irregularities, especially the E_s layer (Mannucci et al., 2011). A better understanding of E_s in RO signals will benefit the improvement of neutral atmosphere retrievals and therefore numerical weather prediction (NWP) and climate change study.

The remainder of the paper is organized as follows. Section 2 shows the examples of multiple-peak E_s layers in the same location. The example of E_s occurring in a broad region simultaneously observed by multiple RO events is given in Sect. 3. We then discuss and conclude in Sects. 4 and 5, respectively.

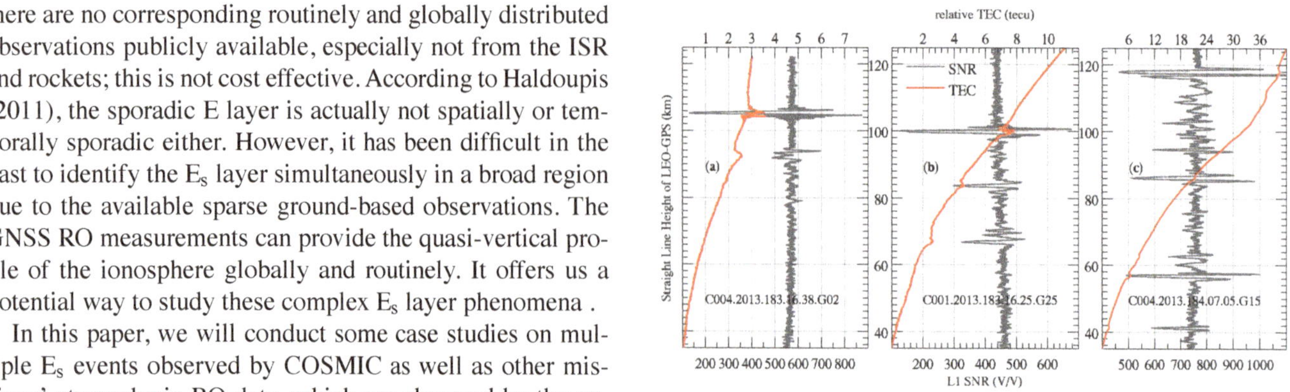

Figure 2. Selected examples of two **(a)**, three **(b)**, and multiple (**c**; in this case, seven) peaks in the L1 SNR and relative slant TEC versus straight line height observed by COSMIC satellites.

2 Case study 1: multiple-peak E_s layer in the same location

The identification of E_s occurrence is different from paper to paper (Arras et al., 2008; Chu et al., 2014; Dou et al., 2010; Hocke et al., 2001; Wu et al., 2005; Yeh et al., 2012; Zeng and Sokolovskiy, 2010). Hocke et al. (2001) and Dou et al. (2010) derived E_s occurrence from the fluctuations of RO inverted electron density profiles. Wu et al. (2005) and Arras et al. (2008) calculated the standard deviation of filtered and normalized SNR and identified E_s occurrence through either the maximum standard deviation or a threshold. Chu et al. (2014) applied a stricter criterion on both the amplitude and phase fluctuations of L1 and L2 signals to avoid potential influences from the instrument and measurement noise. In this study, the 50 Hz amplitude and phase observations made by the COSMIC occultation antennas and archived at the University Corporation for Atmospheric Research (UCAR) COSMIC Data Analysis and Archive Center (CDAAC) level1b atmPhs files will be used. The amplitude scintillation index S4 and relative slant TEC are calculated based on the 50 Hz SNR and excess phase, respectively. Only the cases that have a maximum S4 index calculated from 50 Hz SNR larger than 0.3 are considered. E_s occurrence will then be identified manually from case to case. Only the cases that show visible fluctuations of L1/2 SNR and phases and obvious enhancement of slant TEC around the E region will be considered.

2.1 Multiple-peak E_s examples

As stated, the E_s layer can cause the defocusing and scintillation of the SNR of RO signals. So the E_s occurrence can be identified from the fluctuation of the SNR. In the meantime, the enhanced ionization in the E_s layer also results in an additional phase delay of the radio signal and therefore the enhancement of the slant TEC along the GNSS ray. When we looked at the RO data case by case, we can find many cases

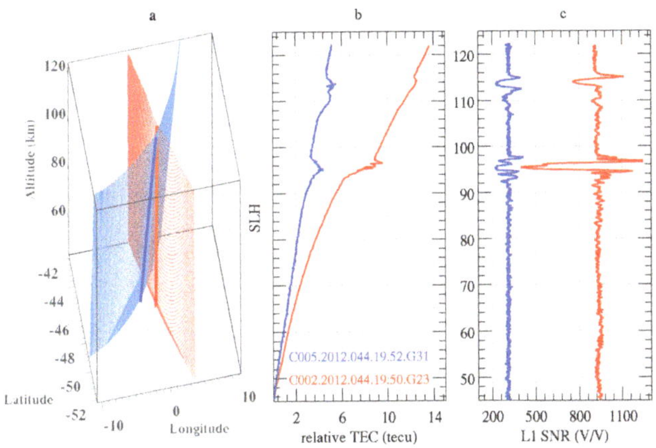

Figure 3. Two colocated RO events, which present two peaks in either relative TEC or L1 SNR versus straight line height, observed by different COSMIC LEO satellites at the same time on day 44 of 2012: **(a)** occultation plane and corresponding tangent points; **(b)** relative slant TEC versus straight line height (SLH); **(c)** L1 SNR versus straight line height. Blue and red represent two different RO events.

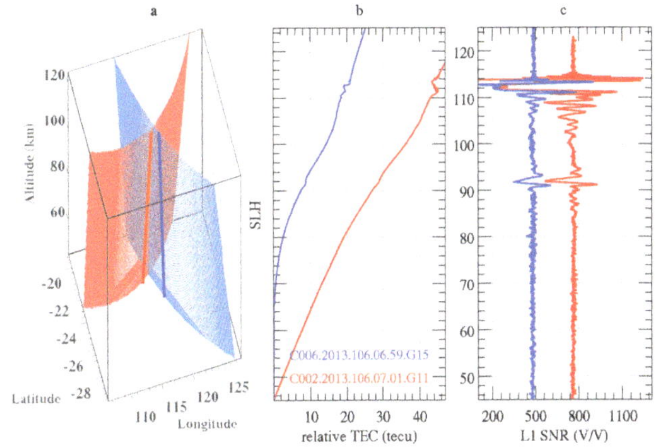

Figure 4. The same as Fig. 3 but for another case on day 106 of 2013.

that have more then one peak in SNR or slant TEC along the tangent height. We give three typical such examples in Fig. 2. The case in Fig. 2a was observed by COSMIC satellite FM4 on day 183 of 2013. Both SNR fluctuation and slant TEC show major and minor peaks around ~ 105 and 93 km. In Fig. 2b case, the SNR/TEC shows one larger peak around ~ 100 km and two smaller peaks around 84 and 67 km, respectively. Figure 2c shows an extreme case with up to seven visible peaks in both SNR fluctuation and slant TEC. Please note that the much larger range of TEC amplitude here makes its enhancements less visible, although the TEC peaks are actually significant in this case. Due to the special observational geometry of RO, either the SNR or the slant TEC actually reflects the integral effect along the whole GNSS ray. When we assign these values to the tangent points of the GNSS ray, the effect of the E_s layer in the higher altitude could map down to the lower tangent point altitude, which might result in pseudo multiple peaks along the tangent point altitude, as shown in Fig. 2 (Zeng and Sokolovskiy, 2010). However, the E_s layer mostly extends horizontally. For a specific GNSS ray, the effect of E_s far from the tangent points will be much smaller than that of E_s in the tangent points area due to the tilt angle between E_s layer and GNSS ray. In the following, we will show two kinds of validation studies on multiple E_s peaks in the same RO event.

2.2 Evaluation by colocated simultaneous RO events

In UCAR/CDAAC, we process most RO data available to the scientific community (Schreiner et al., 2011). This comes from missions including COSMIC, CHAMP, GRACE, SAC-C, TerraSAR-X, C/NOFS, and MetopA/B. These combined missions can generate 3000–4000 RO events daily in terms of atmospheric excess phase (atmPhs). This number varies with time due to either satellite launching or degradation. One potential way to evaluate multiple E_s layers is looking at the colocated RO events. If two colocated RO events with totally different azimuth angles show similar multiple E_s structures at the same time, we can conclude that these structures are realistic structures. We searched the colocated RO events among all CDAAC data during 2012–2013. The Metop data are excluded because Metop usually begins to sample occultation around at an altitude of ~ 80 km. The colocation criteria is that the difference in latitude, longitude, and universal time at an altitude of 100 km of the tangent point between two occultations is less than 1 degree, 2 degree, and 15 min, respectively. This criterion is far smaller than the typical spatial and temporal scales of E_s. In addition, the azimuth angle difference between two occultations at an altitude of 100 km should be larger than 60 degrees to ensure that the areas away from the tangent points of two occultations have no overlap. Those short occultations having no integral coverage of 90–120 km will be discarded too. The excess phase, SNR, and relative slant TEC of these colocated occultations were printed out to be checked further manually. Generally, we can find ~ 6–10 colocated RO pairs per day on average. Unfortunately, some of them either have no E_s occurrence or only one peak E_s, a matter which is beyond the scope of this study. Finally, we found ~ 100 cases manually in which both colocated occultations show visible double peaks in both SNR and slant TEC at the same tangent point altitude. Since two occultations sample a totally different area and intersect with each other in the tangent point area, there is a high possibility that these double-peak E_s layers are the realistic double E_s layers occurring around the tangent points area. As examples, we show two such cases in Figs. 3 and 4. In Fig. 3, two occultations made by COSMIC FM5-GPS31 and FM2-GPS23 within 2 min around the southern middle latitude on day 44 of 2012 at local sunset are displayed. Indicated by SNR and TEC variations, two E_s lay-

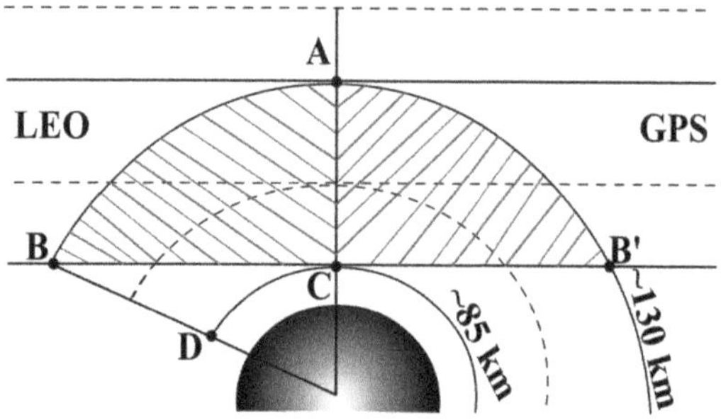

Figure 5. The E_s occurrence area (EOA) is defined by the shaded area in the picture.

ers around 114 and 95 km occurred simultaneously in two occultations. In another case shown in Fig. 4, two simultaneous occultations made by FM6-GPS15 and FM2-GPS11 around the southern middle-latitude area during the afternoon are shown. One major peak at around 113 km and another minor peak at around 92 km can be identified in both occultations. The SNRs around the two peaks have an obvious U shape, which indicates that the tilt angle between E_s direction and GPS ray is insignificant (Zeng and Sokolovkiy, 2010).

2.3 Independent evaluation by lidar observations

According to Gardner et al. (1993), sporadic metal (Fe and Na) layers could be observed simultaneously with the sporadic E layer due to the neutralization of metal ions (Fe^+ and Na^+) in the associated E_s layer. Dou et al. (2010) have compared the E_s derived from COSMIC electron density profiles with the meteors from a meteor radar and the sodium atom layers from a lidar statistically. Well-correlated seasonal variations were detected among three parameters. It implies a link between E_s and metal layers. Another potential way to evaluate these multiple-peak E_s layers is by comparing them with lidar results. In this study, the lidar observations made in Beijing (40.2° N, 116.2° E), Hefei (31.8° N, 117.3° E), Wuhan (30.5° N, 114.4° E), and Haikou (19.5° N, 109.1° E) over China, accessed through the Chinese Meridian Project during 2012, are used (Wang, 2011). Detailed information about these lidar stations can be found in Dou et al. (2013). Based on these lidar observations, Dou et al. (2013) carried out statistical investigations into the occurrence of sporadic sodium layers (SSLs) and the thermosphere enhanced sodium layers (TeSLs) in China during 2011–2012. The TeSLs is defined as the secondary sodium layer in the lower thermosphere, separated from the main SSLs (Dou et al., 2013). We are not going to look at all the lidar observations in this study. Instead, we tried to search the COSMIC RO events nearby the lidar stations for the TeSLs cases listed in Table A2 of Dou et al. (2013). These TeSLs were all observed during local nighttime and their duration ranges from ~1 h up to ~6 h. The colocation criteria are that the difference in latitude and longitude between lidar and RO tangent points at the altitude of 100 km is less than 2.5° and 4.5°, respectively, and the occurrence time of RO is within the TeSLs occurrence. Of all 28 cases, we finally found a perfect case for 29 July 2012 (DOY: 211) over Haikou lidar station. Before showing the results of this case, we would like to define a concept called the E_s occurrence area (EOA) for convenience. As shown in Fig. 5, the peaks of either SNR or TEC in tangent point C could only originate from the E_s layers located above that altitude along the BB' GNSS ray. Usually, the E_s layer occurs in the altitude range of 85–130 km (Haldoupis, 2011). We do not consider the intermediate layer because most RO missions were designed to untertake high-resolution sampling around 130 km (Arras et al., 2009; Li et al., 2013). So the E_s layers in the plots could only come from the area between 85 and 130 km that the GNSS rays pass through. This area is defined as EOA, indicated by the shaded area in Fig. 5. In the TeSLs case over Haikou lidar station from 29 July 2012, the TeSLs begins to separate from the main SSL at around ~15:00 (UT), becomes distinguishable at 15:32, peaks at around 16:00 at an altitude of 113 km, and ends at 18:19. During the whole TeSLs occurrence interval, there exist significant SSLs with peaks around ~93 km. There are a total of four COSMIC colocated ROs observed at 15:20, 15:30, 15:50, and 16:03. The first two ROs are made by COSMIC FM2 and occurred in the initial phase of TeSLs, while the latter two are by FM1 around the peak of TeSLs. Figure 6 shows the geolocation of the Haikou lidar station and the EOA of the four colocated RO events. The EOA of the four ROs have few overlaps, which implies that the E_s observed by the RO events are independent. The SNR/TEC of RO events and the Na density of Haikou lidar corresponding to the RO occurrence time are given in Fig. 7. As indicated by the altitude variations of the SNR/TEC, all four RO events show obvious complex E_s occurrences. In case 1, the occultation is in southwest–northeast direction and move over the Haikou lidar station. The SNR/TEC show simultaneous fluctuations in the altitude range of 85–110 km. While in case 2, the RO event is ~500 km west of Haikou and the SNR/TEC fluctuates at almost all the altitudes. In both cases, the SNR oscillations are not constant and show some minor peaks occasionally. This excludes the potential effects from the F region ionospheric scintillations (Straus et al., 2003). During the occurrence of cases 1 and 2, the TeSLs separates from the main SSL. Since the TeSLs is mainly formed from the neutralization of metal ions, the ionosphere might experience complicated dynamic and chemistry evolutions during the initial phase of TeSLs. RO measurements confirmed that the ionization is complicated and no distinct layered structures are formed at that time, which is consistent with the Na density observations. In cases 3 and 4, the TeSLs are already distinguishable. Both RO events show two peaks at ~100 and 108 km, respectively, in terms of either SNR or slant TEC. In case 3, the higher-altitude peak is smaller than the lower peak, while the situation is reversed in case 4. This might be caused by the different geometry of occultation,

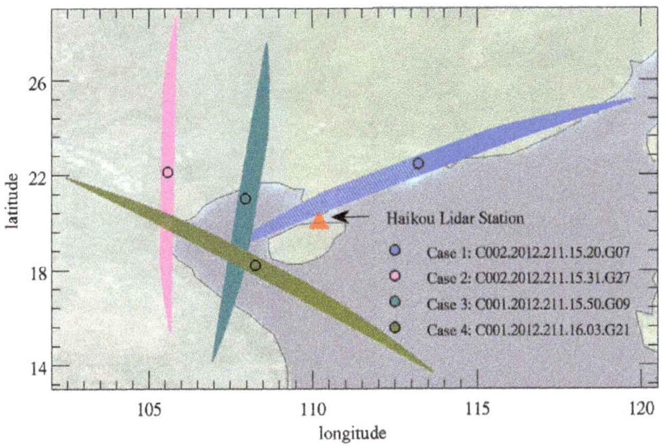

Figure 6. The red triangle represents the location of Haikou lidar station over China. The shaded areas show the E_s occurrence area (EOA) for the four colocated RO events observed by COSMIC, represented by four colors. The points at which the straight line height is 100 km are marked by the black circles.

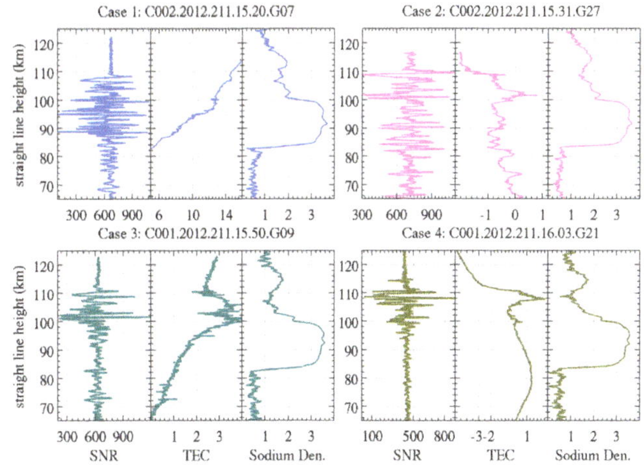

Figure 7. Altitude variations of L1 SNR and relative slant TEC of RO events and Na density of lidar observations corresponding to the four cases shown in Fig. 6. The unit of Na density is cm^{-3} and is represented logarithmically in the plots.

which results in the difference of the tilt angle between E_s and GNSS ray. Please note that the peak altitude and amplitude are not necessarily consistent between RO and lidar observations due to the different dynamic processes involved. The Na density used here mainly serves to demonstrate the potential existence of multiple E_s layers in the ionosphere, which is confirmed by the RO measurements.

3 Case study 2: multiple simultaneous E_s layers in a broad region

E_s was mainly observed by the ionosondes in the early stage. E_s can only be identified by ionosonde when it is larger than a certain value because most ionosonde transmitters have a low limitation on the transmitted wave frequency ($\sim$ 1–1.5 MHz). It makes the appearance of E_s "sporadic" regarding space and time (Haldoupis, 2011). However, the wind shear theory suggests that E_s is closely linked to the atmospheric dynamics. Inherently, this implies that E_s possesses a variable but regular character that reflects both the complexity and repeatability of the atmospheric wind and wave dynamics in the mesosphere–lower-thermosphere (MLT) region. The high-resolution and sensitivity observations made by ISR in Arecibo have confirmed that E_s is present at all times in some cases, which exhibit a non-sporadic behavior (Mathews, 1998). Mathews (1998) even proposed an alternative term "tidal ion layers (TIL)" because "sporadic" is an instrumental limitation rather than a prominent physical property (Haldoupis, 2011). The ISR observations already prove the non-sporadic property of E_s with respect to time. However, due to the lack of global E_s observations, there is still no direct evidence to prove that E_s was also spatially non-sporadic in the past. The global distribution feature of RO technique enables us to find direct evidence for the first time.

Although we can observe 3000–4000 RO events daily and globally by the combined multiple RO missions, it is still impossible to get enough occultations for a certain time and space range. Fortunately, right after the launch, the six COSMIC satellites were clustered together in one orbit and the distances between LEOs increased gradually (Schreiner et al., 2007). It enables tracking multiple RO events in a broad region simultaneously in the initial stage of LEO orbit separation. We looked through all the COSMIC RO data as well as occultations available from other missions during 2006. Finally, we found many cases in which multiple LEOs sample a certain region at a certain time. Some of these cases show that the E_s can occur in a broad region simultaneously, which supports the idea that E_s is non-sporadic spatially too. These cases are mainly found in the middle latitude of Eurasian continent and the Pacific Ocean. This is logical because the relatively higher horizontal component of the geomagnetic field over this region makes the wind shear more efficient (Haldoupis, 2011). To evaluate these results, the raw ionograms accessed from the Digital Ionogram DataBase (DIDB) operated by the Lowell Global Ionospheric Radio Observatory (GIRO) Data Center (LGDC) through the SAO Explorer software are used to make an independent comparison (Galkin et al., 1999; Reinisch and Galkin, 2011). For 2006, a total of about 35–40 ionosonde stations shared their data in DIDB. Generally, Europe and northern America have relatively more and denser ionosonde distributions. We checked the ionograms over these regions during the occurrence of simultaneous E_s over a broad region indicated by RO. For most cases, we can find the visible E_s occurrence in the corresponding ionograms. Figures 8–11 show such an example over Europe.

On day 294 of 2006 (21 October), five satellites of COSMIC, excluding FM5, which was already raised to a higher

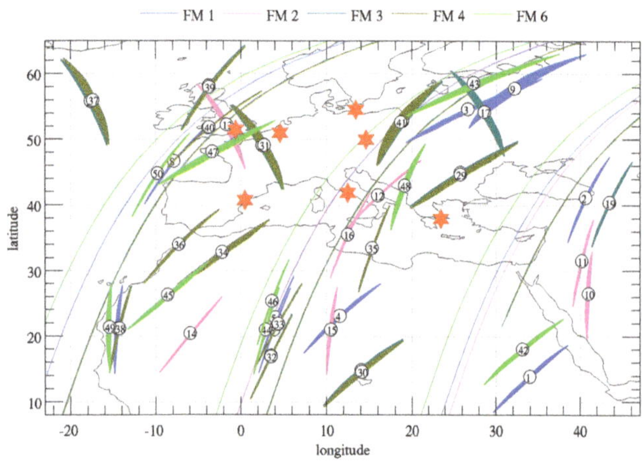

Figure 8. Longitude and latitude distributions of the E_s Occurrence Area (EOA) of 50 RO events observed by COSMIC during 2006.294, 11:00–15:00 UT over and surrounding Europe region. The circle on each EOA is the corresponding tangent point at an altitude of 100 km with the case serial number embedded. The light lines are the orbits of COSMIC LEO satellites during this time interval. Different colors represent different COSMIC FM satellites. The red stars show the location of seven ionosondes used.

orbit, have adjacent orbits. FM3 and FM4 even have almost the same orbit within a distance of 200 km. A total of 62 RO events were observed over 10–60° latitude and −20–45° longitude area during 11:00–15:00 UT interval. Through our manual checking, we found that 50 of them show obvious E_s-related oscillations in SNR. Figure 8 shows the LEOs' orbits and these 50 RO events. The EOA is again shown here for each occultation to demonstrate the independence of the observed E_s between different ROs. Each occultation is labeled by an index number (1–50). The corresponding altitude variations of SNR are shown in Fig. 9. Note that the SNR is normalized to the range of 0–1 for convenience. As indicated in the figure, all the ROs selected here show obvious oscillations related to E_s. Most of the SNRs have a visible U shape, which illustrates that the E_s is significant. At this specific region and time interval, the DIDB has seven ionosondes that have observations available. The geolocation, ionosonde system type, and the operation organization are given in Table 1. The red stars in Fig. 8 represent the geographic distribution of the ionosondes. The hourly ionograms of seven ionosondes during 11:00–15:00 UT are plotted in Fig. 10, with an emphasis on the wave traces below 150 km, indicated by black dots. We can see clearly the wave traces due to the E_s in most ionograms except in 11:00 and 12:00 UT of Athens. Please note the different noise level between the ionograms by different types of ionosondes. In Fig. 11, we plotted the automatically scaled E_s critical frequency (foEs) and the E layer critical frequency (foE) of seven ionosondes and the averaged International Reference Ionosphere (IRI) modeled foE versus local time (Bilitza and Reinisch, 2008). The high occurrence rate of E_s can be identified. Although these seven ionosondes are distributed in a relatively narrow area, they still strongly support the simultaneous occurrence of E_s over a broad region.

4 Discussion

4.1 Multiple E_s layers in one profile

We have summarized some complex E_s layers observed by rockets and ISRs from previous publications (Smith and Miller, 1980; Mathews et al., 2001; Bernhardt, 2002; Cosgrove and Tsunoda, 2003; Wakabayashi and Ono, 2005). In Sect. 2, we reported some examples of RO events that show multiple peaks in the tangent altitude variations of either SNR or relative slant TEC; these manifest the potentially complex multiple layered structures of E_s in the occultation area. In comparison with ISR and rocket measurement, which could manifest either a small or large scale of E_s horizontally, the GNSS-RO-observed E_s layer should have a relatively larger horizontal scale due to the lower horizontal resolution of RO observation (Schreiner et al., 2007; Zeng and Sokolovskiy, 2010). Then we showed some double-peak E_s layers observed by simultaneously colocated RO events with different azimuth angles to prove that these double-peak E_s layers are probably not caused by the localization problem and could be realistic. Furthermore, we also gave an example of a comparison between the complex E_s layers made by four RO events and the colocated lidar-observed double sporadic Na layers (TeSLs). With the development of TeSLs over time, RO measurements accordingly show complex E_s. Actually, Xue et al. (2013) also reported an example showing that the lidar observes TeSLs associated with simultaneous double peaks of SNR made by colocated RO events from TerraSAR-X. According to wind shear theory, at lower (upper) altitudes, the zonal (meridional) wind will dominate the shear due to the altitude variations of the ion-neutral collision frequency regarding the ion gyrofrequency ratio. In comparison with the low altitude, the E_s layers in the upper altitude are more variable and short-lived since ion convergence is much faster there (Haldoupis, 2011). Similarly, an existing E_s layer could be broadened and dissipated by an opposite wind shear. The layers can deform faster (slower) at the altitudes where they form faster (slower). This is the reason why E_s layers remain fairly stable for many hours at lower heights, while their lifetime is much shorter at high altitude. The E_s is controlled by the tidal wind and therefore also shows tidal structures in its formation and descent. Main tidal components can be diurnal, semidiurnal, and even terdiurnal and quarterdiurnal (Arras et al., 2009; Chu et al., 2014; Fytterer et al., 2013; Wu et al., 2005). All these factors make the complex E_s layer structures understandable. However, due to the limited amount of data available from ISR and rockets, the investigations into this issue were mainly

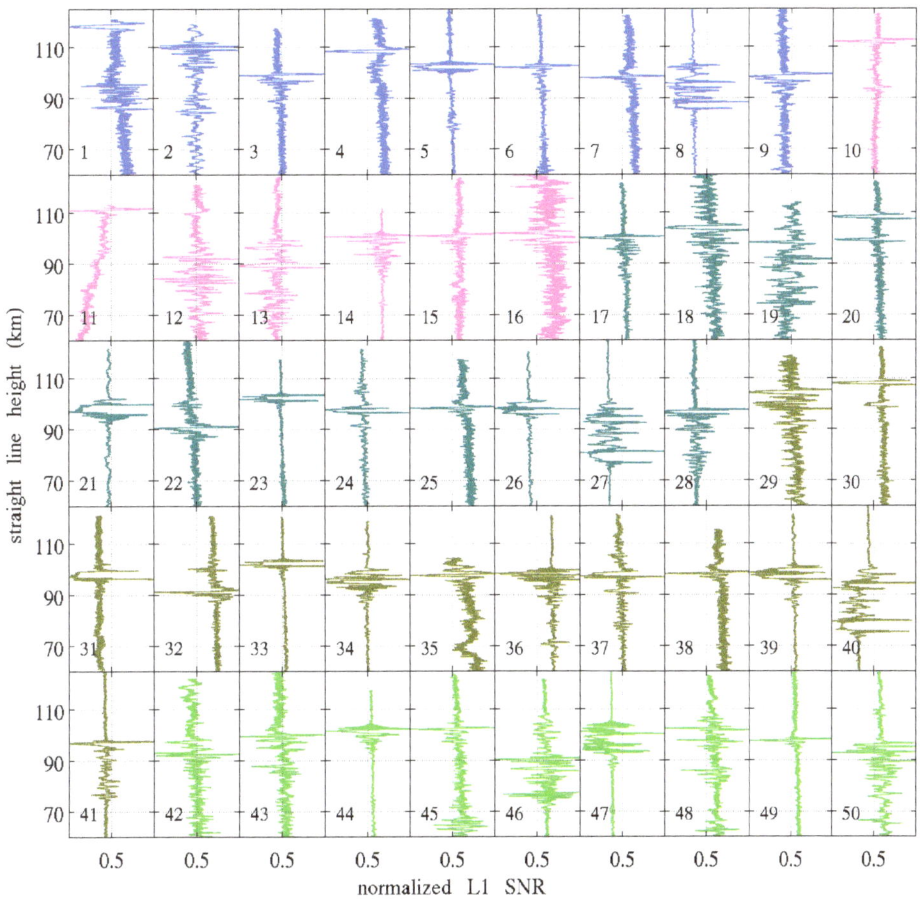

Figure 9. Straight line height variations of the normalized L1 SNR for the 50 COSMIC RO events shown in Fig. 8. Different colors represent different LEO satellites.

Table 1. Latitude, longitude, ionosonde system type, and operation organization/country of the seven ionosondes used in the study. DGS-256: Digisonde-256; DPS-1: Single-Receiver Digisonde Portable Sounder; DPS-4: Four-Receiver Digisonde Portable Sounder.

Station	Latitude	Longitude	Ionosonde	Organization/country
Athens	38.0° N	23.5° E	DPS-4	National Observatory of Athens, Greece
Dourbes	50.1° N	4.6° E	DGS-256	Royal Meteorological Institute, Belgium
Roquetes	40.8° N	0.3° E	DGS-256	Observatori de l'Ebre, Spain
Juliusruh	54.6° N	13.4° E	DPS-1	Institute of Atmospheric Research, University of Rostock, Germany
Pruhonice	50.0° N	14.6° E	DPS-4	Institute of Atmospheric Physics, Czech Republic
Chilton	51.5° N	0.6° W	DPS-1	Rutherford Appleton Laboratory, United Kingdom
Rome	41.9° N	12.5° E	DPS-4	L'Instituto Nazionale di Geofisica, Italy

made through either theoretical analysis or case studies. The huge amount of accumulated RO events enables us to study this phenomenon statistically, which will be done in a further study. This kind of investigation will definitely enhance our understanding of the E_s mechanism.

4.2 Simultaneous E_s occurrence in a broad region

We have found many cases that show simultaneous E_s occurrence over a broad region made by clustered COSMIC satellites in the early stage. One of these cases is displayed in the paper and evaluated by the independent ionosonde observations. Regarding this comparison, we should be aware of two key points. (1) In the RO SNR profile, we can see that some cases show E_s occurrence below even 80 km. However, due to the localization problem of RO geometry, as stated above, these E_s could either happen at the altitude of the tangent point or at a higher altitude beyond the tangent point due to the mapping effect along the GNSS ray. If the E_s occurs below 80 km, it could not be the result of the wind shear theory

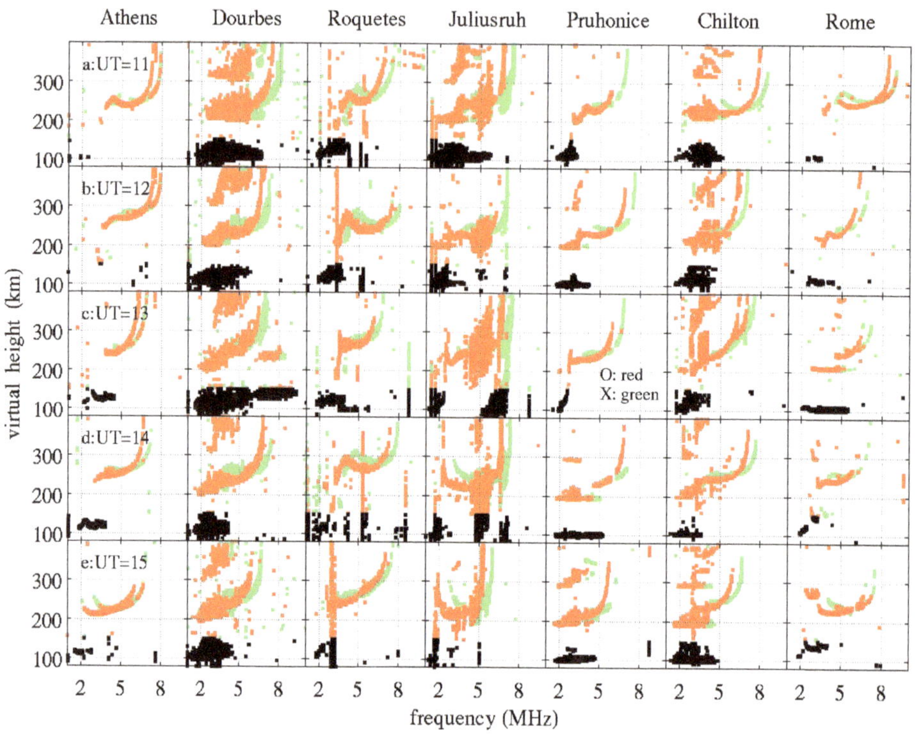

Figure 10. Raw ionogram: graph of the virtual height plotted against frequency. The panels from left to right are for the seven ionosondes, with the name given at the top. Panels a to e are for 11:00 to 15:00 UT at 1 h interval on day 294 of 2006. Ordinary and extraordinary returned waves are represented by the red and green dots, respectively. The wave traces below ~150 km are given by the black dots to highlight the occurrence of sporadic E layer.

because of the high collision frequency between ionization and neutral particles over the lower altitude. (2) According to Whitehead (1989), the thickness of E_s is usually less than 4 km. As indicated in Fig. 10, sometimes the vertical extent of echo trace can be larger than 10 km. This might have additional drivers beyond E_s. The echo trace in ionograms could have a variety of sources (Maruyama et al., 2006). However, we are sure that these irregularities, which can cause echo traces in ionograms, will cause the RO SNR fluctuations too if the horizontal extent is sufficiently long.

We have not seen any cases of the broad E_s occurrence that is reported in the previous literature. One reason is that the available E_s observations from global ionosondes and CSR/ISRs still have far less coverage than necessary to study this phenomenon. Another reason is that the ionosondes mainly used have a lower frequency limitation, as we stated before. By contrast, the high-resolution RO radio wave signal is very sensitive to the layered structure in the ionosphere, which has been proved by the simulation studies by Yeh et al. (2012) and Zeng and Sokolovskiy (2010). During that selected case, the ionosphere is geomagnetically quiet (minimum Dst index is −28 nT). We can exclude the potential effects from the storm time penetration electric field, which can result in broad region E_s occurrence as reported by some studies (Abdu et al., 2014). The region of simultaneous occurrence of E_s has a width of roughly 50° (10–60°) in latitude and 65° (−20–45°) in longitude. We have actually found some other cases which show quasi-simultaneous E_s occurrence in a much larger area, especially in the East Asia region. It has not been selected to be shown here because fewer colocated ionosondes are available to do the evaluation there. Our results strongly support the non-sporadic property of E_s, as proposed by Haldoupis (2011) and Mathews (1998). However, current RO observations are still far less than sufficient to map the global E_s at any time. In the near future, with the dramatic increase of available RO events from different planned RO missions especially COSMIC-2 (Yue et al., 2014), we will be able to map the global E_s occurrence routinely. It will benefit E_s research as well as some applications, such as the long-distance communication at Very High Frequency (VHF) frequencies via reflections from E_s layers.

4.3 Wind shear theory

Wind shear theory has been used to successfully explain the E_s occurrence, especially in the middle latitude regions in the past century (Haldoupis, 2011; Whitehead, 1989). It has been generally accepted by the community and relatively well understood. In this study, we have only shown some cases of RO E_s observations. It is hard to examine the wind shear theory in detail. Multiple observations, statistical anal-

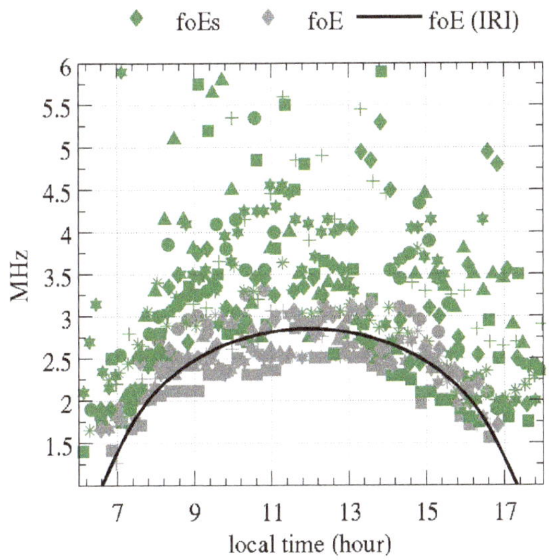

Figure 11. Local time variation of the sporadic E layer critical frequency (foEs, green) and the E layer critical frequency (foE, grey) observed by seven selected ionosondes and the corresponding IRI modeled foE results (black line) during 2006.294. The results of different ionosondes are distinguished by the dot shape. The IRI model results are corresponding to the mean latitude and longitude of seven ionosonde stations.

ysis, and theoretical simulation would be helpful in a further study to address this issue. Yeh et al. (2012) and Chu et al. (2014) have examined the wind shear theory based on RO E_s occurrence in great detail. In Chu et al. (2014), they compared the COSMIC E_s global morphology with the parameters that control the wind shear, including the global horizontal magnetic field intensity, global Fe^+ distribution from the Whole-Atmosphere Community Climate Model (WACCM) model calculation (Feng et al., 2013), and the meridional and zonal winds from the Horizontal Wind Model (HWM) model. Based on these parameters, they also calculated the global distribution of the divergence of the vertical Fe^+ flux. According to their results, the wind shear can explain the global distribution of E_s and their altitude, local time, seasonal, and hemispheric asymmetry quite well, especially at middle latitudes. The tidal wind and geomagnetic field are quite stable. According to Feng et al. (2013), the Fe^+ density is also substantial and distributed globally. We can conclude that the simultaneous occurrence of E_s in a broad region under wind shear theory is possible. Case 2 therefore supports the wind shear theory. It also demonstrates the advantages of RO signals in terms of high sensitivity to sporadic E occurrence.

4.4 Potential improvement on the neutral radio occultation retrievals with a better understanding of E_s occurrence

In this study, we have demonstrated that the GNSS RO signals would be useful to investigate the complex E_s occurrence in the ionosphere. However, on the other hand, the occurrence of E_s will degrade the retrieved data quality in the lower atmosphere, including profiles of bending angle, refractivity, and temperature in the troposphere and stratosphere (Mannucci et al., 2011). Currently, in the RO neutral data retrievals, the most frequently used method to calibrate the ionospheric bending effect is the linear combination of the L1 and L2 bending angle interpolated to the same impact parameter (Schreiner et al., 2011). During the occurrence of E_s, the ionosphere will cause a much larger bending effect and a multipath effect on both L1 and L2, which will make the linear combination of L1 and L2 bending angles less accurate. In addition, the occurrence of E_s itself will cause higher observational uncertainty in the GNSS signals. Small-scale ionospheric irregularity including the E_s and F region has been one of the most significant error sources of neutral RO data retrievals, especially in the upper stratosphere (Schreiner et al., 2011). We have compared the global E_s occurrence and the standard deviation of bending angles between 60 and 80 km derived from COSMIC RO data in detail. It is found that both the E_s occurrence and bending angle standard deviation show a similar global distribution, seasonal variation, and local time variation, especially in the middle- and high-latitude regions. It implies that the occurrence of E_s dominates the bending angle errors in middle- and high-latitude regions. The standard deviation in bending is $\sim 10^{-6}$ radiance, which means it is not negligible in the upper stratosphere. In the equatorial region, the bending angle standard deviations show a maximum at sunset, which is caused by the equatorial F layer scintillation. Having a better understanding of E_s will definitely be helpful in improving the neutral RO retrievals, especially in the upper stratosphere.

5 Conclusions

In this case study, the high-resolution GNSS RO signals from multiple missions processed in UCAR/CDAAC are used to investigate the complex E_s layer structures. The selected cases are evaluated by colocated simultaneous RO events and independent observations from lidar and ionograms. We come to the following key conclusions:

1. The E_s occurrence could be identified from the RO measurements by either the relative slant TEC peak or the SNR fluctuations (or sometimes the U shape).

2. Only the E_s from the defined E_s occurrence area (EOA) can result in E_s-related features in an RO event.

3. When assigning the observations to the tangent height, either SNR or TEC might show multiple E_s layer structures. Based on the evaluations between colocated simultaneous RO events and between RO and lidar observations, we conclude that some of these do manifest the multiple E_s layered structures.

4. We have found some cases that show the occurrence of E_s in a broad region during a certain time interval. This result is validated by independent ionosonde observations.

5. These complex E_s structures could be understood well using the popular wind shear theory. Further studies could benefit both E_s mechanism research and relevant applications.

6. A better understanding of E_s occurrence will be helpful for improving the RO-retrieved data quality in the upper stratosphere.

Acknowledgements. This material is based upon work supported by the National Science Foundation under Cooperative Agreement AGS-1033112. X. Xue acknowledges the support by the Project (KJCX2-EW-J01) of the Chinese Academy of Science. We acknowledge the usage of lidar data from the Chinese Meridian Project and ionosonde data accessed through the Lowell Global Ionospheric Radio Observatory (GIRO) Data Center (LGDC) Digital Ionogram DataBase (DIDB). We appreciate the organizations presented in Table 1 making their ionosondes observations publicly available.

Edited by: J. Y. Liu

References

Abdu, M. A., de Souza, J. R., Batista, I. S., Santos, A. M., Sobral, J. H. A., Rastogi, R. G., and Chandra, H.: The role of electric fields in sporadic E layer formation over low latitudes under quiet and magnetic storm conditions, J. Atmos. Solar-Terr. Phys., 115–116, 95–105, doi:10.1016/j.jastp.2013.12.003, 2014.

Anthes, R. A.: Exploring Earth's atmosphere with radio occultation: contributions to weather, climate and space weather, Atmos. Meas. Tech., 4, 1077–1103, doi:10.5194/amt-4-1077-2011, 2011.

Arras, C., Wickert, J., Beyerle, G., Heise, S., Schmidt, T., and Jacobi, C.: A global climatology of ionospheric irregularities derived from GPS radio occultation, Geophys. Res. Lett., 35, L14809, doi:10.1029/2008GL034158, 2008.

Arras, C., Jacobi, C., and Wickert, J.: Semidiurnal tidal signature in sporadic E occurrence rates derived from GPS radio occultation measurements at higher midlatitudes, Ann. Geophys., 27, 2555–2563, doi:10.5194/angeo-27-2555-2009, 2009.

Axford, W. I.: The formation and vertical movement of dense ionized layers in the ionosphere due to neutral wind shears, J. Geophys. Res., 68, 769–779, 1963.

Bernhardt, P. A.: The modulation of sporadic-E layers by Kelvin-Helmholtz billows in the neutral atmosphere, J. Atmos. Solar-Terr. Phys., 64, 1487–1504, 2002.

Bilitza, D. and Reinisch, B. W.: International Reference Ionosphere 2007: Improvements and new parameters, Adv. Space Res., 42, 599–609, doi:10.1016/j.asr.2007.07.048, 2008.

Chu, Y. H. and Wang, C. Y.: Interferometry observations of three-dimensional spatial structures of sporadic E irregularities using the Chung-Li VHF radar, Radio Sci., 32, 817–832, 1997.

Chu, Y. H., Wang, C. Y., Wu, K. H., Chen, K. T., Tzeng, K. J., Su, C. L., Feng, W., and Plane, J. M. C.: Morphology of sporadic E layer retrieved from COSMIC GPS radio occultation measurements: Wind shear theory examination, J. Geophys. Res., 119, 2117–2136, doi:10.1002/2013JA019437, 2014.

Cosgrove, R. B. and Tsunoda, R. T.: Simulation of the nonlinear evolution of the sporadic-E layer instability in the nighttime midlatitude ionosphere, J. Geophys. Res., 108, 1283, doi:10.1029/2002JA009728, 2003.

Dou, X.-K., Xue, X. H., Li, T., Chen, T. D., Chen, C., and Qiu, S.-C.: Possiblerelations between meteors, enhanced electron density layers, and sporadic sodiumlayers, J. Geophys. Res., 115, A06311, doi:10.1029/2009JA014575, 2010.

Dou, X. K., Qiu, S. C., Xue, X. H., Chen, T. D., and Ning, B. Q.: Sporadic and thermospheric enhanced sodium layers observed by a lidar chain over China, J. Geophys. Res., 118, 6627–6643, doi:10.1002/jgra.50579, 2013.

Feng, W., Marsh, D. R., Chipperfield, M. P., Janches, D., Hoffner, J., Yi, F., and Plane, J. M. C.: A global atmospheric model of meteoric iron, J. Geophys. Res. Atmos., 118, 9456–9474, doi:10.1002/jgrd.50708, 2013.

Foelsche, U., Scherllin-Pirscher, B., Ladstädter, F., Steiner, A. K., and Kirchengast, G.: Refractivity and temperature climate records from multiple radio occultation satellites consistent within 0.05 %, Atmos. Meas. Tech., 4, 2007–2018, doi:10.5194/amt-4-2007-2011, 2011.

Fytterer, T., Arras, C., and Jacobi, C.: Terdiurnal signatures in sporadic E layers at midlatitudes, Adv. Radio Sci., 11, 333–339, doi:10.5194/ars-11-333-2013, 2013.

Galkin, I. A., Kitrosser, D. F., Kecic, Z., and Reinisch, B. W.: Internet access to ionosondes, J. Atmos. Solar-Terr. Phys., 61, 181–186, 1999.

Gardner, C. S., Kane, T. J., Senft, D. C., Qian, J., and Papen, G. C.: Simultaneous observations of sporadic E, Na, Fe, and Ca^+ layers at Urbana, Illinois: Three case studies, J. Geophys. Res., 98, 16865–16873, doi:10.1029/93JD01477, 1993.

Haldoupis, C.: A Tutorial Review on Sporadic E Layers, Aeronomy of the Earth's Atmosphere and Ionosphere, IAGA Special Sopron Book Series 2, 381–394, 2011.

Hocke, K., Igarashi, K., Nakamura, M., Wilkinson, P., Wu, J., Pavelyev, A., and Wikert, J.: Global sounding of sporadic E layers by the GPS/MET radio occultation experiment, J. Atmos. Solar-Terr. Phys., 63, 1973–1980, doi:10.1016/S1364-6826(01)00063-3, 2001.

Hysell, D. L., Nossa, E., Larsen, M. F., Munro, J., Smith, S., Sulzer, M. P., and González, S. A.: Dynamic instability in the lower thermosphere inferred from irregular sporadic E layers, J. Geophys. Res., 117, A08305, doi:10.1029/2012JA017910, 2012.

Li, G., Ning, B., Patra, A. K., Abdu, M. A., Chen, J., Liu, L., and Hu, L.: On the linkage of daytime 150 km echoes and abnormal intermediate layer traces over Sanya, J. Geophys. Res., 118, 7262–7267, doi:10.1002/2013JA019462, 2013.

Liu, J. Y., Lin, C. Y., Lin, C. H., Tsai, H. F., Solomon, S. C., Sun, Y. Y., Lee, I. T., Schreiner, W. S., and Kuo, Y. H.: Artificial plasma cave in the low-latitude ionosphere results from the radio occultation inversion of the FORMOSAT-3/COSMIC, J. Geophys. Res., 115, A07319, doi:10.1029/2009JA015079, 2010.

Mannucci, A. J., Ao, C. O., Pi, X., and Iijima, B. A.: The impact of large scale ionospheric structure on radio occultation retrievals, Atmos. Meas. Tech., 4, 2837–2850, doi:10.5194/amt-4-2837-2011, 2011.

Maruyama, T., Saito, S., Yamamoto, M., and Fukao, S.: Simultaneous observation of sporadic E with a rapid-run ionosonde and VHF coherent backscatter radar, Ann. Geophys., 24, 153–162, doi:10.5194/angeo-24-153-2006, 2006.

Mathews, J. D.: Sporadic E: Current views and recent progress, J. Atmos. Solar-Terr. Phys., 60, 413–435, doi:10.1016/S1364-6826(97)00043-6, 1998.

Mathews, J. D., Machuga, D. W., and Zhou, Q.: Evidence for electrodynamic linkages between spread-F, ion rain, the intermediate layer, and sporadic E: results from observations and simulations, J. Atmos. Solar-Terr. Phys., 63, 1529–1543, 2001.

Pavelyev, A. G., Liou, Y. A., Zhang, K., Wang, C. S., Wickert, J., Schmidt, T., Gubenko, V. N., Pavelyev, A. A., and Kuleshov, Y.: Identification and localization of layers in the ionosphere using the eikonal and amplitude of radio occultation signals, Atmos. Meas. Tech., 5, 1–16, doi:10.5194/amt-5-1-2012, 2012.

Reinisch, B. W. and Galkin, I. A.: Global ionospheric radio observatory (GIRO), Earth Planet Space, 63, 377–381, doi:10.5047/eps.2011.03.001, 2011.

Schreiner, W., Rocken, C., Sokolovskiy, S., Syndergaard, S., and Hunt, D.: Estimates of the precision of GPS radio occultations from the COSMIC/FORMOSAT-3 mission, Geophys. Res. Lett., 34, L04808, doi:10.1029/2006GL027557, 2007.

Schreiner, W., Sokolovskiy, S., Hunt, D., Rocken, C., and Kuo, Y.-H.: Analysis of GPS radio occultation data from the FORMOSAT-3/COSMIC and Metop/GRAS missions at CDAAC, Atmos. Meas. Tech., 4, 2255–2272, doi:10.5194/amt-4-2255-2011, 2011.

Smith, L. G. and Miller, K. L.: Sporadic-E layers and unstable wind shears, J. Atmos. Terr. Phys., 42, 45–50, 1980.

Sokolovskiy, S., Schreiner, W., Rocken, C., and Hunt, D.: Detection of high-altitude ionospheric irregularities with GPS/MET, Geophys. Res. Lett., 29, 1033, doi:10.1029/2001GL013398, 2002.

Straus, P. R., Anderson, P. C., and Danaher, J. E.: GPS occultation sensor observations of ionospheric scintillation, Geophys. Res. Lett., 30, 1436, doi:10.1029/2002GL016503, 2003.

Wakabayashi, M. and Ono, T.: Multi-layer structure of mid-latitude sporadic-E observed during the SEEK-2 campaign, Ann. Geophys., 23, 2347–2355, doi:10.5194/angeo-23-2347-2005, 2005.

Wang, C.: New chains of space weather monitoring stations in China, Space Weather, 8, S08001, doi:10.1029/2010SW000603, 2011.

Whitehead, J. D.: Recent work on midlatitude and equatorial sporadic E, J. Atmos. Terr. Phys., 51, 401–424, doi:10.1016/0021-9169(89)90122-0, 1989.

Wickert, J., Pavelyev, A. G., Liou, Y. A., Schmidt, T., Reigber, C., Igarashi, K., Pavelyev, A. A., and Matyugov, S.: Amplitude variations in GPS signals as possible indicator of ionospheric structures, Geophys. Res. Lett., 31, L24801, doi:10.1029/2004GL020607, 2004.

Wu, D. L., Ao, C. O., Hajj, G. A., de la Torre Juarez, M., and Mannucci, A. J.: Sporadic E morphology from GPS-CHAMP radio occultation, J. Geophys. Res., 110, A01306, doi:10.1029/2004JA010701, 2005.

Xue, X. H., Dou, X. K., Lei, J., Chen, J. S., Ding, Z. H., Li, T., Gao, Q., Tang, W. W., Cheng, X. W., and Wei, K.: Lower thermospheric enhanced sodium layers observed at low latitude and possible formation: Case studies, J. Geophys. Res., 118, 2409–2418, doi:10.1002/jgra.50200, 2013.

Yeh, W.-H., Huang, C.-Y., Hsiao, T.-Y., Chiu, T.-C., Lin, C.-H., and Liou, Y.-A.: Amplitude morphology of GPS radio occultation data for sporadic-E layers, J. Geophys. Res., 117, A11304, doi:10.1029/2012JA017875, 2012.

Yue, X., Wan, W., Liu, L., and Ning, B.: An empirical model of ionospheric foE over Wuhan, Earth Planets Space, 58, 323–330, 2006.

Yue, X., Schreiner, W. S., Kuo, Y.-H., Hunt, D. C., Wang, W., Solomon, S. C., Burns, A. G., Bilitza, D., Liu, J.-Y., Wan, W., and Wickert, J.: Global 3-D Ionospheric Electron Density Reanalysis based on Multi-Source Data Assimilation, J. Geophys. Res., 117, A09325, doi:10.1029/2012JA017968, 2012.

Yue, X., Schreiner, W. S., Pedatella, N., Anthes, R. A., Mannucci, A. J., Straus, P. R., and Liu, J.-Y.: Space Weather Observations by GNSS Radio Occultation: From FORMOSAT-3/COSMIC to FORMOSAT-7/COSMIC-2, Space Weather, 12, 616–621, doi:10.1002/2014SW001133, 2014.

Zeng, Z. and Sokolovskiy, S.: Effect of sporadic E cloud on GPS radio occultation signal, Geophys. Res. Lett., 37, L18817, doi:10.1029/2010GL044561, 2010.

Zhang, K., Fu, E., Silcock, D., Wang, Y., and Kuleshov, Y.: An investigation of atmospheric temperature profiles in the Australian region using collocated GPS radio occultation and radiosonde data, Atmos. Meas. Tech., 4, 2087–2092, doi:10.5194/amt-4-2087-2011, 2011.

12

A six-beam method to measure turbulence statistics using ground-based wind lidars

A. Sathe, J. Mann, N. Vasiljevic, and G. Lea
DTU Wind Energy, Risø campus, Roskilde, Denmark

Correspondence to: A. Sathe (amsat@dtu.dk)

Abstract. A so-called six-beam method is proposed to measure atmospheric turbulence using a ground-based wind lidar. This method requires measurement of the radial velocity variances at five equally spaced azimuth angles on the base of a scanning cone and one measurement at the centre of the scanning circle, i.e.using a vertical beam at the same height. The scanning configuration is optimized to minimize the sum of the random errors in the measurement of the second-order moments of the components (u, v, w) of the wind field. We present this method as an alternative to the so-called velocity azimuth display (VAD) method that is routinely used in commercial wind lidars, and which usually results in significant averaging effects of measured turbulence. In the VAD method, the high frequency radial velocity measurements are used instead of their variances. The measurements are performed using a pulsed lidar (WindScanner), and the derived turbulence statistics (using both methods) such as the u and v variances are compared with those obtained from a reference cup anemometer and a wind vane at 89 m height under different atmospheric stabilities. The measurements show that in comparison to the reference cup anemometer, depending on the atmospheric stability and the wind field component, the six-beam method measures between 85 and 101 % of the reference turbulence, whereas the VAD method measures between 66 and 87 % of the reference turbulence.

1 Introduction

Wind lidars are being used significantly for wind energy applications. They measure mean wind speeds with great accuracy, and are very useful tools in the measurement of wind profiles (Smith et al., 2006; Kindler et al., 2007; Peña et al., 2009; Wagner et al., 2011). New recommended practices are being defined for wind resource assessments (Clifton et al., 2013). However their use in measuring atmospheric turbulence has not yet been established, particularly with the commercial lidars (Sathe et al., 2011b). The main reason is that for a commercial lidar, the measured lidar data is processed using the so-called velocity azimuth display (VAD) method, where the measurements of the radial velocity (also called the line-of-sight velocity) at different azimuth angles are combined to deduce the wind field components. For the mean wind speed estimation, the VAD method produces negligible errors. For turbulence statistics the VAD method produces significant systematic errors (Sathe et al., 2011b; Sathe and Mann, 2012) mainly due to two reasons; one is the filtering of the smaller scales due to the large size of the probe volume within which the radial velocity is measured, and second is the contamination by the two-point correlation between the components of the wind field.

In this article we present a so-called six-beam method that significantly improves the measurement of turbulence in comparison to the VAD method. This method uses the variances of the radial velocities from six different lidar beams, five of which are at equally spaced azimuth angles on the base of a scanning cone and one beam is vertical. These variances are then combined in order to deduce the second-order moments of the wind field. A framework for this method was originally proposed by Lhermitte (1969), which was used by Wilson (1970) and Kropfli (1986) for radar studies, and subsequently by Eberhard et al. (1989) and Mann et al. (2010) for lidar studies of turbulence measurements. In their studies only the covariances were estimated, either by combining several measurements of the radial velocity variances from several lidar beams (Eberhard et al., 1989) at equally spaced azimuth angles and one elevation angle, or by using only two lidar beams (Mann et al., 2010). In the present work, six beams are used, five at an elevation angle of 45° and one vertical that enable us to also deduce the variances.

The ideas to measure turbulence using remote sensing instruments have evolved, albeit slowly, since the pioneering works on radar meteorology (Lhermitte, 1962; Browning and Wexler, 1968). Based on the VAD scanning, Lhermitte (1969) was the first (to our knowledge) to suggest a technique of deducing turbulence using the measurements of the variance of the radial velocity. Subsequently Wilson (1970) was the first to conduct an experiment using a pulsed Doppler radar and deducing turbulence in the convective boundary layer (0.1–1.3 km). Only turbulence scales larger than the pulse volume but smaller than the scanning circle could be measured since all the data from a single scan was used. Also, no comparison with any reference instrument was carried out, and hence, the reliability of the radar measurements could not be verified. Kropfli (1986) extended the study of Wilson (1970) to also include the turbulence scales larger than the scanning circle by using the data from multiple scans. Although the method was developed for Doppler radar studies, it could also be used for Doppler lidar studies. Eberhard et al. (1989) was the first to perform turbulence studies using a lidar following the methods of Wilson (1970); Kropfli (1986). Gal-Chen et al. (1992) also used the variances of the radial velocities to deduce turbulence, but with a different scanning configuration. In all of the aforementioned studies with a Doppler lidar (or radar), the probe length was quite significant (of the order of 100 m), which perhaps was the reason to restrict these studies to the convective boundary layer. However if the turbulence measurements were desired close to the ground then they would be subjected to a significant amount of probe volume averaging. It was perhaps this reason that the focus on turbulence research with lidars shifted to understanding the probe volume averaging effect and providing potential solutions to compensate for it (Frehlich, 1994, 1997; Frehlich et al., 1994, 1998, 2006, 2008; Frehlich and Cornman, 2002; Frehlich and Kelley, 2008; Banakh et al., 1995a, b, 1996, 1999, 2010; Banakh and Smalikho, 1997a, b; Banakh and Werner, 2005; Smalikho, 1995; Smalikho et al., 2005; Mann et al., 2010; Branlard et al., 2013). Even with the development of the modern lidar systems, where the probe lengths have shrunk to about 30 m for a pulsed lidar, significant amount of averaging still remains in the turbulence measurements within the surface layer, where the wind turbines operate (Mann et al., 2009, 2010; Sjöholm et al., 2009; Sathe et al., 2011b; Sathe and Mann, 2012). A detailed review of the state of the art with respect to turbulence measurements using ground-based wind lidars can be found in Sathe and Mann (2013).

Unfortunately within the wind energy sector, turbulence measurements are being deduced using the VAD scanning method, which results in a significant amount of filtering of turbulence, and contamination by the two-point correlation between the components of the wind field (Sathe et al., 2011b). In this work we attempt to significantly improve the turbulence measurements compared to those obtained by the VAD method, by extending the previously developed ideas of using the radial velocity variances (Lhermitte, 1969; Wilson, 1970; Kropfli, 1986; Eberhard et al., 1989; Mann et al., 2010), but restricting them to using only six beams.

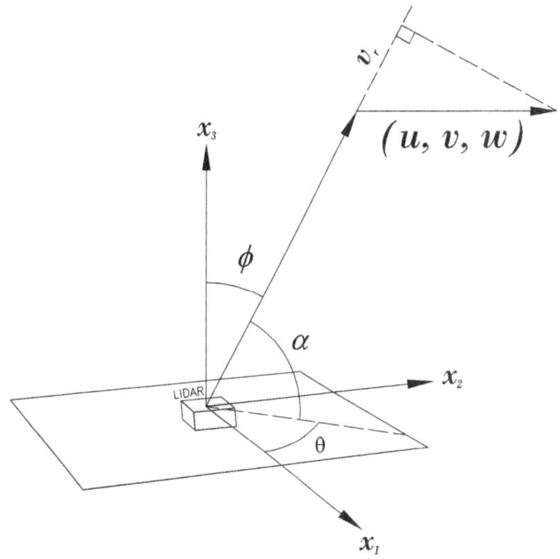

Figure 1. Coordinate system of a lidar.

The structure of this article is divided into the following sections. Section 2 gives a detailed explanation of the six-beam technique. The optimum six-beam configuration, which is one of the main contributions of this article is also described in detail. In order to verify our method, turbulence measurements using the pulsed lidar WindScanner were performed and compared with a reference cup anemometer at a height of 89 m. The site description for the measurements is given in Sect. 3, whereas the results are described in Sect. 4. Discussions and conclusions are made in Sects. 5 and 6, respectively.

2 Theory of six-beam configuration

The instantaneous velocity field is characterized as a vector $\boldsymbol{v} = (u, v, w)$, and turbulence is characterized as the components of the Reynolds stress tensor,

$$\mathbf{R} = \begin{bmatrix} \langle u'^2 \rangle & \langle u'v' \rangle & \langle u'w' \rangle \\ \langle v'u' \rangle & \langle v'^2 \rangle & \langle v'w' \rangle \\ \langle w'u' \rangle & \langle w'v' \rangle & \langle w'^2 \rangle \end{bmatrix}, \quad (1)$$

where the diagonal terms are the variances of the respective wind field components and the off-diagonal terms are the covariances, $\langle\rangle$ denote ensemble average, and $'$ denotes fluctuations around the average.

As shown in Fig. 1, at a given instant of time if we assume that a lidar measures at a point, and that the lidar beam is inclined at a certain zenith angle ϕ (in some literature the complement of ϕ is used, which is called as the elevation angle $\alpha = 90° - \phi$) from the vertical axis, and makes an azimuth angle θ with respect to the axes in the horizontal plane, then the radial velocity (also called as the line-of-sight velocity) can be mathematically written as,

$$v_r(\phi, \theta, d_f) = \boldsymbol{n}(\phi, \theta) \cdot \boldsymbol{v}(\boldsymbol{n}(\phi, \theta) d_f), \tag{2}$$

where v_r is the radial velocity measured at a point, $\boldsymbol{n} = (\cos\theta \sin\phi, \sin\theta \sin\phi, \cos\phi)$ is the unit directional vector for a given ϕ and θ, and d_f is the distance at which the measurement is obtained. In Eq. (2), we have implicitly assumed that v_r is positive for the wind going away from the lidar axis, the coordinate system is right-handed, and u is aligned with the x_1 axis in a horizontal plane, i.e. from west to east. In reality, a lidar never receives backscatter from exactly a point, but from all over the physical space. Fortunately the transverse dimensions of a lidar beam are much smaller than the longitudinal dimensions, and for all practical purposes we can consider that the backscatter is received only along the lidar beam axis. Mathematically the radial velocity can be represented as the convolved signal,

$$\widetilde{v}_r(\phi, \theta, d_f) = \int_{-\infty}^{\infty} \varphi(s) \boldsymbol{n}(\phi, \theta) \cdot \boldsymbol{v}(\boldsymbol{n}(\phi, \theta)(d_f + s)) \mathrm{d}s, \tag{3}$$

where $\widetilde{v}_r$ is the weighted average radial velocity, $\varphi(s)$ is any weighting function integrating to one that depends on the type of lidar, i.e. a continuous wave (c-w) lidar or a pulsed lidar, and s is the distance along the beam from the measurement point of interest. From simple geometrical considerations the radial velocity variance can be written as a function of the components of $\mathbf{R}$ (Lhermitte, 1969; Eberhard et al., 1989),

$$\begin{aligned}\left\langle v_r'^2 \right\rangle =& \left\langle u'^2 \right\rangle \sin^2\phi \cos^2\theta + \left\langle v'^2 \right\rangle \sin^2\phi \sin^2\theta + \left\langle w'^2 \right\rangle \cos^2\phi \\ &+ 2\left\langle u'v' \right\rangle \sin^2\phi \sin\theta \cos\theta + 2\left\langle u'w' \right\rangle \sin\phi \cos\phi \cos\theta \\ &+ 2\left\langle v'w' \right\rangle \sin\phi \cos\phi \sin\theta,\end{aligned} \tag{4}$$

where $\langle v_r'^2 \rangle$ is the radial velocity variance. From Eq. (4) we can see that for a given θ and ϕ, if we have six measurements of $\langle v_r'^2 \rangle$ then there are six unknowns to be determined, which in a matrix form can be written as,

$$\mathbf{M} \underbrace{\begin{bmatrix} \left\langle u'^2 \right\rangle \\ \left\langle v'^2 \right\rangle \\ \left\langle w'^2 \right\rangle \\ \left\langle u'v' \right\rangle \\ \left\langle u'w' \right\rangle \\ \left\langle v'w' \right\rangle \end{bmatrix}}_{\Sigma} = \underbrace{\begin{bmatrix} \left\langle v_{r_1}'^2 \right\rangle \\ \left\langle v_{r_2}'^2 \right\rangle \\ \left\langle v_{r_3}'^2 \right\rangle \\ \left\langle v_{r_4}'^2 \right\rangle \\ \left\langle v_{r_5}'^2 \right\rangle \\ \left\langle v_{r_6}'^2 \right\rangle \end{bmatrix}}_{\boldsymbol{S}}, \tag{5}$$

where Σ is a vector of the components of $\mathbf{R}$ (because $\mathbf{R}$ is symmetric, we only need six components), $\mathbf{M}$ is a 6×6 matrix of the coefficients of Σ that consist of different combinations of θ and ϕ (see Eq. 4), and $\boldsymbol{S}$ is a vector of measurements of $\langle v_r'^2 \rangle$ at different θ and ϕ (where the suffices denote measurements from beam 1 to 6). In principle we can then estimate Σ using the relation $\Sigma = \mathbf{M}^{-1}\boldsymbol{S}$, where $^{-1}$ denotes matrix inverse. It is interesting to know beforehand whether the measurements from the six beams on only one zenith angle are adequate, i.e. whether we can have six θs and only one ϕ.

From fundamental algebra we understand that Eq. (5) will have a finite solution if and only if $\det\mathbf{M} \neq 0$, where det denotes the determinant of a matrix. In other words $\mathbf{M}$ should not be a degenerate matrix. From the properties of determinants we know that if any two rows (or columns) of a matrix are identical then its determinant is zero. Also, if the elements of any row (or column) are increased (or decreased) by equal multiples of the corresponding elements of any other row (or column), the value of determinant is unchanged. If we use only one ϕ at different θ, and add the first two columns of $\mathbf{M}$, we get the first and the third columns of $\mathbf{M}$ to be multiples of each other, which according to the property of determinants implies $\det\mathbf{M} = 0$. Thus $\mathbf{M}$ becomes degenerate if we use only one ϕ, and thus need $\langle v_r'^2 \rangle$ measurements from more than one ϕ.

We are then confronted with the challenge of obtaining an optimum combination of θ and ϕ. Measured $\boldsymbol{S}$ is stochastic, and the random error of Σ will depend on the particular choice of the θs and ϕs. We thus choose the objective function such that the sum of the random errors of the components of Σ are minimized. For simplicity, we neglect the probe volume filtering effect in the derivation of the optimum combination, but including that will not change the optimum configuration.

It is to be noted that in Eq. (5), we have neglected the random error in the lidar estimate of the radial velocity. For the particular lidar used in this work the uncorrelated noise in the velocity estimation is exceedingly small and will not have any effect for the analysis presented in this paper. The very low noise level can be seen from plots of spectra in Figs. 2, 5, and 7 of Sathe and Mann (2012).

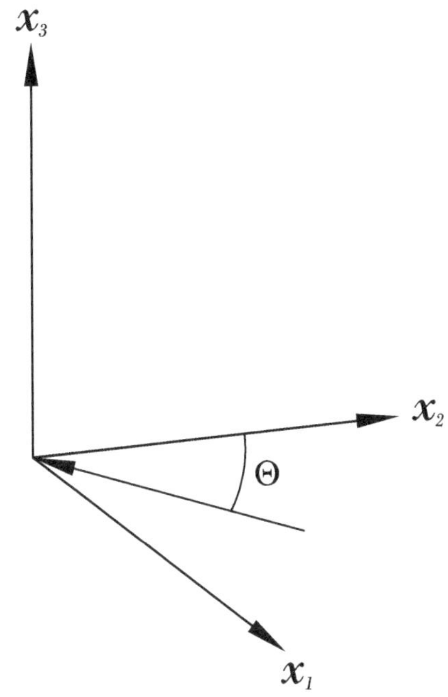

Figure 2. Standard meteorological convention of depicting the mean wind direction.

2.1 Formulation of the objective function

Equation (4) is valid for the mean wind direction aligned in the x_1 direction. Following standard meteorological conventions, let us consider the mean wind direction to be at an angle Θ with respect to the north, i.e. x_2 axis as shown in Fig. 2. At first we derive an objective function for the wind aligned with the x_1 axis, and then extend the derivation to the coordinate system aligned with the mean wind direction.

2.1.1 Mean wind aligned with the x_1 axis

If we consider that $\delta\Sigma$ is the random error on Σ, and $\delta \boldsymbol{S}$ is the random error on $\boldsymbol{S}$, then Eq. (5) can be written as,

$$\begin{aligned} \mathbf{M}(\Sigma+\delta\Sigma) &= \boldsymbol{S}+\delta\boldsymbol{S}, \\ \Sigma+\delta\Sigma &= \mathbf{M}^{-1}(\boldsymbol{S}+\delta\boldsymbol{S}). \end{aligned} \tag{6}$$

We can thus write,

$$\delta\Sigma = \mathbf{M}^{-1}\delta\boldsymbol{S}. \tag{7}$$

If we consider the sum of the error variances $\langle\delta\Sigma^{\mathrm{T}}\delta\Sigma\rangle$, where ^{T} denotes matrix transpose, then the objective is to minimize the sum of the error variances of the components of Σ. Taking the transpose and multiplying by Eq. (7) we get,

$$\begin{aligned} \delta\Sigma\cdot\delta\Sigma = \delta\Sigma^{\mathrm{T}}\delta\Sigma &= (\mathbf{M}^{-1}\delta\boldsymbol{S})^{\mathrm{T}}(\mathbf{M}^{-1}\delta\boldsymbol{S}) \\ &= \delta\boldsymbol{S}^{\mathrm{T}}(\mathbf{M}^{-1\mathrm{T}}\mathbf{M}^{-1})\delta\boldsymbol{S} \end{aligned} \tag{8}$$

The task now is to simplify Eq. (8) such that it can be represented as a function of θ and ϕ only. If we assume that the random errors in the variances of the radial velocities are independent of each other, and that the error variance for each radial velocity variance is $\langle\epsilon_{\mathrm{s}}^2\rangle$, we get,

$$\frac{\langle\delta\Sigma^{\mathrm{T}}\delta\Sigma\rangle}{\langle\epsilon_{\mathrm{s}}^2\rangle} = \mathrm{Tr}(\mathbf{M}^{-1}\mathbf{M}^{-1\mathrm{T}}), \tag{9}$$

where Tr is the trace of a matrix. The objective function is to minimize Eq. (9).

2.1.2 Coordinate system aligned with the mean wind direction

In order to align the coordinate system with the mean wind direction, we need to apply coordinate transformations on any tensors that are defined in the original coordinate system. The vector $\boldsymbol{v}$ rotated in the mean wind direction has to be multiplied by a transformation matrix $\mathbf{T}$ given as,

$$\mathbf{T} = \begin{bmatrix} -\sin\Theta & -\cos\Theta & 0 \\ \cos\Theta & -\sin\Theta & 0 \\ 0 & 0 & 1 \end{bmatrix}. \tag{10}$$

In the coordinate system aligned with the mean wind direction, we then get in matrix form,

$$\mathbf{R}_{\mathrm{r}} = \mathbf{T}\mathbf{R}\mathbf{T}^{\mathrm{T}}, \tag{11}$$

where $\mathbf{R}_{\mathrm{r}}$ is the Reynolds stress tensor in a coordinate system aligned with the mean wind direction. If we denote Σ_{r} as the vector of the components of $\mathbf{R}_{\mathrm{r}}$, then we can write,

$$\Sigma_{\mathrm{r}} = \underbrace{\begin{bmatrix} \sin^2\Theta & \cos^2\Theta & 0 & \sin 2\Theta & 0 & 0 \\ \cos^2\Theta & \sin^2\Theta & 0 & -\sin 2\Theta & 0 & 0 \\ 0 & 0 & 1 & 0 & 0 & 0 \\ -\frac{1}{2}\sin 2\Theta & \frac{1}{2}\sin 2\Theta & 0 & -\cos 2\Theta & 0 & 0 \\ 0 & 0 & 0 & 0 & -\sin\Theta & -\cos\Theta \\ 0 & 0 & 0 & 0 & \cos\Theta & -\sin\Theta \end{bmatrix}}_{\mathbf{N}} \Sigma. \tag{12}$$

Using Eqs. (6) and (7), we can write,

$$\delta\Sigma_{\mathrm{r}} = \mathbf{N}\mathbf{M}^{-1}\delta\boldsymbol{S} \tag{13}$$

Following the same procedure as in Sect. 2.1.1, we get

$$\frac{\langle\delta\Sigma_{\mathrm{r}}^{\mathrm{T}}\delta\Sigma_{\mathrm{r}}\rangle}{\langle\epsilon_{\mathrm{s}}^2\rangle} = \mathrm{Tr}(\mathbf{N}\mathbf{M}^{-1}(\mathbf{N}\mathbf{M}^{-1})^{\mathrm{T}}). \tag{14}$$

Equation (14) states that the error variance is dependent on the mean wind direction. In order to make it independent of the mean wind direction, we assume a uniform distribution of the mean wind direction, and estimate the averaged ratio of the error variance. Thus the directionally averaged ratio is,

$$\begin{aligned} \overline{\frac{\langle\delta\Sigma_{\mathrm{r}}^{\mathrm{T}}\delta\Sigma_{\mathrm{r}}\rangle}{\langle\epsilon_{\mathrm{s}}^2\rangle}}^{\Theta} &= \frac{1}{2\pi}\int_0^{2\pi}\mathrm{Tr}(\mathbf{N}\mathbf{M}^{-1}(\mathbf{N}\mathbf{M}^{-1})^{\mathrm{T}})\mathrm{d}\Theta \\ &= \frac{1}{2\pi}\int_0^{2\pi}\mathrm{Tr}(\mathbf{N}\mathbf{M}^{-1}\mathbf{M}^{-1\mathrm{T}}\mathbf{N}^{\mathrm{T}})\mathrm{d}\Theta, \end{aligned}$$

where $\overline{\cdots}^{\Theta}$ denotes directional average. Using the property of matrix trace that it is invariant under cyclic permutations we get,

$$\frac{\overline{\langle \delta \Sigma_r^T \delta \Sigma_r \rangle}^{\Theta}}{\langle \epsilon_s^2 \rangle} = \frac{1}{2\pi} \int_0^{2\pi} \mathrm{Tr}(\mathbf{N}^T \mathbf{N} \mathbf{M}^{-1} \mathbf{M}^{-1T}) d\Theta \tag{15}$$

We can also switch the order between integration and matrix trace, i.e. either we can estimate the trace first and then the integration or vice-versa. Thus,

$$\frac{\overline{\langle \delta \Sigma_r^T \delta \Sigma_r \rangle}^{\Theta}}{\langle \epsilon_s^2 \rangle} = \mathrm{Tr}\left(\left[\frac{1}{2\pi} \int_0^{2\pi} \mathbf{N}^T \mathbf{N} d\Theta \right] \mathbf{M}^{-1} \mathbf{M}^{-1T} \right). \tag{16}$$

Solving the integral we get,

$$\frac{\overline{\langle \delta \Sigma_r^T \delta \Sigma_r \rangle}^{\Theta}}{\langle \epsilon_s^2 \rangle} = \mathrm{Tr}\left(\begin{bmatrix} \frac{7}{8} & \frac{1}{8} & 0 & 0 & 0 & 0 \\ \frac{1}{8} & \frac{7}{8} & 0 & 0 & 0 & 0 \\ 0 & 0 & 1 & 0 & 0 & 0 \\ 0 & 0 & 0 & \frac{3}{2} & 0 & 0 \\ 0 & 0 & 0 & 0 & 1 & 0 \\ 0 & 0 & 0 & 0 & 0 & 1 \end{bmatrix} \mathbf{M}^{-1} \mathbf{M}^{-1T} \right). \tag{17}$$

The objective is to minimize Eq. (17), subject to the constraints that θ varies between 0 and 360° and ϕ varies between 0 and 45°. The limit of 45° for ϕ is arbitrary, and is based on the considerations of statistical homogeneity in a horizontal plane. Depending on the type of the terrain the range of ϕ could thus be increased or decreased, i.e. if a terrain is horizontally homogeneous over a very large extent, then ϕ could be greater than 45° and vice-versa.

2.2 Optimizing the objective function

Equation (17) represents a non-linear optimization problem with 12 unknown variables, i.e. six unknowns are zenith angles ϕ, and the remaining six are the azimuth angles θ. Owing to the complexity of the optimization problem, an analytical solution of Eq. (17) is not possible. We thus use numerical methods, where either gradient or direct search methods could be used (Rao, 2009). For gradient methods, it is essential that the objective function is differentiable. However we assume that Eq. (17) is a discontinuous function, and hence, we do not use gradient methods. Thus we optimize Eq. (17) using direct search methods only. The main advantage of using direct search methods is that they can be used for discontinuous and non-differentiable functions. The main limitation of such methods is that the found optimum may only be a local optimum.

Different algorithms such as simplex (Nelder and Mead, 1965), simulated annealing (Ingber, 1993) and random search (Rao, 2009) are tested, which result in the optimum angles as given in Table 1, which shows that the optimum configuration consists of five beams equally spaced on the base of a scanning cone and one vertical beam.

Table 1. Optimum six-beam configuration.

Beam no.	1	2	3	4	5	6
θ°	0	72	144	216	288	288
ϕ°	45	45	45	45	45	0

3 Description of the measurements

The six-beam measurements were carried out using the newly developed 1543 nm pulsed coherent Doppler scanning lidar "long-range WindScanner" (henceforth referred to as WindScanner) at the DTU Wind Energy Department in Denmark. The WindScanner is based on the pulsed lidar Windcube 200 from Leosphere and a dual-axis mirror based steerable scanner head designed by DTU Wind Energy and IPU. The WindScanner is intended for radial velocity measurements from the range of distances between 50 and 6000 m. The current maximum measurement rate is 10 Hz. The maximum number of simultaneous radial velocities acquired at any rate along each line-of-sight is 500. The WindScanner can emit either 400 or 200 ns laser pulses, which are streamed with two corresponding pulse repetition frequencies of 10 and 20 kHz respectively. The energy content of 400 ns laser pulses is 100 μJ, while the energy content of the 200 ns laser pulses is half of this value. The scanner head has two rotational degrees of freedom and can rotate around the azimuth and elevation axes, thus it directs the laser pulses into the atmosphere at any combination of azimuth and elevation. The maximum scanner head rotation speed is 50° s^{-1}, while the maximum acceleration is 100° s^{-2}. The scanner head can rotate around both axes from 0 to 360°, and the rotation can be endless. The pointing accuracy of the WindScanner is 0.05°. The WindScanner is operated via a remote "master computer" through a UDP/IP and TCP/IP network using the remote sensing communication protocol (RSComPro) (Vasiljevic, 2014).

The measurements were performed at the Danish National Test Center for Large Wind Turbines at Høvsøre, Denmark. Figure 3 shows the location of the test centre in Denmark (see inset in Fig. 3) and the location of the reference 89 m meteorological (met) mast located at the UTM zone 32 V 447 229 m E and 6 256 195 m N (WGS84 datum). The high frequency measurements from a reference Risø P2564A cup anemometer at 89 m placed on the top of the met mast are combined with the wind direction measurements from a F2919A Vector W200P wind vane placed in the north direction at 86 m to deduce $\langle u'^2 \rangle$ and $\langle v'^2 \rangle$ over a 30 min period. Since it is not possible to measure the w component using a cup anemometer, comparisons of the corresponding second-order statistics were not possible with the WindScanner measurements. The choice of the 30 min averaging period (instead of the standard 10 min statistics prevalent in the wind energy industry) is made based on the considerations

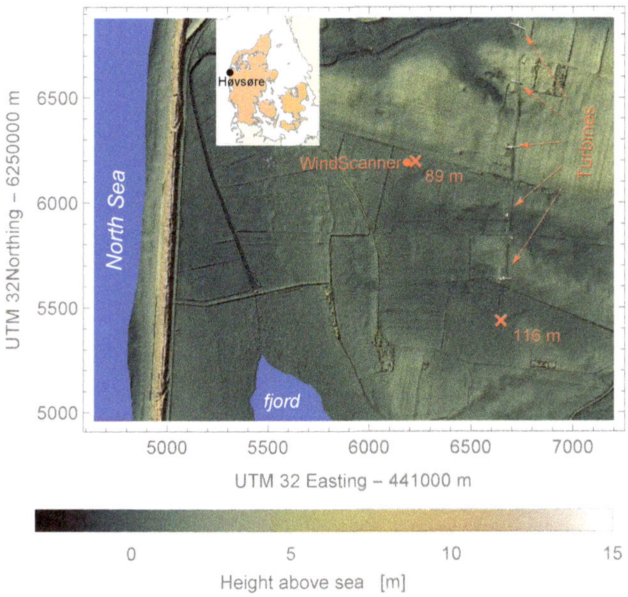

Figure 3. Location of the Høvsøre test centre in Denmark (inset) and details of the site, where the location of the WindScanner and the met masts are shown.

of the random errors in turbulence measurements (Lenschow et al., 1994). The site is about 2 km from the west coast of Denmark. The eastern sector is characterized by flat homogeneous terrain, and to the south is a lagoon. The WindScanner is placed at the UTM zone 32 V 447 188 m E and 6 256 189 m N (WGS84 datum) which is about 41 m away from the met mast in the west direction. Since the wind turbines are to the east of the WindScanner and the met mast, only the measurements from the western sector (225–315°) are used. Owing to the sudden change in the surface roughness from sea to land in the western sector, we expect the turbulence structure to be influenced by the development of the internal boundary layer, particularly under different atmospheric stabilities. However we do not expect a significant influence on the flow homogeneity in the horizontal direction around the scanning circle, which is one of the key assumptions of the six-beam method.

The duration of the full cycle of the six-beam measurements from the WindScanner was about 15 s. The period of measurement was between 1 and 28 July 2013, where 764 30 min periods were measured. After filtering for data availability within each 30 min period, where only those periods were chosen with 95 % data, the number of 30 min periods reduced to 625. Finally filtering for wind directions to avoid wakes from the wind turbines and the met mast rendered 401 30 min periods. The available 30 min ensembles are further classified into different atmospheric stabilities, characterized by Monin–Obukhov length L_{MO} based on the intervals given in Table 2 (Sathe et al., 2011a).

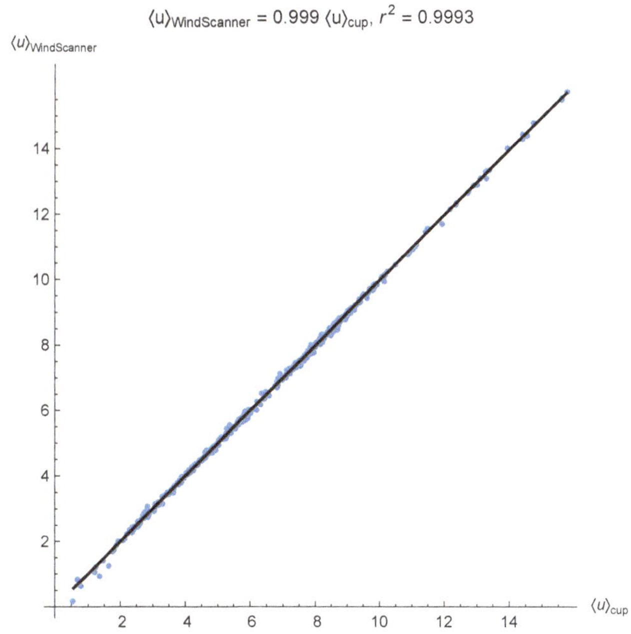

Figure 4. Comparison of the 30 min mean wind speed between the WindScanner and a cup anemometer using two methods.

Table 2. Classification of atmospheric stability according to Monin–Obukhov length intervals.

unstable (u)	$-500 \leq L_{MO} \leq -50$ m
neutral (n)	$\mid L_{MO} \mid \geq 500$ m
stable (s)	$10 \leq L_{MO} \leq 500$ m

L_{MO} is estimated using the eddy covariance method (Kaimal and Finnigan, 1994) from the high frequency (20 Hz) measurements of a sonic anemometer at 80 m, that is mounted on a 116 m tall met mast (UTM zone 32 V 447 647 m E and 6 255 435 m N WGS84 datum) in the south-east direction (see Fig. 3). Mathematically, L_{MO} is given as,

$$L_{MO} = -\frac{{u_*}^3 \theta_v}{\kappa g \overline{w' \theta_v'}}, \tag{18}$$

where u_* is the friction velocity, $\kappa = 0.4$ is the von Kármán constant, g is the acceleration due to gravity, θ_v is the virtual potential temperature, $\overline{\cdots}$ denotes time average, and $\overline{w'\theta_v'}$ (covariance of w and θ_v) is the virtual kinematic heat flux. u_* is estimated as,

$$u_* = \sqrt[4]{\overline{u'w'}^2 + \overline{v'w'}^2}, \tag{19}$$

where $\overline{u'w'}$ (covariance of u and w) and $\overline{v'w'}$ (covariance of v and w) are the vertical fluxes of the horizontal momentum.

As an initial validation of the accuracy and precision of the WindScanner, the 30 min mean wind speeds were compared with those obtained from the cup anemometer. Figure 4 shows that the WindScanner is very accurate (within 0.1 %)

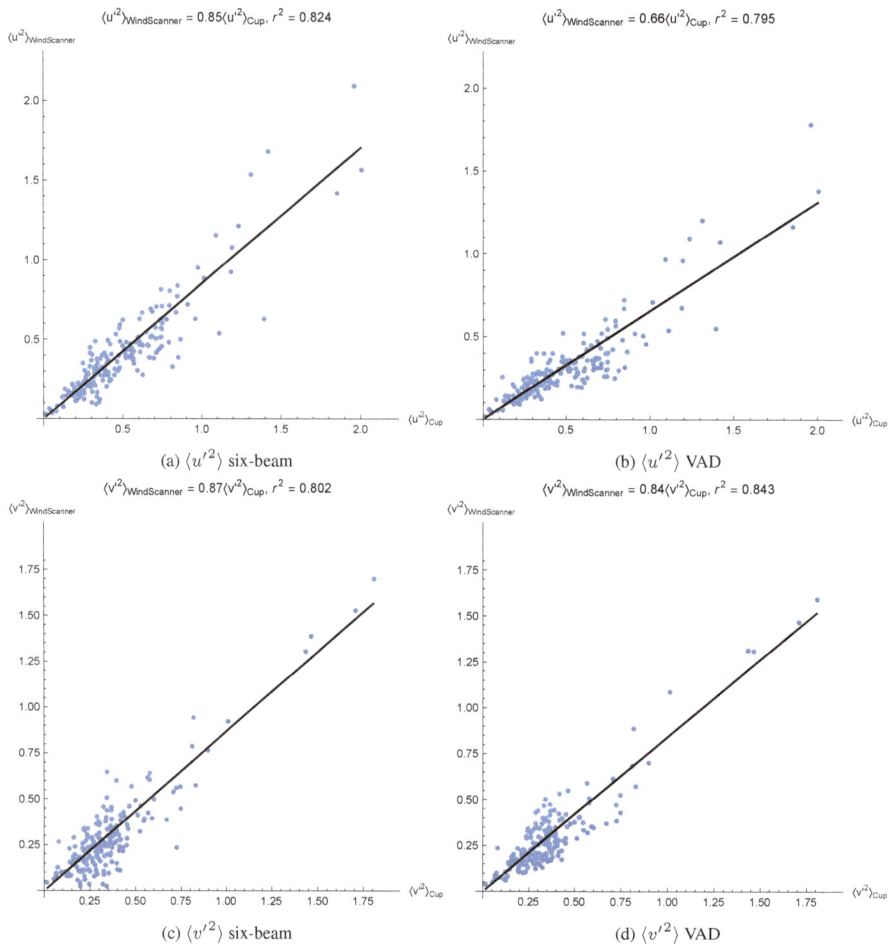

(a) $\langle u'^2 \rangle$ six-beam
(b) $\langle u'^2 \rangle$ VAD
(c) $\langle v'^2 \rangle$ six-beam
(d) $\langle v'^2 \rangle$ VAD

Figure 5. Comparison of the turbulence statistics under unstable conditions between the WindScanner and the cup anemometer using two methods.

and precise (coefficient of determination, $r^2 \approx 0.9993$) in measuring the mean wind speeds. For one 30 min period, the mean radial velocities measured by each of the six beams on the base of the scanning cone were fitted to Eq. (2) in a least squares sense to obtain the 30 min mean wind speed. This procedure was repeated for all 30 min periods. It is to be noted that the mean wind speed obtained using both methods (six-beam and VAD) is identical, since averaging the radial velocity for each beam, and then making a linear fit to obtain the u, v, an w components commute. Such an exercise provided enough confidence to proceed with deducing the turbulence measurements from the WindScanner using both methods.

4 Turbulence measurements

Two methods are used to deduce the turbulence statistics from the WindScanner measurements:

1. Six-beam method – for each 30 min period the measured $\boldsymbol{S}$ vector is used in combination with Eq. (5) to deduce the Σ vector. Finally the Σ vector is rotated in the mean wind direction for the respective 30 min period.

2. VAD method – within each 30 min period the v_r measurements from every single cycle of the six beams are fitted in a least squares sense to Eq. (2) to deduce a 30 min time series of the wind field components. The Σ vector is subsequently computed and rotated in the mean wind direction for the respective 30 min period.

Figure 5 shows the comparison of the turbulence statistics derived from the WindScanner measurements using the six-beam and the VAD methods and those obtained from the cup anemometer under unstable conditions. It is clear that the six-beam method measures more turbulence, about 19 % for $\langle u'^2 \rangle$ and 3 % for $\langle v'^2 \rangle$ than the VAD method, where the orthogonal least-squares regression is used to fit the cup anemometer measurements. The scatter using both methods is comparable to each other, but there is a slightly more scatter using the six-beam method for $\langle v'^2 \rangle$.

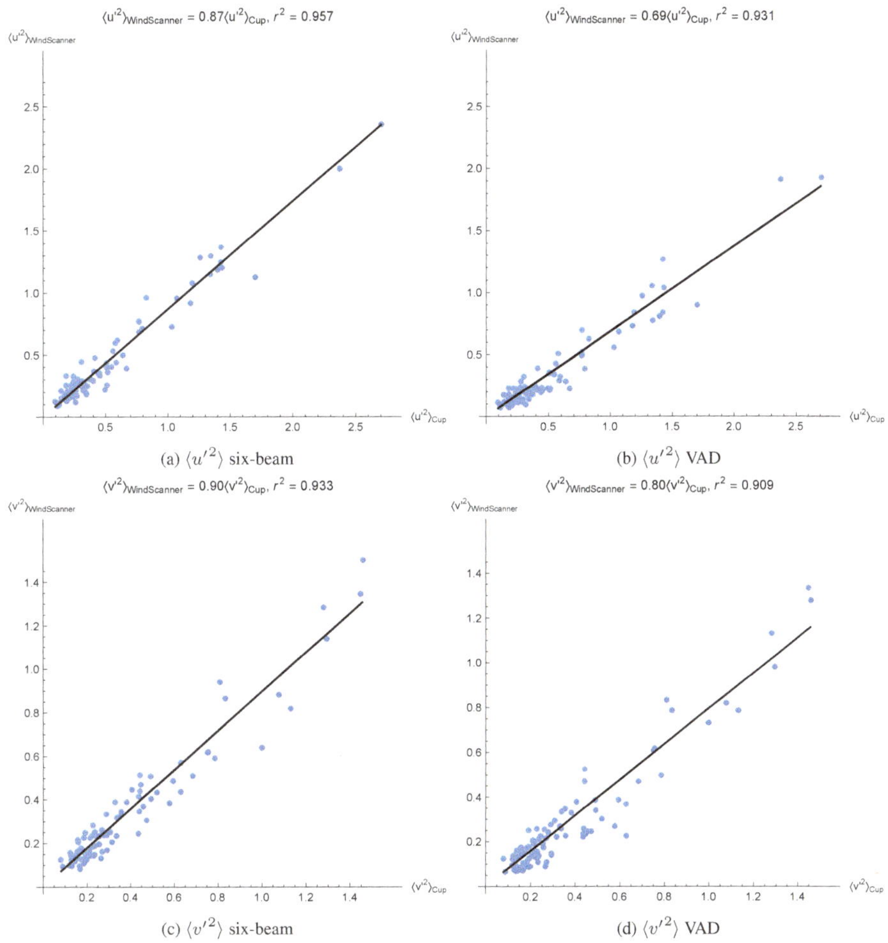

Figure 6. Comparison of the turbulence statistics under neutral conditions between the WindScanner and the cup anemometer using two methods.

Figure 6 shows the same as Fig. 5 but under neutral conditions. As for the unstable conditions, the six-beam method measures more turbulence, about 18 % for $\langle u'^2 \rangle$ and 10 % for $\langle v'^2 \rangle$ than the VAD method. The scatter using both methods is comparable to each other, with the six-beam method giving a slightly reduced scatter than the VAD method.

Figure 7 shows the same as Fig. 5 but under stable conditions. As for the unstable conditions, the six-beam method measures more turbulence, about 19 % for $\langle u'^2 \rangle$ and 4 % for $\langle v'^2 \rangle$ than the VAD method. The scatter using both methods is comparable to each other, but there is slightly more scatter using the six-beam method for $\langle u'^2 \rangle$.

Thus under all stabilities the six-beam method is closer to the turbulence measurements carried out using the reference cup anemometer. There is however some probe volume averaging using both methods, but this is significantly larger for the VAD method. The probe volume averaging can be observed clearly by comparing the radial velocity spectra, which can be observed in Fig. 6 in Mann et al. (2009), and Fig. 4 in Sjöholm et al. (2009).

5 Discussion

From Figs. 5–7 it is clear that using both methods the WindScanner measures more turbulence under stable conditions than under unstable and neutral conditions. This may be contrary to our intuitive understanding, because usually the turbulence scales are much larger under unstable conditions than under stable conditions (Sathe et al., 2013). These results are also contrary to what has been observed by Sathe et al. (2011b) at the same site. However, it is to be noted that Sathe et al. (2011b) used the lidar measurements when the wind was blowing from the eastern direction, whereas in this work we use the measurements when the wind is blowing from the western direction. As described in Sect. 4, in the western sector there is a sudden change of roughness due to the transition from sea to land. As a consequence there is a development of the internal boundary layer (IBL). Also the growth of the IBL depends on atmospheric stability, where under unstable conditions the growth will be faster than under stable conditions. Panofsky and Dutton (1984) state that the growth of the height of the boundary layer is proportional

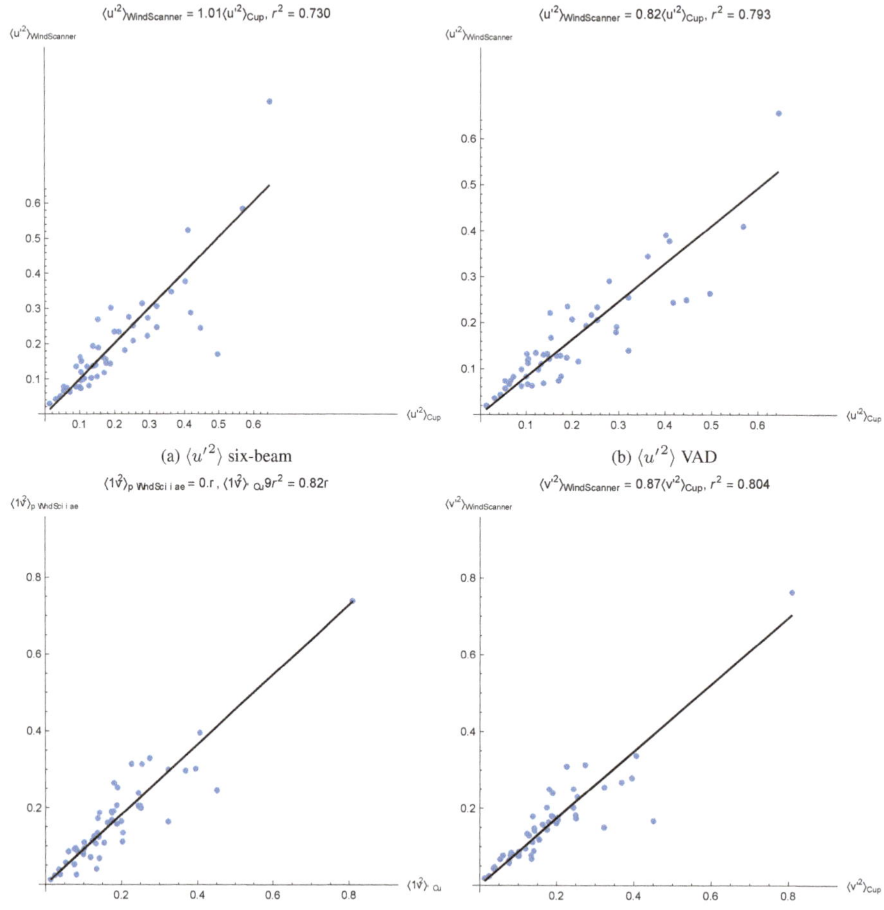

(a) $\langle u'^2 \rangle$ six-beam

(b) $\langle u'^2 \rangle$ VAD

(c) $\langle v'^2 \rangle$ six-beam

(d) $\langle v'^2 \rangle$ VAD

Figure 7. Comparison of the turbulence statistics under stable conditions between the WindScanner and the cup anemometer using two methods.

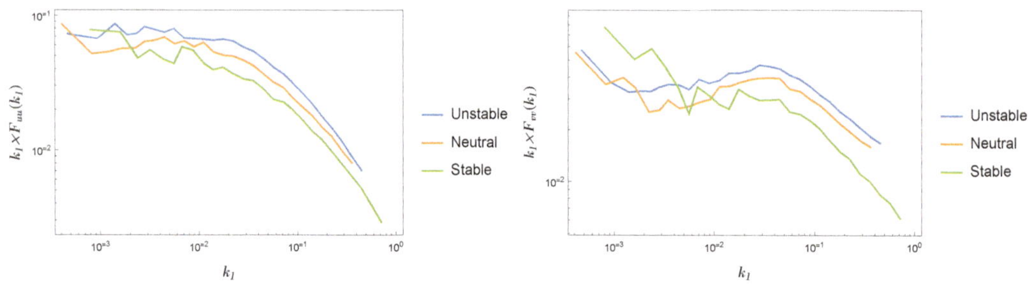

(a) u-spectra

(b) v-spectra

Figure 8. Comparison of the u- and v-spectra derived from high-frequency cup anemometer measurements under different stability conditions

to the drag coefficient $u_*/\langle u \rangle$. And it is well known that the drag coefficient is larger for unstable stratification. Consequently the turbulence scales within the IBL will be smaller as compared to those outside of it. It is then interesting to check whether the WindScanner measures more within the IBL under unstable conditions as compared to the stable conditions.

Figure 8 shows the u and v spectra derived from high-frequency cup anemometer measurements under different stability conditions. If we define the characteristic length scale $\mathcal{L}$ as the length scale corresponding to the maximum spectral energy, it is then clear that the peak of the v spectra is shifted to the right for unstable conditions as compared to the stable conditions. It is not that clear for the u spectra, however the shift of scales to larger wavenumbers under unstable conditions can still be observed. Thus $\mathcal{L}$ appears smaller under unstable conditions than under stable conditions for the measurements from the western sector used in this work. There is thus more probe volume averaging under unstable conditions than under stable conditions. Hence the WindScanner attenuates the turbulence measurements more under unstable conditions than under stable conditions.

Another interesting observation is that using the VAD method the WindScanner does not measure more turbulence than the reference cup anemometer under any stability condition. This, too, does not agree with that observed by Sathe et al. (2011b), even though the same basic pulsed commercial lidar technology was also used in that work. It is likely due to the fact that in Sathe et al. (2011b) only four beams were used as opposed to six beams, and α was 60° compared to 45° used in this work. Therefore the turbulence statistics are not directly comparable with those obtained in Sathe et al. (2011b) even though the same basic commercial lidar was used. Due to the application of the least squares technique on the v_r measurements in this work, there is significant volume averaging around the scanning circle, which is also observed in Sathe et al. (2011b) for a continuous-wave lidar.

6 Conclusions

An alternative so-called six-beam method is proposed in place of the standard VAD method to measure atmospheric turbulence using a ground-based wind lidar. The major difference between the two methods is that the six-beam method uses the measurement of the radial velocity variances, whereas the VAD method uses the high frequency measurement of the radial velocity transformed into Cartesian coordinates to deduce turbulence statistics. The scanning configuration of the six-beam method is optimized to minimize the sum of the random errors in the measurement of the components of the **R** matrix. In comparison to the reference cup anemometer the six-beam method measures between 85 and 101 % of the reference turbulence, whereas the VAD method measures between 66 and 87 % depending on atmospheric stability. The six-beam method thus overcomes partly the problem of significant probe volume averaging that is otherwise observed by the VAD method.

Furthermore two interesting observations have been made in this study. One is that, using both methods the WindScanner measures more turbulence under stable conditions than under unstable conditions, mainly due to the influence of the internal boundary layer (see Sect. 5). The other is that despite using the same underlying pulsed lidar technology as in Sathe et al. (2011b), the VAD method never measures more turbulence than the reference instrument as was observed in Sathe et al. (2011b) (see Sect. 5 for some explanation). It emphasizes the point that the VAD method is highly sensitive to the turbulence structure in the atmosphere, and one must avoid using it to measure atmospheric turbulence.

Future studies must certainly focus on tackling the probe volume averaging effect, which will further strengthen the arguments of using the six-beam method. Smalikho et al. (2005) have provided us with such a framework for pulsed lidars, whereas Mann et al. (2010) have demonstrated it for a continuous-wave lidar.

Acknowledgements. This work is carried out as a part of a research project funded by the Danish Ministry of Science, Innovation and Higher Education – Technology and Production, grant no. 0602-02486B. The resources provided by the Center for Computational Wind Turbine Aerodynamics and Atmospheric Turbulence funded by the Danish Council for Strategic Research grant no. 09-067216 are also acknowledged. We also thank Erik Juul from DTU Wind Energy department for helping with some sketches.

Edited by: W. Ward

References

Banakh, V. A. and Smalikho, I. N.: Determination of the turbulent energy dissipation rate from lidar sensing data, Atmos. Ocean. Opt., 10, 295–302, 1997a.

Banakh, V. A. and Smalikho, I. N.: Estimation of the turbulence energy dissipation rate from the pulsed Doppler lidar data, Atmos. Ocean. Opt., 10, 957–965, 1997b.

Banakh, V. A. and Werner, C.: Computer simulation of coherent Doppler lidar measurement of wind velocity and retrieval of turbulent wind statistics, Opt. Eng., 44, 071205, doi:10.1117/1.1955167, 2005.

Banakh, V. A., Smalikho, I. N., Köpp, F., and Werner, C.: Representativeness of wind measurements with a CW Doppler lidar in the atmospheric boundary layer, Appl. Opt., 34, 2055–2067, doi:10.1364/AO.34.002055, 1995a.

Banakh, V. A., Werner, C., Kerkis, N. N., Köpp, F., and Smalikho, I. N.: Turbulence measurements with a CW Doppler lidar in the atmospheric boundary layer, Atmos. Ocean. Opt., 8, 955–959, 1995b.

Banakh, V. A., Werner, C., Köpp, F., and Smalikho, I. N.: Measurement of turbulent energy dissipation rate with a scanning Doppler lidar, Atmos. Ocean. Opt., 9, 849–853, 1996.

Banakh, V. A., Smalikho, I. N., Köpp, F., and Werner, C.: Measurements of turbulent energy dissipation rate with a CW Doppler lidar in the atmospheric boundary layer, J. Atmos. Ocean. Technol., 16, 1044–1061, doi:10.1175/1520-0426(1999)016<1044:MOTEDR>2.0.CO;2, 1999.

Banakh, V. A., Smalikho, I. N., Pichugina, Y. L., and Brewer, W. A.: Representativeness of measurements of the dissipation rate of turbulence energy by scanning Doppler lidar, Atmos. Ocean. Opt., 23, 48–54, doi:10.1134/S1024856010010100, 2010.

Branlard, E., Pedersen, A. T., Mann, J., Angelou, N., Fischer, A., Mikkelsen, T., Harris, M., Slinger, C., and Montes, B. F.: Retrieving wind statistics from average spectrum of continuous-wave lidar, Atmos. Meas. Tech., 6, 1673–1683, doi:10.5194/amt-6-1673-2013, 2013.

Browning, K. A. and Wexler, R.: The determination of kinematic properties of a wind field using a Doppler radar, J. Appl. Meteorol., 7, 105–113, doi:10.1175/1520-0450(1968)007<0105:TDOKPO>2.0.CO;2, 1968.

Clifton, A., Elliott, D., and Courtney, M.: Ground-Based vertically profiling remote sensing for wind resource assessment, Tech. Rep. 15, Internation Energy Agency, 2013.

Eberhard, W. L., Cupp, R. E., and Healy, K. R.: Doppler lidar measurements of profiles of turbulence and momentum flux, J. Atmos. Ocean. Technol., 6, 809–819, doi:10.1175/1520-0426(1989)006<0809:DLMOPO>2.0.CO;2, 1989.

Frehlich, R.: Coherent Doppler lidar signal covariance including wind shear and wind turbulence, Appl. Opt., 33, 6472–6481, doi:10.1364/AO.33.006472, 1994.

Frehlich, R.: Effects of wind turbulence on coherent Doppler lidar performance, J. Atmos. Ocean. Technol., 14, 54–75, doi:10.1175/1520-0426(1997)014<0054:EOWTOC>2.0.CO;2, 1997.

Frehlich, R. and Cornman, L.: Estimating spatial velocity statistics with coherent Doppler lidar, J. Atmos. Ocean. Techno., 19, 355–366, doi:10.1175/1520-0426-19.3.355, 2002.

Frehlich, R. and Kelley, N.: Measurements of wind and turbulence profiles with scanning Doppler lidar for wind energy applications, IEEE J. Selected Topics Appl. Earth Obs. Remote Sens., 1, 42–47, doi:10.1109/JSTARS.2008.2001758, 2008.

Frehlich, R., Hannon, S. M., and Henderson, S. W.: Performance of a 2-μm coherent Doppler lidar for wind measurements, J. Atmos. Ocean. Technol., 11, 1517–1528, doi:10.1175/1520-0426(1994)011<1517:POACDL>2.0.CO;2, 1994.

Frehlich, R., Hannon, S. M., and Henderson, S. W.: Coherent Doppler lidar measurements of wind field statistics, Bound.-Lay. Meteorol., 86, 233–256, doi:10.1023/A:1000676021745, 1998.

Frehlich, R., Meillier, Y., Jensen, M. L., Balsley, B., and Sharman, R.: Measurements of boundary layer profiles in urban environment, J. Appl. Meteorol. Climatol., 45, 821–837, doi:10.1175/JAM2368.1, 2006.

Frehlich, R., Meillier, Y., and Jensen, M. L.: Measurements of boundary layer profiles with in situ sensors and Doppler lidar, J. Atmos. Ocean. Technol., 25, 1328–1340, doi:10.1175/2007JTECHA963.1, 2008.

Gal-Chen, T., Xu, M., and Eberhard, W. L.: Estimation of atmospheric boundary layer fluxes and other turbulence parameters from Doppler lidar data, J. Geophys. Res., 97, 18409–18423, doi:10.1029/91JD03174, 1992.

Ingber, L.: Simulated Annealing: Practice versus Theory, Math. Comput. Model., 18, 29–57, 1993.

Kaimal, J. C. and Finnigan, J. J.: Atmospheric Boundary Layer Flows – Their Structure and Measurement, chap. Acquisition and processing of atmospheric boundary layer data, 254–281, 7, Oxford University Press, New York, 1994.

Kindler, D., Oldroyd, A., Macaskill, A., and Finch, D.: An eight month test campaign of the QinetiQ ZephIR system: Preliminary results, Meteorol. Z., 16, 479–489, doi:10.1127/0941-2948/2007/0226, 2007.

Kropfli, R. A.: Single Doppler radar measurement of turbulence profiles in the convective boundary layer, J. Atmos. Ocean. Technol., 3, 305–314, doi:10.1175/1520-0426(1986)003<0305:SDRMOT>2.0.CO;2, 1986.

Lenschow, D. H., Mann, J., and Kristensen, L.: How long is long enough when measuring fluxes and other turbulence statistics?, J. Atmos. Ocean. Technol., 11, 661–673, doi:10.1175/1520-0426(1994)011<0661:HLILEW>2.0.CO;2, 1994.

Lhermitte, R. M.: Note on wind variability with Doppler radar, J. Atmos. Sci., 19, 343–346, doi:10.1175/1520-0469(1962)019<0343:NOWVWD>2.0.CO;2, 1962.

Lhermitte, R. M.: Note on the observation of small-scale atmospheric turbulence by Doppler radar techniques, Radio Sci., 4, 1241–1246, doi:10.1029/RS004i012p01241, 1969.

Mann, J., Cariou, J., Courtney, M., Parmentier, R., Mikkelsen, T., Wagner, R., Lindelow, P., Sjöholm, M., and Enevoldsen, K.: Comparison of 3D turbulence measurements using three staring wind lidars and a sonic anemometer, Meteorol. Z., 18, 135–140, doi:10.1127/0941-2948/2009/0370, 2009.

Mann, J., Peña, A., Bingöl, F., Wagner, R., and Courtney, M. S.: Lidar scanning of momentum flux in and above the surface layer, J. Atmos. Ocean. Technol., 27, 792–806, doi:10.1175/2010JTECHA1389.1, 2010.

Nelder, J. A. and Mead, R.: A Simplex Method for Function Minimization, Computer J., 7, 308–313, 1965.

Panofsky, H. A. and Dutton, J. A.: Atmospheric Turbulence, John Wiley & Sons, New York, 1984.

Peña, A., Hasager, C. B., Gryning, S.-E., Courtney, M., Antoniou, I., and Mikkelsen, T.: Offshore Wind Profiling Using Light Detection and Ranging Measurements, Wind Energy, 12, 105–124, doi:10.1002/we.283, 2009.

Rao, S.: Engineering Optimization: Theory and Practice, John Wiley and Sons Inc., Hoboken, New Jersey, 4th Edn., 2009.

Sathe, A. and Mann, J.: Measurement of turbulence spectra using scanning pulsed wind lidars, J. Geophys. Res., 117, D01201, doi:10.1029/2011JD016786, 2012.

Sathe, A. and Mann, J.: A review of turbulence measurements using ground-based wind lidars, Atmos. Meas. Tech., 6, 3147–3167, doi:10.5194/amt-6-3147-2013, 2013.

Sathe, A., Gryning, S.-E., and Peña, A.: Comparison of the atmospheric stability and wind profiles at two wind farm sites over a long marine fetch in the North Sea, Wind Energy, 14, 767–780, doi:10.1002/we.456, 2011a.

Sathe, A., Mann, J., Gottschall, J., and Courtney, M. S.: Can wind lidars measure turbulence?, J. Atmos. Ocean. Technol., 28, 853–868, doi:10.1175/JTECH-D-10-05004.1, 2011b.

Sathe, A., Mann, J., Barlas, T., Bierbooms, W. A. A. M., and van Bussel, G. J. W.: Influence of atmospheric stability on wind turbine loads, Wind Energy, 16, 1013–1032, doi:10.1002/we.1528, 2013.

Sjöholm, M., Mikkelsen, T., Mann, J., Enevoldsen, K., and Courtney, M.: Spatial averaging-effects on turbulence measured by a continuous-wave coherent lidar, Meteorol. Z., 18, 281–287, doi:10.1127/0941-2948/2009/0379, 2009.

Smalikho, I., Kopp, F., and Rahm, S.: Measurement of atmospheric turbulence by 2-μm Doppler lidar, J. Atmos. Ocean. Technol., 22, 1733–1747, doi:10.1175/JTECH1815.1, 2005.

Smalikho, I. N.: On measurement of dissipation rate of the turbulent energy with a CW Doppler lidar, Atmos. Ocean. Opt., 8, 788–793, 1995.

Smith, D. A., Harris, M., Coffey, A. S., Mikkelsen, T., Jørgensen, H. E., Mann, J., and Danielian, R.: Wind Lidar Evaluation at

the Danish Wind Test Site in Høvsøre, Wind Energy, 9, 87–93, doi:10.1002/we.193, 2006.
Vasiljevic, N.: A time-space synchronization of coherent Doppler scanning lidars for 3D measurements of wind fields, Ph.D. Thesis PhD-0027 (EN), Technical University of Denmark, 2014.
Wagner, R., Courtney, M., Gottschall, J., and Lindelöw-Marsden, P.: Accounting for the speed shear in wind turbine power performance measurement, Wind Energy, 14, 993–1004, doi:10.1002/we.509, 2011.
Wilson, D. A.: Doppler Radar Studies of Boundary Layer Wind Profiles and Turbulence in Snow Conditions, in: Proc. 14th Conference on Radar Meteorology, 191–196, Tucson, USA, 1970.

13

A quantitative analysis of the impact of wind turbines on operational Doppler weather radar data

L. Norin

Atmospheric Remote Sensing Unit, Research Department, Swedish Meteorological and Hydrological Institute, Norrköping, Sweden

Correspondence to: L. Norin (lars.norin@smhi.se)

Abstract. In many countries wind turbines are rapidly growing in numbers as the demand for energy from renewable sources increases. The continued deployment of wind turbines can, however, be problematic for many radar systems, which are easily disturbed by turbines located in the radar line of sight. Wind turbines situated in the vicinity of Doppler weather radars can lead to erroneous precipitation estimates as well as to inaccurate wind and turbulence measurements. This paper presents a quantitative analysis of the impact of a wind farm, located in southeastern Sweden, on measurements from a nearby Doppler weather radar. The analysis is based on 6 years of operational radar data. In order to evaluate the impact of the wind farm, average values of all three spectral moments (the radar reflectivity factor, absolute radial velocity, and spectrum width) of the nearby Doppler weather radar were calculated, using data before and after the construction of the wind farm. It is shown that all spectral moments, from a large area at and downrange from the wind farm, were impacted by the wind turbines. It was also found that data from radar cells far above the wind farm (near 3 km altitude) were affected by the wind farm. It is shown that this in part can be explained by detection by the radar sidelobes and by scattering off increased levels of dust and turbulence. In a detailed analysis, using data from a single radar cell, frequency distributions of all spectral moments were used to study the competition between the weather signal and wind turbine clutter. It is shown that, when weather echoes give rise to higher reflectivity values than those of the wind farm, the negative impact of the wind turbines is greatly reduced for all spectral moments.

1 Introduction

As a response to the increasing demand for renewable energy the number of wind turbines is growing rapidly in many countries around the world. The worldwide installed cumulative energy capacity of wind turbines has shown a more-than-13-fold increase during 2001–2013. In Sweden, the wind power capacity has increased even more, 15 times, during the same period (Global Wind Energy Council, 2014). In the coming years many more wind turbines are expected to be built and existing, older ones are likely to be replaced by larger, next-generation turbines. Modern wind turbines are large structures, many reaching 150 m above the ground. Clusters of densely spaced wind turbines, so-called wind farms, are being built both on- and offshore.

The continued deployment of wind turbines and wind farms presents a problem for many radar systems, which are easily disturbed by wind turbines located in the radar line of sight. Due to their rotating blades, interference caused by wind turbines is more severe for radar systems than interference caused by stationary structures (e.g. masts or towers). Many Doppler radars use a clutter filter that suppresses echoes originating from objects with no or little radial velocity, but such filters do not work for moving objects such as the rotating blades of a wind turbine. It has been shown that wind turbines located in the line of sight of Doppler radars can have a detrimental impact on the performance of both military and civilian radar systems (see, e.g., Poupart, 2003; Department of Defense, 2006; Lemmon et al., 2008; Lute and Wieserman, 2011).

Doppler weather radars, employed by meteorological and hydrological services, can also be negatively affected by nearby wind turbines. Weather radars are valuable tools for monitoring precipitation and wind shear as well as for observing hazardous events such as hailstorms, heavy rainfall, and tornadoes. Information from weather radars is also used as input to numerical weather prediction and flood-forecasting models. Errors in weather radar data may propagate to affect the output of such forecast systems (Rossa et al., 2011).

Several studies dedicated to wind turbine impact on weather radar data have presented images of radar composites that convincingly show that wind farms indeed can be detected by weather radars (see, e.g., Burgess et al., 2008; Crum and Ciardi, 2010; Vogt et al., 2011). In other studies, time series of raw radar data have been recorded in order to perform detailed analyses of the impact of wind turbines by spectral analysis. Gallardo et al. (2008) collected raw radar data over a few months from a Spanish C-band weather radar to analyse the impact of a large wind farm, while Isom et al. (2009) used several hours' worth of raw data from two US S-band weather radars to investigate the same phenomenon. Toth et al. (2011) used a mobile X-band Doppler radar to study the impact of wind turbines from close range. However, only a few analyses of wind turbine impact on long time series of operational weather radar data have been published (Haase et al., 2010; Norin and Haase, 2012).

The main objective of this study is to analyse the impact of wind turbines on operational Doppler weather radar volume data. This study thus extends the work by Haase et al. (2010) and Norin and Haase (2012). In this study the wind turbine impact on all three spectral moments (the radar reflectivity factor, radial velocity, and spectrum width) have been investigated in order to improve the understanding of wind turbine impact on Doppler weather radar data. In total, 6 years' worth of operational polar volume data were used for the data analysis.

The structure of the paper is as follows. The technical characteristics of the Swedish radars are described in Sect. 2. The radar data set together with the analysed wind farm are presented in Sect. 3. In Sect. 4.1 the impact of wind turbines on data from a single radar scan is shown by comparing average values of the spectral moments, before and after the construction of the wind farm. In Sect. 4.2 the wind turbine impact on polar volume data is investigated. Section 4.3 presents a detailed analysis of the wind turbine impact on data from a single radar cell, by examining frequency distributions for all spectral moments as a function of reflectivity from a reference radar cell. The competition between the weather signal and wind turbine clutter is investigated in Sect. 4.4, and in Sect. 5 the impact of the wind farm on measurements far above the radar is discussed. Finally, in Sect. 6, a summary of the study is given and conclusions of the analyses are drawn.

2 The Swedish weather radars

The Swedish weather radar network consists of 12 horizontally polarised Ericsson C-band Doppler radars, providing almost complete national coverage. The radars perform azimuthal scans of 360° around a vertical axis for 10 tilt angles, θ, ranging from $\theta = 0.5°$ to $\theta = 40°$. The antenna rotational speed is $v_r = 2$ rpm. Together, these scans make up polar volume data sets which are provided with an update time of 15 min.

The data processing is managed by the radar signal processor. The four lowest scans ($\theta = 0.5°$ to $\theta = 2.0°$) use a different measurement strategy than the highest six scans ($\theta = 2.5°$ to $\theta = 40.0°$). The received signal is digitised using an 8 bit analogue-to-digital converter with a sampling rate of 0.9 (1.8) MHz for the lower (higher) scans, corresponding to 167 (83) m range bins. A radar cell consists of 12 range bins, and thus the range resolution for a radar cell is 2 (1) km for the lower (higher) scans. The lower scans also use a lower set of pulse repetition frequencies (PRFs), 450 Hz or 600 Hz, whereas the higher tilt angles use a higher set of PRFs (900 or 1200 Hz).

For all scans the signal processor uses 32 measurements from each PRF to compute a Fourier transform, which means that the azimuthal resolution depends on the set of PRFs. For the lower scans the azimuthal resolution is $360/\left\{60/\left[v_r\,(32/\mathrm{PRF}_1 + 32/\mathrm{PRF}_2)\right]\right\} \approx 1.49°$, whereas for the higher scans the azimuthal resolution is approximately 0.75°. Data are, however, output in matrices consisting of 120×420 radar cells for every scan. The apparent azimuthal resolution is thus $360/420 \approx 0.86°$ for all scans.

The main radar lobe has a half-power beam width of 0.9°. The first sidelobes appear at approximately 2.5° near −30 dB in the horizontal direction and at approximately 2° near −30 dB in the vertical direction (a vertical cut of the antenna pattern is shown in Sect. 5).

Doppler weather radars measure three spectral moments (see, e.g., Doviak and Zrnić, 2006): the radar reflectivity factor (hereafter referred to as reflectivity), radial velocity, and spectrum width. These spectral moments are used to estimate quantities such as precipitation rate, wind speed, and turbulence.

Reflectivity, Z, is the power of the returned signal and is measured by the radar in units of dBZ. The minimum detectable signal of the Swedish radars is better than −114 dBm. Whenever the radar receives a signal stronger than the minimum detectable signal, a hit is recorded and converted to reflectivity. The Swedish weather radars have a dynamic range of > 85 dB and measure Z between −30 and 71.6 dBZ in steps of 0.4 dBZ. The minimum value (−30 dBZ) represents all measurements ranging from $-\infty$ to −30 dBZ. Such values are classified as undetected measurements.

Radial velocity, V, is obtained as the first moment of the power-normalised Doppler spectrum. Two PRFs are used al-

Table 1. Selected characteristics of Swedish weather radars.

Tilt angles	0.5°, 1.0°, 1.5°, 2.0°	2.5°, 4.0°, 8.0°, 14.0°, 24.0°, 40.0°
Transmit power	250 kW	250 kW
Wavelength	5.35 cm	5.35 cm
Gain	44.7 dB	44.7 dB
Pulse width	0.5 μs	0.5 μs
Beam width	0.9°	0.9°
PRFs	600/450 Hz	1200/900 Hz
Rotational speed	2 rpm	2 rpm
Measurement radius	240 km	120 km
Radial resolution	2 km	1 km
Azimuthal resolution	0.86°	0.86°
Range cells	120	120
Azimuth gates	420	420
Max unambiguous velocity	24 m s^{-1}	48 m s^{-1}

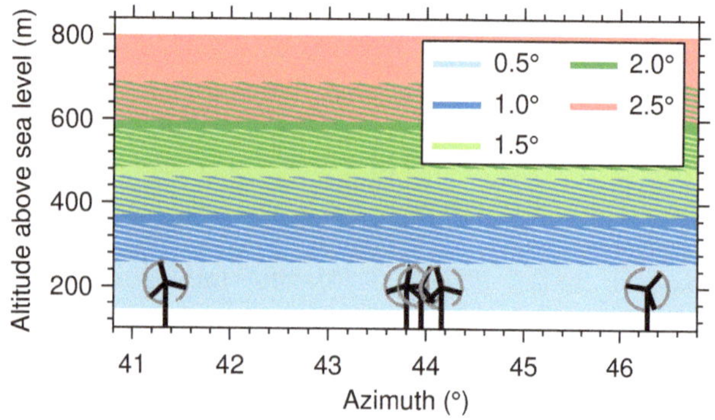

Figure 1. Schematic plot of the Brunsmo wind farm. The positions and heights of the wind turbines are shown together with the altitudes of the half-power beam width of the main radar lobe for the five lowest tilt angles ($0.5° \le \theta \le 2.5°$) at the wind farm.

ternatively to allow real-time dealiasing of the radial velocities. The maximum unambiguous velocity is ± 24 m s^{-1} for the four lowest scans and ± 48 m s^{-1} for the six highest scans. The radial velocity resolution is 0.1875 m s^{-1} (0.375 m s^{-1} for the higher scans). Undetected reflectivity measurements ($Z \le -30$ dBZ) result in unreliable estimates of radial velocity, and such measurements are therefore classified as undetected.

Spectrum width, W, is calculated as the square root of the second moment about the first of the power-normalised spectrum. The measurements of the spectrum width fall into one of four classes: 0–1, 1–2, 2–3, and > 3 m s^{-1} for the four lowest scans and 0–2, 2–4, 4–6, and > 6 m s^{-1} for the six highest scans. As for radial velocity, undetected reflectivity measurements ($Z \le -30$ dBZ) result in unreliable estimates of spectrum width, and such measurements are therefore classified as undetected.

Invalid measurements are produced when the radar does not perform correctly, such as when failing to find a requested angle in azimuth or elevation or if the transmitter is not ready in time. Invalid measurements for any spectral moment are given separate values by the radar.

All radars in the network are equipped with clutter filters which are used to suppress ground echoes. The clutter filter is turned on for all scans. Ground echo suppression is obtained by omitting amplitudes from the three frequency channels closest to 0 in the frequency spectrum (channel 0, 1, and 31). For the lowest four scans (using the lower set of PRFs) this removes echoes with radial velocities less than ± 1 m s^{-1}; for the highest six scans radial velocities with ± 1.5 m s^{-1} are omitted. To protect the radar receiver from overload the signal is damped by 60 dB near the radar, making the data of the first two rows of radar cells unusable.

Relevant radar characteristics are summarised in Table 1.

3 Data set and wind turbines

Operational polar volume data from the Swedish weather radars are available from the year 2005 and later. Some changes were made in the radar scan strategy as well as to the radar hardware during 2007, but from 2008 the radar scan strategy and measurement techniques have remained unaltered. Radar data from 2008 and later hence constitute a homogeneous data set.

In order to investigate the impact of wind turbines on Doppler weather radar data, we have analysed operational polar volume data from weather radar Karlskrona (56.2955° N, 15.6103° E) for a period of 6 years (1 January 2008 to 31 December 2013). Throughout this period no significant data gaps exist, and radar service records do not show any indication of radar malfunctioning occurring during this period.

Brunsmo wind farm is located in southeastern Sweden, approximately 13 km northeast of weather radar Karlskrona. This wind farm consists of five General Electric 2.5 MW wind turbines with total heights of 150 m above the ground. These wind turbines have a rotor diameter of 100 m, a cut-in wind speed of 3.5 m s^{-1}, and a cut-out wind speed of 25 m s^{-1}.

The wind turbines were erected in October and November 2009, and the wind farm became operational in April 2010. The proximity of the Brunsmo wind farm to weather radar Karlskrona together with the date of the wind farm's start of operations (near the middle of the homogeneous radar data set) makes the Brunsmo wind farm well suited for a detailed study.

Figure 1 shows a schematic picture of the locations and heights of the wind turbines of the Brunsmo wind farm together with the altitudes of the radar lobes' half-power beam width for the five lowest tilt angles (0.5, 1.0, 1.5, 2.0, and 2.5°) at the wind farm, assuming standard propagation conditions. From Fig. 1 it is seen that all wind turbines are in

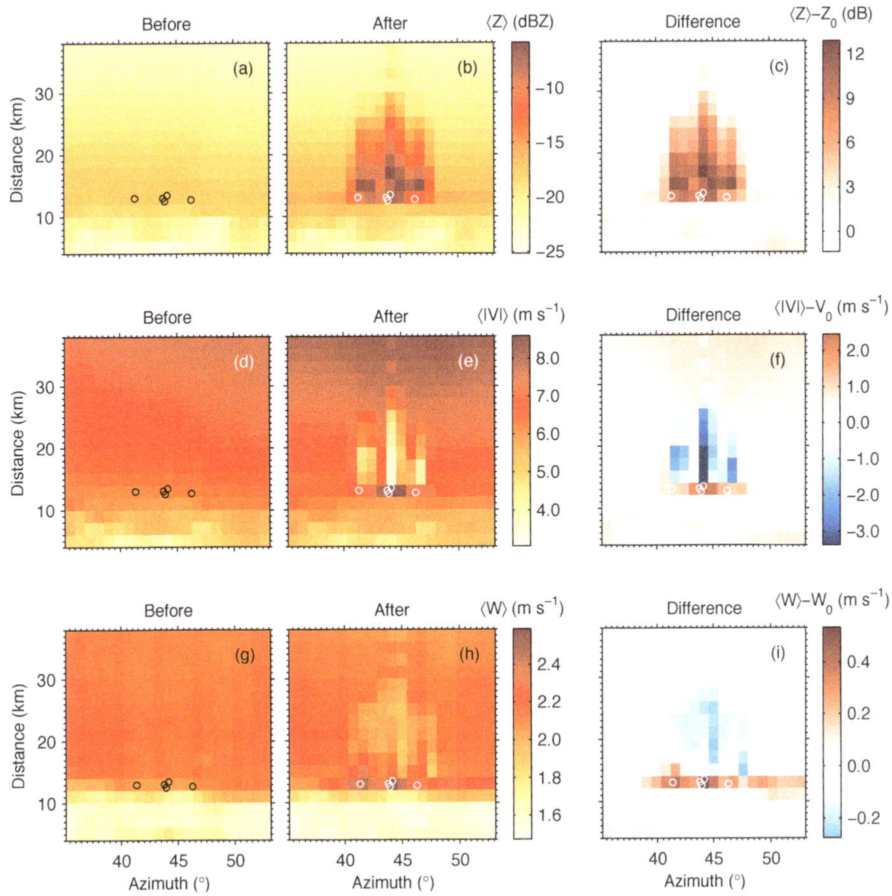

Figure 2. The columns to the left and in the middle show average reflectivity, $\langle Z\rangle$; average absolute radial velocity, $\langle |V|\rangle$; and average spectrum width, $\langle W\rangle$, before and after the construction of the Brunsmo wind farm. The right-hand column shows the difference (after−before). The locations of the wind turbines are shown with black or white circles. The impact of the wind turbines is seen for all spectral moments as changes to their average value. The impact of the wind turbines extends to radar cells both cross- and downrange from the wind farm.

the radar line of sight for the scan with the lowest tilt angle ($\theta = 0.5°$). Three of the five wind turbines are located within the same radar cell (azimuth gate 52, near 44° azimuth).

4 Methods and results

4.1 Wind turbine impact on a single radar scan

One way to investigate the impact of wind turbines on Doppler weather radar data is to compare the average values of the spectral moments before and after the construction of a wind turbine. The differences between these average values provide an estimate of the impact of the wind turbine.

Reflectivity, Z, and spectrum width, W, are well suited for calculating average values, but radial velocity measurements, V, can be either positive or negative, depending on the wind direction relative to the radar. Since we are interested in finding an estimate of the wind turbine impact on V, the influence of the wind direction was minimised by studying the absolute radial velocity, $|V|$.

The average values of the above-described spectral moments (Z, $|V|$, and W) were calculated before the construction of the Brunsmo wind farm (January 2008–March 2010) and after the wind farm's start of operations (May 2010–December 2013). In order to investigate the areal extent of the impact of the wind turbines, multiple radar cells (range: 4–38 km; azimuth: 35–53°) in the vicinity of the Brunsmo wind farm were analysed using data from the scan with the lowest tilt angle ($\theta = 0.5°$).

Undetected reflectivity measurements (see Sect. 2) were included when calculating the average value of Z since these measurements carry actual information. However, as undetected measurements for V and W do not contain reliable information, such measurements were excluded when calculating the average values of $|V|$ and W. Invalid measurements were excluded for all spectral moments.

Figure 2a and b shows the average reflectivity, $\langle Z\rangle$, before and after the construction of the Brunsmo wind farm, respectively. Figure 2a shows that $\langle Z\rangle$ was uniform over the whole analysed area (especially in azimuth) before the con-

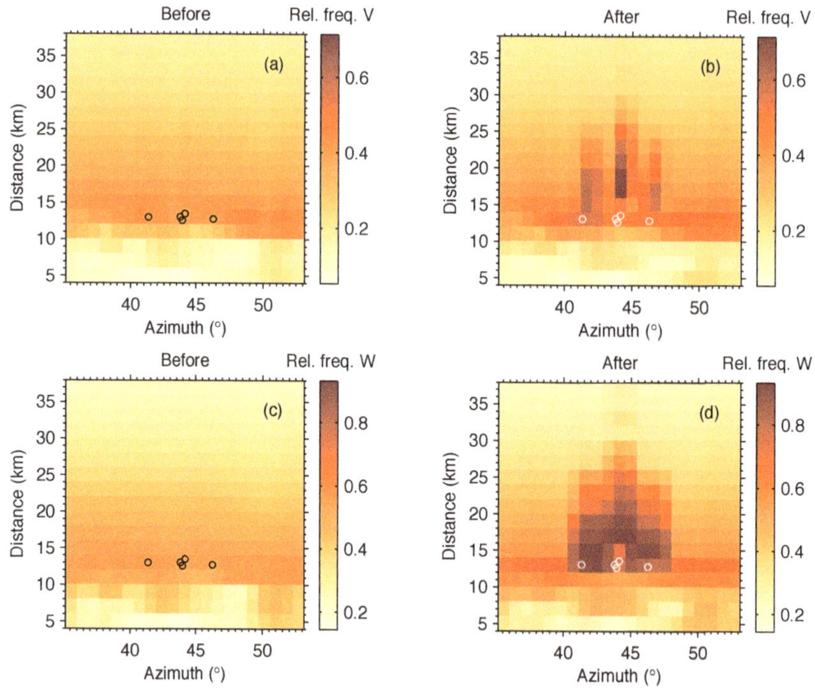

Figure 3. Relative frequency of real (detected and valid) measurements of radial velocity, V, and spectrum width, W, before and after the construction of the Brunsmo wind farm. The locations of the wind turbines are shown with black or white circles. The wind turbines are seen to increase the frequency of real measurements of V and W, both at the location of the wind turbines as well as cross- and downrange from the wind farm.

struction of the wind farm, with only small variations in amplitude. However, after the construction of the wind farm Fig. 2b shows that $\langle Z \rangle$ increases in amplitude, not only in the radar cells in which the wind turbines are located but also in several radar cells downrange of the wind turbines. One of the highest values of $\langle Z \rangle$ is seen in the radar cell in which three wind turbines are located. The average reflectivity in this radar cell is $\langle Z \rangle \approx -6$ dBZ, which should be compared to $\langle Z \rangle \approx -18$ dBZ in the same radar cell before the wind turbines were built. For convenience, Fig. 2c shows the difference between $\langle Z \rangle$ after and before the construction of the wind farm.

Tails of increased reflectivity downrange of wind turbines have been noted in several other works (see, e.g., Crum et al., 2008; Isom et al., 2009; Haase et al., 2010; Vogt et al., 2011; Norin and Haase, 2012). Such tails are believed to be caused by multiple scattering effects (scattering between multiple turbines and/or scattering between turbine and ground) (Isom et al., 2009; Vogt et al., 2011; Kong, 2014). The tails of increased reflectivity seen in Fig. 2b and c extend more than 20 km downrange of the wind turbines. The amplitude of $\langle Z \rangle$ in the tails is seen to reach a maximum shortly behind the wind turbines, after which it decreases with increasing distance.

Figure 2d and e shows the average absolute radial velocity, $\langle |V| \rangle$, before and after the construction of the wind farm, respectively. Before the construction of the wind farm it is seen from Fig. 2d that the amplitude of $\langle |V| \rangle$ increases with range from the radar but is very homogeneous in azimuth. Figure 2e shows that after the wind farm became operational the amplitude of $\langle |V| \rangle$ increased in radar cells containing wind turbines. However, radar cells downrange from the wind turbines show a decrease in amplitude. The largest value of $\langle |V| \rangle$ is found in the radar cell containing three wind turbines, where $\langle |V| \rangle > 8.5\,\mathrm{m\,s^{-1}}$. In the radar cell behind the three turbines the smallest value of $\langle |V| \rangle$ is found, $\langle |V| \rangle \approx 3\,\mathrm{m\,s^{-1}}$. Before the construction of the wind farm the values in both these radar cells were $\langle |V| \rangle \approx 6\,\mathrm{m\,s^{-1}}$. Figure 2f shows the difference between $\langle |V| \rangle$ after and before the construction of the wind farm.

Figure 2g and h shows the average spectrum width, $\langle W \rangle$, before and after the construction of the wind farm, respectively. In these figures it can be seen that after the construction of the wind farm a slight increase in $\langle W \rangle$ appears in the radar cells containing wind turbines, whereas a decrease in $\langle W \rangle$ occurs over a large area cross- and downrange of the wind turbines. The average spectral width increased from $\langle W \rangle \approx 2.0\,\mathrm{m\,s^{-1}}$ to $\langle W \rangle > 2.5\,\mathrm{m\,s^{-1}}$ at the wind turbines, whereas behind the turbines a decrease down to $\langle W \rangle \approx 1.8\,\mathrm{m\,s^{-1}}$ can be seen. Decreased levels of $\langle W \rangle$ can be seen up to 20 km behind the wind turbines. The difference between $\langle W \rangle$ after and before the construction of the wind farm is shown in Fig. 2i.

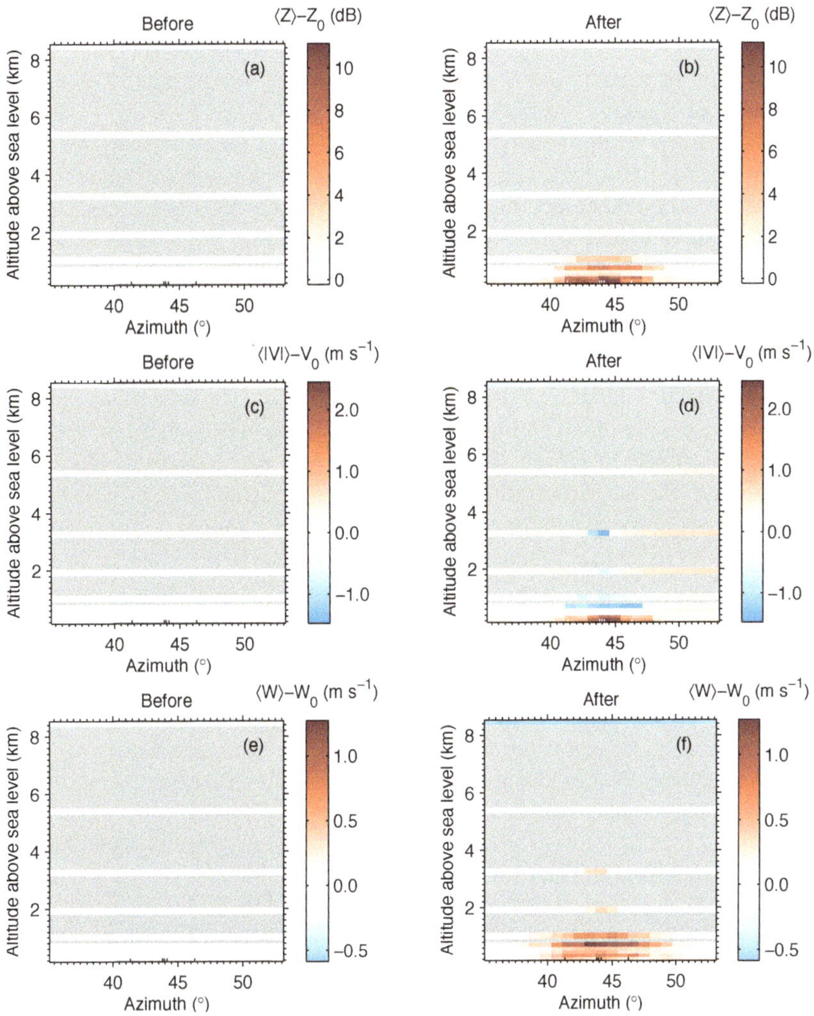

Figure 4. Difference in average reflectivity, $\langle Z\rangle - Z_0$; difference in average absolute radial velocity, $\langle |V|\rangle - V_0$; and difference in average spectrum width, $\langle W\rangle - W_0$, for all tilt angles ($0.5° \leq \theta \leq 40°$) before and after the construction of the Brunsmo wind farm. For each scan, Z_0, V_0, and W_0 were calculated as the mean value in azimuth (for the azimuth gates shown in the figure) of the corresponding spectral moment before the construction of the wind farm. For every scan, the data shown were taken from the range bin corresponding to the location of the wind turbines. For scans with higher tilt angles ($2.5° \leq \theta \leq 40°$) the average values of the two range bins nearest the wind turbines were used since for these scans the range resolution is 2 times that of the scans with lower tilt angles. The locations of the wind turbines are shown on the bottom of each plot as black or white vertical lines. The wind turbines are only in the radar line of sight for the scan with the lowest tilt angle but are nevertheless seen to have an impact on scans with higher tilt angles.

The average values of V and W are based on real measurements, i.e. measurements that are neither undetected nor invalid. In addition to their average values it is also of interest to study their frequency of occurrence. Complementary to the results presented in Fig. 2 the relative frequency of occurrence of V and W is shown in Fig. 3.

Figure 3a and b shows the relative frequency of occurrence of V before and after the construction of the Brunsmo wind farm, respectively. By comparing the two figures, it is seen that after the wind farm was constructed the relative frequency of detected radial velocity measurements increased from approximately 40 % to more than 70 % behind the wind turbines.

Figure 3c and d shows the relative frequency of occurrence of W before and after the construction of the Brunsmo wind farm, respectively. It is seen that the presence of the wind farm led to spectral width being detected more often. The relative frequency of occurrence of W increased from approximately 50 % to more than 90 %, at and behind the wind turbines.

Together, Figs. 2 and 3 show that the construction of the Brunsmo wind farm has led to real (detected) measurements of all spectral moments occurring more frequently. The average value of Z increased at and near the wind turbines, whereas $\langle |V|\rangle$ and $\langle W\rangle$ increased in radar cells located at the wind turbines but decreased behind.

4.2 Wind turbine impact on polar volume data

As was shown in Fig. 1, the wind turbines in the Brunsmo wind farm are in the radar line of sight for the scans with the lowest tilt angle ($\theta = 0.5°$). However, this calculation is based on the assumption of standard atmospheric conditions and further assumes that the extent of the radar lobe is limited by its half-power beam width. In order to investigate whether scans with higher tilt angles also are affected by the Brunsmo wind farm, data from all 10 scans in the polar volume were analysed.

Figure 4 shows the impact of the wind farm on $\langle Z \rangle$, $\langle |V| \rangle$, and $\langle W \rangle$ for all 10 scans in the polar volume. In Fig. 4 every scan is represented by the radar cells in which the wind turbines would be located if a vertical line from the turbines were drawn from the ground. For the higher tilt angles ($2.5° \leq \theta \leq 40°$) the average values of the spectral moments from the two radar cells nearest to the wind turbines were used, since for these scans the range resolution is 2 times that of the lower scans (cf. Table 1). Furthermore, in order to clearly display the change in the spectral moments for all scans in the same figure, the mean value in azimuth (for the extent of the analysed area, 35–53°) for every spectral moment from before the construction of the wind farm were subtracted from the measurements (both before and after the construction of the wind farm). These mean values are denoted by Z_0, V_0, and W_0.

Figure 4a and b shows $\langle Z \rangle - Z_0$ before and after the construction of the wind farm, respectively. Figure 4a shows that before the wind farm was built $\langle Z \rangle - Z_0$ was homogeneous in azimuth for all scans. In Fig. 4b, after the construction of the wind farm, an increase in $\langle Z \rangle - Z_0$ can be seen in the radar cells near the location of the wind turbines for some scans. For the scan with the lowest tilt angle ($\theta = 0.5°$) the increase is approximately 12 dB (cf. Fig. 2b). However, increased levels of $\langle Z \rangle - Z_0$ are seen not only in the lowest scan (in which the wind turbines are in the radar line of sight) but also in higher scans. From Fig. 4b it is seen that the fifth scan ($\theta = 2.5°$) shows an increase of $\langle Z \rangle - Z_0 \approx 8$ dB, which is much larger than that of the fourth scan ($\theta = 2.0°$), where $\langle Z \rangle - Z_0 \approx 2$ dB. This can partly be explained by the change in radar cell resolution that occurs between the four lowest scans and the six highest scans (cf. Sect. 2 and Table 1) and is discussed further in Sect. 5. A slight increase in $\langle Z \rangle - Z_0$ (approximately 2 dB) still exists near 3 km altitude.

Figure 4c and d shows $\langle |V| \rangle - V_0$ before and after the construction of the wind farm, respectively. Before the wind farm was built, $\langle |V| \rangle - V_0$ was uniform in azimuth for all scans, as seen in Fig. 4c. Figure 4d shows that after the wind farm was constructed an increase in $\langle |V| \rangle - V_0$ exists in the radar cells near the wind turbines for the scans with the two lowest tilt angles ($\theta = 0.5°$ and $\theta = 1.0°$). The largest increase in $\langle |V| \rangle - V_0$ is seen in the scan with the lowest tilt angle, where the increase is close to $2.5\,\mathrm{m\,s^{-1}}$. For higher tilt angles no change ($\theta = 1.5°$ and $\theta = 2.0°$) or a decrease ($2.5° \leq \theta \leq 14°$) in $\langle |V| \rangle - V_0$ is seen. A decrease of approximately $-1.5\,\mathrm{m\,s^{-1}}$ can still be seen near 3 km altitude, far above the wind farm.

Figure 4e and f shows $\langle W \rangle - W_0$ before and after the construction of the wind farm, respectively. As for the other two examined spectral moments, $\langle W \rangle - W_0$ is homogeneous in azimuth for all scans before the construction of the wind farm (see Fig. 4e), whereas increased amplitudes in $\langle W \rangle - W_0$ are seen in Fig. 4f, after the wind farm was built. The increase in $\langle W \rangle - W_0$ for the second-lowest scan ($\theta = 1.0°$) is close to $1.0\,\mathrm{m\,s^{-1}}$, higher than the increase for the lowest scan ($\theta = 0.5°$), which is near $0.5\,\mathrm{m\,s^{-1}}$. A sharp increase in $\langle W \rangle - W_0$ can be seen between the fourth scan ($\theta = 2.0°$) and the fifth scan ($\theta = 2.5°$), where the change in radar cell resolution occurs. Increased values of $\langle W \rangle - W_0$ can be seen far above the wind farm, near 3 km altitude.

Figure 5 shows a cross section in range (4–38 km) and height (0–4 km) of the polar volume for azimuth gate 52, the gate in which three wind turbines are located (near 44° azimuth). The same quantities as in Fig. 4 are shown (i.e. $\langle Z \rangle - Z_0$, $\langle |V| \rangle - V_0$, and $\langle W \rangle - W_0$).

Figure 5a and b shows $\langle Z \rangle - Z_0$ before and after the construction of the wind farm, respectively. In Fig. 5b it is seen that, after the construction of the wind farm, increased values of $\langle Z \rangle - Z_0$ extend tens of kilometres downrange of the wind turbines for the lowest two tilt angles ($\theta = 0.5°$ and $\theta = 1.0°$); cf. Fig. 2b. For higher tilt angles no increase in $\langle Z \rangle - Z_0$ downrange of the wind turbines can be seen.

Figure 5c and d shows $\langle |V| \rangle - V_0$ before and after the construction of the wind farm, respectively. Figure 5d shows that, after the wind farm was constructed, increased values of $\langle |V| \rangle - V_0$ exist in the radar cells in which the wind turbines are located for the lowest two tilt angles ($\theta = 0.5°$ and $\theta = 1.0°$). Downrange of the wind turbines for these tilt angles a decrease in $\langle |V| \rangle - V_0$ is seen. For higher tilt angles ($2.5° \geq \theta \leq 14°$) a decrease in $\langle V \rangle - V_0$ exists at the location of the wind turbines.

Figure 5e and f shows $\langle W \rangle - W_0$ before and after the construction of the wind farm, respectively. Figure 5f shows that, after the construction of the wind farm, $\langle W \rangle - W_0$ increased at and above the locations of the wind turbines, for altitudes up to 3 km. Downrange of the wind turbines a slight decrease in $\langle W \rangle - W_0$ is seen for the lowest tilt angle ($\theta = 0.5°$), whereas a slight increase in $\langle W \rangle - W_0$ can be seen for $\theta = 1.0°$.

Supplementary to Figs. 4 and 5, animations of $\langle Z \rangle$, $\langle |V| \rangle$, and $\langle W \rangle$ for all 10 scans are shown in Figs. S1–S3 in the Supplement, respectively. These animations show a 3-D view of the impact the Brunsmo wind farm has on the three spectral moments. From the animations it is seen that tails of changed amplitudes of the spectral moments are only visible for the scans with the lowest two tilt angles.

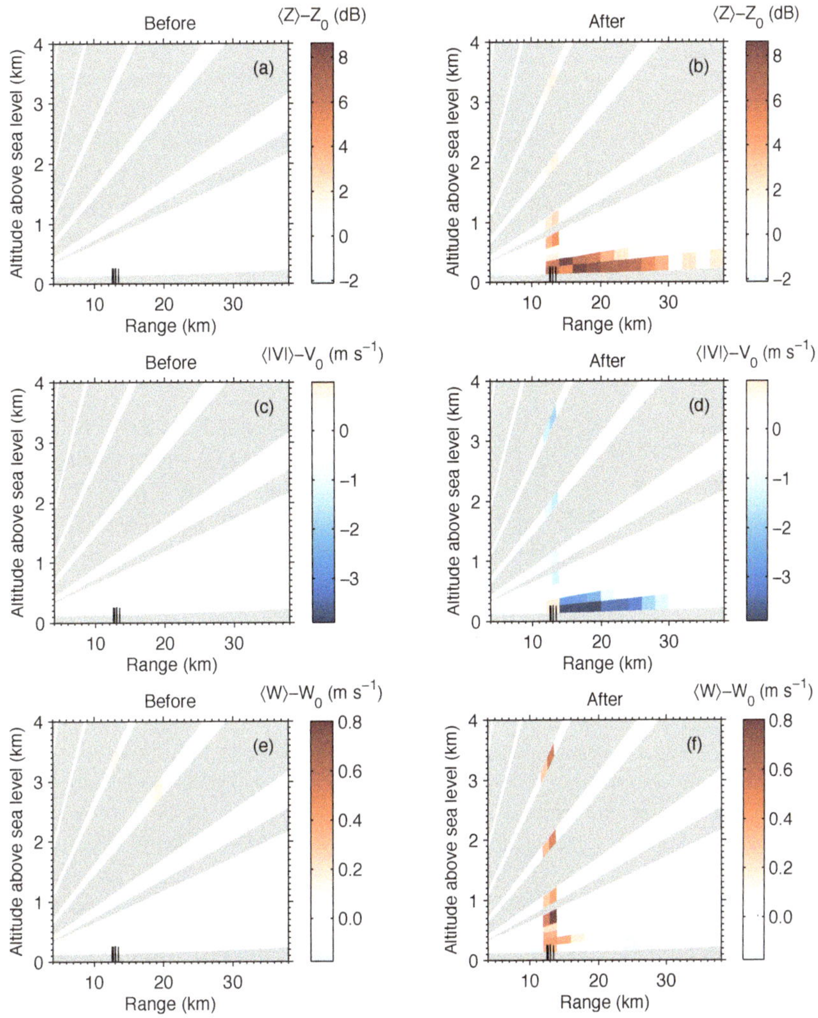

Figure 5. Difference in average reflectivity, $\langle Z\rangle - Z_0$; difference in average absolute radial velocity, $\langle |V|\rangle - V_0$; and difference in average spectrum width, $\langle W\rangle - W_0$, for all tilt angles ($0.5° \le \theta \le 40°$) before and after the construction of the Brunsmo wind farm. For each scan, Z_0, V_0, and W_0 were calculated as the mean value in azimuth (for the azimuth gates shown in Fig. 4) of the corresponding spectral moment before the construction of the wind farm. Data were taken from azimuth gate 52, the gate in which three wind turbines are located (near 44° azimuth). The locations of the wind turbines are shown on the bottom of each plot as black vertical lines. The wind turbines are only in the radar line of sight for the scan with the lowest tilt angle but are nevertheless seen to impact scans with higher tilt angles.

4.3 Wind turbine impact on a single radar cell

In Sects. 4.1 and 4.2 average values of the spectral moments were compared before and after the construction of the Brunsmo wind farm. While this is a robust and simple way to illustrate the impact of wind turbines on weather radar data, these average values do not separate the weather signal from the unwanted wind turbine clutter. Hence, this method tacitly assumes that the weather signal, on average, was similar during the two measurement periods (before and after the construction of the wind farm).

In order to perform a more in-depth analysis of the impact of wind turbines on radar data, one would ideally want to know the actual, unaffected values of the weather signal at all times. A way to simulate this is to use simultaneous measurements from a reference radar cell, a radar cell unaffected by the wind turbines.

This technique was used by Haase et al. (2010) and Norin and Haase (2012) in their works. As reference cells they used radar cells with the same location in range and azimuth as the wind-turbine-affected radar cells, albeit from a scan with a higher tilt angle ($\theta = 2.0°$). Furthermore, in order to remove the influence of weather, they chose to select only those measurements for which the reflectivity in the reference cell was undetected (i.e. $Z_{\text{ref}} = -30$ dBZ). As can be seen from Fig. 4, such a choice of reference radar cells must be treated with caution since radar cells in scans from higher tilt angles may also be affected by the presence of the wind farm.

In this work a different choice of reference radar cell was used: the radar cells one radar cell up-range from the radar

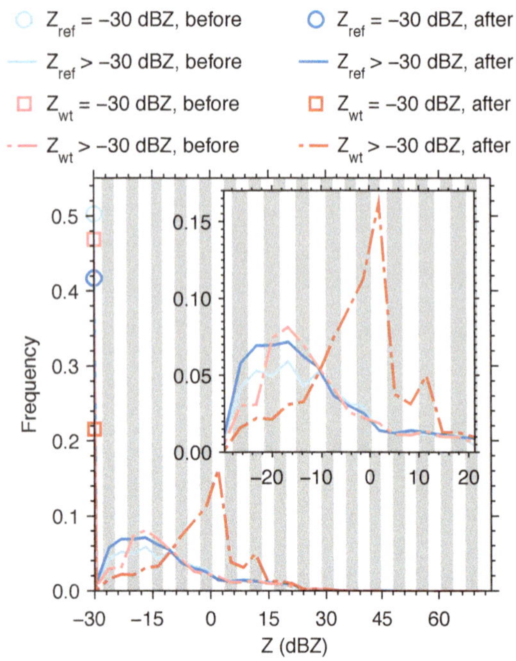

Figure 6. Relative frequency distributions of reflectivity from a reference radar cell, Z_{ref}, and from a radar cell in which three wind turbines are located, Z_{wt}, before and after the construction of a wind farm, using data from the scan with the lowest tilt angle ($\theta = 0.5°$). The grey and white backgrounds represent the bins that were used to create the distributions.

cells in which the wind turbines are located. From Fig. 2 it is seen that these radar cells are not affected by the wind farm.

In order to investigate the impact of wind turbines on a single radar cell, we chose to study data from the radar cell in which three wind turbines are located ($\theta = 0.5°$, range cell 7, azimuth gate 52). The corresponding reference radar cell was selected from the scan with the same tilt angle and azimuth gate albeit one range bin up-range from the wind-turbine-affected cell ($\theta = 0.5°$, range cell 6, azimuth gate 52).

In the analysis frequency distributions of the spectral moments (Z, V, and W) were examined as functions of Z_{ref}. In order to have sufficiently many samples for the analysis, data for Z and V were sorted into 35 bins.

For Z, one bin was used to count all measurement for which $Z = -30$ dBZ (undetected measurements) and one bin counted all invalid measurements. The remaining 33 bins were equally spaced from $Z = -29.6$ dBZ to $Z = 71.6$ dBZ. For V, one bin was used to count all undetected measurements and one bin counted all invalid measurements. The remaining 33 bins were equally spaced from $V = -24\,\text{m}\,\text{s}^{-1}$ to $V = 24\,\text{m}\,\text{s}^{-1}$.

Since measurements of W only consist of four different classes (see Sect. 2) and have separate representations of undetected and invalid measurements, data for W were used as they are.

Let us first examine the validity of the choice of reference radar cell. Figure 6 shows relative frequency distributions of reflectivity from the reference cell, Z_{ref}, and from the wind-turbine-affected radar cell, Z_{wt}, before and after the construction of the Brunsmo wind farm, using data from the scan with the lowest tilt angle ($\theta = 0.5°$). From Fig. 6 it is seen that the distributions of Z_{ref}, both before and after the construction of the wind farm, are similar to the distribution of Z_{wt} from before the wind farm was built. Some small differences exist between these distributions, but they are minor when compared to the distribution of the Z_{wt} after the construction of the wind farm, which is dominated by a prominent peak near $Z \approx 5$ dBZ.

From Fig. 6 it is seen that neither Z_{ref} nor Z_{wt} is normally distributed. This means that the average value of Z, used in Sect. 4.1 and Sect. 4.2, does not coincide with the value of Z at the peak of the frequency distribution.

Distributions of reflectivity from scans with higher tilt angles, using the same choice for reference radar cells (the radar cells one radar cell up-range from the wind-turbine-affected cells), all show similar behaviour. The chosen reference radar cells were therefore judged to provide a good (albeit not perfect) representation of the weather signal in the wind-turbine-affected radar cells.

Having validated the choice of reference radar cell, a more in-depth analysis of the wind-turbine-affected radar cell from the scan with the lowest tilt angle ($\theta = 0.5°$) could be performed. To analyse the impact of the wind turbines, frequency distributions of the spectral moments from the wind-turbine-affected radar cell (Z_{wt}, V_{wt}, and W_{wt}) were created as a function of the simultaneous reflectivity measurements from the reference radar cell (Z_{ref}). In Fig. 7 the relative frequency distributions of Z_{wt}, V_{wt}, and W_{wt} are shown.

Figure 7a shows relative frequency distributions of Z_{wt} as a function of Z_{ref} before the wind farm was constructed. It is seen that, as expected, the measurements of Z_{ref} in general accurately represent the measurements of Z_{wt}. For $Z_{\text{ref}} \lesssim -15$ dBZ the values of Z_{wt} tend to be somewhat higher than Z_{ref} (cf. Fig. 6). In Fig. 7b the same relative frequency distributions are shown, albeit this time for data gathered after the construction of the wind farm. For $Z_{\text{ref}} \gtrsim 5$ dBZ it is seen that Z_{ref} provides a good representation of Z_{wt}, whereas for $Z_{\text{ref}} \lesssim 5$ dBZ the situation is different. For all values of $Z_{\text{ref}} \lesssim 5$ dBZ the distributions of Z_{wt} exhibit a peak at $Z_{\text{wt}} \approx 5$ dBZ. This reflectivity value, $Z \approx 5$ dBZ, is the most frequent reflectivity value generated by the wind turbines. When the reflectivity value from the actual weather is smaller than this value, the wind turbines effectively hide the weather signal. When the weather signal is larger than the wind turbine clutter, the reflectivity value approaches that of the reference cell.

Figure 7c shows the relative frequency distributions of V_{wt} as a function of Z_{ref} before the wind farm was constructed. From Fig. 7c it is seen that the distributions of V_{wt} have their maxima near $V_{\text{wt}} \approx \pm 5\,\text{m}\,\text{s}^{-1}$, regardless of the

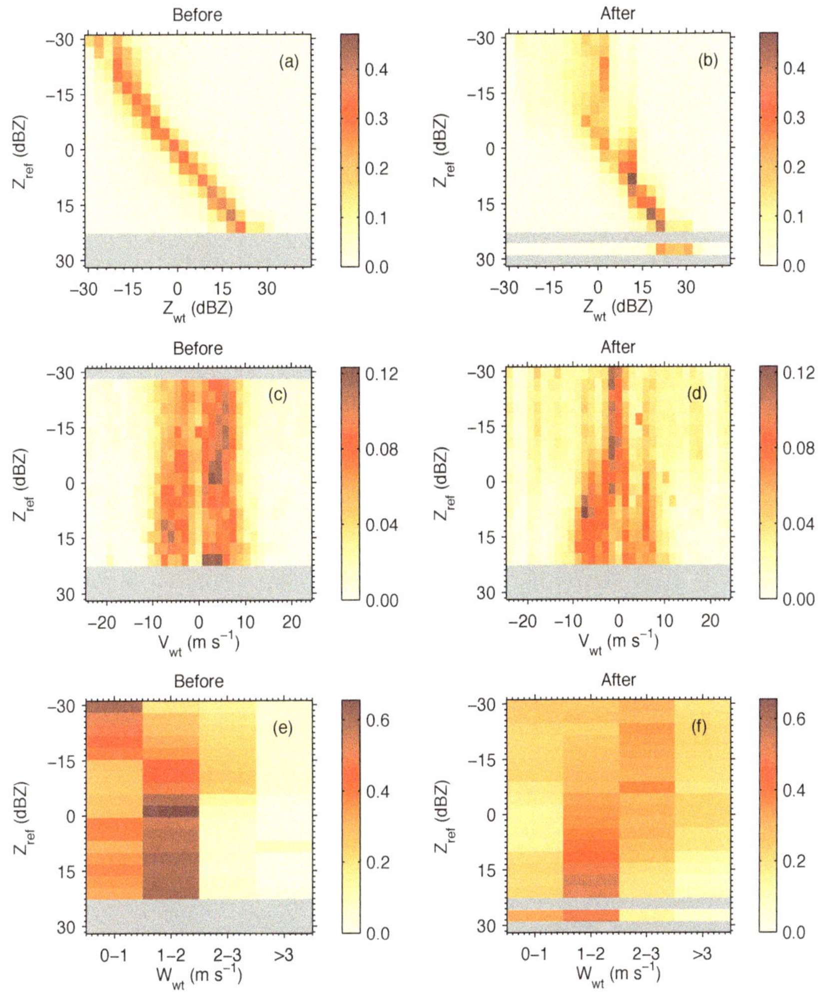

Figure 7. Relative frequency distributions of reflectivity, Z_{wt}; radial velocity, V_{wt}; and spectrum width, W_{wt}, from a radar cell in which three wind turbines are located as a function of reflectivity from a reference radar cell, Z_{ref}, using data from the scan with the lowest tilt angle ($\theta = 0.5°$). Panels on the left-hand side (**a**, **c**, and **e**) show relative frequency distributions before the wind farm was constructed, and panels to the right (**b**, **d**, and **f**) show the corresponding distributions after the wind farm start of operations. For $Z_{\mathrm{ref}} \gtrsim 5$ dBZ the distributions of Z_{wt}, V_{wt}, and W_{wt} are similar before and after the construction of the wind farm. However, for $Z_{\mathrm{ref}} \lesssim 5$ dBZ the distributions before and after the construction of the wind farm look different.

value of Z_{ref}. Figure 7d shows the same distributions, albeit for data recorded after the construction of the wind farm. For $Z_{\mathrm{ref}} \gtrsim 5$ dBZ the distributions of V_{wt} are similar to those shown in Fig. 7c, but for smaller values of Z_{ref} the distributions of V_{wt} look different. For $Z_{\mathrm{ref}} \lesssim 5$ dBZ the distributions of V_{wt} all have their maxima at $V_{\mathrm{wt}} \approx 0\,\mathrm{m\,s^{-1}}$. Much as for Z_{wt}, for $Z_{\mathrm{ref}} \gtrsim 5$ dBZ the distributions of V_{wt} approach the corresponding distributions with data from before the construction of the wind farm.

For $Z_{\mathrm{ref}} \lesssim 5$ dBZ, the frequency distribution of V_{wt} was spread over almost all velocity bins, resulting in an increase in $\langle|V_{\mathrm{wt}}|\rangle$. However, when extending the analysis to radar cells downrange of the wind turbines, it was found that the frequency distribution of V_{wt} was concentrated to the low-velocity bins, resulting in a decrease in $\langle|V_{\mathrm{wt}}|\rangle$. A possible reason for this is discussed in Sect. 5.

For comparison a wind rose for the period of study (2008–2013), together with the distribution of wind speed, is shown in Fig. 10. The measurements of wind speed and wind direction for this figure come from an automatic weather station (56.2619° N, 15.2742° E), located approximately 21 km to the west of the weather radar. The wind speed measurements have a resolution of $1\,\mathrm{m\,s^{-1}}$ and were available at most once per hour during the period of study.

Figure 7e shows the relative frequency distributions of W_{wt} as a function of Z_{ref} before the wind farm was constructed. From Fig. 7e it is seen that the distributions of W_{wt} vary depending on the value of Z_{ref}. For $Z_{\mathrm{ref}} \lesssim -15$ dBZ, the distributions of W_{wt} peak for $W_{\mathrm{wt}} = 0$–$1\,\mathrm{m\,s^{-1}}$, whereas for

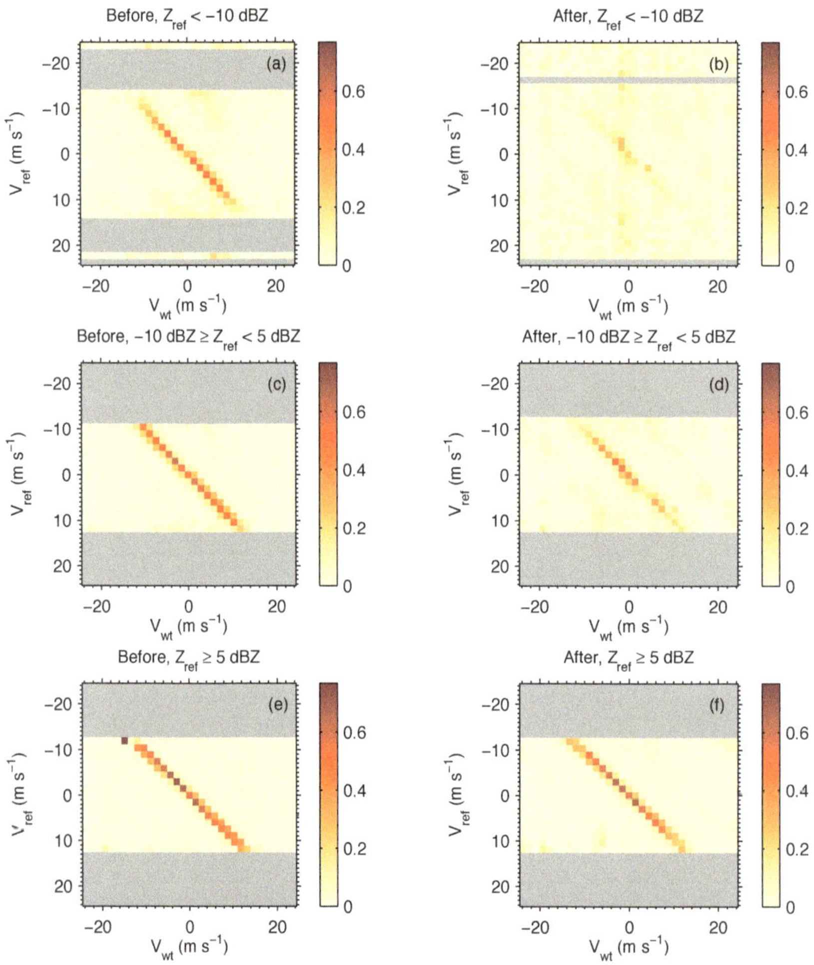

Figure 8. Relative frequency distributions of radial velocity from a radar cell in which three wind turbines are located, V_{wt}; as a function of radial velocity, V_{ref}; and reflectivity, Z_{ref}, from a reference radar cell. Data were taken from the scan with the lowest tilt angle ($\theta = 0.5°$). Panels on the left-hand side (**a**, **c**, and **e**) show relative frequency distributions of V_{wt} for $Z_{ref} < -10$ dBZ, -10 dBZ $\geq Z_{ref} < 5$ dBZ, and $Z_{ref} \geq 5$ dBZ, respectively, before the construction of the wind farm. Panels on the right-hand side (**b**, **d**, and **f**) show the corresponding distributions, using data obtained after the construction of the wind farm. In (**b**), for $Z_{ref} < -10$ dBZ, almost no correlation exists between V_{wt} and V_{ref}, but it is clear that the correlation improves for increasing values of Z_{ref} (**d** and **f**).

$Z_{ref} \gtrsim -15$ dBZ the distributions of W_{wt} have their maxima for $W_{wt} = 1$–$2\,\mathrm{m\,s^{-1}}$. From Fig. 7f it is seen that after the construction of the wind farm the distributions of W_{wt} are evenly spread out for $Z_{ref} \lesssim -5$ dBZ. For larger reflectivity values the distributions of W_{wt} gradually approach those before the wind farm was built.

From Fig. 7 it is clear that competition between the weather signal and the wind turbine clutter exists. Once the strength of the weather echoes exceeds that of the wind turbine clutter, all spectral moments are seen to approach their respective values of the reference cell.

4.4 Improvement of the weather signal

In Fig. 7c and d it was shown that measurements of V_{wt} approached the values of the reference cell for $Z_{ref} \gtrsim 5$ dBZ. Another way to see how V_{wt} approaches its true value for increasing values of Z_{ref} is shown in Fig. 8. In this figure the relative frequency distributions of V_{wt} are shown as functions of V_{ref} and Z_{ref}. Figure 8a, c, and e shows the distributions of V_{wt} as a function of V_{ref} for $Z_{ref} < -10$ dBZ, -10 dBZ $\geq Z_{ref} < 5$ dBZ, and $Z_{ref} \geq 5$ dBZ, respectively, using data from before the construction of the wind farm. From these figures it is seen that before the construction of the wind farm there was a strong correlation between V_{wt} and V_{ref} regardless of the value of Z_{ref}. In Fig. 8b, d, and f the corresponding distributions are shown, albeit for data gathered after the construction of the wind farm. In Fig. 8b, where $Z_{ref} < -10$ dBZ, there is almost no correlation between V_{wt} and V_{ref}, except for $V_{ref} \lesssim \pm 3\,\mathrm{m\,s^{-1}}$. (It is interesting to note that the cut-in speed of the wind turbines is $3.5\,\mathrm{m\,s^{-1}}$ (see Sect. 3), which could explain the correlation

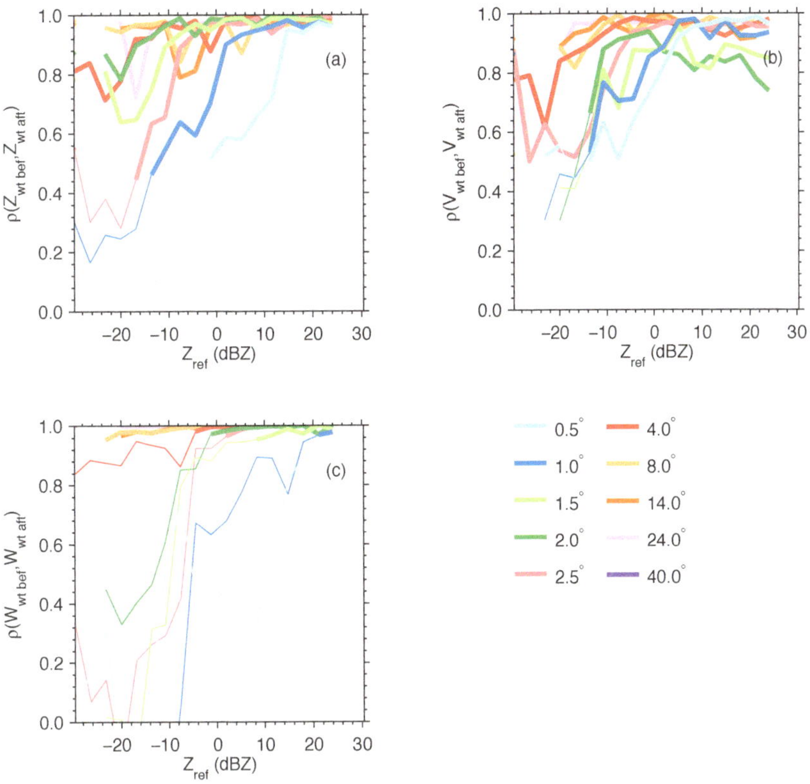

Figure 9. Correlation coefficients, ρ, between frequency distributions of spectral moments from a wind-turbine-affected radar cell before ($Z_{\mathrm{wt\ bef}}$, $V_{\mathrm{wt\ bef}}$, and $W_{\mathrm{wt\ bef}}$) and after ($Z_{\mathrm{wt\ aft}}$, $V_{\mathrm{wt\ aft}}$, and $W_{\mathrm{wt\ aft}}$) the construction of the Brunsmo wind farm, as a function of reflectivity from a reference radar cell, Z_{ref}, for all tilt angles. The correlation of the spectral moments recovers for higher values of Z_{ref} (increasing influence of the weather signal) and for higher tilt angles (decreasing influence of the wind farm). Correlations significant on the $p < 0.05$ level are drawn with thick lines.

for $V_{\mathrm{ref}} \lesssim \pm 3\,\mathrm{m\,s^{-1}}$.) In Fig. 8d the correlation between V_{wt} and V_{ref} is seen to increase for $-10\,\mathrm{dBZ} \geq Z_{\mathrm{ref}} < 5\,\mathrm{dBZ}$, and for $Z_{\mathrm{ref}} \geq 5\,\mathrm{dBZ}$, shown in Fig. 8f, it is as strong as before the construction of the wind farm.

As was shown in Fig. 7, the values of all spectral moments are seen to approach those of the reference cell for increasing values of Z_{ref}. A way to quantitatively analyse this change is to calculate the correlation coefficient between the distributions of the spectral moments before and after the construction of the wind farm, as a function of Z_{ref}. This analysis was performed for all 10 scans in the polar volume. The wind-turbine-affected radar cells were represented by the radar cells with azimuth gate 52 and the range cell in which the wind turbines would be located if a vertical line were drawn from the ground. The corresponding reference cells were selected from the same azimuth gate but one radar cell closer to the radar (cf. Sect. 4.3). Correlation coefficients were only calculated for distributions for which more than 300 measurements existed. The results of the analyses for the three spectral moments (Z, V, and W) for all 10 scans in the polar volume are shown in Fig. 9.

Figure 9a shows the correlation coefficient, ρ, between the distribution of Z at the wind turbines, before ($Z_{\mathrm{wt\ bef}}$) and after ($Z_{\mathrm{wt\ aft}}$) the construction of the wind farm, as a function of Z_{ref} for all scans in the polar volume. The correlation coefficient, $\rho(Z_{\mathrm{wt\ bef}}, Z_{\mathrm{wt\ aft}})$, is seen to increase for increasing values of Z_{ref} (i.e. for increasing influence of the weather signal) and also for higher tilt angles (i.e. for decreasing influence of the wind farm). However, for $Z_{\mathrm{ref}} < -5\,\mathrm{dBZ}$ the correlation coefficient for the fifth scan ($\theta = 2.5°$, the lowest scan using the higher radar cell resolution) is lower than for two scans with lower tilt angles ($\theta = 1.5°$ and $\theta = 2.0°$). This can partly be explained by the changes in radar cell resolution (cf. Sect. 2 and Sect. 5). That the fifth scan is more affected by the wind farm than some of the lower scans can also be seen in Figs. 4 and 5, as well as in Fig. S1 in the Supplement.

Figure 9b shows the correlation coefficient, ρ, between the distribution of V at the wind turbines, before ($V_{\mathrm{wt\ bef}}$) and after ($V_{\mathrm{wt\ aft}}$) the construction of the wind farm, as a function of Z_{ref} for all scans in the polar volume. In Fig. 9b it is again seen that the correlation coefficient, $\rho(V_{\mathrm{wt\ bef}}, V_{\mathrm{wt\ aft}})$, increases for increasing values of Z_{ref} as well as for higher tilt angles. However, the use a higher radar cell resolution ($\theta \geq 2.5°$) seems to have a smaller negative impact on the improvement of V_{wt} than for Z_{wt}.

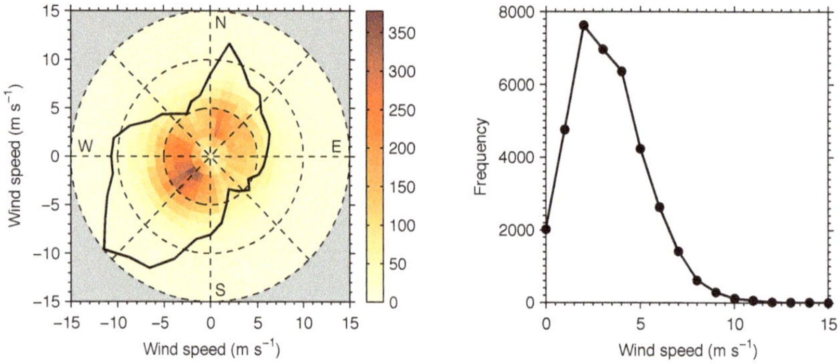

Figure 10. Wind rose showing the wind speed and wind direction using hourly data from an automatic weather station located approximately 21 km to the west of the weather radar for the period April 2010–December 2013. The figure to the left shows the distribution of wind speed and wind direction. The solid black line shows the wind direction for all wind speeds. The figure to the right shows the distribution of wind speed for all wind directions.

Figure 9c shows the correlation coefficient, ρ, between the distribution of W at the wind turbines, before ($W_{\text{wt bef}}$) and after ($W_{\text{wt aft}}$) the construction of the wind farm, as a function of Z_{ref} for all scans in the polar volume. As for Z and V, the correlation coefficient for W, $\rho(W_{\text{wt bef}}, W_{\text{wt aft}})$, increases for increasing values of Z_{ref} and also for higher tilt angles. For the scans with the lowest tilt angle ($\theta = 0.5°$) a correlation greater than 0.9 is only achieved for $Z_{\text{ref}} > 15$ dBZ. Since W is limited to four classes (see Sect. 2), fewer values of the correlation coefficient are statistically significant compared to the other spectral moments.

Figures 7 and 8 demonstrate that the values of all spectral moments approach those of the reference cell when the weather signal is stronger than the wind turbine clutter. Figure 9 shows that this effect can be seen for all tilt angles. However, in order to quantitatively determine the impact of a radar cell for a certain tilt angle, data should be examined more carefully, for example as described in Sect. 4.3.

5 Discussion

Some of the results presented in the previous sections require further discussion. Here, we first examine the changes in the average values of the spectral moments. Secondly, the sudden increase in reflectivity occurring between the fourth and the fifth scan is investigated. Thirdly, possible explanations of the impact seen on spectral moments far above the wind farm are discussed.

Let us first examine the changes in the average values of the spectral moments. Figures 2 and 3 showed that the impact of the Brunsmo wind farm on spectral moments from the scan with the lowest tilt angle extends up to 20 km behind the wind turbines. As mentioned in Sect. 4.1, tails of increased reflectivity downrange of wind turbines are believed to be caused by multiple scattering effects (scattering between multiple turbines and/or scattering between turbine and ground) (Isom et al., 2009; Vogt et al., 2011; Kong, 2014). A wind-turbine-scattered radar signal, shown to occur behind the wind turbines, has evidently experienced a time delay, likely caused by multiple scattering, whereas a signal shown to occur at the location of the wind farm is the result of direct scattering off the turbines. At the location of the wind farm the impact on $\langle Z\rangle$ should therefore always be positive (or negligible if in the presence of a strong weather signal). Behind the wind farm, competition may exist between blockage, leading to a decrease in $\langle Z\rangle$, and multiple scattering effects, which would lead to an increase in $\langle Z\rangle$. For the Brunsmo wind farm, examined in this work, it is clear that the net effect is an increase in $\langle Z\rangle$.

In Fig. 2 it was shown that both average absolute radial velocity, $\langle|V|\rangle$, and average spectrum width, $\langle W\rangle$, show an increase at the location of the wind turbines. If the received radar signal is dominated by scattering off one or more moving wind turbine blades, a large and variable value in $|V|$ would be expected, consistent with the observations (cf. Fig.7d). A signal composed of several frequency components of similar strength, for example generated by scattering off different locations on a single blade, would give rise to a large value of W.

Behind the wind turbines, however, $|V|$ and W show a decrease in their average values. If the main contribution to the received radar signal originated from multiple scattering off stationary parts of the wind turbine (e.g. scattering between the tower and the ground), the resulting frequency spectrum would have a narrow peak, centred around zero frequency shift. For such a frequency spectrum an increase in $\langle Z\rangle$ together with small values in $\langle|V|\rangle$ and $\langle W\rangle$ would follow, consistent with the observations. To validate this hypothesis, in-phase (I) and quadrature (Q) measurements would need to be sampled so that individual frequency spectra could be generated and examined. Unfortunately, the Swedish weather radars lack such capability.

The second result from the previous sections worth discussing is the sudden increase in $\langle Z\rangle$ occurring between the fourth and the fifth scan, seen in Figs. 4 and 5. This increase can be explained by the way the Swedish radars are configured to perform their measurements. As described in Sect. 2, the radar cells from the four lowest scans use a lower resolution in both range and azimuth. A wind-turbine-affected radar cell (consisting of 12 range bins) can have one or more unaffected range bins. Since radar cells from the four lowest scans have a lower range resolution, the impact on these cells is expected to be lower than (or at most equal to) those from the higher scans. For example, a wind turbine located in the centre of a 2 km wide radar cell would have an impact on the last six range bins of that radar cell. However, all 12 range bins from a 1 km wide radar cell would be affected, and that radar cell would register a higher value in $\langle Z \rangle$ compared to the radar cell using the lower range resolution.

The changes in azimuthal resolution of the radar cells can also contribute to the observed sudden increase in $\langle Z \rangle$. A lower resolution in azimuth (used by the lower scans) means that fewer measurements in a frequency spectrum will be affected by the wind turbine, leading to a smaller impact compared to measurements from the higher scans, using the higher azimuthal resolution. These hypotheses could be tested by examining individual frequency spectra from wind-turbine-affected radar cells using different resolutions or by changing the radar cell resolution so that it is the same for all tilt angles.

Finally, let us investigate why the wind turbines in the Brunsmo wind farm are seen to have an impact on radar measurements far above the turbines (cf. Figs. 4 and 5). That a wind farm can have a negative impact on radar data at altitudes far above the wind turbines can be due to several mechanisms. For example, Isom et al. (2009) showed that wind farms can be detected by radar sidelobes. It is, however, also conceivable that the impact could be due to anomalous propagation, i.e., bending of the radar beam due to changes in the atmospheric refractive index. A third explanation of the observed increase in reflectivity is that it is due to scattering off increased levels of dust and turbulence, generated by the wind farm. Let us examine these mechanisms one at a time.

Figure 11 shows the increase in $\langle Z \rangle$ for the different tilt angles from the radar cells in and below which three wind turbines are located. In the same figure is also shown a vertical cut of the antenna directivity. In Fig. 11 it is seen that $\langle Z \rangle$ follows the antenna pattern for the lowest four tilt angles, after which an increase in $\langle Z \rangle$ can be seen. This increase, as discussed above, is likely a result of the Swedish radars' measurement technique. For the higher tilt angles $\langle Z \rangle$ does not decrease as much as the antenna pattern does. Also, the received power (which is proportional to reflectivity) is proportional to the directivity squared. Thus the antenna directivity, shown in Fig. 11, should be squared in order provide a fair comparison to the observed impact in $\langle Z \rangle$. From Fig. 11 it is seen that detection by the radar sidelobes does not provide a complete explanation of the impact seen on the higher scans.

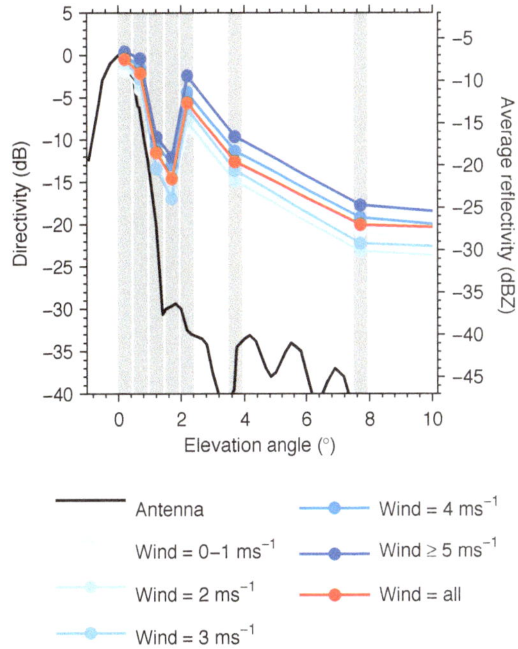

Figure 11. Increase in average reflectivity, $\langle Z \rangle$, from radar cells in and below which three wind turbines are located for different tilt angles and wind speeds. The wind speed measurements come from an automatic weather station, located approximately 21 km to the west of the weather radar. The black solid line shows a vertical cut of the antenna pattern. The grey areas show the angular extent of the wind turbine blades for the different tilt angles.

Another possible reason for the observed impact far above the wind farm is anomalous propagation. Anomalous propagation can sometimes result in scattering from objects on the ground, not normally seen by the radar. Such propagation conditions can be caused by temperature inversions or by a rapid decrease in water vapour content with height (see, e.g., Patterson, 2008). Temperature inversions can exist at any time of the day in Sweden during the cool season (Devasthale and Thomas, 2012), whereas such inversions most commonly occur in early morning during the warm season. A rapid decrease in water vapour content with height can occur during afternoons in summer (Bodine et al., 2011).

In order to investigate the effect of anomalous propagation, ray tracing of the radar beam was performed for the period April 2010–December 2013 (after the wind farm was built). Ray tracing of the radar beam was calculated every hour using the method proposed by Zeng et al. (2014). Vertical profiles of refractivity, needed for the calculations, were generated using data obtained from Sweden's operational numerical weather prediction model. The results, shown in Fig. 12, reveal that the changes in beam height were very small at the distance of the wind turbines. For the Brunsmo wind farm, anomalous propagation is unlikely to explain the impact seen on the higher tilt angles.

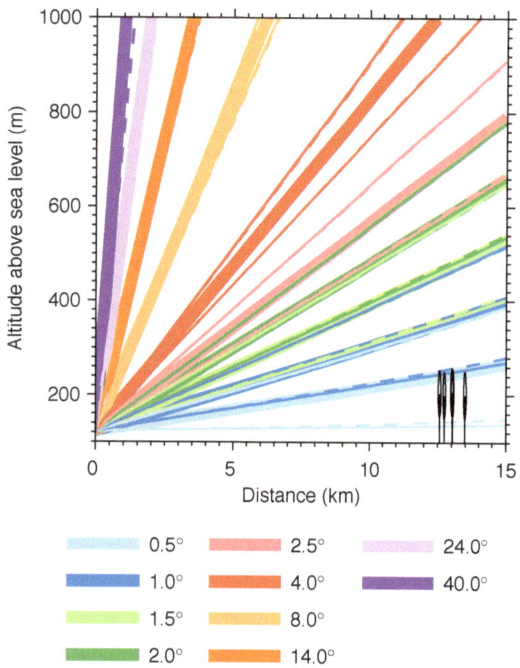

Figure 12. Beam propagation for different tilt angles calculated using ray tracing. Vertical profiles of refractivity were obtained using data from Sweden's operational NWP model. Thick solid lines show the median centre of the radar beams. Thin solid lines show the 1st and the 99th percentile of the half-power beam widths for the calculated beams. Thin dashed lines show half-power beam widths using standard propagation conditions. Wind turbines are shown in black.

The third possibility, scattering off increased levels of dust and turbulence, generated by the wind farm is difficult to examine directly. However, it is conceivable that levels of wind-turbine-generated dust and turbulence could be higher for higher wind speeds. Higher wind speeds let the turbine blades rotate faster, which in turn could lead to increased levels of dust and turbulence. Measurements of wind speed from an automatic weather station, located approximately 21 km to the west of weather radar Karlskrona, was shown in Fig. 10. The wind speed at the location of the automatic weather station does not necessarily coincide with the wind speed at the location of the wind farm, but the general tendency should be the same (higher wind speeds at the automatic weather station should correspond to higher wind speed at the wind farm). The wind speed measurements were divided into five bins: 0–1, 2, 3, 4, and 5–15 $\mathrm{m\,s^{-1}}$. All bins contained similar amounts of data. The average values of $\langle Z \rangle$, corresponding to the different wind speeds, are shown in Fig. 11. From this figure it is seen that higher wind speeds lead to higher values of $\langle Z \rangle$, supporting the hypothesis of increased scattering. However, the difference in $\langle Z \rangle$ between low (0–1 $\mathrm{m\,s^{-1}}$) and high (5–15 $\mathrm{m\,s^{-1}}$) wind speeds is only of the order of 5 dB.

In summary, the observed impact on higher scans are most likely to be caused by detection of the radar sidelobes, although this does not provide a complete explanation. High wind speeds, which could lead to increased levels of dust and turbulence, also lead to an increase in reflectivity, whereas the effects of anomalous propagation seem to have a limited effect on the measurements at the Brunsmo wind farm.

6 Summary and conclusions

Wind turbines located in the vicinity of a Doppler weather radar can have a detrimental impact on the radar's performance. The objective of this study was to investigate the impact of wind turbines on operational Doppler weather radar data. Doppler weather radars measure three spectral moments: the radar reflectivity factor, Z; radial velocity, V; and spectrum width, W. In this work we have analysed the impact of wind turbines on all spectral moments, using two different approaches.

First, the average values for the three spectral moments (reflectivity, $\langle Z \rangle$; absolute radial velocity, $\langle |V| \rangle$; and spectrum width, $\langle W \rangle$) were computed. In this study these averages were calculated using operational radar data from long periods of time before (approximately 2.5 years) and after (approximately 3.5 years) the construction of a wind farm near radar Karlskrona in southeastern Sweden. By comparing the average values before and after, the impact of the wind farm on all spectral moments was estimated. For the wind farm studied in this work the average values of the spectral moments in the radar cells in which wind turbines are located, from the scan with the lowest tilt angle, were shown to change after the wind turbines were built; reflectivity increased from $\langle Z \rangle \approx -18$ dBZ to $\langle Z \rangle \approx -6$ dBZ, the absolute radial velocity increased from $\langle |V| \rangle \approx 6\,\mathrm{m\,s^{-1}}$ to $\langle |V| \rangle > 8.5\,\mathrm{m\,s^{-1}}$, and the spectrum width increased from $\langle W \rangle \approx 2.0\,\mathrm{m\,s^{-1}}$ to $\langle W \rangle > 2.5\,\mathrm{m\,s^{-1}}$. However, not only those radar cells in which the wind turbines are located were affected by the presence of the wind farm. Several radar cells both cross- and downrange were also affected.

Downrange from the wind turbines, tails with increased values of $\langle Z \rangle$ were seen, visible for more than 20 km behind the turbines. This is believed to be caused by multiple scattering of the radar signal between rotor blades and the ground. For $\langle |V| \rangle$ and $\langle W \rangle$, tails with decreased values were observed. A possible explanation of these changes is that the dominating contribution to the radar signal received behind the wind farm originates from stationary parts of the wind turbines, but this needs to be investigated further.

In addition to the changes in values of all spectral moments in data from the scan with the lowest tilt angle it was shown that data from scans with higher tilt angles, far above the wind farm, were also affected. Increased values in $\langle Z \rangle$ and $\langle W \rangle$ and decreased values in $\langle |V| \rangle$ were seen up to 3 km above the wind farm.

That a wind farm can have an impact on radar data at altitudes far above the wind turbines can be due to several reasons. Wind farms not in the line of sight of a radar can still be detected by the radar's sidelobes. Scattering off increased levels of dust and turbulence above the wind farm and anomalous propagation, i.e., bending of the radar beam, are also conceivable explanations of changed values in the spectral moments. In this work all three mechanisms were investigated. It was shown that anomalous propagation is unlikely to explain the observed changes, whereas detection by the radar sidelobes and scattering off increased levels of dust and turbulence can, at least in part, explain the observed changes.

One of the problems with wind turbine clutter is that their impact on the spectral moments is similar to that of real weather signals. To analyse the competition between the weather signal and wind turbine clutter, a second approach, using data from reference radar cells, was used. Ideally, data from a reference radar cell should, for any given measurement, show what the unaffected weather signal should be in the wind-turbine-affected radar cell. In this work the reference cells were chosen as the radar cells one radar cell up-range from the affected radar cells. Data from all three spectral moments, before and after the construction of the wind farm, were then visualised using relative frequency distributions of the spectral moment under study, as a function of Z_{ref}.

The results using this method revealed that the values of all spectral moments from wind-turbine-affected radar cells approach those of the reference cell (i.e. the value of the weather signal) for increasing values of Z_{ref}. When precipitation gives rise to reflectivity values stronger than those caused by the wind turbines, the negative impact of the wind turbines is greatly reduced for all spectral moments.

These results suggest that the weather information from wind-turbine-affected radar cells is not always lost. One way to mitigate the impact of wind turbines could be to create conditional filters. If reflectivity from a wind-turbine-affected radar cell is larger than that caused by the wind turbine, the radar data could still be used. However, when the reflectivity is similar to or lower than the reflectivity caused by the wind turbine, caution should be applied when using radar data from any spectral moment from the wind-turbine-affected radar cells.

The Supplement related to this article is available online at doi:10.5194/amt-8-593-2015-supplement.

Acknowledgements. This work was financed by the Swedish Energy Agency, under the project VINDRAD+.

Edited by: M. Kulie

References

Bodine, D., Michaud, D., Palmer, R. D., Heinselman, P. L., Brotzge, J., Gasperoni, N., Cheong, B. L., Xue, M., and Gao, J.: Understanding Radar Refractivity: Sources of Uncertainty, J. Appl. Meteorol. Clim., 50, 2543–2560, doi:10.1175/2011JAMC2648.1, 2011.

Burgess, D. W., Crum, T. D., and Vogt, R. J.: Impacts of Wind Farms on WSR-88D Radars, in: 24th International Conference on Interactive Information and Processing Systems for Meteorology, Oceanography, and Hydrology, New Orleans, LA, United States, American Meteorological Society, paper 6B.3, 2008.

Crum, T. and Ciardi, E.: Wind Farms and the WSR-88D: An Update, Nextrad Now, 20, 17–22, 2010.

Crum, T., Ciardi, E., and Sandifer, J.: Wind Farms: Coming Soon to a WSR-88D Near You, Nexrad Now, 18, 1–7, 2008.

Department of Defense: The Effect of Windmill Farms on Military Readiness, Report to the congressional defense committees, Office of the Director of Defense Research and Engineering, 2006.

Devasthale, A. and Thomas, M. A.: An investigation of statistical link between inversion strength and carbon monoxide over Scandinavia in winter using AIRS data, Atmos. Environ., 56, 109–114, doi:10.1016/j.atmosenv.2012.03.042, 2012.

Doviak, R. J. and Zrnić, D. S.: Doppler Radar and Weather Observations, Dover Publications, Mineola, New York, 2nd Edn., 2006.

Gallardo, B., Pérez, F., and Aguado, F.: Characterization Approach of Wind Turbine Clutter in the Spanish Weather Radar Network, in: Proceedings of the Fifth European Conference on Radar in Meteorology and Hydrology, Helsinki, Finland, ERAD, 2008.

Global Wind Energy Council: Global Wind Report, Annual market update 2013, Global Wind Energy Council, 2014.

Haase, G., Johnson, D., and Eriksson, K.-Å.: Analyzing the impact of wind turbines on operational weather radar products, in: Proceedings of the Sixth European Conference on Radar in Meteorology and Hydrology, Sibiu, Romania, 276–281, ERAD, 2010.

Isom, B. M., Palmer, R. D., Secrest, G. S., Rhoton, R. D., L., D. S. T., Allmon, Reed, J., Crum, T., and Vogt, R.: Detailed Observations of Wind Turbine Clutter with Scanning Weather Radars, J. Atmos. Oceanic Technol., 26, 894–910, doi:10.1175/2008JTECHA1136.1, 2009.

Kong, F.: Wind Turbine Clutter in Weather Radar: Characterization and Mitigation, Ph.D. thesis, University of Oklahoma, Norman, OK, United States, 2014.

Lemmon, J. J., Caroll, J. E., Sanders, F. H., and Turner, D.: Assessment of the Effects of Wind Turbines on Air Traffic Control Radars, Tech. Rep. TR-08-454, US Department of Commerse, National Telecommunications & Information Administration, 2008.

Lute, C. and Wieserman, W.: ASR-11 radar performance assessment over a wind turbine farm, in: Proceedings of the 2011 IEEE Radar Conference, Kansas City, MO, United States, 226–230, IEEE, 2011.

Norin, L. and Haase, G.: Doppler Weather Radars and Wind Turbines, in: Doppler Radar Observations – Weather Radar, Wind Profiler, Ionospheric Radar, and Other Advanced Applications, edited by: Bech, J. and Chau, J. L., InTech, Croatia, doi:10.5772/39029, 2012.

Patterson, W. L.: The Propagation Factor, F_p, in the Radar Equation, in: Radar Handbook, edited by Skolnik, M., McGraw-Hill, United States, 3rd Edn., 2008.

Poupart, G. J.: Wind farms impact on radar aviation interests, Tech. Rep. FES W/14/00614/00/REP, DTI PUB URN 03/1294, QinetiQ, 2003.

Rossa, A., Liechti, K., Zappa, M., Bruen, M., Germann, U., Haase, G., Keil, C., and Krahe, P.: The COST 731 Action: A review on uncertainty propagation in advanced hydro-meteorological forecast systems, Atmos. Res., 100, 150–167, doi:10.1016/j.atmosres.2010.11.016, 2011.

Toth, M., Jones, E., Pittman, D., and Solomon, D.: DOW Radar Observations of Wind Farms, B. Am. Meteorol. Soc., 92, 987–995, doi:10.1175/2011BAMS3068.1, 2011.

Vogt, R. J., Crum, T. D., Greenwood, W., Ciardi, E. J., and Guenther, R. G.: New Criteria for Evaluating Wind Turbine Impacts on NEXRAD Radars, in: Proceedings of WINDPOWER 2011, Anaheim, CA, United States, American Wind Energy Association, 2011.

Zeng, Y., Blahak, U., Neuper, M., and Jerger, D.: Radar Beam Tracing Methods Based on Atmospheric Refractive Index, J. Atmos. Oceanic Technol., 31, 2650–2670, doi:10.1175/JTECH-D-13-00152.1, 2014.

14

Empirical validation and proof of added value of MUSICA's tropospheric δD remote sensing products

M. Schneider[1], **Y. González**[2], **C. Dyroff**[1], **E. Christner**[1], **A. Wiegele**[1], **S. Barthlott**[1], **O. E. García**[2], **E. Sepúlveda**[2], **F. Hase**[1], **J. Andrey**[3,*], **T. Blumenstock**[1], **C. Guirado**[2], **R. Ramos**[2], and **S. Rodríguez**[2]

[1]Institute for Meteorology and Climate Research (IMK-ASF), Karlsruhe Institute of Technology, Karlsruhe, Germany
[2]Izaña Atmospheric Research Center, Agencia Estatal de Meteorología (AEMET), Izaña, Spain
[3]Area de Investigación e Instrumentación Atmosférica, INTA, Torrejón de Ardoz, Spain
[*]now at: CNRM-GAME, Météo France and CNRS, Toulouse, France

Correspondence to: M. Schneider (matthias.schneider@kit.edu)

Abstract. The project MUSICA (MUlti-platform remote Sensing of Isotopologues for investigating the Cycle of Atmospheric water) integrates tropospheric water vapour isotopologue remote sensing and in situ observations. This paper presents a first empirical validation of MUSICA's H_2O and δD remote sensing products, generated from ground-based FTIR (Fourier transform infrared), spectrometer and space-based IASI (infrared atmospheric sounding interferometer) observation. The study is made in the area of the Canary Islands in the subtropical northern Atlantic. As reference we use well calibrated in situ measurements made aboard an aircraft (between 200 and 6800 m a.s.l.) by the dedicated ISOWAT instrument and on the island of Tenerife at two different altitudes (at Izaña, 2370 m a.s.l., and at Teide, 3550 m a.s.l.) by two commercial Picarro L2120-i water isotopologue analysers.

The comparison to the ISOWAT profile measurements shows that the remote sensors can well capture the variations in the water vapour isotopologues, and the scatter with respect to the in situ references suggests a δD random uncertainty for the FTIR product of much better than 45 ‰ in the lower troposphere and of about 15 ‰ for the middle troposphere. For the middle tropospheric IASI δD product the study suggests a respective uncertainty of about 15 ‰. In both remote sensing data sets we find a positive δD bias of 30–70 ‰.

Complementing H_2O observations with δD data allows moisture transport studies that are not possible with H_2O observations alone. We are able to qualitatively demonstrate the added value of the MUSICA δD remote sensing data. We document that the δD–H_2O curves obtained from the different in situ and remote sensing data sets (ISOWAT, Picarro at Izaña and Teide, FTIR, and IASI) consistently identify two different moisture transport pathways to the subtropical north eastern Atlantic free troposphere.

1 Introduction

The water cycle (continuous evaporation, transport and condensation of water) is closely linked to the global energy and radiation budgets and has thus fundamental importance for climate on global as well as regional scales. Understanding the different, strongly coupled, and often competing processes that comprise the tropospheric water cycle is of prime importance for reliable climate projections. Since different water cycle processes differently affect the isotopic composition of atmospheric water, respective measurements are potentially very useful for tropospheric water cycle research. For instance, they can help to investigate cloud processes (Webster and Heymsfield, 2003; Schmidt et al., 2005), rain recycling and evapotranspiration (Worden et al., 2007), or the processes that control upper tropospheric humidity (Risi et al., 2012).

In the following, we express $H_2^{16}O$ and $HD^{16}O$ as H_2O and HDO, respectively; and $\frac{HD^{16}O}{H_2^{16}O}$ in the δ-notation (with the Vienna standard mean ocean water, VSMOW = 3.1152×10^{-4}, Craig, 1961):

$$\delta D = \frac{HD^{16}O/H_2^{16}O}{VSMOW} - 1. \quad (1)$$

Tropospheric water isotopologue data obtained by space- and ground-based remote sensing techniques are particularly useful, since they can be produced continuously and at global scale. Schneider et al. (2012) present tropospheric H_2O and δD remotely sensed from the ground at 10 globally distributed FTIR (Fourier transform infrared) stations of the NDACC (Network for the Detection of Atmospheric Composition Change, http://www.acd.ucar.edu/irwg/). Space-based remote sensing observations of tropospheric δD are possible by thermal nadir sensors (Worden et al., 2006; Schneider and Hase, 2011; Lacour et al., 2012) and sensors detecting reflected solar light in the near infrared (Frankenberg et al., 2009, 2013; Boesch et al., 2013).

While there have been efforts for theoretically assessing the quality of the remotely sensed δD products (e.g. Schneider et al., 2006b; Worden et al., 2006; Schneider and Hase, 2011; Lacour et al., 2012; Schneider et al., 2012; Boesch et al., 2013), so far no rigorous empirical quality assessment has been presented. The few existing studies consist of very few indirect comparisons (Worden et al., 2011) or use a δD reference that itself is not comprehensively validated (Schneider and Hase, 2011; Boesch et al., 2013). This situation is rather unsatisfactory: since δD remote sensing measurements are very difficult and the nature of the data is very complex, there is urgent need to support the theoretical quality assessment studies by elaborated empirical validation exercises.

In this context the objective of our paper is to empirically document the quality and prove the scientific value of tropospheric water vapour isotopologue remote sensing products for investigating moisture transport pathways. To do so we use tropospheric δD reference data obtained by continuously calibrated in situ instruments. The study is made for the Izaña observatory and the surroundings of the island of Tenerife, where tropospheric water isotopologues have been measured in coincidence from different platforms and by different techniques: (1) since 1999 by ground-based FTIR remote sensing within NDACC, (2) since 2007 by space-based remote sensing using IASI aboard METOP, (3) since 2012/2013 by commercial Picarro in situ instruments from ground at two different altitudes, and (4) in July 2013 during six aircraft profile measurements using the dedicated ISOWAT in situ instrument. All these data have been generated within the project MUSICA (www.imk-asf.kit.edu/english/musica). The Izaña observatory and the surroundings of Tenerife comprise the principal water vapour isotopologue reference area of MUSICA and allow a first empirical validation with dedicated aircraft campaign data as well as with long-term data sets. We are not aware of any other site where free tropospheric water vapour isotopologue observations have been made continuously and simultaneously by so many different observational techniques.

Section 2 briefly presents the in situ observations that we use as reference. In Sect. 3 we briefly describe the investigated ground- and space-based remote sensing data. In Sects. 4 and 5 the inter-comparisons are shown and discussed. Section 6 resumes the study.

2 The in situ reference observations

Compared to the remote sensing measurements, the in situ techniques have the great advantage that they can be continuously calibrated against an isotopologue standard. In the following we briefly describe the in situ instruments as operated in the Tenerife area.

2.1 Aircraft-based ISOWAT

The ISOWAT instrument is a tunable diode laser spectrometer specifically designed for the use aboard research aircraft (Dyroff et al., 2010). The first prototype has been successfully used within the project CARIBIC (Civil Aircraft for the Regular Investigation of the atmosphere Based on an Instrument Container, http://www.caribic-atmospheric.com/). For the validation campaign, a second prototype of ISOWAT was installed aboard a CASA C-212 aircraft of INTA (Instituto Nacional de Técnica Aeroéspacial, http://www.inta.es/).

ISOWAT measures δD in water vapour by means of laser-absorption spectroscopy. Ambient air is extracted from the atmosphere and is pumped through a multi-reflection measurement cell. The beam of a diode laser is guided into the cell where it is reflected back and forth between two mirrors to form an optical path of 32 m. The emission wavelength of the laser is continuously tuned across individually resolved absorption lines of the $H_2^{16}O$ and $HD^{16}O$ isotopologues at a frequency of 10 Hz. Absorption spectra are averaged over 1 s, which defines the maximum temporal resolution of the measurements.

During the campaign ISOWAT was calibrated before and after each flight performing measurements of a known isotope standard. The standard was provided by a custom-designed bubbler system that was used to humidify dry synthetic air to δD of about -130 ‰ at humidity ranging from about 200 to > 25000 ppmv. One of the key strengths of ISOWAT is its in-flight calibration. While substantial calibration measurements were performed on the ground, we perform in-flight calibrations in order to confirm the instrumental performance during the flight. The in-flight calibration source provided a known isotope standard at a humidity of 3000–5000 ppmv. Thereby, ambient air was first dried (< 5 ppmv) in a cartridge containing a molecular sieve. The

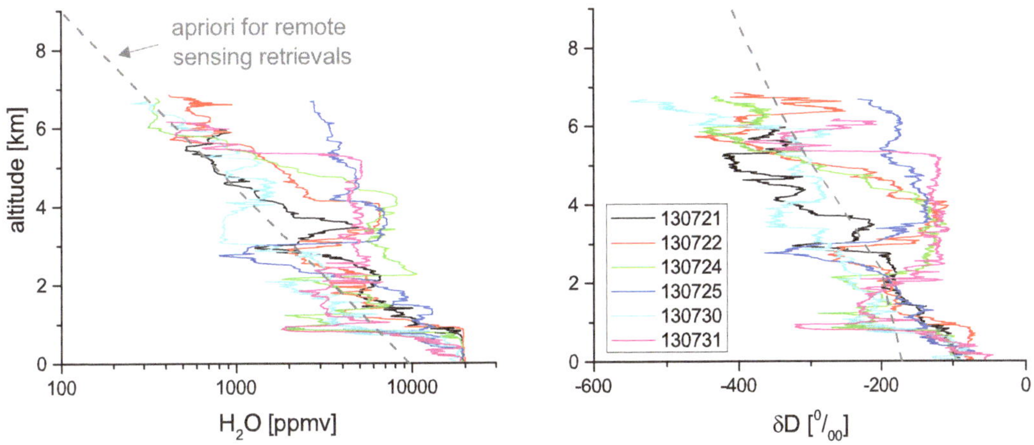

Figure 1. H_2O (left panel) and δD profiles (right panel) as measured by ISOWAT above the subtropical Atlantic close to Tenerife between 10:00 and 13:00 UT on 6 days at the end of July 2013. The depicted data are 10 m vertical averages. For comparison we also depict the a priori profiles used for the MUSICA NDACC/FTIR and METOP/IASI remote sensing retrievals (grey dashed lines).

dry air was then humidified in a small temperature-stabilized bubbler system. Calibrations were performed 2–5 times per flight at varying altitudes in order to check the performance of ISOWAT under all environmental conditions. By combining the ground-based and in-flight calibration measurements we were able to provide a robust accuracy estimate for the ISOWAT in-flight data. For the majority of conditions encountered, ISOWAT provided profile measurements with an accuracy for δD of about 10 ‰. Below 2000 ppmv the accuracy increases and can exceed 35 ‰ for a humidity below 500 ppmv.

The ISOWAT humidity data are calibrated against a commercial hygrometer (Edge Tech model Dew Master). We estimate a precision and accuracy of about 4 %.

ISOWAT measures the isotopologues with a temporal resolution of 1 s. This translates to a horizontal spatial resolution of about 80 m depending on the aircraft velocity. During the MUSICA campaign, profile flights consisted in a relatively fast ascent followed by a slower descent with intermediate level flights at certain selected altitudes and we achieved an average vertical resolution of about 1.2 m. The good temporal resolution of ISOWAT allows observing the rather sharp vertical gradients encountered during the flights (see Fig. 1).

For a more detailed report about the aircraft campaign, the ISOWAT measurements, and the calibration procedures please refer to Dyroff et al. (2015).

2.2 Surface-based Picarro

Similar to ISOWAT, the Picarro water isotopologue analysers work with a tunable diode laser. The laser light enters a cavity filled with the air mass to be analysed. The absorption signal of the air mass is determined by the ring-down time of the laser light intensity. In order to get an absorption spectrum the ring-down time is measured for different laser frequencies between 7183.5 and 7184 cm^{-1}. For more details see the Picarro website (http://www.picarro.com).

In March 2012 we installed a Picarro L2120-i water isotopologue analyser at the Izaña Observatory at 2370 m a.s.l. The measured H_2O and δD time series is depicted in the left graphs of Fig. 2. In June 2013 we installed an additional Picarro L2120-i analyser at a small observatory located close to the cable car summit station of the Teide mountain (at 3550 m a.s.l.). The Teide H_2O and δD time series is depicted in the right graphs of Fig. 2. We only depict data measured during the second half of each night (between midnight and 1 hour after sunrise). By this measure we avoid that the measurements are affected by local upslope flow, which can bring boundary layer air to the mountain observatories. The measurement gaps are due to instrumental problems when the instruments were sent to the manufacturer for repair. The time series document that the atmospheric water vapour fields are rather variable already on hourly timescales.

The δD calibration procedure for the Picarro instruments is based on a standard delivery module (SDM). It consists of two syringe pumps and allows injecting two different standards into the vaporizer where a constant dry air flow sustains immediate evaporation of the liquid in the air stream (for more details see also Aemisegger et al., 2012). The calibration is made every 8 h. It uses two different δD standards (at −142.2 and −245.3 ‰) that are analysed by the instrument at three different humidity levels covering the typical humidity range at Izaña and at Teide (during summer).

The precision of our 0.6 Hz measurements was 13.5 and 2 ‰ at 500 and 4500 ppmv, respectively. This noise almost completely averages out in the 10 min averages (our final data product). The random uncertainty of the individual calibrations (performed every 8 h) was 0.5 and 0.2 ‰ at 500 and 4500 ppmv, respectively, which also averages out on longer timescales. Systematic uncertainties of the used standard

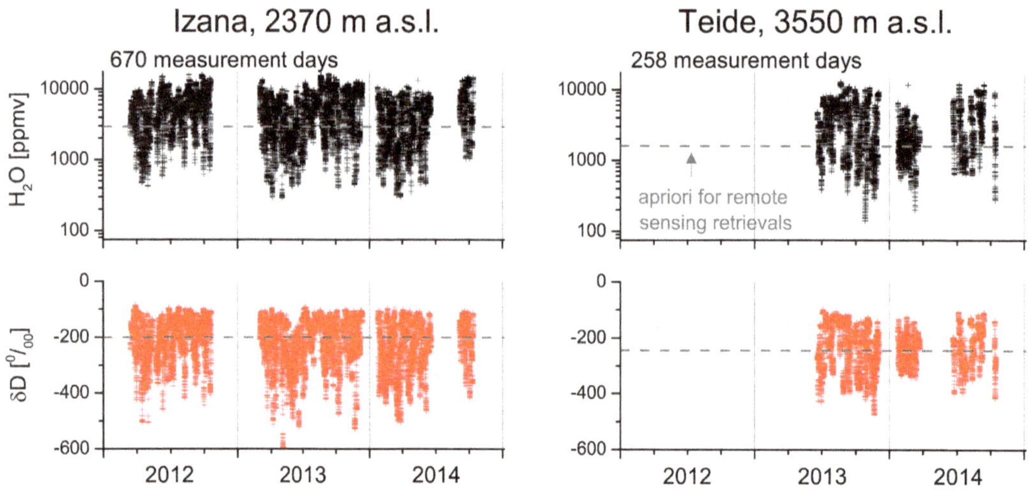

Figure 2. H_2O (upper panel) and δD data (bottom panel) as measured by the Picarro instruments on Tenerife between midnight and 1 hour after sunrise. The depicted data are 10 min averages obtained during a large amount of different days (numbers as given in the panel). For comparison we also depict the a priori values used for the MUSICA NDACC/FTIR and METOP/IASI remote sensing retrievals (grey dashed lines).

water and uncorrected humidity dependence were 1.7 and 0.7 ‰ at 500 and 4500 ppmv, respectively.

Our Picarro humidity data are calibrated with respect to humidity data measured by standard meteorological sensors. The humidity values measured by the Picarro and the meteorological sensors are very strongly correlated and for the calibrated data we estimate a precision and accuracy of better than 2 %.

A more detailed description of the MUSICA Picarro in situ measurements at Tenerife island is the subject of a dedicated paper, which is currently in preparation.

3 The investigated remote sensing products

The characteristics of the water vapour isotopologue remote sensing products are rather complex. Schneider et al. (2012) introduce a proxy state concept that allows for reasonably characterizing the interesting products, namely humidity (or H_2O) and δD (or HDO/H_2O). They identify two product types. Product type 1 is a vertical H_2O profile and product type 2 is for isotopologue studies. Product type 2 provides H_2O and δD data that are representative for the same air mass and assures that the δD product is optimally independent on H_2O. In the following we use this proxy state concept for briefly discussing the characteristics of the MUSICA remote sensing products that are available for the area of Tenerife island.

3.1 Ground-based NDACC/FTIR

Ground-based FTIR remote sensing measurements have been performed at Izaña since 1999 (e.g. García et al., 2012). The experiment is situated in its own measurement laboratory (a shipping container equipped with electricity, air-conditioning, internet, etc.) and just about 50 m away from the Observatory's main building. Typically we measure on about 2–3 days per week. The number of measurement days is mainly limited by manpower, which is needed for operating the instrument. Ground-based FTIR systems measure high-resolution mid-infrared solar absorption spectra. The spectra contain absorption signatures of many different atmospheric trace gases. Therefore, the technique can simultaneously monitor many different atmospheric trace gases. Continuous developments of sophisticated retrieval algorithms allow also the retrieval of tropospheric water vapour isotopologues (for retrieval details please refer to Schneider et al., 2006b, 2010b).

Using the measured spectra for estimating the atmospheric trace gas distributions is generally an ill-posed problem. For this purpose the solution state is constraint towards an a priori state. The H_2O and δD a priori values applied for the isotopologue retrievals are depicted in Fig. 1 as grey line. The MUSICA FTIR retrieval works with several spectral microwindows between 2655 and 3025 cm^{-1} and with HITRAN 2008 (Rothman et al., 2009) spectroscopic parameters, and the H_2O and HDO parameters have been adjusted for speed-dependent Voigt line shapes (Schneider et al., 2011), which affects pressure broadening parameters but also line intensities. For H_2O the adjustment has been made taking coincident H_2O profile observations as the reference. For HDO the adjustment is more uncertain, since the atmospheric HDO state was calculated from the measured H_2O state assuming standard δD profiles (there were no HDO measurements available).

The averaging kernel is an important output of the retrieval. It describes the response of the retrieval on actual

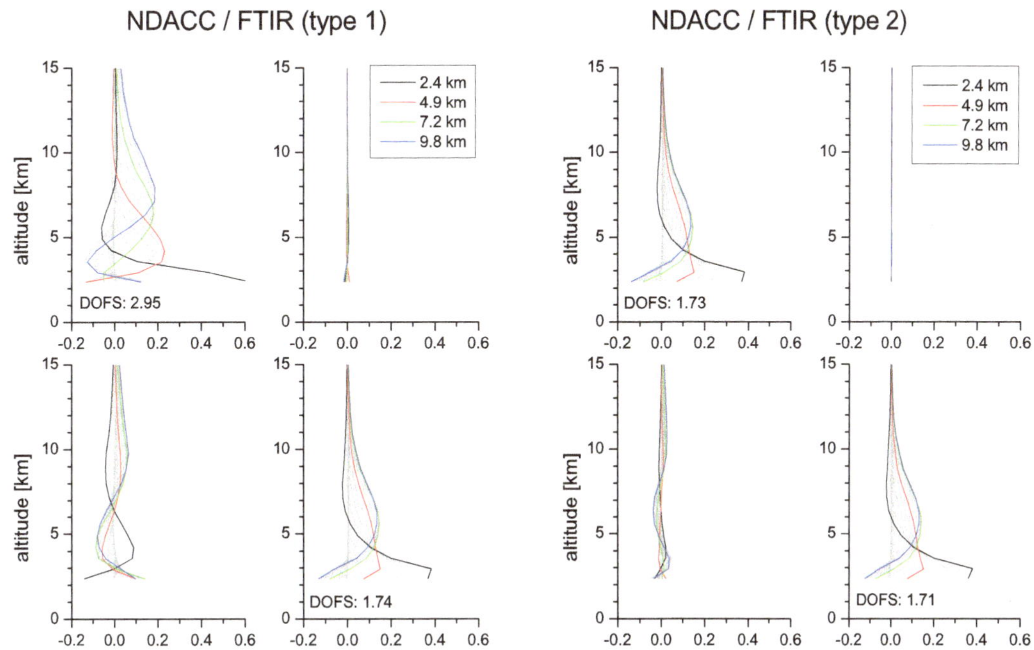

Figure 3. Typical row kernels for the {H_2O, δD}-proxy states as retrieved from Izaña's ground-based NDACC/FTIR spectra. Left panel of graphs: product type 1 kernels; Right panel of graphs: product type 2 kernels. The upper graphs of each panel display how the retrieved H_2O is affected by actual H_2O variations (left graph) and by actual δD variations (right graph). The lower graphs of each panel display the same for the retrieved δD. For more details on the proxy state kernels see Schneider et al. (2012).

atmospheric variability. The left panel of graphs in Fig. 3 shows typical averaging kernels for the product type 1 {H_2O, δD}-state. The panel consists of four graphs. The upper left graph demonstrates how actual atmospheric H_2O affects retrieved H_2O, the upper right graph how actual atmospheric δD affects retrieved H_2O, the bottom right graph how actual atmospheric H_2O affects retrieved δD, and the bottom left graph how actual atmospheric δD affects retrieved δD. Product type 1 offers H_2O profiles (FWHM, full width at half maximum, of kernels is 2 km in the lower troposphere, 4 km in the middle troposphere, and 6–8 km in the upper troposphere, Schneider et al., 2012), which have been empirically validated in a variety of previous studies (e.g. Schneider et al., 2006a; Schneider and Hase, 2009; Schneider et al., 2010a). A further validation of product type 1 is not the subject of this paper.

Product type 2 is for isotopologue studies and shall be investigated in this paper. The type 2 product is calculated with the a posteriori correction method as presented in Schneider et al. (2012). The respective {H_2O, δD}-state averaging kernels are depicted as the graphs in the right panel of Fig. 3. The a posteriori correction significantly reduces the vertical resolution of the retrieved H_2O state (compare upper left graphs between type 1 and type 2 panels). This sensitivity reduction is mandatory in order to assure that the H_2O product is representative for the same air mass as the δD product (compare upper left and bottom right graphs in the type 2 panel). In addition, the a posteriori correction reduces the dependency of the retrieved δD on humidity, i.e. the cross-dependency on humidity (compare the bottom left graphs between the type 1 and type 2 panels).

The Izaña FTIR product type 2 offers a degree of freedom for the signal (DOFS, trace of the averaging kernel matrix quantifying the retrieval's independency on the a priori assumptions) for H_2O and δD of typically 1.75. The lower tropospheric kernels have an FWHM (full width at half maximum) of about 2 km and the middle tropospheric kernels a FWHM of about 5–6 km. In Schneider et al. (2012) the lower and middle tropospheric random uncertainty has been estimated to be smaller than 2 % for H_2O and 25 ‰ for δD, respectively, whereby measurement noise, spectral baseline uncertainties, and assumed tropospheric a priori temperature are the leading error sources. The systematic uncertainty can reach 10 % for H_2O and 150 ‰ for δD. It is dominated by uncertainties in the spectroscopic parameters (intensity and pressure broadening). The a posteriori correction method reduces the δD cross-dependency on humidity to less than 10–15 ‰.

3.2 Space-based METOP/IASI

IASI is a Fourier transform spectrometer flown aboard the METOP satellites. It combines high temporal, horizontal, and spectral resolution (covers the whole globe twice a day, measures nadir pixels with a diameter of only 12 km, and records thermal radiation between 645 and 2760 cm^{-1} with a

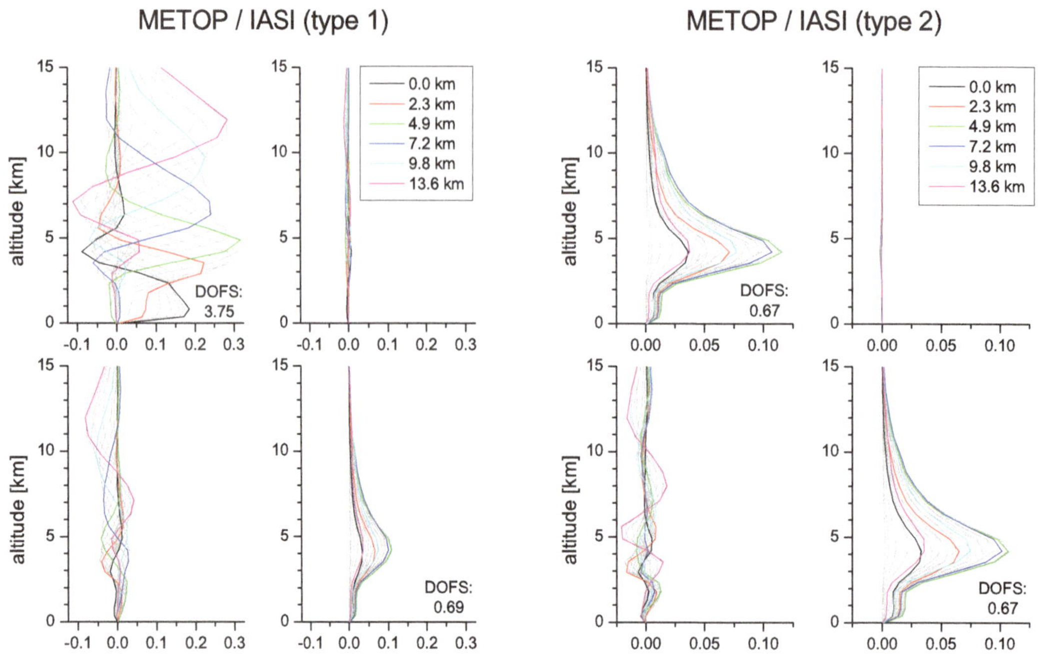

Figure 4. As Fig. 3, but for the $\{H_2O, \delta D\}$-proxy states as retrieved from METOP/IASI spectra measured over the subtropical ocean. Please note the different scales in the x axis for the type 1 and type 2 kernels. For more details on the METOP/IASI proxy state kernels see Schneider et al. (2012) and Wiegele et al. (2014).

resolution of 0.5 cm^{-1}). IASI measurements are assured until 2020 on three different METOP satellites, whereby the first two are already in orbit: MetOp-A has been launched in October 2006, MetOp-B in September 2012. For our study we consider spectra measured by the IASI instruments aboard both satellites. Each IASI instrument has two Tenerife overpasses a day (about 10:30 and 22:30 UT). For MUSICA we work so far only with overpasses that observe clear sky scenes. We use very strict cloud-screening conditions and therefore work with only about 10 % of all possible overpass spectra.

The retrieval uses the spectral window between 1190 and 1400 cm^{-1} and HITRAN 2008 (Rothman et al., 2009) spectroscopic line parameters. The applied a priori values are the same as for the FTIR retrievals and are depicted in Fig. 1 as grey line. Schneider and Hase (2011) present our IASI retrieval algorithm as well as a first theoretical error assessment of the respective H_2O and δD products.

The left panel of graphs in Fig. 4 shows typical averaging kernels for the product type 1 $\{H_2O, \delta D\}$-state of MUSICA IASI retrievals corresponding to spectra measured over the subtropical northern Atlantic. Product type 1 offers vertical H_2O profiles with the FWHM of the kernels of about 2 km in the lower and middle troposphere and 3 km in the upper troposphere (see upper left graph in type 1 panel of Fig. 4 and Schneider and Hase, 2011; Wiegele et al., 2014, for more a more detailed characterization). The MUSICA product type 1 IASI H_2O profiles are empirically validated in Schneider and Hase (2011) by comparison to meteorological radiosondes, ground-based FTIR, and the EUMETSAT METOP/IASI product and in Wiegele et al. (2014) by comparison to ground-based FTIR at three different locations (polar, mid-latitudinal and subtropical) and a further validation of this type 1 product is not the subject of this paper.

Interesting for isotopologue studies is product type 2 which shall be further investigated in this paper. The respective $\{H_2O, \delta D\}$-state kernels are depicted in the right panel of graphs in Fig. 4. The sensitivity is typically limited to between 2 and 8 km altitude, with a maximum in the middle troposphere around 4–5 km. The random uncertainty for type 2 H_2O and δD is estimated to be about 5 % and 20 ‰, respectively, whereby measurement noise, surface emissivity, assumed tropospheric a priori temperature, and unrecognized thin elevated clouds are the leading error sources. The systematic uncertainty is very likely dominated by errors in the spectroscopic line parameters and can reach 2–5 % for H_2O in the case of a 2–5 % error in the line intensity parameter. For δD the systematic uncertainty can reach 50 ‰ in the case of inconsistencies of 5 % between the spectroscopic intensity parameters of H_2O and HDO. Despite the a posteriori correction there remains a δD cross-dependency on humidity, which can cause an additional δD uncertainty of 30–50 ‰ (Wiegele et al., 2014).

4 Assessing random uncertainty and bias

Remote sensing data do not represent a single altitude. Instead they reflect the atmospheric situation as averaged out over a range of different altitudes (the kind of averaging is documented by the averaging kernels of Figs. 3 and 4). Such data can only be quantitatively compared to reference data that are available as vertical profiles. Here we compare the NDACC/FTIR and METOP/IASI products (product type 2, see Schneider et al., 2012) to the in situ profiles as measured by ISOWAT between 200 and 6800 m a.s.l.

For a quantitative comparison we have to adjust the vertically highly resolved ISOWAT profiles (x_{IW} as depicted in Fig. 1) to the modest vertical resolution of the remote sensing profiles. For this purpose we first perform a smoothing towards the altitude grid of the remote sensing retrieval's forward model. It can be described by $\mathbf{M}x_{\mathrm{IW}}$, whereby $\mathbf{M}$ is a $v \times w$ matrix with v and w being the numbers of the grid points of the forward model and of the vertically highly resolved ISOWAT profile, respectively. This smoothing works in a way that assures a conservation of the partial columns between two forward model grid points (i.e. the respective partial columns are very similar, before and after the smoothing process). In a second step we convolve $\mathbf{M}x_{\mathrm{IW}}$ with the remote sensing averaging kernel $\hat{\mathbf{A}}$:

$$\hat{x}_{\mathrm{IW}} = \hat{\mathbf{A}}\,(\mathbf{M}x_{\mathrm{IW}} - x_{\mathrm{a}}) + x_{\mathrm{a}}. \tag{2}$$

Here x_{a} is the a priori profile used by the remote sensing retrieval. When using the {H_2O, δD}-proxy state averaging kernels as depicted in Figs. 3 and 4 we have to make sure that $\mathbf{M}x_{\mathrm{IW}}$ in Eq. (2) is in the {H_2O, δD}-proxy state. The result of this convolution is an ISOWAT profile ($\hat{x}_{\mathrm{IW}}$) with the same vertical resolution as the remote sensing profile. In the following we will use the "hat"-index for marking parameters or amounts that are retrieved from measured spectra or obtained by convolution calculations according to Eq. (2).

The ISOWAT profiles are limited to 200–6800 m altitude, whereas the FTIR and IASI remote sensing instruments are also sensitive to altitudes above 7000 m (and IASI is also very weakly sensitive for altitudes below 200 m). For the comparison study we extend the ISOWAT profiles above the aircraft's ceiling altitude by the values we use as a priori for the remote sensing retrievals (the a priori values are the same for FTIR and IASI and depicted as grey dashed line in Fig. 1). Below 200 m altitude we use the values as measured during the first 100 m of available aircraft data.

4.1 NDACC/FTIR

When comparing two atmospheric measurements we have to take care that the different instruments detect the same (or at least very similar) air masses. Large differences in the vertical resolution can be accounted for by Eq. (2). In addition we have to be aware that ISOWAT measures the profile aboard a descending and ascending aircraft. Each flight, i.e. each ISOWAT profile measurement, takes about 3 hours and is made over the ocean (see flight tracks of Fig. 5). This is in contrast to the FTIR remote sensing measurements, which need only about 10 min and are made at Izaña observatory at 2370 m a.s.l. on a mountain ridge of an island. It is well documented (e.g. Rodríguez et al., 2009) that the immediate surroundings of the island at this altitude are affected by mixing from the marine boundary layer (MBL). This mixing is mostly thermally driven: it is weak during the night and starts increasing during the morning hours and the air mass measured at or very close to the island becomes more and more affected by the MBL. At the end of the morning this air mass is not well representative for the situation encountered at the same altitude over the ocean.

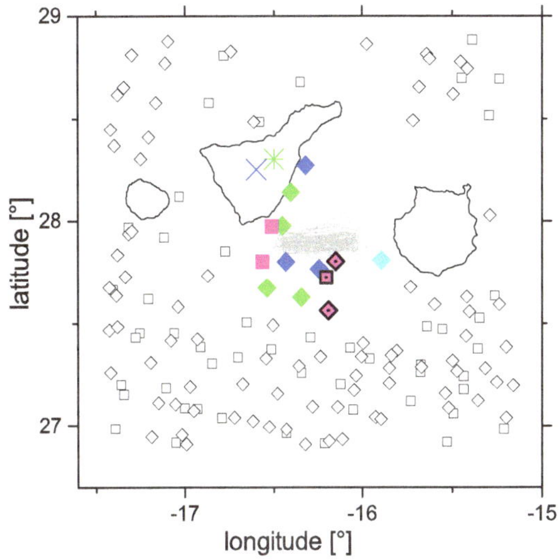

Figure 5. Site map indicating the location of the different instruments and ground pixels during the aircraft campaign on 6 days in July 2013. Green star: Izaña observatory (location of the first Picarro and the FTIR); blue star: Teide observatory (location of the second Picarro); grey lines: aircraft flight track during ISOWAT measurements; black squares and diamonds: cloud-free ground pixels of IASI-A and -B, respectively, during the six aircraft flights; coloured filled squares and diamonds: pixels that fulfil our coincidence criteria for IASI-A and -B, respectively (the colours mark the different days in analogy to Fig. 1).

For this study we compare the ISOWAT profile with the FTIR data obtained from the spectra measured in the morning with a solar elevation angle of 30°. This elevation assures that the FTIR has good sensitivity (absorption pathes are short enough and water vapour lines are not too heavily saturated) and it occurs only about 2:30 h after sunrise, when the influence of the MBL on the immediate surroundings of the island is still moderate.

In Fig. 6 we compare the H_2O measurements. The data are depicted as difference to the a priori, i.e. we discuss here the variability that is actually introduced by the remote sensing measurement and observed by ISOWAT. Here and in the

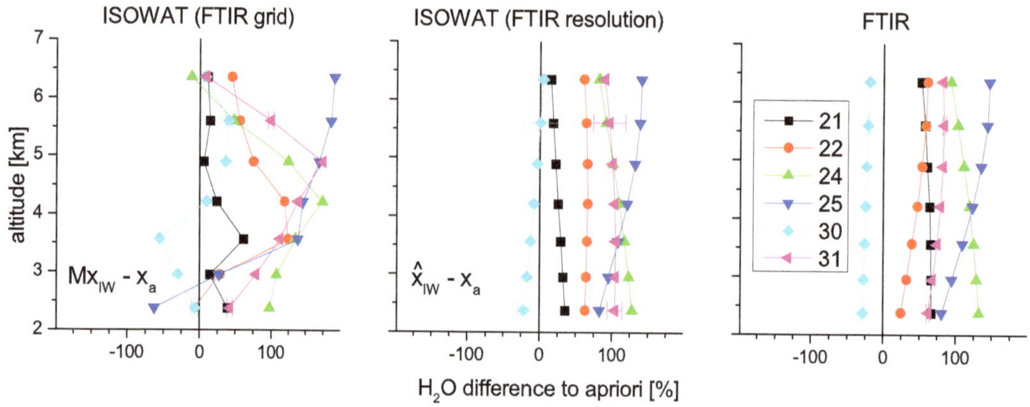

Figure 6. Comparison of ISOWAT and FTIR H_2O measurements (one coincidence for each of the days 21, 22, 24, 25, 30, and 31 of July 2013). Shown are differences with respect to the a priori profiles. Left panel: ISOWAT data represented for the FTIR altitude grid; central panel: ISOWAT data convolved with the FTIR kernels, i.e. representative for the vertical resolution of the FTIR; right panel: FTIR data measured at the end of the morning when solar elevation is still $< 30°$. Depicted are also lower and middle tropospheric error bars for day 130730 (dry conditions) and 130731 (humid conditions), which are hardly discernible, since the H_2O errors are only a few percent.

following we calculate the H_2O differences as differences between the logarithms of the H_2O concentrations. Since $\Delta \ln X \approx \frac{\Delta X}{X}$ we can interpret these differences between the logarithms of the H_2O concentrations as relative differences. The graphs also show the predicted uncertainties for the lower and middle troposphere and for dry (day 130730) and humid conditions (day 130731). The respective error bars are hardly discernible, since the H_2O errors are only a few percent. Please note that the uncertainties as shown in the central graph take into account the assumption of climatologic a priori values for altitudes above the aircraft's ceiling altitude (no measurements!), which is the reason for the relative large uncertainty as estimated for the middle troposphere for the ISOWAT data smoothed by the FTIR kernels.

The comparison between the left and the central panels of Fig. 6 gives insight into the vertical structures that can be resolved by the remote sensing measurements. The profiles depicted in the central and the right panels represent the same vertical structures and their comparison allows conclusions about the quality of the remote sensing data.

For this purpose we statistically analyse the differences between the remote sensing product ($\hat{x}_{RS}$) and the convolved ISOWAT product ($\hat{x}_{IW}$, see Eq. 2). We determine the mean difference $\hat{b}_{DIFF}$ from the n numbers of coincidences:

$$\hat{b}_{DIFF} = \frac{1}{n}\sum_{i=1}^{n}\left(\hat{x}_{RS,i} - \hat{x}_{IW,i}\right) \tag{3}$$

and the standard deviation of the differences (the scatter between both data sets):

$$\hat{\sigma}_{DIFF} = \sqrt{\frac{1}{n}\sum_{i=1}^{n}\left(\hat{x}_{RS,i} - \hat{x}_{IW,i} - \hat{b}_{DIFF}\right)^2}. \tag{4}$$

Furthermore, we calculate the typical/mean scatter as predicted from the combined uncertainties of the remote sensing and the convolved ISOWAT data, $\hat{\epsilon}_{RS}$ and $\hat{\epsilon}_{IW}$, respectively:

$$\hat{\epsilon}_{DIFF} = \frac{1}{n}\sum_{i=1}^{n}\sqrt{\hat{\epsilon}^2_{RS,i} + \hat{\epsilon}^2_{IW,i}}. \tag{5}$$

For comparison we also calculate a parameter, which is the scatter in the convolved ISOWAT data:

$$\hat{\sigma}_{IW} = \sqrt{\frac{1}{n}\sum_{i=1}^{n}\left(\hat{x}_{IW,i} - \hat{\mu}_{IW}\right)^2} \tag{6}$$

whereby $\hat{\mu}_{IW} = \frac{1}{n}\sum_{i=1}^{n}\hat{x}_{IW,i}$. The $\hat{\sigma}_{IW}$ parameter informs about the random uncertainty of the water vapour state when no measurements are available (a priori uncertainty).

The values that result from these calculations for the comparison of FTIR and ISOWAT H_2O data are depicted in Fig. 7. In the left graph, the scatter between FTIR and ISOWAT (parameter $\hat{\sigma}_{DIFF}$ according to Eq. 4) is shown as black triangles, together with the predicted scatter ($\hat{\epsilon}_{DIFF}$ according to Eq. (5), red dashed line). We observe a scatter of about 28 and 20 % for the lower and middle troposphere, respectively. In the lower troposphere the observed scatter is significantly larger than the scatter predicted from the combined uncertainties, which might be caused by the fact that the two instruments observe similar, but not the same air masses. The large inhomogeneity in the tropospheric water vapour fields becomes clearly visible in Fig. 1 (spatial inhomogeneity) and Fig. 2 (temporal inhomogeneity). In this context the observed scatter has to be interpreted as a rather conservative empirical estimation of the FTIR's random uncertainty. The blue solid line depicts the scatter between the six ISOWAT profiles when smoothed with the FTIR kernels

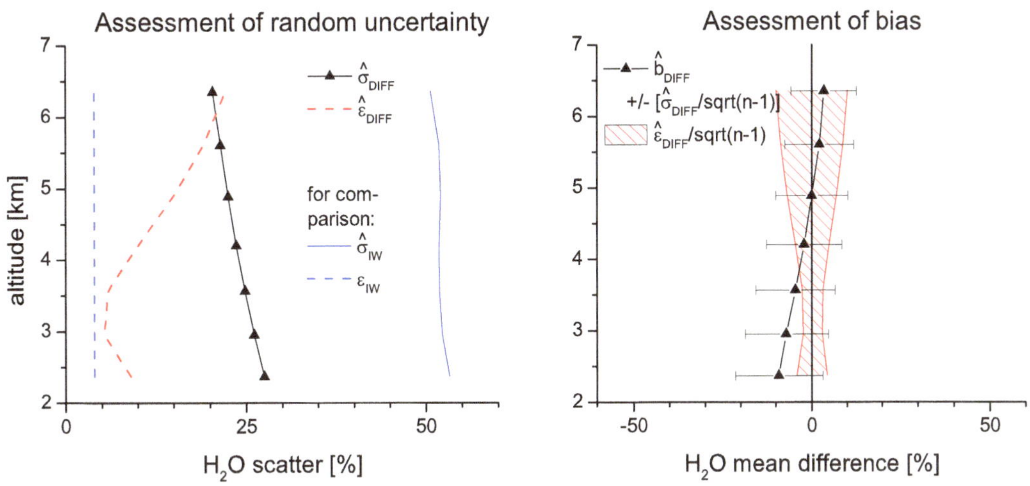

Figure 7. Left graph, assessment of the random uncertainty. Black symbols: the FTIR-ISOWAT scatter ($\hat{\sigma}_{\mathrm{DIFF}}$ value according to Eq. 4). Red dashed line: the predicted scatter ($\hat{\epsilon}_{\mathrm{DIFF}}$ value according to Eq. 5). Blue solid line: the scatter in the convolved ISOWAT data ($\hat{\sigma}_{\mathrm{IW}}$ value according to Eq. 6). Blue dashed line: the uncertainty in the not convolved ISOWAT data (ϵ_{IW}). Right graph, assessment of the bias. Black symbols and error bars: the FTIR-ISOWAT mean difference and standard error of the mean ($\hat{b}_{\mathrm{DIFF}}$ and $\hat{\sigma}_{\mathrm{DIFF}}/\sqrt{n-1}$ values according to Eqs. (3) and (4), respectively). Red area: the predicted uncertainty range (calculated as $\hat{\epsilon}_{\mathrm{DIFF}}/\sqrt{n-1}$).

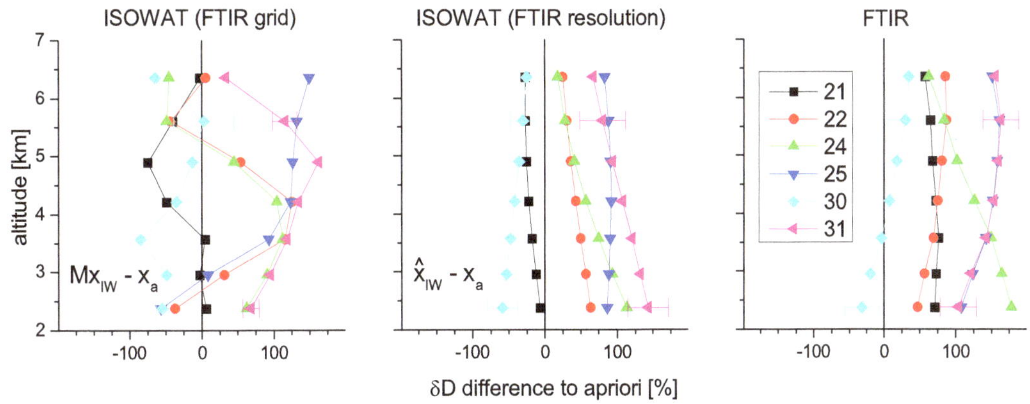

Figure 8. As Fig. 6, but for δD. Here the error bars are clearly discernible.

(the $\hat{\sigma}_{\mathrm{IW}}$ values according to Eq. 6). It is much larger than the scatter in the difference with the FTIR profiles meaning that the FTIR well follows the ISOWAT reference, which confirms previous empirical validation exercises made for H_2O (e.g. Schneider et al., 2006a; Schneider and Hase, 2009; Schneider et al., 2010a). Please note that the depicted red dashed line represents the combined uncertainty of the convolved ISOWAT data and the FTIR data. The pure ISOWAT uncertainty (uncertainty of ISOWAT if not convolved with the FTIR kernel, ϵ_{IW}) is much smaller and depicted as blue dashed line.

The mean difference between FTIR and ISOWAT and the respective standard error of the mean ($\hat{b}_{\mathrm{DIFF}}$ and $\hat{\sigma}_{\mathrm{DIFF}}/\sqrt{n-1}$, according to Eqs. (3) and (4), respectively) are shown in the right graph of Fig. 7 as black triangles and error bars. The red area depicted around the zero line indicates the zone where a bias might be observed accidently due to the combined random uncertainties of ISOWAT and FTIR (this area is calculated as $\hat{\epsilon}_{\mathrm{DIFF}}/\sqrt{n-1}$, see Eq. 5). The observed differences overlap well with this area meaning that we cannot observe a significant bias between the FTIR and the ISOWAT data.

The coincident δD profiles are shown in Fig. 8. Here the predicted uncertainties are clearly discernible. The results of the statistical analysis of the δD comparison are presented in Fig. 9. In the left graph we observe a scatter ($\hat{\sigma}_{\mathrm{DIFF}}$) of about 45 and 15 ‰ for the lower and middle troposphere, respectively, which can be interpreted as a conservative empirical estimation of the FTIR's random uncertainty. In the middle troposphere this observed scatter is clearly smaller than the predicted one ($\hat{\epsilon}_{\mathrm{DIFF}}$) indicating that our theoretical ISOWAT and/or FTIR error estimations might be too conservative. In particular the ISOWAT estimates are rather conservative since we assumed that on a single day all δD errors

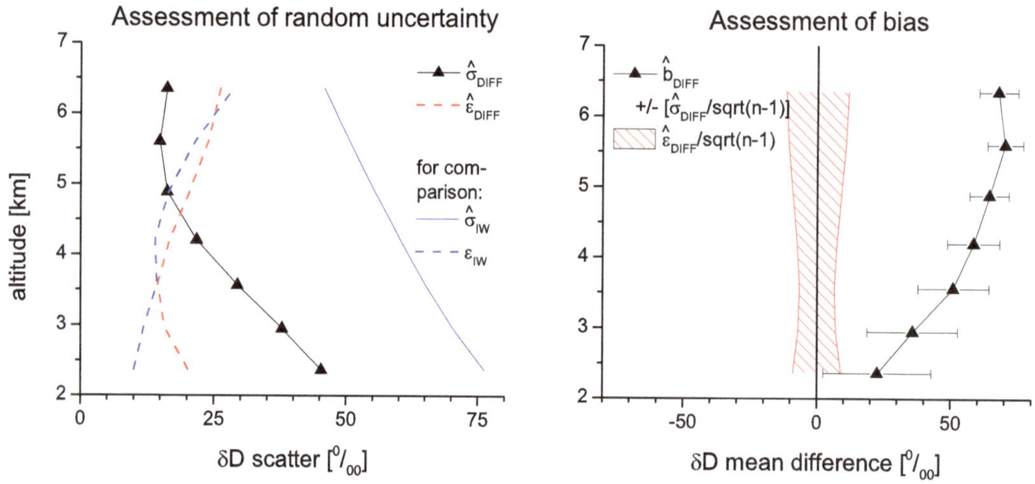

Figure 9. As Fig. 7, but for δD.

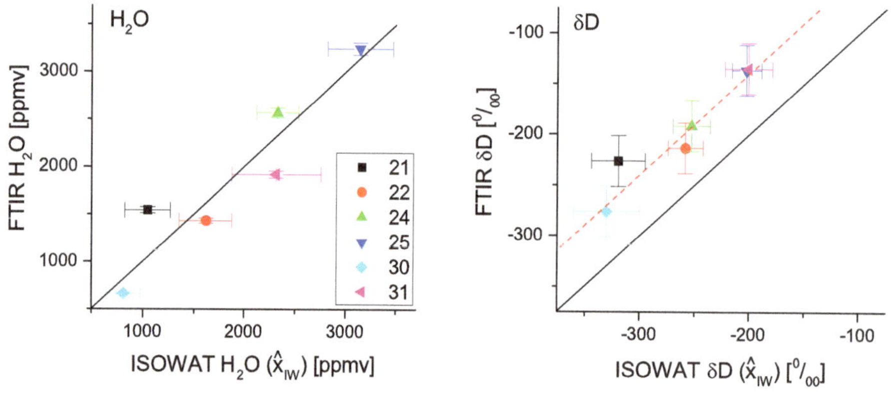

Figure 10. Correlation between ISOWAT data (smoothed with FTIR kernels) and FTIR data for 5 km altitude (left for H_2O and right for δD). Black line is the 1 : 1 diagonal and red dotted line is the 1:1 diagonal shifted by +60 ‰. Error bars represent the ISOWAT and FTIR uncertainty estimations. Note that a large part of the uncertainty in the smoothed ISOWAT data is due to the fact that there are no ISOWAT measurements above the aircraft's ceiling altitude.

that occur at different altitudes and during ascent and descent are fully correlated. Allowing for δD errors that are not fully correlated (for instance positive errors at 2 km altitude and negative errors at 5 km altitude) would strongly decrease the error as estimated for the ISOWAT data, since through the convolution with the kernel a large part of the error would cancel out. The comparison of the observed scatter with the blue solid line (scatter between the six convolved ISOWAT profiles, $\hat{\sigma}_{IW}$) demonstrates that the FTIR δD product well captures the atmospheric variations as seen by the ISOWAT reference data.

The right graph of Fig. 9 reveals a clear systematic difference between the FTIR and the ISOWAT δD data.

At and above 3 km altitude this difference lies outside the zone where a bias might be observed accidentally due to the combined random uncertainties of ISOWAT and FTIR (red area calculated according to Eq. 5), meaning that we are able to detect a significant bias. It is about 25 ‰ in the lower troposphere and 70 ‰ in the middle troposphere.

In Fig. 10 we show a correlation plot for the H_2O and δD data as observed in the middle troposphere (altitude of 5 km) by ISOWAT and FTIR. This plot can serve as a summary of the comparison exercise: (1) H_2O and δD signals are similarly observed by both instruments (good correlation), (2) there is no significant difference between the H_2O data, and (3) there is a significant positive bias in the FTIR δD data.

4.2 METOP/IASI

The black squares and diamonds in Fig. 5 show all the IASI-A and IASI-B pixels for morning overpasses that have been declared cloud free (according to EUMETSAT level 2 data) within a 250 km × 250 km area around the aircraft's flight tracks. In order to assure that IASI and ISOWAT observe

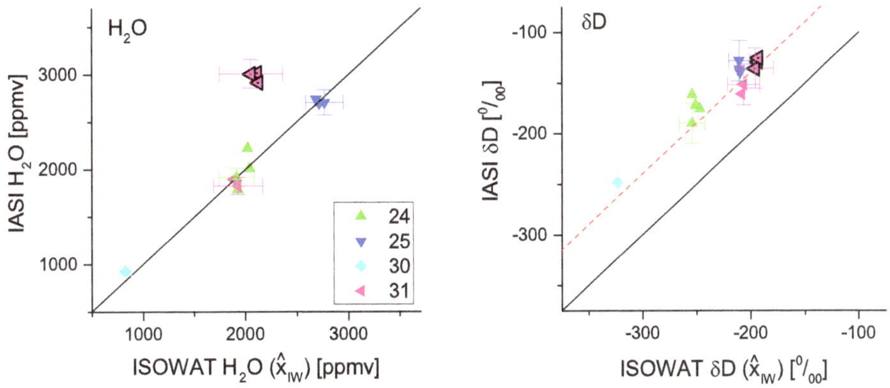

Figure 11. As Fig. 10, but for correlation between ISOWAT data (here smoothed with IASI kernels, i.e. this ISOWAT data are different to the ISOWAT data as depicted in Fig. 10) and IASI data for 5 km altitude. Three H_2O outliers on day 130731 (blue triangles) are marked by black edges. For clarity the error bars are only given for one observation per day and on day 130731 in addition for one of the three outliers.

similar air masses we only work with IASI observations that are made not further away than 50 km from the typical location of the aircraft. This yields 13 coincidences (three for IASI-A and ten for IASI-B) made on four different days: 130724, 130725, 130730, and 130731. The respective pixels are marked by the filled squares and diamonds in Fig. 5 (the different colours are for the different days).

IASI can hardly measure the vertical distribution of δD (see right panel of Fig. 4). For this reason we can limit the validation exercise to a single altitude level where IASI has typically good sensitivity. This is the case for the altitude of 5 km, which typically represents the atmospheric situation between 2 and 8 km altitude.

Figure 11 shows the same as Fig. 10 but for the IASI instead of the FTIR data. It depicts the correlation between the ISOWAT data (convolved with IASI kernels) and the IASI data.

For H_2O the data group nicely around the 1 : 1 diagonal. For three of the five IASI observations on day 130731 the agreement is unusually poor. These data points are marked by black edges and they correspond to the one IASI-A pixel and the two IASI-B pixels marked by the black edges in Fig. 5. The reason for their relatively poor agreement is that for these coincidences the two IASI and ISOWAT instruments observe different air masses. For more details see Appendix A. For δD we observe a good correlation but there is a significant systematic difference. All the IASI data show rather consistently about 60 ‰ less HDO depletion than ISOWAT and this bias is outside the predicted uncertainty range, i.e. it is significant. In this context it is interesting that for the FTIR data we observe a similar systematic difference, which confirms the work of Schneider and Hase (2011) and Wiegele et al. (2014), where no systematic differences between IASI and FTIR are reported.

The good correlations as seen in Fig. 11 demonstrate that IASI can well capture the variations of the middle tropospheric water vapour isotopologues. In order to document this quantitatively, we compare the scatter observed in the difference between IASI and ISOWAT with the scatter observed in the ISOWAT profiles as well as with the scatter predicted from our ISOWAT and IASI uncertainty estimations. These scatter values are collected in Table 1. First, we see that the scatter in the IASI – ISOWAT difference (the value $\hat{\sigma}_{DIFF}$ according to Eq. 4) is much smaller than the scatter in the convolved ISOWAT data (the value $\hat{\sigma}_{IW}$ according to Eq. 6), i.e. IASI well tracks the H_2O and δD variations as observed by ISOWAT. From the $\hat{\sigma}_{DIFF}$ values we empirically assess an IASI random uncertainty of better than 6–16 % for H_2O and about 13 ‰ for δD. For the latter the value is not larger than the combined estimated errors of the convolved ISOWAT and the IASI δD data ($\hat{\epsilon}_{DIFF}$ according to Eq. 5), which is also about 13 ‰. Since ISOWAT and IASI actually do not detect the same air mass, part of the observed scatter should be due to the observation of a different air mass. Consequently our combined uncertainty estimations are very likely too conservative, which might be explained by the aforementioned conservative ISOWAT uncertainty assumption (assumption of fully correlated errors).

Table 1. Scatter values as obtained from IASI and ISOWAT coincidences that fulfil our coincidence criteria (presents the same as the left panels of Figs. 7 and 9, but for IASI instead of FTIR and at 5 km altitude only). Predicted: combined ISOWAT and IASI estimated uncertainties ($\hat{\epsilon}_{DIFF}$); IASI – ISOWAT: scatter as observed in the difference between ISOWAT and IASI ($\hat{\sigma}_{DIFF}$); ISOWAT: scatter as observed in the ISOWAT data ($\hat{\sigma}_{IW}$). The values in parenthesis are for calculations without considering the three outliers on 130731.

	H_2O [%]	δD [‰]
Predicted ($\hat{\epsilon}_{DIFF}$)	9.9 (8.5)	13.2 (12.3)
IASI – ISOWAT ($\hat{\sigma}_{DIFF}$)	16.0 (6.0)	12.0 (13.2)
ISOWAT ($\hat{\sigma}_{IW}$)	29.9 (34.4)	37.1 (36.7)

4.3 Bias in the remote sensing data

When comparing the remote sensing and ISOWAT H2O data we find a systematic difference of up to 10 %. However, the mean difference is clearly smaller than the respective standard error of the mean, making this empirical bias estimation statistically insignificant (see black triangles and error bars in the right panel of Fig. 7). Since the lower/middle tropospheric H_2O fields are temporally and spatially very variable (variability in the range of 50–150 %), it is difficult to identify a bias in the H_2O remote sensing product using a limited number of nearby in situ reference observations.

Tropospheric δD variability is in the range of 50–150 ‰ and a remote sensing δD bias of a similar magnitude might be detected even by using a small number of coinciding reference measurements. In fact the bias we observe between the few ISOWAT reference data and our remote sensing δD product is statistically significant. It is between 10 and 70 ‰ for the FTIR (see right panels in Figs. 9 and 10) and about 60 ‰ for the IASI data (see right panel of Fig. 11). Wiegele et al. (2014) found consistency between a large amount of IASI and FTIR data obtained at three rather different locations (subtropics, mid-latitudes and Arctic), which suggests that the biases are similar and stable for a wide range of different locations. However, in order to investigate the consistency of the bias for different locations more quantitatively further ISOWAT profile measurements at different sites and during different seasons would be desirable.

5 Proofing the added value of δD

5.1 The added value of δD observations

In Fig. 12 we plot all the in situ data that have been measured so far in the lower or middle troposphere within the project MUSICA. These data represent different altitudes and two different locations (subtropics and mid-latitudes). We observe a strong correlation between the logarithm of H_2O concentrations and δD (for linear correlation we get an R^2 of about 80 %), which suggests that δD and H_2O contain similar information thereby relativizing the results as presented in Sect. 4: in Fig. 9 and Table 1 we show that the remote sensing measurements can correctly detect a large part of the actual δD variations. However, we do not show if the remote sensing δD measurements are precise enough for detecting the small part of the lower/middle tropospheric δD variations that are complementary to H_2O.

In this context, we have to be aware that the remote sensing retrievals provide δD data that are slightly dependent on the atmospheric H_2O state (see the bottom left graphs in the type 1 and 2 panels of Figs. 3 and 4), meaning that the remote sensing δD data might be correlated to the in situ δD data only via its dependency on H_2O and not because δD is correctly retrieved. The fact that in Sect. 4 we convolve

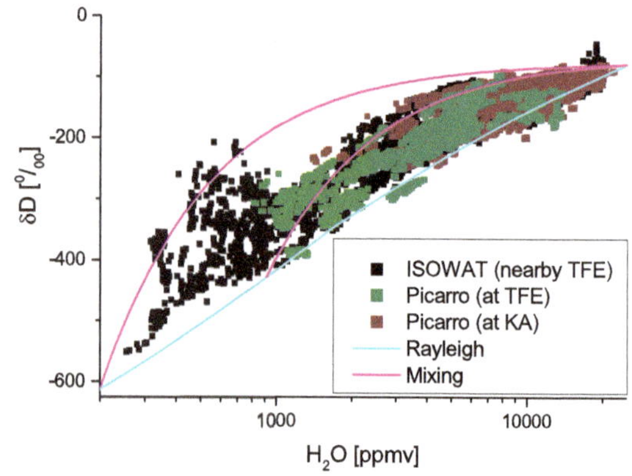

Figure 12. δD–H_2O plot using in situ observations made from different platforms (aircraft/ISOWAT and surface/Picarro) and at different sites (mid-latitudes at Karlsruhe, KA, and subtropics at Tenerife, TFE). Cyan line: Rayleigh curve for initialization with $T = 25\,°C$ and $RH = 80\,\%$. Magenta lines: two mixing lines. First for mixing between $H_2O[1] = 25\,000$ ppmv; $\delta D[1] = -80$ ‰ and $H_2O[2] = 900$ ppmv; $\delta D[2] = -430$ ‰. Second for mixing between $H_2O[1] = 25\,000$ ppmv; $\delta D[1] = -80$‰ and $H_2O[2] = 200$ ppmv; $\delta D[2] = -610$ ‰.

the in situ reference data with the remote sensing averaging kernels further increases this risk of misinterpreting the δD remote sensing data quality: by convolving the in situ data with the averaging kernels we create a δD in situ reference that has a similar dependency on H_2O as the remote sensing data and the good correlation between the remote sensing and the convolved in situ δD data might partly be a result of the convolution calculations according to Eq. (2).

For these reasons it is rather important to complement the quantitative validation exercises as shown in Sect. 4 by a qualitative study demonstrating that the δD remote sensing measurements add new information to the H_2O remote sensing data. This added value of δD can be examined by analysing δD–H_2O diagrams (we face a two-dimensional validation problem!). We have to demonstrate that our combined $\delta D/H_2O$ remote sensing products are really useful for studying atmospheric moisture transport processes.

In this section we work with pure in situ data (not convolved with the remote sensing averaging kernels). This does not allow a quantitative study, but it ensures that our in situ reference remains completely independent from the remote sensing data.

5.2 Dehydration via condensation versus vertical mixing from the boundary layer

The downwelling branch of the Hadley circulation plays a dominating role for the free troposphere (FT) of the subtropics. It is responsible for a strong subsidence inversion layer which hinders mixing of planetary boundary layer (PBL) air into the FT. Under such situation the FT air mass is transported from higher latitudes and/or higher altitudes

(Galewsky et al., 2005; Barnes and Hartmann, 2010; Cuevas et al., 2013).

In addition, the summertime FT of the northern subtropical Atlantic is frequently affected by air masses that are advected from the African continent (e.g. Prospero et al., 2002; Rodríguez et al., 2011; Andrey et al., 2014). At that time of the year there are strong and extended heat lows over the African continent at 20–35° N. These heat lows modify the transport pathways to the FT and enable mixing from the PBL to the FT (e.g. Valero et al., 1992).

These two different transport pathways can be observed at our reference site (island of Tenerife and surroundings), and we can use these circumstances for validating the added value of the δD signal and for demonstrating the usefulness of the remote sensing data for studying atmospheric moisture transport processes.

5.2.1 Aerosol data as proxy for transport pathways

Saharan dust aerosols measured in the FT around Tenerife can be used for identifying an air mass that has experienced vertical transport over the African continent. This shall be demonstrated by Fig. 13, which shows that at Izaña and in summer high aerosol concentrations ($PM_{10} > 25\,\mu g\,m^{-3}$) are clearly correlated to air masses that have been advected from the African continent. The PM_{10} aerosol data show the aerosol mass concentration considering all particles with aerodynamic diameter smaller than $10\,\mu m$, and at Izaña, PM_{10} is, by far, dominated by Saharan dust particles (Rodríguez et al., 2011). For the calculation of the backward trajectories we use HYSPLIT (Hybrid Single-Particle Lagrangian Integrated Trajectory – some examples and explanations of the HYSPLIT calculations are given in Cuevas et al., 2013).

In the following we use low aerosol concentrations as proxy for an FT air mass that has been mainly transported from higher latitudes and/or higher altitudes (clean conditions, no vertical mixing from the PBL to the FT). High aerosol concentrations are our proxy for air masses influenced by vertical mixing over the African continent (dust conditions). We investigate the isotopologue data measured by the different instruments under the different conditions. For each instrument we need aerosol data that are representative for a similar air mass as the isotopologue data.

For ISOWAT we use the V_{10} aerosol data observed in situ during the flights. The V_{10} aerosol data is the aerosol volume concentration considering particles with a diameter smaller than $10\,\mu m$. The vertical resolution of these measurements is 100 m. The ISOWAT data are much higher resolved. We calculate 100 m averages from the ISOWAT data in order to make them representative for the same layers as the aerosol data. An air mass with $V_{10} < 1\,\mu m^3\,cm^{-3}$ is defined as a clean air mass and an air mass with $V_{10} > 10\,\mu m^3\,cm^{-3}$ is defined as dust air mass. For more details on the aircraft aerosol data see Appendix B and Fig. B1.

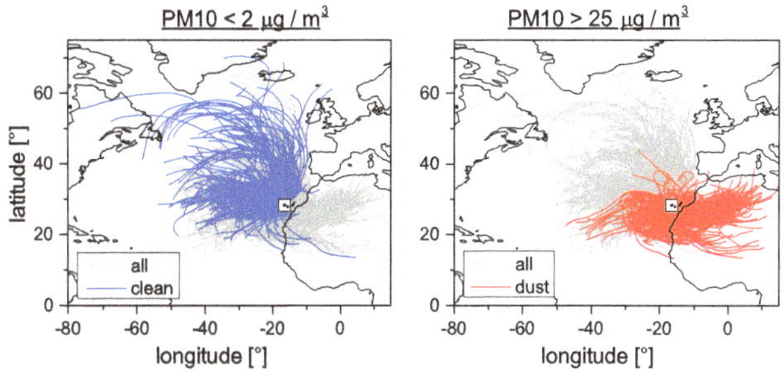

Figure 13. 72 h backward trajectories between May and October (2012 and 2013) for Izaña classified with respect to the PM10 value as measured in situ at Izaña. Grey lines: all backward trajectories; blue lines: for $PM_{10} < 2\,\mu g\,m^{-3}$; red lines: for $PM_{10} > 25\,\mu g\,m^{-3}$.

The Izaña Picarro data are paired with Izaña's PM_{10} in situ aerosol measurements. As definition for clean conditions we use $PM_{10} < 2\,\mu g\,m^{-3}$ and for dust conditions $PM_{10} > 25\,\mu g\,m^{-3}$. More information about these data is given in Appendix B (Fig. B2). We only work with nighttime data (observed between midnight to 1 hour after sunrise). Thereby we can minimize the risk of data affected by local mixing between the MBL and the FT.

At Teide there are no continuous in situ aerosol observations and we pair the Teide Picarro data with the Izaña AERONET (http://aeronet.gsfc.nasa.gov) remote sensing data. The AERONET photometers analyse the direct sunlight and report the aerosol optical depth (AOD) for the whole atmosphere above Izaña, i.e. they are reasonably representative for the altitudes around Teide. For an AOD value at 500 nm below 2×10^{-2} we assume clean conditions and above 5×10^{-2} dust conditions. In order to avoid that the Picarro observations are affected by local vertical mixing we only work with nighttime data (as for the Izaña Picarro we only use data obtained between midnight and one our after sunrise). Note that we have to interpolate to AERONET data during the nighttime, since the AERONET photometer observations work with direct sunlight.

Izaña's ground-based FTIR remote sensing system has the capability to resolve the first few kilometres above Izaña. The Izaña AERONET data are representative for similar air masses and can be used for classifying the FTIR data. We use the same threshold levels as for the classification of the Teide Picarro data.

For the classification of the IASI data we also use Izaña's AERONET data. However, we have to consider that IASI is most sensitive at around 5 km altitude, i.e. an altitude significantly above the location of the Izaña AERONET photometer. In order to have a reasonable proxy for aerosol amounts around 5 km altitude we work with higher AOD threshold values. We use a value below 5×10^{-2} for clean conditions and above 1×10^{-1} for dust conditions.

A brief overview of the Izaña AERONET data is given in Appendix B (Fig. B3).

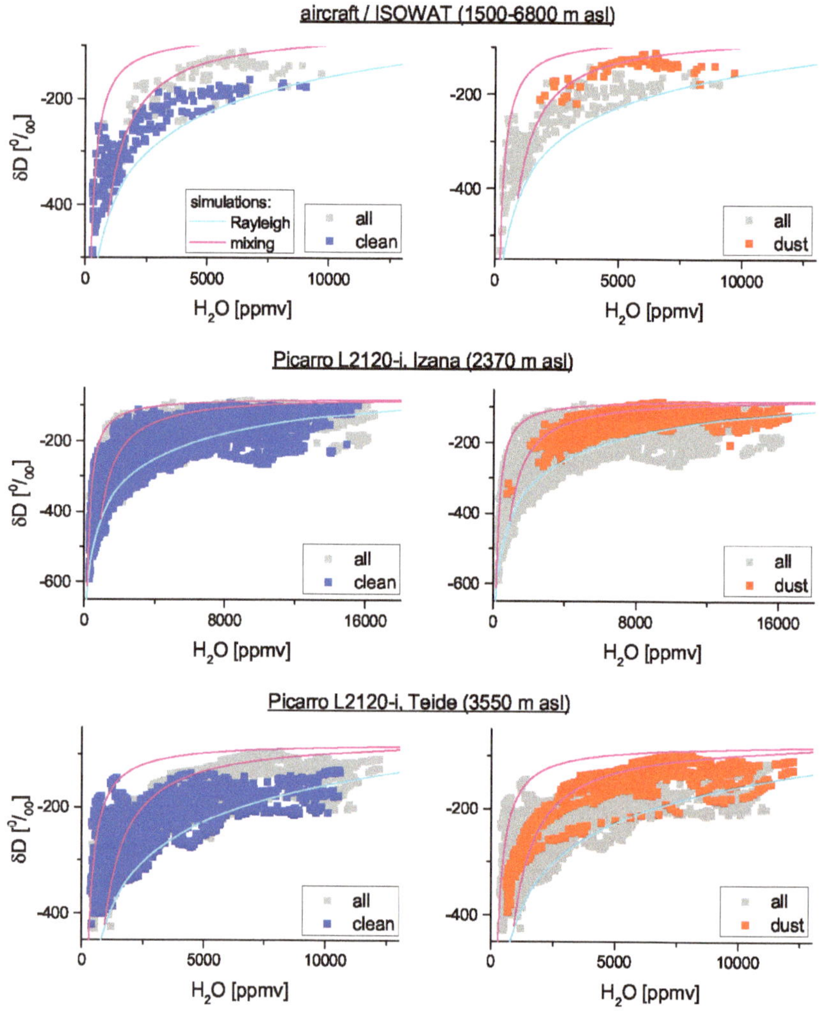

Figure 14. δD–H_2O plot observed from March 2012 to October 2014 at and close to Tenerife by different in situ instruments at different altitudes. Grey dots: all data; blue dots: for air mass not affected by the African continent; red dots: for air mass affected by strong vertical mixing over the African continent. In addition all panels show the same simulated curves as depicted in Fig. 12. Cyan line: Rayleigh curve for initialization with $T = 25\,°C$ and $RH = 80\,\%$. Magenta lines: two mixing lines. First for mixing between $H_2O[1] = 25\,000$ ppmv; $\delta D[1] = -80$ ‰ and $H_2O[2] = 900$ ppmv; $\delta D[2] = -430$ ‰. Second for mixing between $H_2O[1] = 25\,000$ ppmv; $\delta D[1] = -80$ ‰ and $H_2O[2] = 200$ ppmv; $\delta D[2] = -610$ ‰.

5.2.2 Identifying the transport pathways in the water vapour isotopologue data

Figure 14 shows the δD–H_2O plots for the different in situ instruments (from the top to the bottom: ISOWAT, Picarro at Izaña, and Picarro at Teide). The plot is made with the whole data set as shown in Figs. 1 and 2, i.e. it uses among others the 10 min averaged nighttime data measured almost continuously between March 2012 and October 2014. Grey dots represent all measurements. The blue dots (left panels) represent the observations made for clean conditions, i.e. when there is weak/no influence from the African continent. The red dots (right panels) are for observations made for dust conditions, i.e. when the air mass is clearly affected by vertical mixing over Africa.

The top graphs show the ISOWAT data together with simulated δD–H_2O curves. The magenta lines represents values that occur for mixing of two air masses: a humid air mass with MBL values (H_2O: 25 000 ppmv; δD: −80 ‰) and a dry FT background air mass (first mixing line with H_2O of 900 ppmv and δD of −430 ‰; second mixing line with H_2O of 200 ppmv and δD of −610 ‰). The cyan line represents a Rayleigh distillation curve for an initial atmospheric temperature of 25 °C, a relative humidity at the surface of 80 %, and an averaged thermodynamic profile. Rayleigh distillation means that all water that condenses during the transport pathway is immediately removed from the air mass. There is a strong correlation between the ISOWAT water vapour isotopologue data and the aerosol proxy: (1) when vertical mixing over Africa is clearly discernible in the aerosol data, the

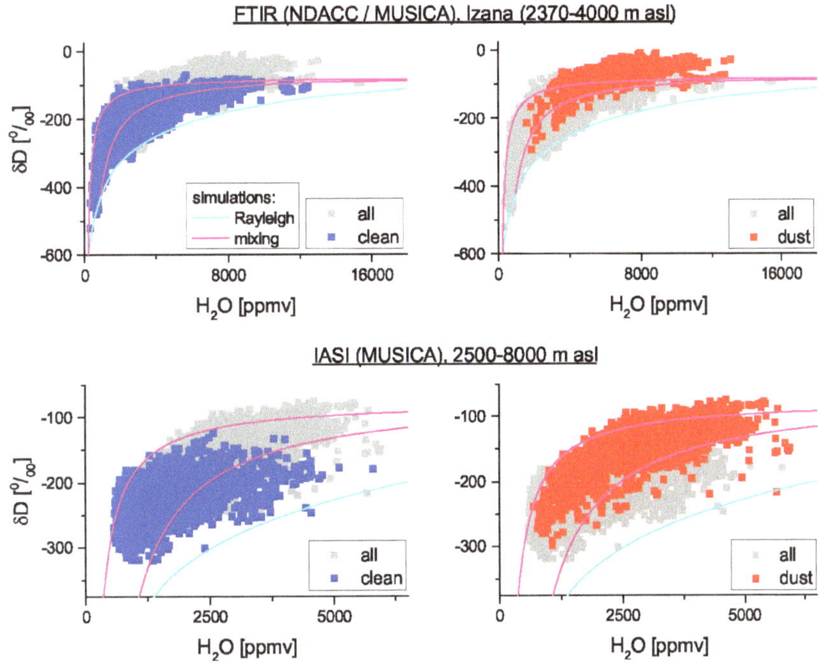

Figure 15. As Fig. 14, but for MUSICA's remote sensing observations between 2005 and 2014.

ISOWAT δD data group around a mixing line. (2) When the aerosol data reveal no significant vertical mixing over Africa the δD data approximate towards a Rayleigh distillation line (blue dots and cyan line in the top graphs of Fig. 14).

These different transport pathways are also well observed in the in situ isotopologue data sets as collected continuously by the two Picarros at Izaña and Teide. In particular the Picarro at Teide, which is less affected by the local airflows on the island's mountains, provides very distinct isotopologue fingerprints for the two different situations. The good consistency between the different in situ reference data sets makes this finding very robust.

Figure 15 shows the respective δD–H_2O plots for the MUSICA's remote sensing data (ground-based FTIR, and IASI). These data are representative for several years. We work with FTIR data measured between 2005 and 2014 and we use IASI-A and IASI-B data measured between October 2007 and the end of July 2013. If compared to the in situ data of Fig. 14 we can again observe the positive bias in the remote sensing δD values. However, the interesting point here is that the δD–H_2O distribution in the remote sensing data sets reveals the different transport pathways similarly to the in situ data sets. This proves that the MUSICA FTIR and IASI δD remote sensing data add scientifically useful information to the respective H_2O measurement and can give valuable insight into tropospheric moisture transport pathways.

A more profound scientific examination of these different transport regimes is out of the scope of this technical validation paper and shall be addressed in an extra paper which is currently in preparation.

6 Conclusions

Global remote sensing of tropospheric δD in addition to H_2O can open up novel opportunities for tropospheric water cycle research. However, the interesting signal of the added value of δD is small. Its detection by remote sensing techniques is a difficult task and prior to a scientific application of the data, their characteristics and uncertainties should be well understood. In this context Schneider et al. (2012) and Wiegele et al. (2014) show detailed feasibility and error assessment studies for MUSICA's NDACC/FTIR and METOP/IASI water vapour isotopologue products. In this paper those efforts are complemented by an empirical validation exercise taking high-quality in situ measurements as the reference. The study is made in the surroundings of the Canary Islands in the subtropical northern Atlantic.

For a quantitative validation of the random uncertainty and the bias of the remote sensing data we need in situ observations of well-resolved vertical profiles. During an aircraft campaign in July 2013 we produce six of such profiles using the dedicated aircraft instrument ISOWAT, whose extensive pre-, post-, and in-flight calibration modes ensure data of high accuracy and a very high vertical resolution. With this unprecedented data set and based on the scatter between the coincident ISOWAT and remote sensing observations we can empirically assess the random uncertainty of the NDACC/FTIR δD product to be smaller than 45 and 15 ‰ for the lower and middle troposphere, respectively. For the middle tropospheric METOP/IASI δD product we assess a value of 13 ‰. For the bias assessment we examine systematic differences with respect to ISOWAT and find that there is a systematic positive difference of about 30–70‰, which is similarly observed for the NDACC/FTIR and the METOP/IASI δD product. For global and more reliable bias and empirical uncertainty estimations we need more reference profiles measured at a variety of globally representative sites (currently we only have six coinciding reference profiles for one single location).

The quantitative comparison to the ISOWAT data does not demonstrate whether the remote sensing δD data is complementary to the remote sensing H_2O data. Such complementarity has to be investigated by δD–H_2O plots. For these plots we need a large number of data points, thus they are hardly possible by means of only six ISOWAT profiles. Instead we use all the data available from the different instruments since 2005 and pair them with aerosol data that we use as proxy for classifying situations of different moisture pathways. We are able to demonstrate that all the instruments involved in our study (ISOWAT, Picarro at Izaña and Teide, NDACC/FTIR, and METOP/IASI) are able to consistently identify two different free tropospheric water vapour transport pathways. Transport from the PBL via vertical mixing can be identified by isotopologue data that approximate a mixing line on the δD-H_2O surface. Such mixing lines occur in all data sets primarily in the presence of large aerosol amounts, which serve as proxy for large scale vertical mixing over the African continent. In contrast, if there are small aerosol amounts we observe primarily δD–H_2O curves that approximate a Rayleigh curve, meaning that the respective air masses have been dried by condensation. This is in agreement with a strong inversion layer that hinders mixing from the PBL and prevailing moisture transport from higher latitudes and altitudes.

Our study proves that the MUSICA H_2O and δD remote sensing data are a reliable tool for water vapour isotopologue studies. However, we have to be aware that there is very likely a bias in the δD values. For a better understanding of this bias further research on the spectroscopic line parameters as well as more high-quality aircraft profiles and thus further campaigns and validation efforts would be desirable.

We would like to propose using the approaches as presented in this paper, in Schneider et al. (2012), and in Wiegele et al. (2014) as guideline for best practice when validating water vapour isotopologue remote sensing data sets.

Appendix A: Discussion of day 130731

On day 130731 there are five IASI observations that fulfil our temporal and spatial coincidence criteria. The respective IASI pixels are depicted in Fig. 5 in magenta as filled symbols. Square symbols are for IASI-A pixels and diamonds are for IASI-B pixels. They group at two different locations. The first group (two IASI-A pixels) is located west of the aircraft's flight track and the second group (one IASI-A pixel and two IASI-B pixels) is located southeast of the flight track. The latter pixels are marked by black edges. The magenta symbols as shown in Fig. 11 have the same marking. The three data points where the ISOWAT versus IASI agreement for H_2O is poor belong to the marked pixels (left panel of Fig. 11), i.e. to the observation made southeast of the flight track. There is consistency between the IASI-A observation and the two IASI-B observations, but all three observations are in disagreement with the ISOWAT observation. Such outliers are not seen in the ISOWAT versus IASI plot for δD (right panel of Fig. 11). There all five IASI observations on day 130731 agree well with the ISOWAT observations.

The issue can be understood by a look at Fig. A1. It shows MODIS (Moderate Resolution Imaging Spectroradiometer) images for the Canary Islands on the three consecutive days between day 130730 and 130801. On day 130730 (left image) the air is rather clean, which is the typical situation for an air mass travelling from the Atlantic northwest of the islands. In contrast, on day 130801 (right image) there is a lot of dust in the air, indicating that the free tropospheric air mass over the Canaries has its origin in the boundary layer over the Sahara desert (located just a few hundred kilometres east of the islands). Day 130731, when we observe the H_2O outliers, is a transition day. During this day the Sahara boundary layer air mass is mixing with the free tropospheric Atlantic air mass. The central panel of Fig. A1 documents that there is a strong gradient in the dust concentration from southeast of ISOWAT's flight track – where the air has mainly the dust-laden Saharan characteristics – to west of the flight track – where the air is still rather clean.

The two groups of IASI pixels on day 130731 are located in areas with different atmospheric situations. West of ISOWAT's flight track the mixing between dry and clean air with humid and dust-laden air has just begun, meaning that H_2O started to increase and δD has strongly increased (during a mixing from dry to humid δD is rapidly determined by the humid end member, see also mixing lines in Figs. 12, 14, and 15). Southeast of the flight track the mixing process is already more advanced and the air mass has almost the characteristics of the humid mixing end member, i.e. both H_2O and δD are strongly increased.

The three outliers as seen in Fig. 11 are due to spatial inhomogeneities in the atmospheric water vapour state on day 130731. We show that the high temporal and spatial resolution of IASI allows resolving these inhomogeneities and that IASI-A and IASI-B can observe the responsible moisture transport pathway in a consistent manner.

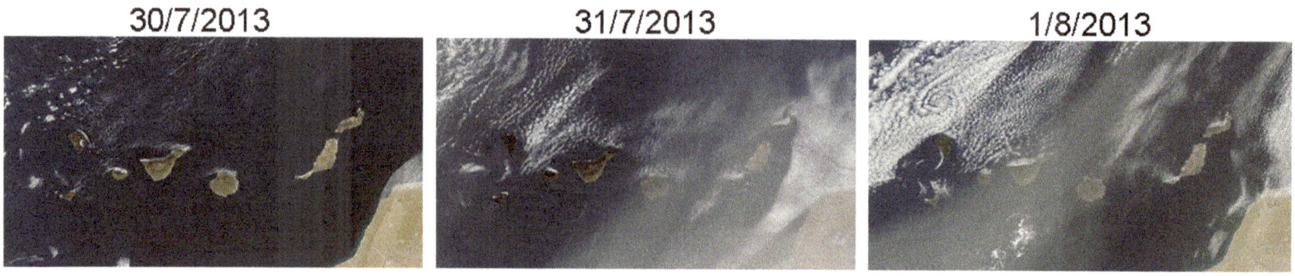

Figure A1. Images for the Canary Islands on three consecutive days (130730, 130731, and 130801). Images are daily composites of data of the Moderate Resolution Imaging Spectroradiometer (MODIS) aboard the NASA Terra and Aqua satellites. Source: https://earthdata.nasa.gov/labs/worldview/.

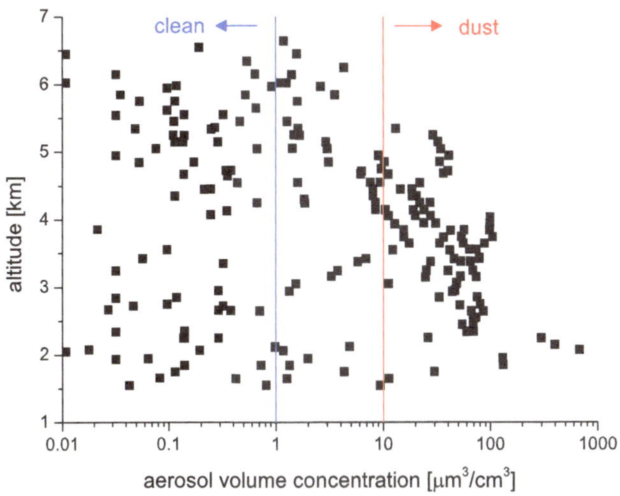

Figure B1. Vertical distribution of aerosol volume concentration (V_{10}) measured above the boundary layer during the six aircraft flights.

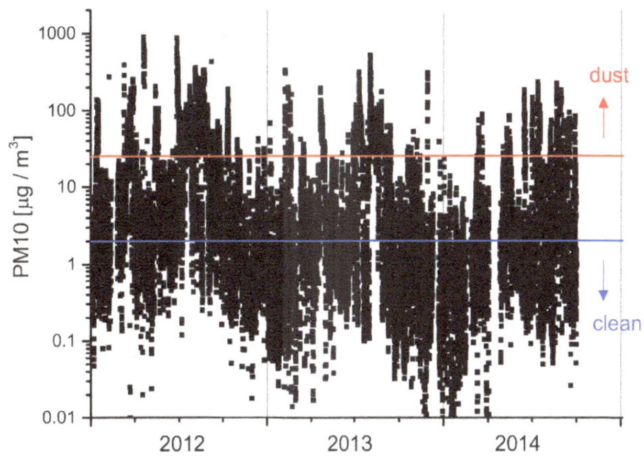

Figure B2. Time series of PM_{10} in situ data as measured continuously at Izaña (2012–2014).

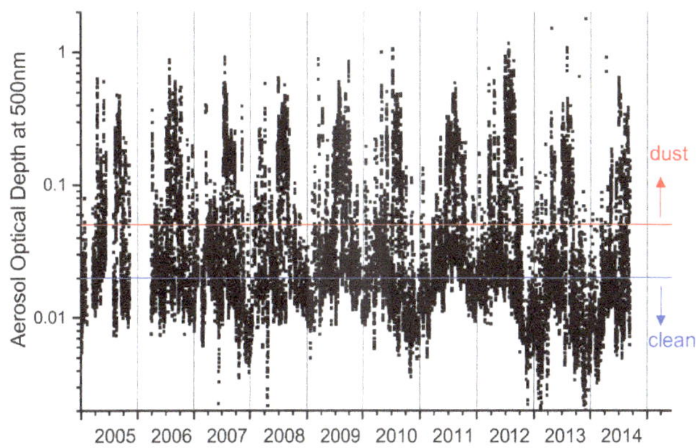

Figure B3. Time series of the AERONET aerosol optical depth at 500 nm reported for Izaña (2005–2012: level 2.0; 2013 and 2014: level 1.5).

Appendix B: The aerosol data sets

The INTA aerosol instrument provides the aerosol number density for different particle sizes, whereby different particle sizes ranging from 0.5 to 50 μm diameter can be distinguished. From these data we calculate the aerosol volume concentration for all the aerosols with a diameter smaller than 10 μm (this volume concentration is called V_{10}), whereby we assume spherical aerosols. The respective data measured during the six flights above the boundary layer (above 1500 m a.s.l.) are plotted in Fig. B1. Please note that there are many occasions with $V_{10} = 0\,\mu\text{m}^3\,\text{cm}^{-3}$ (aerosol number density below the detection limit) that are not visible in the graph.

The Izaña PM_{10} data are hourly averaged mass concentration data and given as aerosol mass per air volume. Figure B2 shows the 2012–2014 time series. The mass concentrations are determined from the continuously measured aerosol size distributions (applying the instrument APS, TSI™, model 3321). First the aerosol volume concentrations (V_{10}) are calculated. From V_{10} the mass concentration (PM_{10}) is obtained by using an empirically estimated aerosol effective density using the methods described by Rodríguez et al. (2012). This density estimation consists in correlating the measured aerosol volume concentrations to mass concentrations measured by the gravimetric method. For the gravimetric method PM_{10} aerosols are collected in microfibre quartz filters. This collection needs several hours and requires a careful treatment of the filters, hence the gravimetric method cannot be performed continuously. The Izaña programme for in situ aerosol characterization was audited by the World Calibration Centre for Aerosol Physics (WCCAP; Institute for Tropospheric Research, Leipzig) in November 2006. For more details on these PM_{10} data please refer to Rodríguez et al. (2011).

The Aerosol Robotic Network (AERONET) provides atmospheric aerosol properties at several wavelengths (Holben et al., 1998). The Cimel CE-318 sun photometer is the standard instrument in the network. It measures direct sun irradiance (every 15 min) with a 1.2° full field of view that is used to compute aerosol optical depth (AOD) for each wavelength available on each photometer type. Three quality levels of AERONET data are available: level 1.0 (real time raw data), level 1.5 (real time cloud screened data), and level 2.0 (quality assured data). Figure B3 shows the time series of AOD at 500 nm as measured at the Izaña observatory.

Acknowledgements. This study has been conducted in the framework of the project MUSICA which is funded by the European Research Council under the European Community's Seventh Framework Programme (FP7/2007-2013)/ERC Grant agreement number 256961.

E. Sepúlveda is supported by the Ministerio de Economía and Competitividad of Spain for the project NOVIA (CGL2012-37505).

The aircraft campaign has been co-funded by the project MUSICA and the Spanish national project AMISOC (CGL2011-24891).

We are grateful to INTA Aerial Platforms, a branch of the Spanish ICTS program, and the Spanish Air Force for their efforts in maintaining and operating the aircraft.

The Izaña aerosol in situ measurements are part of the project POLLINDUST (CGL2011-26259) funded by the Minister of Economy and Competitiveness of Spain.

The AERONET sun photometer at Izaña (PI: Emilio Cuevas) has been calibrated within AERONET EUROPE TNA supported by the European Community Research Infrastructure Action under the FP7 Capacities program for Integrating Activities, ACTRIS grant agreement number 262254.

We acknowledge the support by the Deutsche Forschungsgemeinschaft and the Open Access Publishing Fund of the Karlsruhe Institute of Technology.

The service charges for this open access publication
have been covered by a Research Centre of the
Helmholtz Association.

Edited by: U. Platt

References

Aemisegger, F., Sturm, P., Graf, P., Sodemann, H., Pfahl, S., Knohl, A., and Wernli, H.: Measuring variations of $\delta^{18}O$ and δ^2H in atmospheric water vapour using two commercial laser-based spectrometers: an instrument characterisation study, Atmos. Meas. Tech., 5, 1491–1511, doi:10.5194/amt-5-1491-2012, 2012.

Andrey, J., Cuevas, E., Parrondo, M. C., Alonso-Pérez, S., Redondas, A., and Gil-Ojeda, M.: Quantification of ozone reductions within the Saharan air layer through a 13-year climatologic analysis of ozone profiles, Atmos. Environ., 84, 28–34, doi:10.1016/j.atmosenv.2013.11.030, 2014.

Barnes, E. A. and Hartmann, D. L.: Influence of eddy-driven jet latitude on North Atlantic jet persistence and blocking frequency in CMIP3 integrations, Geophys. Res. Lett., 37, L23802, doi:10.1029/2010GL045700, 2010.

Boesch, H., Deutscher, N. M., Warneke, T., Byckling, K., Cogan, A. J., Griffith, D. W. T., Notholt, J., Parker, R. J., and Wang, Z.: HDO/H_2O ratio retrievals from GOSAT, Atmos. Meas. Tech., 6, 599–612, doi:10.5194/amt-6-599-2013, 2013.

Craig, H.: Standard for Reporting concentrations of Deuterium and Oxygen-18 in natural waters, Science, 13, 1833–1834, doi:10.1126/science.133.3467.1833, 1961.

Cuevas, E., González, Y., Rodríguez, S., Guerra, J. C., Gómez-Peláez, A. J., Alonso-Pérez, S., Bustos, J., and Milford, C.: Assessment of atmospheric processes driving ozone variations in the subtropical North Atlantic free troposphere, Atmos. Chem. Phys., 13, 1973–1998, doi:10.5194/acp-13-1973-2013, 2013.

Dyroff, C., Fütterer, D., and Zahn, A.: Compact diode-laser spectrometer ISOWAT for highly sensitive airborne measurements of water-isotope ratios, Appl. Phys. B, 98, 537–548, doi:10.1007/s00340-009-3775-6, 2010.

Dyroff, C., Sanati, S., Christner, E., Zahn, A., Balzer, M., Bouquet, H., McManus, J. B., González-Ramos, Y., and Schneider, M.: Airborne in situ vertical profiling of $HDO/H_2^{16}O$ in the subtropical troposphere during the MUSICA remote sensing validation campaign, Atmos. Meas. Tech. Discuss., 8, 121–155, doi:10.5194/amtd-8-121-2015, 2015.

Frankenberg, C., Yoshimura, K., Warneke, T., Aben, I., Butz, A., Deutscher, N., Griffith, D., Hase, F., Notholt, J., Schneider, M., Schrejver, H., and Röckmann, T.: Dynamic processes governing lower-tropospheric HDO/H_2O ratios as observed from space and ground, Science, 325, 1374–1377, doi:10.1126/science.1173791, 2009.

Frankenberg, C., Wunch, D., Toon, G., Risi, C., Scheepmaker, R., Lee, J.-E., Wennberg, P., and Worden, J.: Water vapor isotopologue retrievals from high-resolution GOSAT shortwave infrared spectra, Atmos. Meas. Tech., 6, 263–274, doi:10.5194/amt-6-263-2013, 2013.

Galewsky, J., Sobel, A., and Held, I.: Diagnosis of Subtropical Humidity Dynamics Using Tracers of Last Saturation Journal of the Atmospheric Sciences, Atmos. Sci., 62, 3353–3367, doi:10.1175/jas3533.1, 2005.

García, O. E., Schneider, M., Redondas, A., González, Y., Hase, F., Blumenstock, T., and Sepúlveda, E.: Investigating the long-term evolution of subtropical ozone profiles applying ground-based FTIR spectrometry, Atmos. Meas. Tech., 5, 2917–2931, doi:10.5194/amt-5-2917-2012, 2012.

Holben, B. N., Eck, T. F., Slutsker, I., Tanre, D., Buis, J., Setzer, A., Vermote, E., Reagan, J., Kaufman, Y., Nakajima, T., Lavenu, F., Jankowiak, I., and Smirnov, A.: AERONET – A federated instrument network and data archive for aerosol characterization, Remote Sens. Environ., 66, 1–16, 1998.

Lacour, J.-L., Risi, C., Clarisse, L., Bony, S., Hurtmans, D., Clerbaux, C., and Coheur, P.-F.: Mid-tropospheric δD observations from IASI/MetOp at high spatial and temporal resolution, Atmos. Chem. Phys., 12, 10817–10832, doi:10.5194/acp-12-10817-2012, 2012.

Prospero, J. M., Ginoux, P., Torres, O., Nicholson, S. E., and Gill, T. E.: Environmental characterizacion of global sources of dust with the Nimbus 7 total ozone mapping spectrometer (TOMS) absorbing aerosol product, Rev. Geophys., 40, 2–1–2–31, doi:10.1029/2000RG000095, 2002.

Risi, C., Noone, D., Worden, J., Frankenberg, C., Stiller, G., Kiefer, M., Funke, B., Walker, K., Bernath, P., Schneider, M., Bony, S., Lee, J., Brown, D., and Sturm, C.: Process-evaluation of tropospheric humidity simulated by general circulation models using water vapor isotopic observations. Part 2: an isotopic diagnostic to understand the mid and upper tropospheric moist bias in the tropics and subtropics, J. Geophys. Res., 117, D05304, doi:10.1029/2011JD016623, 2012.

Rodríguez, S., González, Y., Cuevas, E., Ramos, R., Romero, P. M., Abreu-Afonso, J., and Redondas, A.: Atmospheric nanoparticle observations in the low free troposphere during upward orographic flows at Izaña Mountain Observatory, Atmos. Chem. Phys., 9, 6319–6335, doi:10.5194/acp-9-6319-2009, 2009.

Rodríguez, S., Alastuey, A., Alonso-Pérez, S., Querol, X., Cuevas, E., Abreu-Afonso, J., Viana, M., Pérez, N., Pandolfi, M., and de la Rosa, J.: Transport of desert dust mixed with North African industrial pollutants in the subtropical Saharan Air Layer, Atmos. Chem. Phys., 11, 6663–6685, doi:10.5194/acp-11-6663-2011, 2011.

Rodríguez, S., Alastuey, A., and Querol, X.: A review of methods for long term in situ characterization of aerosol dust, Aeolian Res., 6, 55–74, doi:10.1016/j.aeolia.2012.07.004, 2012.

Rothman, L. S., Gordon, I. E., Barbe, A., Chris Benner, D., Bernath, P. F., Birk, M., Boudon, V., Brown, L. R., Campargue, A., Champion, J.-P., Chance, K., Coudert, L. H., Dana, V., Devi, V. M., Fally, S., Flaud, J.-M., Gamache, R. R., Goldman, A., Jacquemart, D., Kleiner, I., Lacome, N., Lafferty, W. J., Mandin, J.-Y., Massie, S. T., Mikhailenko, S. N., Miller, C. E., Moazzen-Ahmadi, N., Naumenko, O. V., Nikitin, A. V., Orphal, J., Perevalov, V. I., Perrin, A., Predoi-Cross, A., Rinsland, C. P., Rotger, M., Simecková, M., Smith, M. A. H., Sung, K., Tashkun, S. A., Tennyson, J., Toth, R. A., Vandaele, A. C., and Vander-Auwera, J.: The HITRAN 2008 molecular spectroscopic database, J. Quant. Spectrosc. Ra., 110, 533–572, doi:10.1016/j.jqsrt.2009.02.013, 2009.

Schmidt, G. A., Hoffmann, G., Shindell, D. T., and Hu, Y.: Modeling atmospheric stable water isotopes and the potential for constraining cloud processes and stratosphere-troposphere water exchange, J. Geophys. Res., 110, D21314, doi:10.1029/2005JD005790, 2005.

Schneider, M. and Hase, F.: Ground-based FTIR water vapour profile analyses, Atmos. Meas. Tech., 2, 609–619, doi:10.5194/amt-2-609-2009, 2009.

Schneider, M. and Hase, F.: Optimal estimation of tropospheric H_2O and δD with IASI/METOP, Atmos. Chem. Phys., 11, 11207–11220, doi:10.5194/acp-11-11207-2011, 2011.

Schneider, M., Hase, F., and Blumenstock, T.: Water vapour profiles by ground-based FTIR spectroscopy: study for an optimised retrieval and its validation, Atmos. Chem. Phys., 6, 811–830, doi:10.5194/acp-6-811-2006, 2006a.

Schneider, M., Hase, F., and Blumenstock, T.: Ground-based remote sensing of HDO/H_2O ratio profiles: introduction and validation of an innovative retrieval approach, Atmos. Chem. Phys., 6, 4705–4722, doi:10.5194/acp-6-4705-2006, 2006b.

Schneider, M., Romero, P. M., Hase, F., Blumenstock, T., Cuevas, E., and Ramos, R.: Continuous quality assessment of atmospheric water vapour measurement techniques: FTIR, Cimel, MFRSR, GPS, and Vaisala RS92, Atmos. Meas. Tech., 3, 323–338, doi:10.5194/amt-3-323-2010, 2010a.

Schneider, M., Toon, G. C., Blavier, J.-F., Hase, F., and Leblanc, T.: H_2O and δD profiles remotely-sensed from ground in different spectral infrared regions, Atmos. Meas. Tech., 3, 1599–1613, doi:10.5194/amt-3-1599-2010, 2010b.

Schneider, M., Hase, F., Blavier, J.-F., Toon, G. C., and Leblanc, T.: An empirical study on the importance of a speed-dependent Voigt line shape model for tropospheric water vapor profile remote sensing, J. Quant. Spectrosc. Ra., 112, 465–474, doi:10.1016/j.jqsrt.2010.09.008, 2011.

Schneider, M., Barthlott, S., Hase, F., González, Y., Yoshimura, K., García, O. E., Sepúlveda, E., Gomez-Pelaez, A., Gisi, M., Kohlhepp, R., Dohe, S., Blumenstock, T., Wiegele, A., Christner, E., Strong, K., Weaver, D., Palm, M., Deutscher, N. M., Warneke, T., Notholt, J., Lejeune, B., Demoulin, P., Jones, N., Griffith, D. W. T., Smale, D., and Robinson, J.: Ground-based remote sensing of tropospheric water vapour isotopologues within the project MUSICA, Atmos. Meas. Tech., 5, 3007–3027, doi:10.5194/amt-5-3007-2012, 2012.

Valero, F., Luna, Y., Martín, M., and Sancho, P.: Tropospheric ozone concentrations related to atmospheric conditions at Izaña BAPMoN weather station, Canary Islands, Il Nuovo Cimento C, 15, 159–172, doi:10.1007/BF02507610, 1992.

Webster, C. R. and Heymsfield, A. J.: Water Isotope Ratios H/D, $^{18}O/^{16}O$, $^{17}O/^{16}O$ in and out of Clouds Map Dehydration Pathways, Science, 302, 1742–1745, doi:10.1126/science.1089496, 2003.

Wiegele, A., Schneider, M., Hase, F., Barthlott, S., García, O. E., Sepúlveda, E., González, Y., Blumenstock, T., Raffalski, U., Gisi, M., and Kohlhepp, R.: The MUSICA MetOp/IASI H_2O and δD products: characterisation and long-term comparison to NDACC/FTIR data, Atmos. Meas. Tech., 7, 2719–2732, doi:10.5194/amt-7-2719-2014, 2014.

Worden, J., Noone, D., Bowman, K., Beer, R., Eldering, A., Fisher, B., Gunson, M., Goldman, A., Herman, R., Kulawik, S. S., Lampel, M., Osterman, G., Rinsland, C., Rodgers, C., Sander, S., Shephard, M., Webster, R., and Worden, H.: Importance of rain evaporation and continental convection in the tropical water cycle, Nature, 445, 528–532, doi:10.1038/nature05508, 2007.

Worden, J., Noone, D., Galewsky, J., Bailey, A., Bowman, K., Brown, D., Hurley, J., Kulawik, S., Lee, J., and Strong, M.: Estimate of bias in Aura TES HDO/H_2O profiles from comparison of TES and in situ HDO/H_2O measurements at the Mauna Loa observatory, Atmos. Chem. Phys., 11, 4491–4503, doi:10.5194/acp-11-4491-2011, 2011.

Worden, J. R., Bowman, K., Noone, D., Beer, R., Clough, S., Eldering, A., Fisher, B., Goldman, A., Gunson, M., Herman, R., Kulawik, S. S., Lampel, M., Luo, M., Osterman, G., Rinsland, C., Rodgers, C., Sander, S., Shephard, M., and Worden, H.: TES observations of the tropospheric HDO/H_2O ratio: retrieval approach and characterization, J. Geophys. Res., 11, D16309, doi:10.1029/2005JD006606, 2006.

15

Hygroscopic growth of atmospheric aerosol particles based on active remote sensing and radiosounding measurements: selected cases in southeastern Spain

M. J. Granados-Muñoz[1,2], F. Navas-Guzmán[1,2,*], J. A. Bravo-Aranda[1,2], J. L. Guerrero-Rascado[1,2], H. Lyamani[1,2], A. Valenzuela[1,2], G. Titos[1,2], J. Fernández-Gálvez[1,2], and L. Alados-Arboledas[1,2]

[1]Dpt. Applied Physics, Faculty of Sciences, University of Granada, Fuentenueva s/n, 18071, Granada, Spain
[2]Andalusian Institute for Earth System Research (IISTA-CEAMA), Avda. del Mediterráneo s/n, 18006, Granada, Spain
[*]now at: Institute of Applied Physics (IAP), University of Bern, Bern, Switzerland

Correspondence to: M. J. Granados Muñoz (mjgranados@ugr.es)

Abstract. A new methodology based on combining active and passive remote sensing and simultaneous and collocated radiosounding data to study the aerosol hygroscopic growth effects on the particle optical and microphysical properties is presented. The identification of hygroscopic growth situations combines the analysis of multispectral aerosol particle backscatter coefficient and particle linear depolarization ratio with thermodynamic profiling of the atmospheric column. We analyzed the hygroscopic growth effects on aerosol properties, namely the aerosol particle backscatter coefficient and the volume concentration profiles, using data gathered at Granada EARLINET station. Two study cases, corresponding to different aerosol loads and different aerosol types, are used for illustrating the potential of this methodology. Values of the aerosol particle backscatter coefficient enhancement factors range from 2.1 ± 0.8 to 3.9 ± 1.5, in the ranges of relative humidity 60–90 and 40–83 %, being similar to those previously reported in the literature. Differences in the enhancement factor are directly linked to the composition of the atmospheric aerosol. The largest value of the aerosol particle backscatter coefficient enhancement factor corresponds to the presence of sulphate and marine particles that are more affected by hygroscopic growth. On the contrary, the lowest value of the enhancement factor corresponds to an aerosol mixture containing sulphates and slight traces of mineral dust. The Hänel parameterization is applied to these case studies, obtaining results within the range of values reported in previous studies, with values of the γ exponent of 0.56 ± 0.01 (for anthropogenic particles slightly influenced by mineral dust) and 1.07 ± 0.01 (for the situation dominated by anthropogenic particles), showing the convenience of this remote sensing approach for the study of hygroscopic effects of the atmospheric aerosol under ambient unperturbed conditions. For the first time, the retrieval of the volume concentration profiles for these cases using the Lidar Radiometer Inversion Code (LIRIC) allows us to analyze the aerosol hygroscopic growth effects on aerosol volume concentration, observing a stronger increase of the fine mode volume concentration with increasing relative humidity.

1 Introduction

The influence of atmospheric aerosols in the Earth's climate is still affected by a high uncertainty. Scientific knowledge on the interaction between atmospheric aerosol and solar radiation is quite low compared to other atmospheric constituents mainly due to atmospheric aerosol high spatiotemporal variability (IPCC, 2013). As it is well known, atmospheric aerosol can affect the Earth's atmosphere energy budget by means of direct effects (by scattering or absorbing solar radiation) and indirect effects (mainly by changes in cloud properties). Therefore, changes in aerosol properties can highly influence the Earth's climate. Aerosol particles size may increase due to water uptake (hygroscopic growth) altering their size distribution and their associated optical and microphysical properties under high relative humidity conditions. Therefore, hygroscopic growth affects the direct scattering of radiation (Hänel, 1976; Hegg et al., 1996; Titos et al., 2014a; Zieger et al., 2013). Also to be considered are especially the indirect effects, as the affinity of atmospheric aerosols for water vapor is highly related to their ability to act as cloud condensation nuclei (CCN) (Charlson et al., 1992; Feingold and Morley, 2003; Padró et al., 2010).

In the past years there has been an increasing interest in the hygroscopic growth effects on the aerosol optical and microphysical properties, and many studies have already been performed (e.g., Veselovskii et al., 2009; Zieger et al., 2013; Titos et al., 2014a, c). Much of the recent research was performed by means of humidified nephelometers (Covert et al., 1972; Fierz-Schmidhauser et al., 2010a, and references therein) or humidified tandem differential mobility analyzers (Massling et al., 2007; Wu et al., 2013, and references therein). Nonetheless, these instruments present two main problems. Firstly, due to experimental set-up limitations, it is difficult to provide accurate results above RH of 85 % (Wulfmeyer and Feingold, 2000). Secondly, they modify the ambient conditions by drying the air sample and then humidifying it again up to a certain value of RH, altering thus the aerosol properties and being also subject to aerosol losses in the sampling lines.

Opposite to in situ measurements, the main advantage of lidar systems is that they can provide vertically resolved measurements without modifying the aerosol sample or its surroundings. They also detect RH close to saturation, which is of great importance since the range between 85 and 100 % RH is where the particles are more affected by hygroscopic growth (Feingold and Morley, 2003). Therefore, they are adequate to provide useful information about aerosol hygroscopic properties if favorable atmospheric conditions occur. However, they present the major drawback that the sample and conditions are not controlled in any way, and the number of cases with adequate conditions is usually low compared to in situ measurements. In the past years, some studies have already been performed by using lidar systems to detect aerosol hygroscopic growth with promising results (Ferrare et al., 1998; Wulfmeyer and Feingold, 2000; Feingold and Morley, 2003; Veselovskii et al., 2009; DiGirolamo et al., 2012). However, in many of these studies the experimental RH profiles were not available and assumptions were made to obtain RH data (Wulfmeyer and Feingold, 2000; Feingold and Morley, 2003; Feingold et al., 2006; Pahlow et al., 2006). In some other cases, RH profiles were taken from quite distant radiosounding measurements (Veselovskii et al., 2009). The availability of collocated radiosounding data allows the reduction of assumptions needed. In this work, a methodology to study aerosol hygroscopic growth based on lidar data and collocated radiosounding RH profiles is presented and applied at Granada EARLINET (European Aerosol Research Lidar Network) experimental site in order to study aerosol hygroscopic growth under unperturbed ambient conditions.

2 Experimental site and instrumentation

This study was performed at the Andalusian Institute for Earth System Research, IISTA-CEAMA, located in Granada (37.16° N, 3.61° W, 680 m a.s.l.). The station is described in more detail in (Lyamani et al., 2010; Titos et al., 2012; Valenzuela et al., 2012). Granada is affected by continental climate conditions. Temperature presents large seasonal variations, with cool winters and hot summers, and a strong diurnal thermal oscillation (up to 20 °C), which lead to strong RH variations along the day. Seasonal variations in RH are also quite important with mean monthly values ranging between 41 % in summer and 76 % in winter (source: State Meterological Agency, www.aemet.es). An analysis of RH profiles based on measurements of water vapor mixing ratio from Raman lidar and temperature from microwave radiometry for 1-year period, performed for 500 m-layers at different altitudes revealed that over Granada the 60 % of the data present RH values in the range between 20 and 60 %, whereas only 25 % of these layers present values larger than 60 % (Navas-Guzmán et al., 2014).

Due to its proximity to the African continent, the experimental site is usually affected by mineral dust events (Valenzuela et al., 2012), reaching altitudes of up to 5500 m a.s.l. (Guerrero-Rascado et al., 2008, 2009; Navas-Guzmán et al., 2013). In addition, Europe acts as an important source of anthropogenic pollution (Lyamani et al., 2006a, b). The experimental site is also affected by local and regional sources. In a recent source apportionment study of fine and coarse particulate matter at surface level, Titos et al. (2014b) pointed out that the major aerosol sources at Granada are road traffic (dominated by diesel engines) and mineral dust as well as fuel-based domestic heating during winter time. The levels of mineral matter increase considerably from winter to summer due to the dryness conditions, which favor re-suspension processes and the higher frequency of Saharan outbreaks. Traffic related sources increase its contribution during winter compared with summer (Titos et al., 2014b) as a consequence of higher emissions and lower mixing layer heights (Granados-Muñoz et al., 2012).

The experimental station is equipped with a multiwavelength Raman lidar system model LR331D400 (Raymetrics S.A.). Such an instrument, called MULHACEN, is described in detail in Guerrero-Rascado et al. (2008, 2009). The lidar system emits at 355, 532 and 1064 nm. The detection system records the elastic signals at these three wavelengths and Raman signals in three additional channels: 387 and 607 nm, corresponding to the nitrogen Raman channel, and 408 nm, corresponding to water vapor Raman channel. In addition, the perpendicular and parallel components of the visible channel are detected, allowing us to study the depolarization properties of the atmospheric aerosol (Bravo-Aranda et al., 2013, and references therein). Uncertainties associated to the elastic lidar signals are around 15 %, considering the statistical uncertainties retrieved with Monte

Carlo techniques according to Pappalardo et al. (2004) and Guerrero-Rascado et al. (2008). The Raman lidar is part of EARLINET (Pappalardo et al., 2014). It has been part of the EARLINET-ASOS (European Aerosol Research Lidar Network – Advanced Sustainable Observation System) project and currently is included in the ACTRIS (Aerosols, Clouds, and Trace gases Research InfraStructure Network) European project.

In addition, the radiometric station is equipped with a sun photometer CIMEL CE-318-4. The sun photometer provides column-integrated atmospheric aerosol properties. The automatic tracking sun and sky scanning radiometer makes sun direct measurements with a 1.2° full field of view every 15 min at 340, 380, 440, 500, 675, 870, 940 and 1020 nm (nominal wavelengths). These solar extinction measurements are used to compute aerosol optical depth (τ_λ) at each wavelength except for the 940 nm channel, which is used to retrieve total column water vapor (or precipitable water). The estimated uncertainty in computed τ_λ, due primarily to calibration uncertainty, is around 0.010–0.021 for field instruments (which is spectrally dependent, with the higher errors in the UV) (Eck et al., 1999). The sun photometer located in Granada is included in the AERONET network (Holben et al., 1998). Uncertainty of AERONET inversion products is described in detail in Dubovik et al. (2000). The study showed that the uncertainty in the retrieved single scattering albedo, $\omega(\lambda)$, is ± 0.03 for high aerosol load ($\tau_{440\,\mathrm{nm}} > 0.4$) and solar zenith angle $\theta > 50°$. In cases with low aerosol load ($\tau_{440\,\mathrm{nm}} < 0.2$), the retrieval accuracy of $\omega(\lambda)$, drops down to 0.02–0.07 (Dubovik et al., 2000). For $\tau_{440\,\mathrm{nm}} > 0.4$ and $\theta > 50°$ the imaginary part of the refractive index is about 20–50 %. The aerosol size distribution retrieval depends on aerosol particle sizes, types and actual values of the size distribution. For particles in the size range $0.1 < r < 7\,\mu$m, the uncertainty is around 10–35 %, whereas for sizes lower than 1 μm and higher than 7 μm the uncertainty increases up to 80–100 %. The AERONET Version 2 Level 1.5 data are used in this study for the characterization of the aerosol properties and for the retrieval of the aerosol microphysical properties profiles in combination with backscattered elastic lidar signals by means of the Lidar Radiometer Inversion Code (LIRIC) (Chaikovsky et al., 2012; Wagner et al., 2013; Granados-Muñoz et al., 2014).

For the analysis of the aerosol hygroscopic properties at the station, specific radiosounding launch campaigns are performed in order to obtain the RH humidity profiles since 2011. Radiosoundings (DFM-09 from GRAW Radiosondes) are launched simultaneously and collocated to the lidar measurements. They provide temperature (resolution 0.01 °C, accuracy 0.2 °C), pressure (resolution 0.1 hPa, accuracy 0.5 hPa), humidity (resolution 1 %, accuracy 2 %) and wind speed (resolution 0.1 m s^{-1}, accuracy 0.2 m s^{-1}) profiles. Data acquisition and processing are performed by Grawmet5 software and a GS-E ground station from the same manufacturer. GRAW Radiosondes showed good performance at previous intercomparison studies, especially for the RH in the lower troposphere (deviations below 2 %) (Sun et al., 2013).

In addition, analysis of backwards trajectories is performed in this study by means of the HYSPLIT model (Hybrid Single-Particle Lagrangian Integrated Trajectory) (Draxler and Rolph, 2003). Five-day backwards trajectories of air masses arriving at the experimental site at different height levels depending on the region of interest are computed using the model including vertical wind information. The trajectories analysis allows the interpretation of the source regions of air masses reaching the experimental site. Under low wind conditions, trajectories can have relative error ~ 40 % (Stunder, 1996). The Global Data Assimilation System (GDAS) database is used as input meteorological database for the computations.

3 Methodology

3.1 Retrieval of aerosol optical and microphysical properties and relative humidity profiles

From the inversion of the lidar data, the aerosol particle backscatter coefficient profiles ($\beta_\lambda^{\mathrm{P}}$) are obtained by applying the Klett–Fernald inversion algorithm (Fernald, 1984; Fernald et al., 1972; Klett, 1981, 1985). This algorithm assumes a reference height free of aerosol particles and a height-independent aerosol particle lidar ratio (extinction-to-backscatter ratio) for each wavelength. More details can be found in (Bravo-Aranda et al., 2013, and references therein) and (Guerrero-Rascado et al., 2009, 2011). Aerosol particle lidar ratios assumed when applying the Klett–Fernald algorithm to the lidar data are obtained by minimizing the difference between the integral of the aerosol particle backscatter coefficient profile multiplied by the particle lidar ratio and the aerosol optical depth provided by AERONET for each wavelength (Guerrero-Rascado et al., 2008; Landulfo et al., 2003). The aerosol particle backscatter related Angstöm exponent profiles between 355 and 532 nm (β-AE(355–532 nm)), related to the aerosol particles size, are also obtained. In addition, the particle linear depolarization ratio profiles ($\delta_{532\,\mathrm{nm}}^{\mathrm{P}}$) are also calculated as explained in (Bravo-Aranda et al., 2013) in order to analyze variations in the shape of the particles. The assumption of a height-independent lidar ratio in Klett retrieval might be a source of error in cases of non-homogenous aerosol layers, but even incorrect assumed backscatter-to-extinction ratios can yield rather accurate results (Kovalev, 1995). In general, the errors in the profiles obtained with the Klett method are less than 20 % for $\beta_\lambda^{\mathrm{P}}$ and less than 25 % for $\delta_{532\,\mathrm{nm}}^{\mathrm{P}}$ (Franke et al., 2001; Alados-Arboledas et al., 2011; Preissler et al., 2011).

The volume concentration profiles are retrieved by means of Lidar Radiometer Inversion Code (LIRIC) (Chaikovsky et al., 2012). This algorithm provides vertical profiles of microphysical properties from a combined set of sun photometer and lidar data. LIRIC inputs are column-integrated optical and microphysical properties retrieved from the sun photometer measurements using AERONET code (Dubovik et al., 2006) and measured lidar elastic backscatter signals at three different wavelengths (355, 532, and 1064 nm). The depolarization information from lidar data can be optionally used. From AERONET code, input variables used by LIRIC are columnar integrated volume concentration values for each mode, refractive index, single scattering albedo, integrated backscatter coefficients, the first and second diagonal elements of the scattering matrices at 180°, the fraction of the spherical particles and the aerosol optical depth. In addition, based also on AERONET code, an aerosol model, defined by the columnar integrated volume concentrations of each mode (fine and coarse modes) and based on a mixture of randomly oriented spherical and spheroid particles, is assumed (Dubovik and King, 2000; Dubovik et al., 2006). These data, together with the elastic lidar raw signals at 355, 532 and 1064 nm, are used to obtain the volume concentration profiles for fine and coarse particles, distinguishing between coarse spherical and coarse spheroid mode when 532 nm cross-polarized lidar channel is used. The mathematical procedure used to retrieve the volume concentration profiles is described in detail in Chaikovsky et al. (2008). It is worthy to point out here that the refractive index, percentage of sphericity and size distribution for each mode are assumed to be height-independent in LIRIC retrievals. In cases of different aerosol types in the atmospheric column or variation of these properties with height, this assumption may introduce some uncertainties. Therefore, it is necessary to consider the limitations of LIRIC under these specific situations when analyzing LIRIC results. Ancillary information, such as the lidar retrieved optical properties profiles, has to be used in order to guarantee the reliability of the results. In cases of hygroscopic growth, it is clear that the refractive index is height-dependent, since it varies with increasing/decreasing relative humidity due to the changes in the aerosol composition. However, we performed some tests with LIRIC by varying the refractive index provided by AERONET, which is the one used by LIRIC, and keeping all the other parameters in LIRIC code unvaried. The volume concentration profiles retrieved for the different values of refractive index were almost the same (with differences below 7.5 %). Therefore, it can be assumed that LIRIC's sensitivity to variations of the refractive index is very low and the associated uncertainty is quite small.

The RH profiles are directly measured by the radiosounding simultaneously and collocated to the lidar measurements, therefore no assumptions concerning the RH profile are needed as in previous studies (Ferrare et al., 1998; Wulfmeyer and Feingold, 2000; Feingold and Morley, 2003).

3.2 Procedure for selection of hygroscopic growth case studies

For the retrieval of the aerosol hygroscopic properties from lidar data, very specific conditions need to be fulfilled. Aerosol water uptake is associated to the increase in aerosol optical properties such as the aerosol particle backscatter coefficient $\beta^{\mathrm{P}}_{\lambda}$ and aerosol particle microphysical properties such as the volume concentration. Therefore, for the study of these hygroscopic properties, we need to observe an increase in the aerosol particle backscatter at a certain aerosol layer. This increase has to occur simultaneously to an increase in RH in this aerosol layer in order to consider hygroscopic growth as the possible cause of the changes in the aerosol particles properties. Those cases fulfilling the previously described conditions are considered as potential cases of hygroscopic growth.

β-AE and $\delta^{\mathrm{P}}_{532\,\mathrm{nm}}$ profiles are analyzed as ancillary information. In general, a simultaneous decrease of these two properties is indicative of larger and more spherical particles, which is a good indicator of aerosol hygroscopic growth. This positive correlation between these two aerosol properties usually occurs only in cases of aerosol hygroscopic growth or aging processes. Nonetheless, during our analysis it was observed that the decrease in the β-AE, although existing, is quite smooth, and variations are within the uncertainty limits. Therefore, a more detailed analysis with more case studies and based on Raman analysis will be needed to reinforce this conclusion.

Once the potential cases of hygroscopic growth are detected, it is necessary to verify that in the analyzed aerosol layer the atmospheric aerosol particles present a certain degree of homogeneity. In this way, we can corroborate that the variations in the aerosol particle properties such as $\beta^{\mathrm{P}}_{\lambda}$ and the volume concentration, are caused by the increase in the aerosol size due to water uptake and not to changes in the aerosol composition or load in the analyzed layer. That means that the same aerosol type or mixture must be present along the analyzed height range and almost no variations in the aerosol load must exist. For this purpose, ancillary information such as the backward trajectories of the air masses, the profiles of potential temperature, θ, or the water vapor mixing ratio profiles, r, are used to ensure that the aerosol layer under study is affected by hygroscopic growth. In this sense, if the origin and the trajectory of the air masses are independent of the altitude in the layer analyzed, it is considered that the same aerosol type might have been advected and, therefore, a homogenous aerosol composition might be expected in the analyzed layer. Otherwise, variations in the aerosol composition are expected and the case is not considered for analysis of hygroscopic growth. These air mass backward trajectories analyses are mainly used as a first approach selection criterion. In addition, good mixing is required as a boundary condition in order to guarantee the homogeneity of the atmospheric aerosol in the investigated layer. In general, constant profiles of θ and r are indicators of well mixed conditions within the atmosphere. In our analysis, θ and r are calculated in the analyzed layer to check the mixing conditions. Both atmospheric variables are calculated from the radiosounding temperature and relative humidity profiles. Only those cases with almost constant values of θ and r in the analyzed layer (variations lower than 2° and $2\,\mathrm{g\,kg^{-1}}$, respectively) will be considered for the analysis. Good mixed conditions cannot be guaranteed in any other case.

The backward trajectories analysis and the height independency of r criteria for identifying hygroscopic growth were already used in the study by Veselovskii et al. (2009). However, the height independency of θ is introduced for the first time in this study in order to provide more robust boundary conditions and guarantee the occurrence of hygroscopic growth within a well-mixed layer (Stull, 1988).

If these requirements for the homogeneity on the atmospheric aerosol layer are fulfilled, these cases are selected for a more detailed analysis of atmospheric aerosol hygroscopic properties. For the analysis of the hygroscopic growth, the enhancement factor is defined as

$$f_\zeta\,(\mathrm{RH}) = \frac{\zeta\,(\mathrm{RH})}{\zeta\,(\mathrm{RH_{ref}})}, \tag{1}$$

where ζ(RH) represents an aerosol property at a certain RH. $\mathrm{RH_{ref}}$ is the so-called reference RH. This $\mathrm{RH_{ref}}$ is chosen as the lowest value of RH in the analyzed layer. For this study, f_ζ(RH) is obtained for β_λ^P profiles (f_β(RH)) and the volume concentration profiles (f_VC(RH)).

The enhancement factor total uncertainty is very difficult to determine since it is highly dependent on the uncertainties of the aerosol properties and the RH, the uncertainty in the vertical and temporal coordinates, on the range of RH considered as well as the hygroscopic growth of the particle itself and therefore it is not well characterized yet. There is still a lack of studies about the characterization of the enhancement factor uncertainty that should be fulfilled in future studies. Adam et al. (2012) provided estimations based on a sensitivity test and Mie calculations. According to their study, this uncertainty varies between 4 % (for RH < 40 %) and 38 % (at RH > 95 %).

For different cases, the ranges of RH and, as a consequence, $\mathrm{RH_{ref}}$ values are different. Therefore, in order to make the different results comparable it is necessary to use a common $\mathrm{RH_{ref}}$ value. For this purpose, the Hänel model (Hänel, 1976) is used to parameterize the experimental enhancement factor curves. The general form of the Hänel equation is expressed as

$$f_\zeta\,(\mathrm{RH}) = \left(\frac{1-\mathrm{RH}}{1-\mathrm{RH_{ref}}}\right)^{-\gamma}, \tag{2}$$

where γ is an indicator of the hygroscopicity of the particles. Larger γ values are related to more hygroscopic aerosol types.

Since atmospheric aerosol hygroscopic properties are highly dependent on the aerosol chemical composition information, NAAPS (Navy Aerosol Analysis and Prediction System) (Christensen, 1997) model is also considered in this study to support the data interpretation with respect to the aerosol type.

To summarize, for the detection of aerosol hygroscopic growth the following procedure is followed: (1) detection of the aerosol particle backscatter coefficient increase with simultaneous increase in RH in a certain aerosol layer, (2) confirmation that the atmospheric aerosol layer under study is homogenous. For this purpose, homogeneity boundary conditions are implemented: the origin and the trajectory of the air masses at the analyzed layer has to be independent of the altitude and the profiles of r and θ has to be constant with height in this layer as an indicative of well mixed conditions.

4 Results and discussion

Data corresponding to several radiosounding launch campaigns performed at the experimental site during the period 2011 to 2013 were analyzed following the methodology described in the previous section. During this period, lidar measurements were always running in coincidence with the radiosounding launches. These lidar measurements have been exhaustively analyzed together with the RH profiles provided by the radiosoundings in order to detect cases of aerosol hygroscopic growth. Two case studies corresponding to 22 July 2011 and 22 July 2013 are presented here to show the potential of the technique described in Sect. 3.2. In these case studies, atmospheric conditions are highly supportive for aerosol hygroscopic growth at certain height ranges. A detailed analysis of these cases is presented in the following paragraphs.

Case I corresponds to the 22 July 2011. On that day, a radiosounding was launched at 20:30 UTC in coincidence with night-time lidar measurements. According to NAAPS model, 22 July 2011 at 18:00 UTC is characterized by the presence of mineral dust above Granada (Fig. S1a in the Supplement). However, this is not in agreement with the experimental data at the time of the measurements. Mineral dust was observed just as a remaining small layer at high altitude from a previous dust event above the station, but it was not the predominant aerosol type. No presence of mineral dust was detected in the aerosol layer studied for hygroscopic growth, as we will show later on. A second case of hygroscopic growth is detected on 22 July 2013 (Case II) during the summer radiosounding campaign. For that specific day, the radiosounding was launched at 20:00 UTC in coincidence with simultaneous lidar measurements. On that day, the NAAPS model indicates the presence of sulphates and smoke above the experimental site (Fig. S1b).

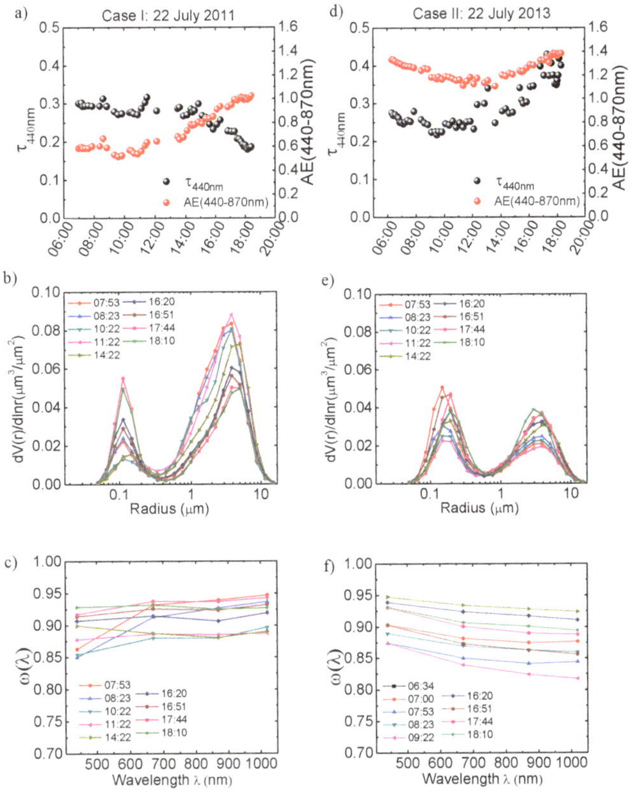

Figure 1. **(a)** AERONET τ_{440} and AE(440–870) daily time series for Case I. **(b)** AERONET retrieved volume size distributions for Case I. **(c)** $\omega(\lambda)$ for Case I. **(d)** AERONET τ_{440} and AE(440–870) daily time series for Case II. **(e)** AERONET retrieved volume size distributions for Case II. **(f)** $\omega(\lambda)$ for Case II.

Sun photometer experimental data are also used for this analysis, since they provide information about the aerosol properties. In addition, sun photometer data are required for the retrieval of the volume concentration profiles with LIRIC. For Case I, sun photometer data suggest the presence of a Saharan dust plume that dissipates at late afternoon (Fig. 1a), when our measurements took place. The aerosol optical depth at 440 nm ($\tau_{440\,\mathrm{nm}}$) decreases during the afternoon, changing from 0.30 in the morning to values of 0.20 at 18:30 UTC. The Angström exponent between 440 and 870 nm (AE(440–870 nm)) increases from 0.5 to 1.1, indicating an enhancement in the contribution of fine particles from 15:00 UTC. The aerosol size distributions retrieved during the day also indicate a decrease in the coarse mode and an increase of the fine mode from midday onwards. At 18:30 UTC (Fig. 1b) the aerosol size distributions indicate a balanced presence of both fine and coarse particles. This is confirmed by the fine mode fraction, determined through the SDA (spectral deconvolution algorithm, not shown) (O'Neill et al., 2003) that increases from 0.35 in the early morning up to 0.55 in the late evening. In the retrievals of single scattering albedo, $\omega(\lambda)$, corresponding to the morning hours it is observed a strong influence of mineral dust (high $\omega(\lambda)$ values and increasing $\omega(\lambda)$ with wavelength) (Fig. 1c). However, in the late afternoon, $\omega(\lambda)$ values around 0.93, and its neutral spectral dependence suggest the presence in the atmospheric column of aerosol from anthropogenic origin with influence of residual mineral dust (Dubovik et al., 2002; Lyamani et al., 2006; Valenzuela et al., 2012), which will be confirmed later with the lidar data and the backward trajectories analysis (Fig. 1c).

For the second case, $\tau_{440\,\mathrm{nm}}$ values indicate high aerosol loads reaching values above 0.40 at 17:19 UTC (Fig. 1d). The AE(440–870 nm) exhibits values larger than 1.2 during the whole day reaching 1.4 at 18:30 UTC, thus indicating a predominance of fine particles. The AERONET inversion retrievals during the whole day show bimodal size distributions with predominance of fine particles mode (Fig. 1e). Both, the ω(440 nm) values close to 0.9, and their spectral dependence, with a decreasing trend with wavelength, evidence the presence of anthropogenic pollution and/or smoke over Granada (Lyamani et al., 2006) (Fig. 1f), in agreement with NAAPS forecast model.

On both case studies, lidar measurements were running from 20:00 to 22:00 UTC. On 22 July 2011, lidar range corrected signal (RCS) time series (Fig. 2a) indicate the presence of atmospheric aerosol up to 3000 m a.s.l. Moreover, a strong increase of the RCS is observed inside the height range around 2400 m a.s.l. between 20:30 and 21:00 UTC. On the other hand, the time series of the lidar RCS on 22 July 2013 (Fig. 2b) indicate that the atmospheric aerosol reach altitudes up to 3500 m a.s.l. with the strongest backscattered lidar signal around 3000 m a.s.l. For this second case, some clouds were observed from 21:30 UTC. The occurrence of these clouds might be related to the ability of hygroscopic aerosol to act as CCN.

The analysis of lidar data by means of the Klett–Fernald inversion algorithm for both cases is shown in Fig. 3. Mean profiles of $\beta^{\mathrm{P}}_{532\,\mathrm{nm}}$, β-AE(355–532 nm) and $\delta^{\mathrm{P}}_{532\,\mathrm{nm}}$ corresponding to the period from 20:30 to 21:00 UTC for Case I and to 20:00–20:30 UTC for Case II are presented in this figure. On both cases, we observe a marked increase with altitude in $\beta^{\mathrm{P}}_{532\,\mathrm{nm}}$ in the range between 1330 and 2330 m a.s.l. for Case I and 1300 and 2700 m a.s.l. for Case II (Fig. 3a). Simultaneous to this increase in $\beta^{\mathrm{P}}_{532\,\mathrm{nm}}$, the RH also increase with altitude in both layers (Fig. 3b). Water vapor mixing ratio, r, profiles retrieved from the radiosounding data are shown in Fig. 3c. Opposite to $\beta^{\mathrm{P}}_{532\,\mathrm{nm}}$, β-AE(355–532 nm) decreases with altitude in the both layers (Fig. 3d, Table 1). The decrease in β-AE(355–532 nm) with altitude indicates an increase in aerosol particles size at higher altitudes, even though it is necessary to be aware that the decrease is very smooth and the variations are within the uncertainty limits in both cases. The $\delta^{\mathrm{P}}_{532\,\mathrm{nm}}$ also decreases with altitude for both cases in the corresponding aerosol layers (Fig. 3e). The decrease in $\delta^{\mathrm{P}}_{532\,\mathrm{nm}}$ is stronger than the one observed in β-AE(355–532 nm) and it indicates that there is an increase

Table 1. Values of the different aerosol properties at the lowest and highest altitudes of the analyzed layers for Case I and Case II.

	Case I		Case II	
	1330 m a.s.l.	2330 m a.s.l	1300 m a.s.l.	2700 m a.s.l.
$\beta^{aer}_{532\,nm}(Mm^{-1}\cdot sr^{-1})$	2.17	4.20	1.11	3.84
RH (%)	60	90	40	85
β-AE(355–532 nm)	1.3	0.8	2.0	1.0
$\delta^{P}_{532\,nm}$	0.10	0.05	0.07	0.03

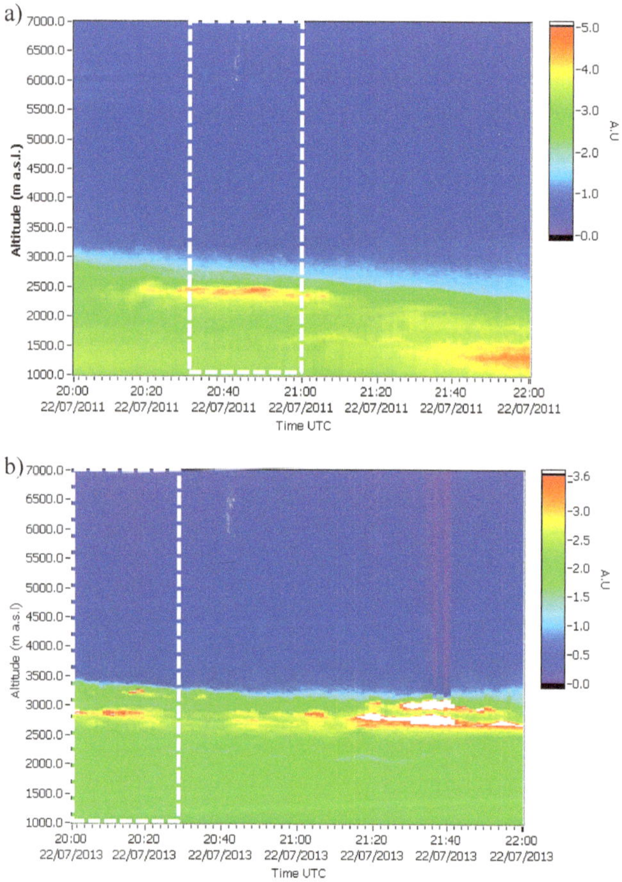

Figure 2. (a) Lidar RCS time series at 532 nm (arbitrary units) on 22 July 2011 from 20:00 to 22:00 UTC. Features shown here are aerosol related. **(b)** Lidar RCS time series at 532 nm (arbitrary units) on 22 July 2013 from 20:00 to 22:00 UTC. For this case, some clouds were present between 21:35 and 24:45, leading to a strong attenuation of the RCS. On both cases data were cloud-screened before proceeding to the investigation of hygroscopic growth effects.

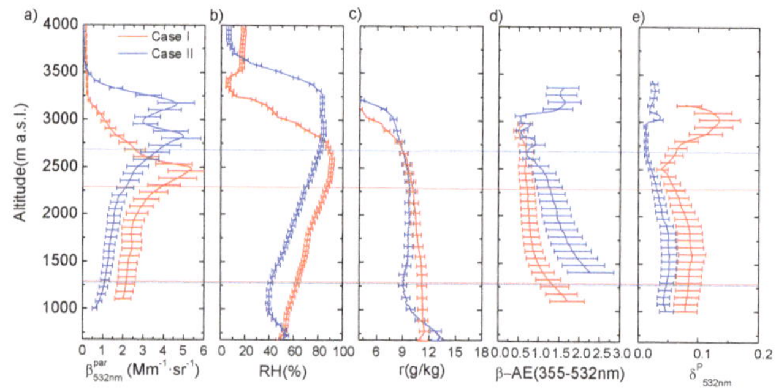

Figure 3. (a) $\beta^{P}_{532\,nm}$ retrieved with Klett–Fernald algorithm (LR = 65 sr for Case I and LR = 70 for Case II) from 20:30 to 21:00 UTC on Case I and 20:00 to 20:30 UTC on Case II **(b)** RH profiles from the radiosounding launched at 20:30 UTC on Case I and at 20:00 UTC on Case II. **(c)** r profiles from the radiosounding launched at 20:30 UTC on Case I and at 20:00 UTC on Case II. **(d)** β-AE(355–532 nm) retrieved with Klett–Fernald algorithm for the same periods. **(e)** $\delta^{P}_{532\,nm}$ retrieved from lidar data for the same periods. Horizontal lines represent the height limits of the aerosol layers selected for the analysis in Case I (red line) and Case II (blue lines).

in particle sphericity, which is a typical behavior of hygroscopic growth. The observed positive correlation between β-AE(355–532 nm) and $\delta^{P}_{532\,nm}$ might be a very good indicator of aerosol hygroscopic growth and, up to our knowledge, it has never been presented in previous studies. However, due to the uncertainty in β-AE(355–532 nm), it has to be carefully read in this particular study, and a more comprehensive study with more case studies will be necessary (as previously stated). On Case I, $\delta^{P}_{532\,nm}$ profiles retrieved from lidar data in the analyzed aerosol layer are rather low ($\delta^{P}_{532\,nm} \sim 0.07$), indicating predominance of spherical particles in the analyzed layer. However, at higher altitudes, $\delta^{P}_{532\,nm}$ reach values of 0.15 and β-AE(355–532 nm) ~ 0.6 with RH values lower than 40 %, indicating that the influence of mineral dust observed in the column with the sun photometer data is more important in this upper layer than in our region of interest.

In order to corroborate that the variations with height in the aerosol properties are due to water uptake and not to inhomogeneities in the aerosol layer, the boundary conditions established in Sect. 3.2 are checked (Fig. 4). The 5-day backward trajectories analysis performed with the HYSPLIT model reveals that the origin and trajectories of the air masses are almost identical for the different altitudes within the aerosol layer considered in each case. Therefore, the same aerosol type might have been advected over Granada in the investigated aerosol layers. For Case I, air masses come from the Northwest of Europe, going through the Northern Iberian Peninsula, finally arriving over Granada after overpassing the

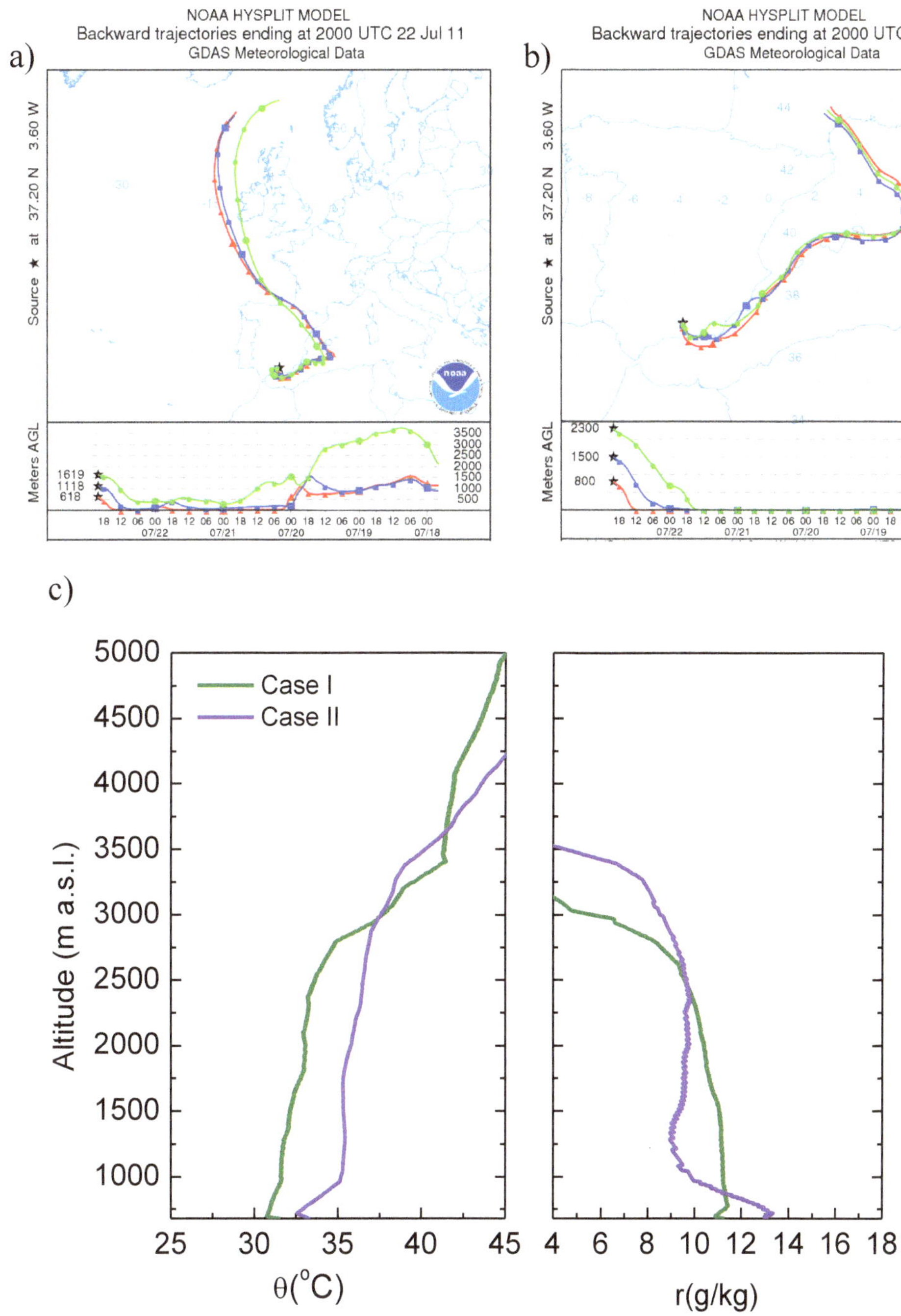

Figure 4. **(a)** 5-day back trajectories of the air masses ending on 22 July 2011 at 20:00 at Granada at three altitude heights within the 1330–2330 m a.s.l. height range. **(b)** 5-day back trajectories of the air masses ending on 22 July 2013 at 20:00 at Granada at three altitude heights within the 1300–2700 m a.s.l. height range. **(c)** Vertical profiles of θ(in °C) and r (in $g\,kg^{-1}$) from radiosounding data on 22 July 2011 at 20:30 UTC (Case I) and 22 July 2013 at 20:00 UTC (Case II).

Iberian Peninsula Mediterranean Coast (Fig. 4a). These air masses might have transported anthropogenic aerosol from Europe to the experimental site, especially considering that they were traveling at very low altitudes, located within the planetary boundary layer according to HYSPLIT retrieved boundary layer heights. The backward trajectories analysis also revealed that the air masses were traveling very close to the sea surface above the Mediterranean Sea and marine aerosols might be probably present in the aerosol mixture (Fig. 4a). For Case II, the air masses come from the Mediterranean region at the three altitude levels considered (Fig. 4b), and they were traveling within the marine boundary layer before reaching Granada station; therefore, they are likely loaded with marine aerosol from the western Mediterranean Sea together with sulphates and smoke from Europe, as indicated by the NAAPS model. The trajectories are very similar to those in Case I, but in this case they traveled more slowly above the Mediterranean Sea, and are therefore likely loaded with more marine aerosol than in the previous case.

Vertical profiles of θ and r measured with radiosounding data are also checked in order to corroborate good mixing conditions within the analyzed aerosol layers. Both θ and r profiles present almost constant values in the analyzed layers in both case studies (Fig. 4c) and thus it can be inferred that the analyzed layers are well mixed in both cases. Once these conditions are fulfilled, vertical homogeneity in the analyzed layers can be assumed. Therefore, hygroscopic growth is foreseen for these cases since there is a high likelihood that changes in the aerosol properties are due to water uptake.

According to all the previous results, these cases are considered representative of hygroscopic growth since there is an enhancement in $\beta^{P}_{532\,\mathrm{nm}}$ in coincidence with an increase in RH in the selected aerosol layers. In addition, the positive correlation between the β-AE(355–532 nm) and the $\delta^{P}_{532\,\mathrm{nm}}$ values suggests hygroscopic growth, since aerosol particles become larger and more spherical due to water uptake. Backward trajectories analysis with HYSPLIT and the height independency of θ and r in the analyzed height range corroborates that the enhancement of $\beta^{P}_{532\,\mathrm{nm}}$ is due to water uptake because of the homogeneity of the aerosol layer.

Following the methodology described in Sect. 3.2, from the combination of $\beta^{P}_{532\,\mathrm{nm}}$ and RH profiles in Fig. 3, the aerosol particle backscatter coefficient enhancement factor f_β(RH) is obtained as indicated in Eq. (1).

In Case I, $\mathrm{RH_{ref}} = 60\,\%$, which is the lowest value measured in the investigated layer. The dependence of f_β(RH) with the RH is shown in the resultant humidogram in Fig. 5a. From this figure, it is evident that $\beta^{P}_{532\,\mathrm{nm}}$ increases 2.5 times ($f_\beta\,(90\,\%) = 2.5$) in the range of humidity between 60 and 90 % . The humidogram in Case II shows that $f_\beta\,(83\,\%) = 3.5$, in the range of RH between 40 and 83 %. For Case II, $\mathrm{RH_{ref}}$ is established at 40 %, since it is the lowest RH value reached in the analyzed layer for this case.

In a similar study performed by Veselovskii et al. (2009) in the East Coast of the United States, they attained a value of

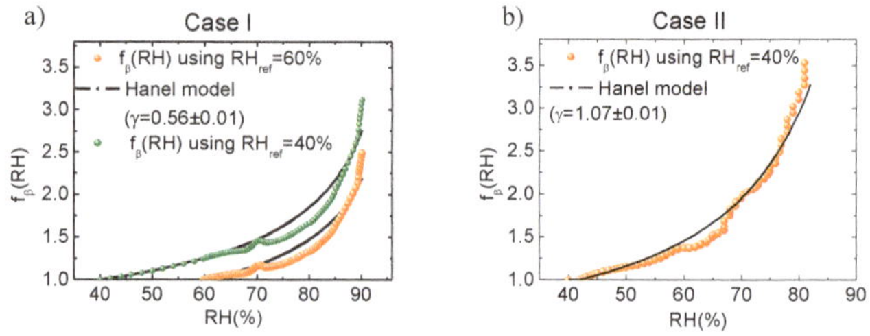

Figure 5. (a) f_β (RH) retrieved on 22 July 2011 (Case I) from 20:30 to 21:00 UTC for the height range between 1330 and 2330 m a.s.l. (yellow dots for $\mathrm{RH_{ref}} = 60\,\%$ and green dots for $\mathrm{RH_{ref}} = 40\,\%$). **(b)** f_β(RH) retrieved on 22 July 2013 (Case II) from 20:00 to 20:30 UTC for the layer corresponding to heights between 1300 and 2700 m a.s.l. using $\mathrm{RH_{ref}} = 40\,\%$.

the aerosol particle extinction coefficient enhancement factor ($f_\alpha\,(85\,\%) = 2.3$) in the presence of the typical continental haze using $\mathrm{RH_{ref}} = 60\,\%$. It is necessary to take into account that Veselovskii et al. (2009) used the aerosol particle extinction coefficient profile; thus, results are comparable only in a contextual way, since it would be necessary to know the influence of the aerosol hygroscopic growth on the aerosol particle lidar ratio to perform a quantitative comparison. For Case I, the value obtained here for $f_\beta\,(85\,\%)$ is much lower than the one provided by Veselovskii et al. (2009) ($f_\beta\,(85\,\%) = 1.5$). However, for Case II $f_\beta\,(85\,\%) = 2.6$ using $\mathrm{RH_{ref}} = 60\,\%$, which is very similar to the one obtained by Veselovskii et al. (2009).

A qualitative comparison with in situ studies can be done in order to contextualize our results. However, when making this comparison it is necessary to take into account the differences between both techniques. In addition, in situ analyses are usually performed under controlled conditions, whereas lidar data are measured under real and unperturbed conditions. In addition, in situ studies are frequently based on the retrieval of the aerosol particle light-scattering coefficient enhancement factor f_σ(RH) and not on the f_β(RH) used here. They usually provide values for $f_\sigma\,(85\,\%)$ using $\mathrm{RH_{ref}}$ values of 40 % or lower (dry conditions). In order to compare our results to these in situ studies using a $\mathrm{RH_{ref}}$ of 40 %, the Hänel parameterization is applied to our data in Case I (Fig. 5a). For Case II, the Hänel parameterization is necessary to obtain $f_\beta\,(85\,\%)$, since RH values above 83 % are not reached. Values of $f_\beta\,(80\,\%)$, $f_\beta\,(85\,\%)$ using $\mathrm{RH_{ref}} = 40\,\%$ and γ obtained are summarized for both cases in Table 2.

As it can be inferred from Table 2 and Fig. 5, the atmospheric aerosol presents a stronger hygroscopic growth for Case II. According to the experimental AERONET and lidar data and the ancillary information of the model, this may be to the larger contribution of sulphates (in the fine mode) and marine aerosol (in the coarse mode) during Case II than during Case I in the analyzed layers. In addition, a minor influence of the residual mineral dust particles (which presents very low hygroscopicity) from the morning hours in Case I

Table 2. Values of f_β (80 %), f_β (85 %) and γ for the two cases of hygroscopic growth corresponding to the 22 July of 2011 and 2013, respectively. The uncertainties in f_β (RH) are obtained by error propagation applied to Eq. (1). Only the uncertainty introduced by the aerosol particle backscatter coefficient is considered.

	Case I	Case II
f_β (80 %)	1.6 ± 0.6	3.0 ± 1.0
γ	0.56 ± 0.22	1.1 ± 0.4
f_β (85%)	2.1 ± 0.8	3.9 ± 1.5

may have led to low hygroscopic growth of the aerosol mixture (especially for the coarse mode). f_β (80 %) values obtained (1.60 ± 0.03 for Case I and 3.00 ± 0.02 for Case II) are in agreement with those presented in Titos et al. (2014a), that range between 1.4 and 3.4, being larger in those cases with marine influence. f_β (85 %) (2.10 ± 0.06 for Case I and 3.90 ± 0.03 for Case II) values are similar to those obtained in previous in situ studies within measurement differences limitations, ranging between 1.2 and 3.4 (Kim et al., 2006; Fierz-Schmidhauser et al., 2010a, b; Adam et al., 2012; Zieger et al., 2013). The γ values obtained from the Hänel parameterization in these case studies are in the range of values obtained in previous studies for the scattering coefficient using tandem of nephelometers (between 0.1 and 1.35) (Raut and Chazette, 2007; Gasso et al., 2000; Randriamiarisoa et al., 2006; Titos et al., 2014a, c).

The availability of the AERONET inversion retrieval data at late afternoon during both cases as shown in Fig. 1 allows us to retrieve volume concentration profiles by means of LIRIC algorithm, assuming no drastic temporal change in aerosol properties between the last AERONET retrieval and lidar measurements (Chaikovsky et al., 2008, 2012; Wagner et al., 2013; Granados-Muñoz et al., 2014). Figure 6a represents these volume concentration profiles obtained by the combination of lidar data from 20:30 to 21:00 UTC and the closest sun photometer retrieval (at 18:30 UTC) on Case I. Figure 6b shows the volume concentration profiles for Case II retrieved from AERONET data at 18:10 UTC and lidar data from 20:00 to 20:30 UTC. As it can be observed, there is a combination of coarse and fine particles along the whole profile for both cases. In both cases, the fine mode volume concentration increases with altitude in both analyzed layers. The total volume concentration for Case I increases with height reaching a maximum around 2500 m a.s.l. Figure 6c shows the volume concentration enhancement factor (f_{VC}(RH)) for the fine mode and the total volume concentration against RH for Case I. In this case, a threshold value of the volume concentration has been established at

$10\,\mu m^3\,cm^{-3}$ since f_{VC}(RH) is a ratio and lower values may induce a significant overestimation of f_{VC}(RH). Because of this, RH_{ref} for the fine mode is around 73 %. The fine mode volume concentration presents a strong increase with

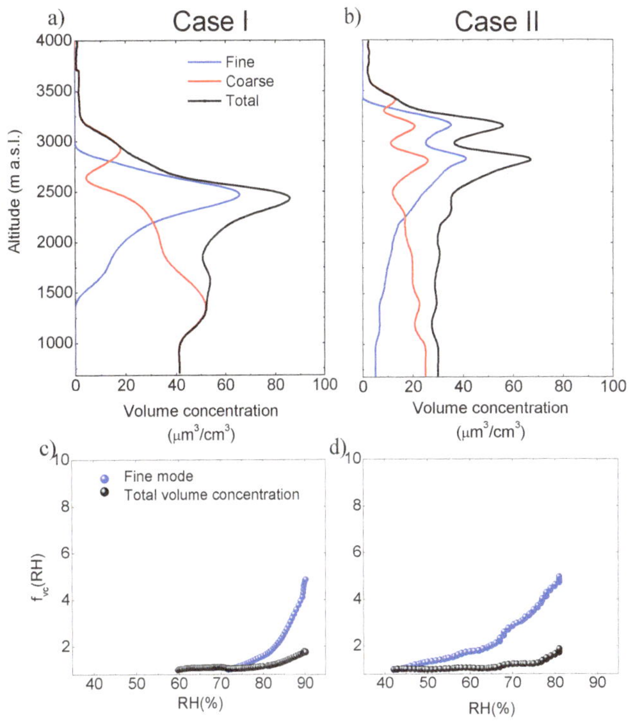

Figure 6. (a) LIRIC retrieved volume concentration (fine mode, coarse mode and total concentration) profiles on 22 July 2011 from 20:30 to 21:00 UTC. The shaded area indicates the height range where hygroscopic growth was investigated. **(b)** LIRIC retrieved volume concentration profiles on 22 July 2013 from 20:00 to 20:30 UTC. The shaded area indicates the investigated height range. **(c)** f_{VC}(RH) versus RH for the fine mode and the total volume concentrations for 22 July 2011 and the layer corresponding to heights between 1330 and 2330 m a.s.l. **(d)** f_{VC}(RH) vs. RH for the fine mode and the total volume concentrations for 22 July 2013 and the layer corresponding to heights between 1300 and 2700 m a.s.l.

RH, being f_{VC} (80 % = 1.57). The total volume concentration smoothly increases with RH, mainly due to the increase in the fine mode (f_{VC} (80 %) = 1.16 with RH_{ref} = 60 %). For Case II, Fig. 6d shows f_{VC}(RH) versus RH for the fine mode and the total volume concentration. An increase of f_{VC}(RH) with RH for the fine mode is observed, slightly smoother than in the previous case, with f_{VC} (80 %) = 1.28. Using RH_{ref} = 60 % in order to make a comparison with Case I, f_{VC} (80 %) for Case II is 1.57 which is larger than in Case I. According to these results, the fine mode is the one dominating the hygroscopic growth in the analyzed layers in both cases. Nonetheless, in Case II there is a larger increase of the total volume concentration with RH than in Case I, indicating that the coarse mode is more hygroscopic for Case II. This can be attributed to the higher influence of the marine aerosol advected from the Mediterranean Sea in the analyzed layer in Case II and the minor influence of the residual mineral dust in the analyzed layer in Case I, evidencing the influence of the chemical composition on the hygroscopic growth. Larger values of f(RH) are usually obtained for fine mode particles (Di Girolamo et al., 2012; Titos et al., 2014a). In our study, it seems that the fine mode is clearly more dominated by more hygroscopic particles whereas the coarse mode is dominated by substances with very low hygroscopic growth, especially for Case I (possible influence of mineral dust in the aerosol mixture). Di Girolamo et al. (2012) observed similar behavior analyzing aged dust particles partially mixed with maritime, urban and organic aerosols. However, according to Zieger et al. (2013), the relative contribution of the fine and the coarse modes and the specific chemical composition for each mode are very important for determining f(RH).

5 Conclusions

A new methodology to detect aerosol particle hygroscopic growth is implemented at Granada EARLINET experimental site. Aerosol hygroscopic properties are analyzed using a multispectral lidar system in combination with radiosounding data obtained during specific campaigns within the period 2011–2013. In the proposed method, an increase of the aerosol particle backscatter coefficient with relative humidity is used to detect aerosol hygroscopic growth. In addition, results point out that there is an associated decrease in the Angström exponent and in the particle depolarization ratio. The hypothesis about the positive correlation between the Angström exponent and the particle linear depolarization ratio is presented here for the first time as an indicator of the aerosol hygroscopic growth. However, the decrease in the Angström exponent is not significant in the case studies presented here since the variation is within the uncertainty limits. A further analysis with additional and more reliable data (e.g., Raman measurements) is needed in order to corroborate this hypothesis.

The height independency of the air masses arriving at the station and of the water vapor mixing ratio and the potential temperature profiles are also used as constraints in the method presented here in order to provide more strictness to the identification of the hygroscopic growth cases. The method proved to be reliable for the identification and analysis of hygroscopic growth situations based on the analysis of the aerosol particle properties profiles. The methodology is also applied to the analysis of the hygroscopic growth effects on the volume concentration profiles retrieved by means of LIRIC algorithm, which have never been done before.

Two cases of hygroscopic growth within the available data set (one on 22 July 2011 and another on 22 July 2013) are presented in this study to illustrate the potential of the exposed methodology. Different conditions were observed in these two cases allowing us to analyze the hygroscopic behavior of different aerosol types. In order to compare the two analyzed cases, the Hänel parameterization was used and data were recalculated using a common RH_{ref} value of 40 %. From this comparison, it was observed that the atmospheric aerosol presented a stronger hygroscopic growth during the case study corresponding to 22 July 2013, as indicated by f_β (80 %), f_β (85 %) and γ values. This can be explained by the different atmospheric conditions in the two cases. The case corresponding to the 22 July 2011 was affected by a mixture of atmospheric aerosols dominated by anthropogenic pollution and slightly affected by mineral dust and marine aerosol, whereas the case detected on 22 July 2013 was influenced mainly by sulphates with a stronger influence of marine aerosol than Case I, as indicated by AERONET data and the NAAPS model. The values obtained for the backscatter enhancement factor f_β (85 %), considering the associated uncertainties, are within the range obtained with in situ studies for the case on 22 July 2011, and slightly larger on 22 July 2013. Uncertainties presented here are related only to the aerosol backscatter coefficient. Since the experimental determination of the enhancement factor uncertainty by both remote sensing and in situ instrumentation is not straightforward and presents similar problems regarding its determination, further studies should point in this direction and would be a significant contribution to this research field.

The analysis of the volume concentration profiles reveals an increase of the total volume concentration with relative humidity, dominated in our particular cases by an increase in the fine mode fraction. The increase in the total volume concentration is larger in Case II corresponding to the 22 July 2013, due to the larger influence of the marine aerosol for this case and to the slight influence of mineral dust (non hygroscopic particles) on Case I.

Acknowledgements. This work was supported by the Andalusia Regional Government through projects P12-RNM-2409 and P10-RNM-6299, by the Spanish Ministry of Economy and Competitiveness through projects CGL2013-45410-R, CGL2011-13580-E/CLI and CGL2011-16124-E; by the EU through ACTRIS project (EU INFRA-2010-1.1.16-262254); and by the University of Granada through the contract "Plan Propio. Programa 9. Convocatoria 2013". CIMEL Calibration was performed at the AERONET-EUROPE calibration center, supported by ACTRIS (European Union Seventh Framework Program (FP7/2007-2013) under grant agreement no. 262254. M. J. Granados-Muñoz was funded under grant AP2009-0552. The authors thankfully acknowledge the computer resources, technical expertise, and assistance provided by the Barcelona Supercomputing Center for the BSC-DREAM8b model dust data. The authors express gratitude to the NOAA Air Resources Laboratory for the HYSPLIT transport and dispersion model. We also thank those at the NRL-Monterey that helped in the development of the NAAPS model.

Edited by: U. Wandinger

References

Adam, M., Putaud, J. P., Martins dos Santos, S., Dell'Acqua, A., and Gruening, C.: Aerosol hygroscopicity at a regional background site (Ispra) in Northern Italy, Atmos. Chem. Phys., 12, 5703–5717, doi:10.5194/acp-12-5703-2012, 2012.

Alados-Arboledas, L., Müller, D., Guerrero-Rascado, J. L., Navas-Guzmán, F., Pérez-Ramírez, D., and Olmo, F. J.: Optical and microphysical properties of fresh biomass burning aerosol retrieved by Raman lidar, and star- and sun-photometry, Geophys. Res. Lett., 38, L01807, doi:10.1029/2010GL045999, 2011.

Bravo-Aranda, J. A., Navas-Guzmán, F., Guerrero-Rascado, J. L., Pérez-Ramírez, D., Granados-Muñoz, M. J., and Alados-Arboledas, L.: Analysis of lidar depolarization calibration procedure and application to the atmospheric aerosol characterization, Int. J. Remote. Sens., 34, 3543–3560, 2013.

Chaikovsky, A., Dubovik, O., Goloub, P., Balashevich, N., Lopatsin, A., Karol, Y., Denisov, S., and Lapyonok, T.: Software package for the retrieval of aerosol microphysical properties in the vertical column using combined lidar/photometer data (test version), Tech. Rep., Institute of Physics, National Academy of Sciences of Belarus, Minsk, Belarus, 2008.

Chaikovsky, A., Dubovik, O., Goloub, P., Tanre, D., Pappalardo, G., Wandinger, U., Chaikovskaya, L., Denisov, S., Grudo, Y., Lopatsin, A., Karol, Y., Lapyonok, T., Korol, M., Osipenko, F., Savitski, D., Slesar, A., Apituley, A., Arboledas, L. A., Binietoglou, I., Kokkalis, P., Granados Muñoz, M. J., Papayannis, A., Perrone, M. R., Pietruczuk, A., Pisani, G., Rocadenbosch, F., Sicard, M., De Tomasi, F., Wagner, J., and Wang, X.: Algorithm and software for the retrieval of vertical aerosol properties using combined lidar/radiometer data: Dissemination in EARLINET,, paper presented at 26th International Laser and Radar Conference, Porto Heli, Greece, 2012.

Charlson, R. J., Schwartz, S. E., Hales, J. M., Cess, R. D., Coakley Jr., J. A., Hansen, J. E., and Hofmann, D. J.: Climate forcing by anthropogenic aerosols, Science, 255, 423–430, 1992.

Christensen, J. H.: The Danish Eulerian hemispheric model – A three-dimensional air pollution model used for the Arctic, Atmos. Environ., 31, 4169–4191, 1997.

Covert, D. S., Charlson, R. J., and Ahlquist, N. C.: A study of the relationship of chemical composition and humidity to light scattering by aerosols, J. Appl. Meteorol., 11, 968–976, 1972.

DiGirolamo, P., Summa, D., Bhawar, R., Di Iorio, T., Cacciani, M., Veselovskii, I., Dubovik, O., and Kolgotin, A.: Raman lidar observations of a Saharan dust outbreak event: characterization of the dust optical properties and determination of particle size and microphysical parameters, Atmos. Environ., 50, 66–78, 2012.

Draxler, R. R. and Rolph, G. D.: HYSPLIT (HYbrid Single-Particle Lagrangian Integrated Trajectory) model access via NOAA ARL READY website (http://ready.arl.noaa.gov/HYSPLIT.php), last access: 26 January 2015, NOAA Air Resources Laboratory, Silver Spring, 2003.

Dubovik, O. and King, M. D.: A flexible inversion algorithm for retrieval of aerosol optical properties from Sun and sky radiance measurements, J. Geophys. Res., 105, 20673–20696, doi:10.1029/2000JD900282, 2000.

Dubovik, O., Holben, B., Eck, T. F., Smirnov, A., Kaufman, Y. J., King, M. D., Tanré, D., and Slutsker, I.: Variability of absorption and optical properties of key aerosol types observed in worldwide locations, J. Atmos. Sci., 59, 590–608, 2002.

Dubovik, O., Sinyuk, A., Lapyonok, T., Holben, B. N., Mishchenko, M., Yang, P., Eck, T. F., Volten, H., Muñoz, O., and Veihelmann, B.: Application of spheroid models to account for aerosol particle nonsphericity in remote sensing of desert dust, J. Geophys. Res.-Atmos., 111, D11208, doi:10.1029/2005JD006619, 2006.

Eck, T. F., Holben, B. N., Reid, J. S., Dubovik, O., Smirnov, A., O'Neill, N. T., Slutsker, I., and Kinne, S.: Wavelength dependence of the optical depth of biomass burning, urban, and desert dust aerosols, J. Geophys. Res.-Atmos., 104, 31333–31349, 1999.

Feingold, G. and Morley, B.: Aerosol hygroscopic properties as measured by lidar and comparison with in situ measurements, J. Geophys. Res.-Atmos., 108, 4327, doi:10.1029/2002JD002842, 2003.

Feingold, G., Furrer, R., Pilewskie, P., Remer, L. A., Min, Q., and Jonsson, H.: Aerosol indirect effect studies at Southern Great Plains during the May 2003 intensive operations period, J. Geophys. Res., 111, D05S14, doi:10.1029/2004JD005648, 2006.

Fernald, F. G.: Analysis of atmospheric lidar observations- Some comments, Appl. Optics, 23, 652–653, 1984.

Fernald, F. G., Herman, B. M., and Reagan, J. A.: Determination of aerosol height distributions by lidar, J. Appl. Meteorol., 11, 482–489, 1972.

Ferrare, R. A., Melfi, S. H., Whiteman, D. N., Evans, K. D., Poellot, M., and Kaufman, Y. J.: Raman lidar measurements of aerosol extinction and backscattering: 2. Derivation of aerosol real refractive index, single-scattering albedo, and humidification factor using Raman lidar and aircraft size distribution measurements, J. Geophys. Res.-Atmos., 103, 19673–19689, 1998.

Fierz-Schmidhauser, R., Zieger, P., Wehrle, G., Jefferson, A., Ogren, J. A., Baltensperger, U., and Weingartner, E.: Measurement of relative humidity dependent light scattering of aerosols, Atmos. Meas. Tech., 3, 39–50, doi:10.5194/amt-3-39-2010, 2010a.

Fierz-Schmidhauser, R., Zieger, P., Gysel, M., Kammermann, L., DeCarlo, P. F., Baltensperger, U., and Weingartner, E.: Measured and predicted aerosol light scattering enhancement factors at the high alpine site Jungfraujoch, Atmos. Chem. Phys., 10, 2319–2333, doi:10.5194/acp-10-2319-2010, 2010b.

Franke, K., Ansmann, A., Müller, D., Althausen, D., Wagner, F., and Scheele, R.: One-year observations of particle lidar ratio over the tropical Indian Ocean with Raman lidar, Geophys. Res. Lett., 56, 1766–1782, doi:10.1029/2001GL013671, 2001.

Gasso, S., Hegg, D. A., Covert, D. S., Collins, D., Noone, K. J., Öström, E., Schmid, B., Russell, P. B., Livingston, J. M., and Durkee, P. A.: Influence of humidity on the aerosol scattering coefficient and its effect on the upwelling radiance during ACE-2, Tellus B, 52, 546–567, 2000.

Granados-Muñoz, M. J., Navas-Guzmán, F., Bravo-Aranda, J. A., Guerrero-Rascado, J. L., Lyamani, H., Fernández-Gálvez, J., and Alados-Arboledas, L.: Automatic determination of the planetary boundary layer height using lidar: one-year analysis over southeastern Spain, J. Geophys. Res.-Atmos., 117, D18208, doi:10.1029/2012JD017524, 2012.

Granados-Muñoz, M. J., Guerrero-Rascado, J. L., Bravo-Aranda, J. A., Navas-Guzmán, F., Valenzuela, A., Lyamani, H., Chaikovsky, A., Wandinger, U., Ansmann, A. Dubovik, O., Grudo, J. O., and Alados-Arboledas, L.: Retrieving aerosol microphysical properties by Lidar-Radiometer Inversion Code (LIRIC) for different aerosol types, J. Geophys. Res.-Atmos., 119, 4836–4858, 2014.

Guerrero-Rascado, J. L., Ruiz, B., and Alados-Arboledas, L.: Multispectral Lidar characterization of the vertical structure of Saharan dust aerosol over southern Spain, Atmos. Environ., 42, 2668–2681, 2008.

Guerrero-Rascado, J. L., Olmo, F. J., Avilés-Rodríguez, I., Navas-Guzmán, F., Pérez-Ramírez, D., Lyamani, H., and Alados Arboledas, L.: Extreme Saharan dust event over the southern Iberian Peninsula in september 2007: active and passive remote sensing from surface and satellite, Atmos. Chem. Phys., 9, 8453–8469, doi:10.5194/acp-9-8453-2009, 2009.

Guerrero-Rascado, J. L., Andrey, J., Sicard, M., Molero, F., Comerón, A., Pujadas, M., and Alados-Arboledas, L.: Aerosol closure study by lidar, Sun photometry, and airborne optical counters during DAMOCLES field campaign at El Arenosillo sounding station, Spain, J. Geophys. Res.-Atmos., 116, D02209, doi:10.1029/2010JD014510, 2011.

Hänel, G.: The Properties of Atmospheric Aerosol Particles as Functions of the Relative Humidity at Thermodynamic Equilibrium with the Surrounding Moist Air, in: Advances in Geophysics, edited by: Landsberg, H. E. and Mieghem, J. V., 73–188, Elsevier, 1976.

Hegg, D. A., Covert, D. S., Rood, M. J., and Hobbs, P. V.: Measurements of aerosol optical properties in marine air, J. Geophys. Res.-Atmos., 101, 12893–12903, 1996.

Holben, B. N., Eck, T. F., Slutsker, I., Tanre, D., Buis, J. P., Setzer, A., Vermote, E., Reagan, J.A., Kaufman, Y. J., Nakajima, T., Lavenus, F., Jankowiak,, I., and Smirnov, A.: AERONET – A Federated Instrument Network and Data Archive for Aerosol Characterization, Remote Sens. Environ., 66, 1–16, 1998.

IPCC: Contribution of Working Group I to the Fifth Assessment Report of the Intergovernmental Panel on Climate Change, in: Summary for Policymakers in Climate Change, edited by: Stocker, T. F., Qin, D., Plattner, G. K., Tignor, M., Allen, S., Boschung, J., Nauels, A., Xia, Y., Bex, V., and Midgley, P., Cambrigde University Press, 2013.

Kim, J., Yoon, S. C., Jefferson, A., and Kim, S. W.: Aerosol hygroscopic properties during Asian dust, pollution, and biomass burning episodes at Gosan, Korea in April 2001, Atmos. Environ., 40, 1550–1560, 2006.

Klett, J. D.: Stable analytical inversion solution for processing lidar returns, Appl. Optics, 20, 211–220, 1981.

Klett, J. D.: Lidar inversion with variable backscatter/extinction ratios, Appl. Optics, 24, 1638–1643, 1985.

Kovalev, V. A.: Sensitivity of the Lidar Solution to Errors of the Aerosol Backscatter-to-Extinction Ratio – Influence of a Monotonic Change in the Aerosol Extinction Coefficient, Appl. Optics, 34, 3457–3462, 1995.

Landulfo, E., Papayannis, A., Artaxo, P., Castanho, A. D. A., de Freitas, A. Z., Souza, R. F., Vieira Junior, N. D., Jorge, M. P. M. P., Sánchez-Ccoyllo, O. R., and Moreira, D. S.: Synergetic measurements of aerosols over São Paulo, Brazil using LIDAR, sunphotometer and satellite data during the dry season, Atmos. Chem. Phys., 3, 1523–1539, doi:10.5194/acp-3-1523-2003, 2003.

Lyamani, H., Olmo, F. J., Alcántara, A., and Alados-Arboledas, L.: Atmospheric aerosols during the 2003 heat wave in southeastern Spain I: Spectral optical depth, Atmos. Environ., 40, 6453–6464, 2006a.

Lyamani, H., Olmo, F. J., Alcántara, A., and Alados-Arboledas, L.: Atmospheric aerosols during the 2003 heat wave in southeastern Spain II: microphysical columnar properties and radiative forcing, Atmos. Environ., 40, 6465–6476, 2006b.

Lyamani, H., Olmo, F. J., and Alados-Arboledas, L.: Physical and optical properties of aerosols over an urban location in Spain: seasonal and diurnal variability, Atmos. Chem. Phys., 10, 239–254, doi:10.5194/acp-10-239-2010, 2010.

Massling, A., Leinert, S., Wiedensohler, A., and Covert, D.: Hygroscopic growth of sub-micrometer and one-micrometer aerosol particles measured during ACE-Asia, Atmos. Chem. Phys., 7, 3249–3259, doi:10.5194/acp-7-3249-2007, 2007.

Navas-Guzmán, F., Bravo-Aranda, J. A., Guerrero-Rascado, J. L., Granados-Muñoz, M. J., and Alados-Arboledas, L.: Statistical analysis of aerosol optical properties retrieved by Raman lidar over Southeastern Spain, Tellus B, 65, 21234, doi:10.3402/tellusb.v65i0.21234, 2013.

Navas-Guzmán, F., Fernández-Gálvez, J., Granados-Muñoz, M. J., Guerrero-Rascado, J. L., Bravo-Aranda, J. A., and Alados-Arboledas, L.: Tropospheric water vapour and relative humidity profiles from lidar and microwave radiometry, Atmos. Meas. Tech., 7, 1201–1211, doi:10.5194/amt-7-1201-2014, 2014.

O'Neill, N. T., Eck, T. F., Smirnov, A., Holben, B. N., and Thulasiraman, S.: Spectral discrimination of coarse and fine mode optical depth, J. Geophys. Res., 108, 4559, doi:10.1029/2002JD002975, 2003.

Padró, L. T., Tkacik, D., Lathem, T., Hennigan, C. J., Sullivan, A. P., Weber, R. J., Huey, L. G., and Nenes, A.: Investigation of cloud condensation nuclei properties and droplet growth kinetics of the water-soluble aerosol fraction in Mexico City, J. Geophys. Res., 115, D09204, doi:10.1029/2009JD013195, 2010.

Pahlow, M., Feingold, G., Jefferson, A., Andrews, E., Ogren, J. A., Wang, J., and Turner, D. D.: Comparison between lidar and nephelometer measurements of aerosol hygroscopicity at the Southern Great Plains Atmospheric Radiation Measurement site, J. Geophys. Res.-Atmos., 111, D05S15, doi:10.1029/2004JD005646, 2006.

Pappalardo, G., Amodeo, A., Pandolfi, M., Wandinger, U., Ansmann, A., Bösenberg, J., Matthias, V., Amiridis, V., De Tomasi, F., Frioud, M., Iarlori, M., Komguem, L., Papayannis, A., Rocadenbosch, F., and Wang, X.: Aerosol lidar intercomparison in the framework of the EARLINET project. 3. Raman lidar algorithm for aerosol extinction, backscatter, and lidar ratio, Appl. Opt., 43, 5370–5385, 2004.

Pappalardo, G., Amodeo, A., Apituley, A., Comeron, A., Freudenthaler, V., Linné, H., Ansmann, A., Bösenberg, J., D'Amico, G., Mattis, I., Mona, L., Wandinger, U., Amiridis, V., Alados-Arboledas, L., Nicolae, D., and Wiegner, M.: EARLINET: towards an advanced sustainable European aerosol lidar network, Atmos. Meas. Tech., 7, 2389–2409, doi:10.5194/amt-7-2389-2014, 2014.

Preissler, J., Wagner, F., Pereira, S. N., and Guerrero-Rascado, J. L.: Multiinstrumental observation of an exceptionally strong Saharan dust outbreak over Portugal, J. Geophys. Res., 116, D24204, doi:10.1029/2011JD016527, 2011.

Randriamiarisoa, H., Chazette, P., Couvert, P., Sanak, J., and Mégie, G.: Relative humidity impact on aerosol parameters in a Paris suburban area, Atmos. Chem. Phys., 6, 1389–1407, doi:10.5194/acp-6-1389-2006, 2006.

Raut, J.-C. and Chazette, P.: Retrieval of aerosol complex refractive index from a synergy between lidar, sunphotometer and in situ measurements during LISAIR experiment, Atmos. Chem. Phys., 7, 2797–2815, doi:10.5194/acp-7-2797-2007, 2007.

Stull, R. B.: An Introduction to Boundary Layer Meteorology, Kluwer Acad., Dordrecht, the Netherlands, 1988.

Stunder, B. J. B.: An assessment of the quality of forecast trajectories, J. Appl. Meteorol., 35, 1319–1331, 1996.

Sun, B., Reale, A., Schroeder, S., Seidel, D. J., and Ballish, B.: Toward improved corrections for radiation-induced biases in radiosonde temperature observations, J. Geophys. Res.-Atmos., 118, 4231–4243, 2013.

Titos, G., Foyo-Moreno, I., Lyamani, H., Querol, X., Alastuey, A., and Alados-Arboledas, L.: Optical properties and chemical composition of aerosol particles at an urban location: An estimation of the aerosol mass scattering and absorption efficiencies, J. Geophys. Res.-Atmos., 117, D04206, doi:10.1029/2011JD016671, 2012.

Titos, G., Jefferson, A., Sheridan, P. J., Andrews, E., Lyamani, H., Alados-Arboledas, L., and Ogren, J. A.: Aerosol light-scattering enhancement due to water uptake during the TCAP campaign, Atmos. Chem. Phys., 14, 7031–7043, doi:10.5194/acp-14-7031-2014, 2014a.

Titos, G., Lyamani, H., Pandolfi, M., Alastuey, A., and Alados-Arboledas, L.: Identification of fine (PM1) and coarse (PM10-1) sources of particulate matter in an urban environment, Atmos. Environ., 89, 593–602, 2014b.

Titos, G., Lyamani, H., Cazorla, A., Sorribas, M., Foyo-Moreno, I., Wiedensohler, A., and Alados-Arboledas, L.: Study of the relative humidity dependence of aerosol light-scattering in southern Spain, Tellus B, 66, 24536, doi:10.3402/tellusb.v66.24536, 2014c.

Valenzuela, A., Olmo, F. J., Lyamani, H., Antón, M., Quirantes, A., and Alados-Arboledas, L.: Classification of aerosol radiative properties during African desert dust intrusions over southeastern Spain by sector origins and cluster analysis, J. Geophys. Res.-Atmos., 117, D06214, doi:10.1029/2011JD016885, 2012.

Veselovskii, I., Whiteman, D. N., Kolgotin, A., Andrews, E., and Korenskii, M.: Demonstration of aerosol property profiling by multiwavelength lidar under varying relative humidity conditions, J. Atmos. Ocean. Tech., 26, 1543–1557, 2009.

Wagner, J., Ansmann, A., Wandinger, U., Seifert, P., Schwarz, A., Tesche, M., Chaikovsky, A., and Dubovik, O.: Evaluation of the Lidar/Radiometer Inversion Code (LIRIC) to determine microphysical properties of volcanic and desert dust, Atmos. Meas. Tech., 6, 1707–1724, doi:10.5194/amt-6-1707-2013, 2013.

Wu, Z., Birmili, W., Poulain, L., Wang, Z., Merkel, M., Fahlbusch, B., van Pinxteren, D., Herrmann, H., and Wiedensohler, A.: Particle hygroscopicity during atmospheric new particle formation events: implications for the chemical species contributing to particle growth, Atmos. Chem. Phys., 13, 6637–6646, doi:10.5194/acp-13-6637-2013, 2013.

Wulfmeyer, V. and Feingold, G.: On the relationship between relative humidity and particle backscattering coefficient in the marine boundary layer determined with differential absorption lidar, J. Geophys. Res.-Atmos., 105, 4729–4741, 2000.

Zieger, P., Fierz-Schmidhauser, R., Weingartner, E., and Baltensperger, U.: Effects of relative humidity on aerosol light scattering: results from different European sites, Atmos. Chem. Phys., 13, 10609–10631, doi:10.5194/acp-13-10609-2013, 2013.

16

Generation of a bending angle radio occultation climatology (BAROCLIM) and its use in radio occultation retrievals

B. Scherllin-Pirscher[1], S. Syndergaard[2], U. Foelsche[1], and K. B. Lauritsen[2]

[1]Wegener Center for Climate and Global Change (WEGC) and Institute for Geophysics, Astrophysics, and Meteorology/Institute of Physics (IGAM/IP), University of Graz, Graz, Austria
[2]Danish Meteorological Institute, Copenhagen, Denmark

Correspondence to: B. Scherllin-Pirscher (barbara.pirscher@uni-graz.at)

Abstract. In this paper, we introduce a bending angle radio occultation climatology (BAROCLIM) based on Formosat-3/COSMIC (F3C) data. This climatology represents the monthly-mean atmospheric state from 2006 to 2012. Bending angles from radio occultation (RO) measurements are obtained from the accumulation of the change in the ray-path direction of Global Positioning System (GPS) signals. Best quality of these near-vertical profiles is found from the middle troposphere up to the mesosphere. Beside RO bending angles we also use data from the Mass Spectrometer and Incoherent Scatter Radar (MSIS) model (modified for RO purposes) to expand BAROCLIM in a spectral model, which (theoretically) reaches from the surface up to infinity. Due to the very high quality of BAROCLIM up to the mesosphere, it can be used to detect deficiencies in current state-of-the-art analysis and reanalysis products from numerical weather prediction (NWP) centers. For bending angles derived from European Centre for Medium-Range Weather Forecasts (ECMWF) analysis fields from 2006 to 2012, e.g., we find a positive bias of 0.5 to 1 % at 40 km, which increases to more than 2 % at 50 km. BAROCLIM can also be used as a priori information in RO profile retrievals. In contrast to other a priori information (i.e., MSIS) we find that the use of BAROCLIM better preserves the mean of raw RO measurements. Global statistics of statistically optimized bending angle and refractivity profiles also confirm that BAROCLIM outperforms MSIS. These results clearly demonstrate the utility of BAROCLIM.

1 Introduction

Global data sets of the lower and middle atmosphere (troposphere to upper mesosphere) provide important information to understand atmospheric dynamics of the Earth's climate system. Observational data as well as analysis/reanalysis data and atmospheric models are used, e.g., to study specific atmospheric phenomena such as MJO (Madden Julian Oscillation), ENSO (El Niño–Southern Oscillation), or the QBO (Quasi Biennial Oscillation). Long-term observational records, reanalysis data sets, and atmospheric models can also be used to investigate atmospheric climate change.

Simple empirical atmospheric models are of high utility if a quick but reasonable estimate of the atmospheric state is of main interest. This is of importance, e.g., for simulation studies in the field of atmospheric remote sensing or within the retrieval of atmospheric parameters from remote sensing measurements. For this purpose several research communities use early empirical models like CIRA (Committee on Space Research (COSPAR) International Reference Atmosphere) (Fleming et al., 1990) or MSIS (Mass Spectrometer and Incoherent Scatter Radar) (Hedin, 1991; Picone et al., 2002).

Published in the early 1960s, the CIRA model was the earliest comprehensive climatological model of the atmosphere to contain information up to the thermosphere. This model is based on observational data such as radiosondes, rocket data, and satellite observations. Its fourth version, CIRA-86, includes thermosphere models as well as tables of monthly-mean zonal-mean temperature, pressure, geopotential height, and zonal wind from the surface to an altitude of 120 km. The most recent CIRA version (CIRA-2012) contains updated versions of empirical models of the Earth's upper atmosphere (above 120 km) (CIRA, 2012).

Recent MSIS model versions (MSIS-90 and NRLMSIS-00) provide information on atmospheric composition, total mass density, and temperature from the ground up to the exosphere also using observations from ground, rockets, and satellites. The MSIS model output depends on time (day of year, universal time, local solar time), location (altitude, latitude, longitude), geomagnetic activity (represented by the magnetic index A_p), and solar activity (represented by the 10.7 cm solar radiation flux, $F_{10.7}$).

An overview on principal features of a number of global and regional models of the middle atmosphere and thermosphere is given by AIAA (2004), summarizing their model content, uncertainties, and limitations. Randel et al. (2004) compared middle-atmosphere climatologies from approximately 10 to 80 km – including historical measurements from rocketsonde winds and temperatures (1970 to 1989), lidar temperature data (1990s), global meteorological analyses, and satellite data – and found notable differences between these data sets. Some biases found in atmospheric analyses were caused by the low vertical resolution of these data as well as the low vertical resolution of some assimilated satellite measurements.

Since 2004 research initiatives have tried to understand and eliminate errors of previous middle-atmosphere models, building new, state-of-the-art analysis and reanalysis products. However, there are still uncertainties and differences in current reanalysis products, and the SPARC (Stratosphere-troposphere Processes And their Role in Climate) Reanalysis/analysis Intercomparison Project (S-RIP; Fujiwara et al., 2012) aims at understanding and reasonably interpreting these differences and contributing to future reanalysis improvements in the middle atmosphere.

In this study, we aim at compiling and investigating a global climatological model from recent high-resolution radio occultation (RO) bending angles.

The RO method (Melbourne et al., 1994; Kursinski et al., 1997) is an active limb sounding technique that utilizes radio signals continuously broadcast by Global Positioning System (GPS) satellites. The measured quantity is the phase change of the GPS signal as a function of time, which is a measure for physical atmospheric parameters, in particular bending angle and radio refractivity, from which density, pressure, geopotential height, temperature, and humidity profiles can be derived with very high accuracy (though in the lower moist troposphere auxiliary information is necessary to separate dry-air and moist-air contributions to the refractivity). Precise and stable oscillators aboard the GPS satellites ensure measurement stability and consistency between various RO missions, without the need for instrument-dependent calibrations (Hajj et al., 2004; Schreiner et al., 2007; Foelsche et al., 2011).

The RO bending angle can be used as a climate variable (Ringer and Healy, 2008). In the upper troposphere and above (where moisture is negligible), it mainly depends on atmospheric density, which again depends on pressure and temperature. This means that a change in bending angle can be caused by changes of all these parameters. This has to be kept in mind when analyzing atmospheric variability in terms of bending angle. However, the use of bending angle as a climate variable is superior to other RO-derived atmospheric quantities that are sensitive to the additional use of a priori information used in the retrieval (see below).

First RO profiles of the Earth's atmosphere were provided by the GPS/MET experiment in intermittent periods from 1995 and 1997 (Rocken et al., 1997). Since 2001, RO data have been continuously available from a number of RO satellite missions: CHAMP RO measurements from 2001 to 2008 are supplemented with measurements from SAC-C (launched in 2000), GRACE (2002), Metop-A (2006), TerraSAR-X (2007), C/NOFS (2008), OceanSat-2 (2009), Tandem-X (2010), SAC-D (2011), Megha-Tropiques (2011), Metop-B (2012), FY-3C (2013), and KOMPSAT-5 (2013) as well as from the first six-satellite RO constellation, Formosat-3/COSMIC (F3C; launched in 2006).

Atmospheric profiles from RO are used for data assimilation in numerical weather prediction (NWP) (e.g., Healy and Thépaut, 2006; Cucurull and Derber, 2008) and in atmospheric and climate research (see Anthes, 2011; Steiner et al., 2011, for reviews). Most studies utilize RO profiles in the upper troposphere and lower stratosphere (UTLS) region, i.e., the altitude range from approximately 5 to 35 km, where RO profiles of pressure, geopotential height, and temperature are known to be of best quality (Scherllin-Pirscher et al., 2011b).

The top altitude of RO measurements depends on the instrument settings but is at least 80 km (e.g., Metop-A and Metop-B). For F3C it is usually about 120 km or higher. However, at these altitudes, individual RO profiles are dominated by measurement noise and ionospheric disturbances, because neutral atmospheric density gradients are small. To derive refractivity profiles via the Abel integral transform (Fjeldbo et al., 1971), and since noisy and erroneous information from high altitudes propagates downwards in the retrieval process, bending angles are combined with a priori information, e.g., from a climatology, usually applying statistical optimization (Rodgers, 2000; Healy, 2001; Gorbunov, 2002; Gobiet and Kirchengast, 2004; Lohmann, 2005).

Formally, the upper limit in the Abel integral transform is infinity (Fjeldbo et al., 1971). If a climatology is used for statistical optimization, it is therefore necessary that it is able to provide a good measure of the bending angle to very high altitudes. Usually, the climatology is used as an extrapolation above the altitude range where the statistical optimization is performed (or some other form of extrapolation is necessary for high-accuracy purposes). However, biases in the a priori information are inevitable. At high enough altitudes biases are non-negligible (in a fractional sense) and affect the quality of derived atmospheric profiles (Ho et al., 2012; Steiner et al., 2013). Therefore high-quality a priori information is of particular importance.

Ao et al. (2012) and Gleisner and Healy (2013) introduced a new approach to derive GPS RO climatological products. Instead of averaging individual refractivity profiles, they calculated monthly-mean bending angles and computed mean refractivity as the Abel inversion of the mean bending angle. Ao et al. (2012) showed that the maximum altitude through which the F3C measurements are useful increased substantially, which leads to a reduced bias in climatological averages of refractivity. The main drawback of this approach, however, is that it can only be used to obtain mean atmospheric fields and is not applicable to individual RO profiles.

The generation of the bending angle radio occultation climatology (BAROCLIM) described in this paper is mainly based on the work by Foelsche and Scherllin-Pirscher (2012) and Scherllin-Pirscher (2013). BAROCLIM is based on measurements from 2006 to 2012. Since this period is much shorter than the standard period for climate normals (30 years), BAROCLIM does not represent a long-term mean. However, it is of very high utility to obtain a quick and reasonable estimate of the atmospheric mean state as required, e.g., within the retrieval of atmospheric parameters from remote sensing measurements.

Section 2 gives a detailed description of the RO data set as well as of reference data used in this study. In Sect. 3 we present the method used to construct BAROCLIM, and in Sect. 4 we discuss potential systematic errors and evaluate the model by comparing it to reference data from NWP analysis fields provided by the European Centre for Medium-Range Weather Forecasts (ECMWF). In Sect. 5 we show that BAROCLIM can further be used as a priori information for statistical optimization in RO profile retrievals. Conclusions are drawn in Sect. 6.

2 Data

For the generation of BAROCLIM, we used ionosphere-corrected bending angles as a function of impact altitude (impact parameter with the local radius of curvature and geoid undulation subtracted). These profiles were retrieved with the Wegener Center for Climate and Global Change (WEGC) Occultation Processing System version 5.6 (OPSv5.6) (Schwärz et al., 2013) and interpolated on a regular 200 m grid.

Input data to the WEGC processing system are profiles of excess phase and amplitude as well as precise orbit information of GPS and low Earth orbit (LEO) satellites (level 1a data) provided by other data centers. Currently WEGC uses level 1a data provided by University Corporation for Atmospheric Research (UCAR)/COSMIC Data Analysis and Archive Center (CDAAC) for all RO satellite missions. Since recent UCAR/CDAAC processing versions vary for different missions and GPS receivers used for RO measurements are not of the same quality (Foelsche et al., 2011, e.g., found different bending angle noise for different missions), we decided to use only data from the F3C constellation. The UCAR/CDAAC processing version of F3C data used in this study was 2010.2640 for the entire time period starting in 2006.

The processing system at UCAR/CDAAC relies on the Bernese software package to obtain precise orbits of F3C satellites and applies single differencing to remove F3C clock offsets (Schreiner et al., 2009). In the inversion retrieval, WEGC first corrects orbits for the Earth's oblateness (Syndergaard, 1998), removes outliers from L1 and L2 excess phase profiles, and smooths data with a regularization filter (Syndergaard, 1999). Doppler shift profiles are then calculated from excess phase profiles via numerical differentiation. In the upper troposphere and above, bending angles are computed from Doppler shift profiles based on geometric optics (Melbourne et al., 1994). Wave optics retrieval is applied in the lower and middle troposphere (Gorbunov, 2002; Gorbunov et al., 2004; Gorbunov and Lauritsen, 2004). To remove the ionospheric contribution to the bending angle, WEGC applies ionospheric correction on bending angle level following Vorob'ev and Krasil'nikova (1994) and Hocke et al. (2003). Except for the use of F3C open-loop data (Sokolovskiy et al., 2006) and the wave optics retrieval in the lower and middle troposphere, the WEGC OPSv5.6 bending angle retrieval is very similar to the OPSv5.4 processing described by Ho et al. (2012) and Steiner et al. (2013).

We used F3C data from August 2006 to December 2012 (more than 6 years of data). The number of F3C measurements increased from 2006 to 2007 until all F3C spacecraft were in their final orbits (F3C/flight model no. 3 (FM-3) did not reach its final orbit altitude because solar panels were stuck, which limited the power and payload operation of this spacecraft). From 2007 to 2009 F3C always tracked more than 60 000 RO events per month (exception: June 2009, UCAR/CDAAC provided approximately 55 000 level 1a profiles). Since 2010 the number of measurements decreased again due to battery degradation of all spacecraft. Furthermore, F3C/FM-3 has been out of contact since 1 August 2010. However, the minimum number of level 1a data provided by UCAR/CDAAC per month in the period between 2006 and 2012 was always larger than 30 000.

Since RO profiles usually do not reach higher than about 120 km (only about 80 km for Metop) and because the true neutral atmospheric bending angle decreases nearly exponentially with height (and therefore measurement noise dominates in a fractional sense in the mesosphere and above), there is an upper limit to which RO data are useful for the generation of a climatology. To extend BAROCLIM to higher altitudes, we used bending angles based on a modified version of the MSIS-90 empirical model of the neutral atmosphere (Hedin, 1991).

The MSIS-90 model is based on data from several satellites, incoherent scatter radar stations, and rocket probes as well as from tabulations from the Middle Atmosphere Program (MAP) Handbook 16 (Hedin, 1983, 1987, 1991). Høeg et al. (1995) modified the original MSIS-90 model by smoothing out a discontinuity at 72.5 km and fixing the local apparent solar time at 0 h, the solar radio flux at $F_{10.7} = 150 \times 10^{-22}\,\mathrm{W\,m^{-1}\,Hz^{-1}}$, and the magnetic index at $A_p = 4$.

Subsequently, this modified version of the MSIS model was used to generate spectral models of refractivity and bending angle for the use of fast and efficient modeling and inversion of RO data. The spectral model of refractivity, N, was obtained from the MSIS pressure, p, and temperature, T ($N = 77.6\, p/T$), for the 15th of every month. The spectral model of bending angle was obtained from the spectral refractivity model via the Abel integral transform. Thus, the MSIS-based bending angles and refractivities contain no information on atmospheric humidity and are given as a function of month, impact height, latitude, and longitude. Both spectral models are based on Chebychev polynomials in the vertical and spherical harmonics in the horizontal and were constructed using an approach similar to the one we now use for BAROCLIM described in Sect. 3.4. The MSIS spectral models were later incorporated into the End-to-end Generic Occultation Performance Simulator (EGOPS) (Fritzer et al., 2012) and the Radio Occultation Processing Package (ROPP) (Culverwell, 2013).

To validate BAROCLIM, we used operational ECMWF analysis fields at T42 horizontal resolution (comparable to RO horizontal resolution) and 91 vertical levels. Profiles were extracted at mean RO event location of those F3C measurements, which were used to construct BAROCLIM. We applied a forward model to derive ECMWF bending angles from refractivity (above the ECMWF model top, refractivities were extended with MSIS profiles scaled to fit the ECMWF model at high altitudes).

We will show in Sect. 5 that BAROCLIM can be used as a priori information for the statistical optimization in the processing of RO measurements. To investigate the performance of BAROCLIM as a priori information, we used level 1a data from CHAMP, GRACE-A, SAC-C, and F3C from January 2008 and July 2008 (level 1a data were again provided by UCAR/CDAAC; CHAMP data version was 2009.2650; other satellites had consistent data version 2010.2640) and applied the WEGC OPSv5.6 retrieval using BAROCLIM and MSIS as a priori information for bending angle initialization. Retrieved atmospheric profiles were then compared to operationally retrieved OPSv5.6 profiles that used ECMWF short-range forecasts as a priori in the statistical optimization (Schwärz et al., 2013).

3 BAROCLIM generation

Figure 1 summarizes the steps of the BAROCLIM generation. Using ionosphere-corrected bending angles as a function of impact altitude on a regular 200 m grid, we first applied a quality control (QC) to identify and exclude bad profiles before averaging bending angles in latitude and altitude and calculating monthly-mean 10° zonal-mean bending angles. These monthly means include data from 6 or 7 years. Since we aimed at generating a BAROCLIM spectral model, which (theoretically) reaches from the surface to infinity, we extended mean RO profiles with MSIS. We refer to these bending angles as the BAROCLIM discrete model. The BAROCLIM spectral model was then constructed by expansion into Chebychev polynomials and zonal harmonics. Below we describe these steps in detail.

3.1 Quality control of individual profiles

Some bending angles are very noisy and/or contain unphysical values. They could strongly affect the quality of the bending angle climatology if they were not properly excluded. To avoid entering these profiles in BAROCLIM, we only used OPSv5.6 profiles that passed the external QC performed at WEGC (Schwärz et al., 2013). This external QC comprises bending angle, refractivity, and temperature profiles, which are compared to co-located profiles from ECMWF analysis fields. The profile is flagged as bad if the difference between the RO and the ECMWF profile is larger than 20 % in bending angle, 10 % in refractivity, and/or 25 K in temperature. However, since these quality checks are only performed in the upper troposphere and lower stratosphere region up to 35 km, profiles passing the external QC can still be very noisy in the upper stratosphere and above.

The inspection of individual bending angle profiles indeed revealed that it is imperative to perform an additional QC. We therefore introduced a threefold approach for an additional outlier rejection. In a first step we rejected all profiles with bending angles smaller than $-40\,\mu$rad or larger than $+40\,\mu$rad in the altitude range from 50 to 80 km. In this altitude range, neutral atmospheric bending angles are usually smaller than $20\,\mu$rad, and bending angles smaller/larger than $\pm 40\,\mu$rad result from very strong (unphysical) noise, possibly sometimes related to ionospheric scintillations (Zeng and Sokolovskiy, 2010).

To also detect very bad profiles below 50 km, we rejected all profiles with bending angles smaller than $-20\,\mu$rad below 50 km in a second step. Profiles that were flagged as bad by these QC mechanisms were outlier profiles, which could degrade the quality of BAROCLIM. Inspecting remaining bending angle profiles after application of these first checks, we still found some very noisy bending angle profiles (top panel of Fig. 2) and therefore applied an additional QC.

We used all remaining profiles and calculated mean bending angles and standard deviations for 10° latitude bands for multiyear monthly means and rejected all profiles with bending angles outside of 4 standard deviations (4σ) from the mean in the altitude range from the surface to 100 km. Figure 2 shows F3C profiles for the month of January before and after application of this 4σ criterion. This figure reveals that

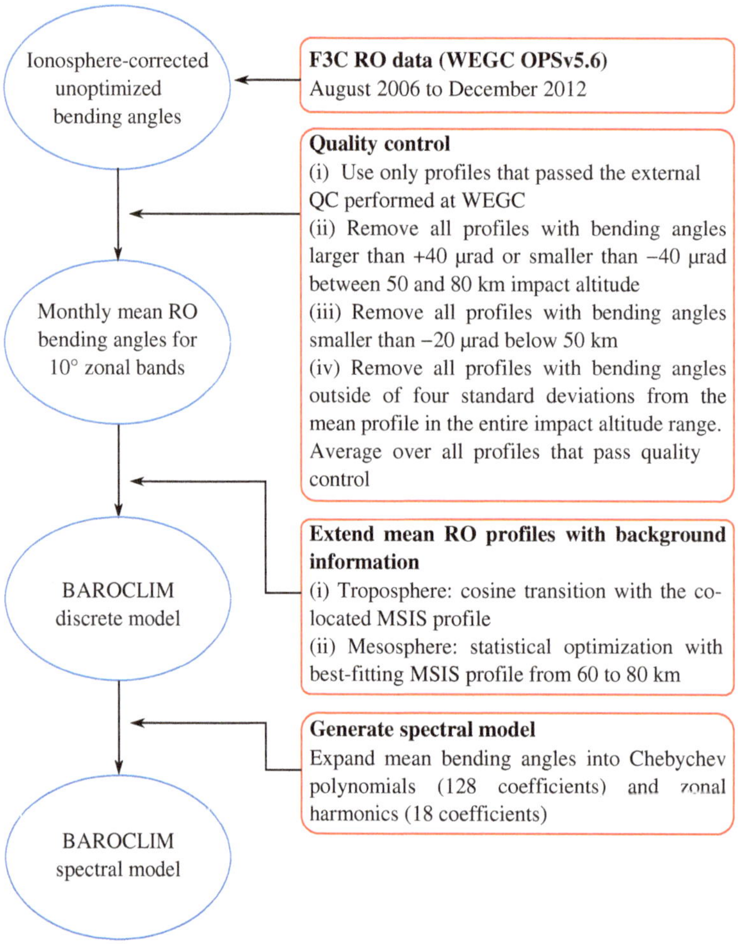

Figure 1. Flowchart of the BAROCLIM generation.

application of this final QC results in a considerable decrease in standard deviation.

Table 1 gives an overview on the number of profiles provided by UCAR/CDAAC, retrieved at WEGC, and passing BAROCLIM QC. The number of bending angle profiles for a given month used to generate BAROCLIM is always larger than 200 000, except for June, when it is slightly below. In some months, however, the number is even larger than 260 000. These numbers are large enough and sufficient to obtain a smooth BAROCLIM up to about 60 km.

3.2 Average over high-quality profiles

The optimal horizontal extent of the regions to calculate a typical climatological mean from high-quality measurements is a trade-off between a sufficiently large number of profiles and atmospheric variability. Our experience of building atmospheric climatologies utilizing RO data (e.g., Foelsche et al., 2008; Scherllin-Pirscher et al., 2011a) showed that 10° zonal bands were a reasonable choice for calculating mean atmospheric profiles from RO data. These bands range from 90° S to 90° N, resulting in 18 zonal bands.

The mean number of profiles per 10° latitudinal band varied between 11 000 (June) and 15 000 (October). However, the latitudinal distribution of RO events is not uniform, which is due to the orbit characteristics of the GPS and F3C satellites (orbit inclinations of 55° and 72°, respectively). The largest number of RO measurements per latitude band is obtained at mid-latitudes, where more than 20 000 profiles enter BAROCLIM. A smaller number of measurements per latitude band is obtained at low (approximately 8000 profiles) and high latitudes (between 2000 and 4000 profiles). We judged that 8000 tropical profiles are sufficient to calculate mean profiles because atmospheric variability at high altitudes is small at low latitudes. We also considered 2000 high-latitude profiles to be sufficient due to the decrease of zonal surface area with increasing latitude. Thus, 2000 profiles are enough to represent a small area.

Multiyear monthly-mean bending angles for 10° zonal bands for January and July are shown in Fig. 3 together with their standard error of the mean $\sigma_{\text{mean}} = \sigma/\sqrt{N}$ (σ is the standard deviation, and N is here the number of observations used to estimate the mean – not to be confused with refrac-

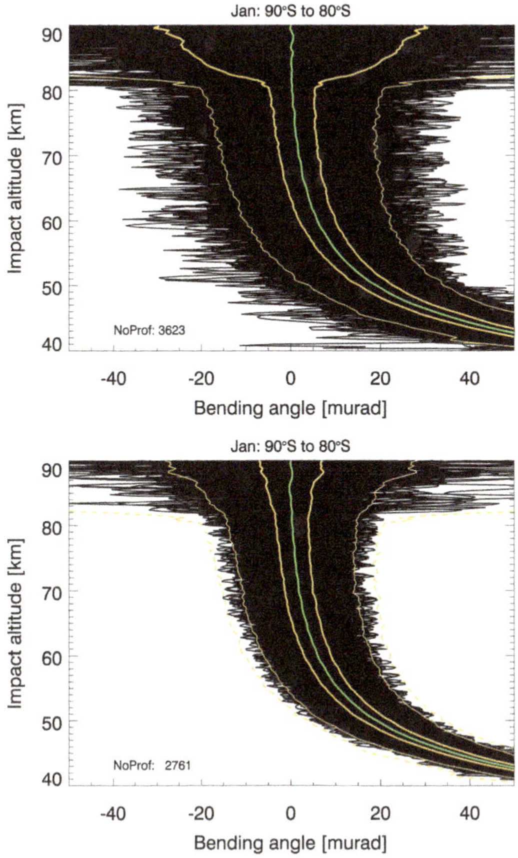

Figure 2. F3C bending angles (in μrad) in January in the latitude band from 80 to 90° S before (top) and after (bottom) the application of the 4σ-outlier rejection criterion. The thick green line shows the mean of all profiles, the thick yellow lines 1 standard deviation, and the thin yellow lines 4 standard deviations. The dashed lines in the lower panel indicate the thin yellow lines in the upper panel.

tivity). Bending angles are negative (white areas) somewhat above an altitude of 80 km, and the standard error of the mean is larger than 10 % above 70 km in the winter hemisphere at high latitudes. While negative mean bending angles might be caused by residual systematic ionospheric errors (see Danzer et al., 2013), the high standard error of the mean (in a fractional sense) is a result of the decreasing bending angle with altitude. Below 60 to 70 km, however, mean bending angles are rather smooth and the standard error of the mean generally does not exceed 2 %.

3.3 BAROCLIM discrete model

Because of the generally decreasing bending angle with altitude, the mean bending angle (Fig. 3) is error-dominated above 80 km. Therefore we combined the mean bending angles with a priori information to generate a model that is useful also above the mesosphere. A priori information profiles can be obtained from already-existing climatological models or profile data sets. Current state-of-the-art analysis, reanalysis, or forecast products from NWP centers do not reach high enough in the atmosphere (the ECMWF model top, e.g., is at 0.01 hPa, corresponding approximately to 80 km) and can therefore not be used for extension to higher altitudes.

Table 1. Number of level 1a F3C data provided by UCAR/CDAAC (first column), number of bending angles profiles retrieved at WEGC (second column), and number of profiles that passed BAROCLIM quality control (last column). Numbers are shown for every month, adding up the data from August 2006 to December 2012.

	Level 1a	Retrieved OPS	After QC
Jan	415 105	307 422	227 815
Feb	396 271	300 670	229 848
Mar	406 435	314 581	230 580
Apr	425 836	336 706	262 513
May	396 528	296 595	222 382
Jun	378 260	273 553	198 177
Jul	422 674	327 990	238 146
Aug	453 726	357 431	268 882
Sep	429 306	335 284	264 264
Oct	435 833	345 237	270 381
Nov	411 974	314 380	240 132
Dec	485 333	360 227	263 235

Since it is readily available, we decided to use the modified MSIS climatology as a priori information. In order to make maximum use of the information content of the RO data, and since the MSIS-90 climatology might be biased at high altitudes, we did not necessarily take the MSIS profile as representative for a given latitude band and month. Instead we performed a search in the spectral model of the MSIS bending angle climatology on a regular 5° × 10° latitude–longitude grid through all months from January to December, and we found the best match to the RO data using a least-squares fit to the RO mean bending angle profile in the altitude range from 60 to 80 km, where RO data quality is still high. To correct remaining background biases, the best-fitting MSIS profile was then multiplied with a fit factor obtained from regression with respect to the mean RO bending angle profile at high altitudes (least-squares adjustment from 60 to 80 km). We found fit coefficients being close to unity (0.99 to 1.01), with exceptions only in Southern Hemisphere winter at high latitudes (80 to 90° S), where fit coefficients were as small as 0.96, 0.88, and 0.91 in May, June, and July, respectively.

To combine the mean RO bending angle profile with the corresponding MSIS profile, we applied statistical optimization by inverse covariance weighting (Gobiet and Kirchengast, 2004) between 60 and 80 km using an error correlation length of 2 km for the RO profile and an error correlation length of 15 km for the adjusted MSIS model profile. Furthermore, we assumed the MSIS background error to increase linearly from 0 % at 80 km to 15 % at 78 km, kept it constant at 15 % between 78 and 62 km, and then increased linearly

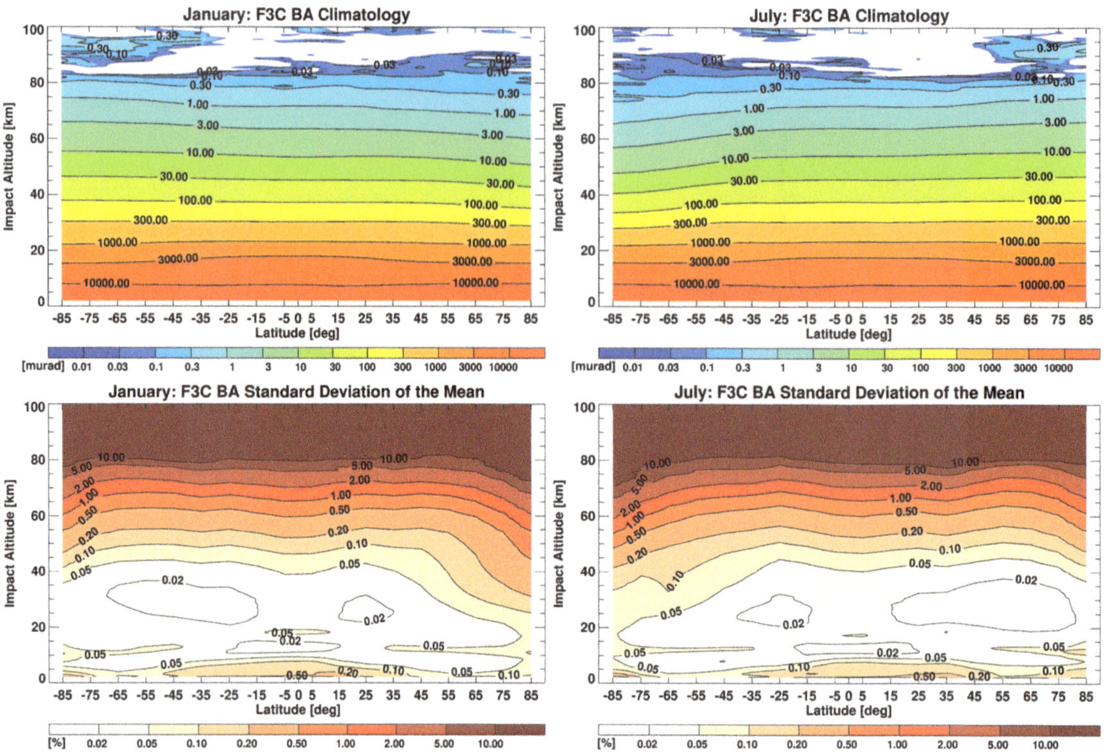

Figure 3. Mean F3C bending angles (in μrad) (top) and their standard deviation (in percent) of the mean (bottom) in January (left) and July (right) as a function of latitude and impact altitude up to 100 km.

again from 15 % at 62 km to 100 % at 60 km. All these percent values refer to the absolute values of the MSIS bending angle at the respective impact altitude level. While the linear increase of the relative error at the top and the bottom end of statistical optimization avoids too sharp transitions, the constant fraction of 15 % was determined empirically by Gobiet and Kirchengast (2004).

The observational error was set to the mean background error between 62 and 78 km and was constant with height (in absolute value, not percentagewise). Using these settings, we obtained smooth statistically optimized bending angles, for which the height where the retrieval to a priori error ratio (Gobiet et al., 2007) equals 50 % is 67.2 km for all profiles. Outside of the transition region from 60 to 80 km, the statistically optimized bending angle equals that of MSIS (above 80 km) or that of the mean RO profiles (below 60 km).

Even though Fig. 3 indicates that the mean bending angles reach down to the surface (2 km impact altitude approximately corresponds to 0 km altitude), mean bending angle quality also decreases in the lower troposphere. This is mainly because of the strongly decreasing number of measurements in the lower troposphere (see also Anthes et al., 2008) but is also due to reduced quality of the measurements. Besides this, averaging bending angles at low impact altitudes is tricky, since the lowest impact altitude in individual profiles (even if profiles are tracked all the way to the surface) depends on the bending angle. For the ray grazing the surface in one occultation event with a large bending angle, the impact altitude is larger than for the ray grazing the surface in another occultation event with smaller bending angle. It thus becomes dubious to talk about bending angle at the lowest impact altitude for the mean profile.

Being aware that MSIS is a dry-air climatology (no humidity is included in this model) and accepting that BAROCLIM will not reflect real atmosphere conditions at the lowest altitudes, we decided to use this model to extend BAROCLIM down to the surface. BAROCLIM is therefore, like MSIS, a dry-air model, being clearly wrong in regions were moisture is usually abundant, but for technical reasons smooth bending angles in the lower troposphere close to the surface are necessary when generating the BAROCLIM spectral model.

To extend mean RO bending angles down to the surface, we first extracted the MSIS profile for the given month and latitude and searched for the best fit in longitude. We then applied a gradual transition using a cosine weighting function from the mean RO bending angle α_{RO} to the MSIS bending angle α_{MSIS}. This weighting function was defined as $w(z) = 1/2\left(1+\cos\left(\pi(z_{\mathrm{top}}-z)/\Delta z\right)\right)$, and the tropospheric bending angle α_{trop} was obtained from $\alpha_{\mathrm{trop}} = w(z)\alpha_{\mathrm{RO}}(z) + (1-w(z))\alpha_{\mathrm{MSIS}}$.

Since the amount of water vapor in the lower troposphere depends on latitude, we performed RO–MSIS transition between 5 and 10 km from 60° S/N to 90° S/N, between 7 and 12 km from 30° S/N to 60° S/N, and between 10 and 15 km from 30° S to 30° N.

To sum up, our BAROCLIM discrete model is available for every month (January to December) and has a horizontal resolution of 10° zonal bands and a vertical gridding of 200 m. It relies 100 % on RO data from the upper troposphere up to 60 km. Above 80 km and below 5, 7, or 10 km (depending on latitude), it consists of data-driven adjusted MSIS profiles.

3.4 BAROCLIM spectral model

For fast and easy access to BAROCLIM at any latitude and impact altitude, and to make the functionality similar to the MSIS bending angle and refractivity spectral models in EGOPS and ROPP, we expanded the BAROCLIM discrete model in Chebychev polynomials and zonal harmonics. Since the bending angle scale height is more finely structured than the smooth, almost exponentially decreasing bending angle, we expanded a function into Chebychev polynomials, which depends on the bending angle scale height.

First we introduced the variable $z = h - h_{\mathrm{surf}}$ $(z \geq 0)$, where h is impact altitude and h_{surf} is the lowest possible impact altitude corresponding to a hypothetical ray grazing the surface. The lowest impact altitude was estimated from the MSIS climatology using $h_{\mathrm{surf}} = N_{\mathrm{surf}}^{\mathrm{MSIS}} \times 10^{-6} R_{\mathrm{E}}$, where $R_{\mathrm{E}} = 6371$ km is the mean radius of the Earth and $N_{\mathrm{surf}}^{\mathrm{MSIS}}$ is MSIS refractivity at the surface extracted at the specific month and latitude and at longitude $\lambda = 0°$. The bending angle $\alpha(z)$ was then extracted from the BAROCLIM discrete model by interpolation to a number of discrete impact heights evaluating $z(x) = 100(\ln 2 - \ln(1-x))$ at $k_{\max}$ values of x given by $x_k = \cos(\pi(k - \frac{1}{2})/k_{\max})$ $(k = 1, \ldots, k_{\max})$. This mapping yields a finer vertical spacing at low altitudes and coarser vertical spacing at higher altitudes. Having $\alpha(z)$ at these discrete impact heights, the bending angle scale height $H_{\mathrm{S}}(z)$ was calculated as

$$H_{\mathrm{S}}(z) = \frac{z}{\ln(\alpha_{\mathrm{surf}}/\alpha(z))}, \tag{1}$$

where α_{surf} is the bending angle at $z = 0$ (also extracted from the BAROCLIM discrete model).

Chebychev coefficients, c_j, were obtained from

$$c_j = \frac{2}{k_{\max}} \sum_{k=1}^{k_{\max}} G(x_k) \cos\left(\frac{\pi(j-1)\left(k-\frac{1}{2}\right)}{k_{\max}}\right), \tag{2}$$

where

$$G(x) = H_{\mathrm{S}}(z(x)) - (m\,z(x) + b) \tag{3}$$

and m and b are slope and intercept, respectively, of a straight line fit to the scale height at high altitudes. Finally, $j = 1, \ldots, k_{\max}$, and $k_{\max}$ is the number of extracted Chebychev coefficients for $k_{\max} - 1$ Chebychev polynomials (Press et al., 1986).

The Chebychev coefficients were then expanded into zonal harmonics. Besides the Chebychev coefficients, also h_{surf}, α_{surf}, m, and b were expanded into zonal harmonics.

In general, zonal harmonics coefficients A_n are obtained from a given function $f(y)$, where normally $y = \cos\theta$ (θ is co-latitude (polar distance)) as (see, e.g., Spiegel, 1979)

$$A_n = \frac{2n+1}{2} \int_{-1}^{+1} f(y) P_n(y)\,\mathrm{d}y, \tag{4}$$

where P_n are Legendre polynomials, $n = 1, \ldots, n_{\max}$, and $n_{\max}$ is the number of extracted zonal harmonics coefficients. In our case $f(y)$ was c_j, h_{surf}, α_{surf}, m, or b. Thus, the final output of the BAROCLIM spectral model was $n_{\max}$ zonal harmonics coefficients of $k_{\max}$ Chebychev coefficients, surface impact altitude, surface bending angle, and slope and intercept of the straight line.

To reconstruct the bending angle from the BAROCLIM spectral model for a given impact altitude and latitude, we first applied Clenshaw's recurrence formula (Press et al., 1986) for zonal harmonics to obtain the Chebychev coefficients c_j as well as h_{surf}, α_{surf}, m, and b. We also applied Clenshaw's recurrence formula to reconstruct $G(x)$ (where $x = 1 - 2\exp(-z/100)$) before reconstructing bending angles using

$$\alpha(z) = \alpha_{\mathrm{surf}} \exp\left(-\frac{z}{H_{\mathrm{S}}(z)}\right). \tag{5}$$

More details on the expansion of BAROCLIM into Chebychev polynomials and zonal harmonics as well as their reconstruction can be found in Scherllin-Pirscher (2013).

To settle on the order of the Chebychev polynomials and the degree of the zonal harmonics, we calculated differences between the bending angles from the BAROCLIM discrete model and the BAROCLIM spectral model for different choices of $k_{\max}$ and $n_{\max}$, aiming at minimizing these differences. Since the BAROCLIM discrete model has a horizontal resolution of 10° zonal bands (18 zonal bands), we found minimum differences for 18 zonal harmonics coefficients. Concerning the order of the Chebychev polynomials, we found reasonably good agreement between the BAROCLIM discrete model and the BAROCLIM spectral model for 64 Chebychev coefficients. When using 128 Chebychev coefficients, the spectral model even reproduces the sharp tropical tropopause. For this reason we decided to use 128 Chebychev coefficients when constructing the spectral model and reconstructing the bending angle, but lower vertical resolution bending angles can be reconstructed using a smaller number of Chebychev coefficients. This could be useful for applications where computational speed is more important than high vertical resolution.

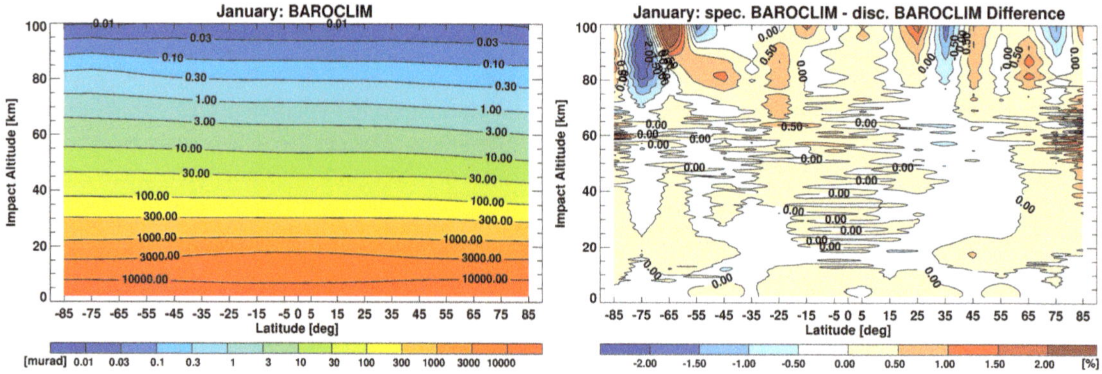

Figure 4. BAROCLIM spectral model (calculated with 128 Chebychev coefficients and 18 zonal harmonics coefficients) as a function of latitude and impact altitude for January (left) and difference between the BAROCLIM spectral model and the BAROCLIM discrete model (right).

The BAROCLIM spectral coefficients (stored in a NetCDF-file) as well as the Fortran-90 code needed to reconstruct bending angles from these coefficients are designed to be included in a future release of the ROPP software. ROPP is free of charge after registration at http://www.romsaf.org.

Figure 4 shows (as an example for the month of January) the BAROCLIM spectral model for 128 Chebychev coefficients and 18 zonal harmonics coefficients. The right panel of Fig. 4 shows that differences between the BAROCLIM discrete model and the BAROCLIM spectral model are, in general, within 0.5 % up to 60 km (a closer inspection of the differences reveals that it is even within 0.3 % in most places). Larger differences are found close to the 60 km altitude level (transition height of RO-only data and statistically optimized RO data) and above 80 km where the absolute amount of the bending angle is so small that even very small differences yield a noticeable percentage value.

4 Evaluation of BAROCLIM

4.1 Error sources

Atmospheric climatological fields of RO data are affected by (i) random statistical errors, (ii) systematic errors, and (iii) sampling errors (Scherllin-Pirscher et al., 2011a). Random statistical errors include, e.g., receiver thermal noise, clock stability/differencing errors, ionospheric noise, and statistical velocity errors (see e.g., Ramsauer and Kirchengast, 2001). Random statistical errors diminish by averaging over a large number of profiles. Since BAROCLIM is based on a very large number of quality-controlled RO soundings, all contributions from statistical errors are negligible, except at the highest altitudes (cf. Fig. 3).

Systematic errors are more important for BAROCLIM. From the RO measurement and retrieval perspective, these errors include systematic errors in orbit determination, local multipath, residual ionospheric errors, and errors due to assumptions in the RO retrieval. Systematic errors of BAROCLIM also include contributions due to the additional use of MSIS at high and low altitudes.

Schreiner et al. (2009) investigated the uncertainty of UCAR/CDAAC precise orbit determination (POD) of F3C satellites and found a velocity error of 0.17 $\mathrm{mm\,s^{-1}}$, which approximately corresponds to an F3C bending angle error of 0.05 μrad (1 % at 60 km). Due to the lack of an alternative measurement system onboard F3C, Schreiner et al. (2009) could not give an estimate of the orbit accuracy. If we assume that half of the F3C velocity error is attributable to a systematic error component, the corresponding BAROCLIM error will be 0.5 % at 60 km.

Errors due to local multipath depend on the spacecraft size and on the reflection coefficient (Rocken et al., 2008). For F3C these local multipath errors are estimated to be smaller than 0.05 $\mathrm{mm\,s^{-1}}$ (Rocken et al., 2008), which corresponds to 0.015 μrad in bending angle when using velocity-error-to-bending-angle-error conversion given by Schreiner et al. (2009). This error corresponds to 0.3 % at 60 km.

Systematic residual ionospheric errors are important for BAROCLIM. In general, ionospheric residual errors depend on the level of ionization at high altitudes, which again depends on local time (i.e., day- versus nighttime conditions due to solar insolation) as well as on solar activity (Kursinski et al., 1997; Schreiner et al., 2011; Danzer et al., 2013). Danzer et al. (2013) showed that this error rarely exceeds 0.1 μrad from mid-2006 to the end of 2011. When averaging over this time period, it is even smaller than 0.05 μrad. Giving a conservative estimate and assuming a 0.1 μrad residual ionospheric error from mid-2006 to 2012 (solar activity increased in 2012), which corresponds to 2 % at 60 km, this error is the most important systematic error source of BAROCLIM.

Systematic errors due to assumptions in the retrieval process (such as spherical symmetry) are assumed to be small at high altitudes (Rocken et al., 2008).

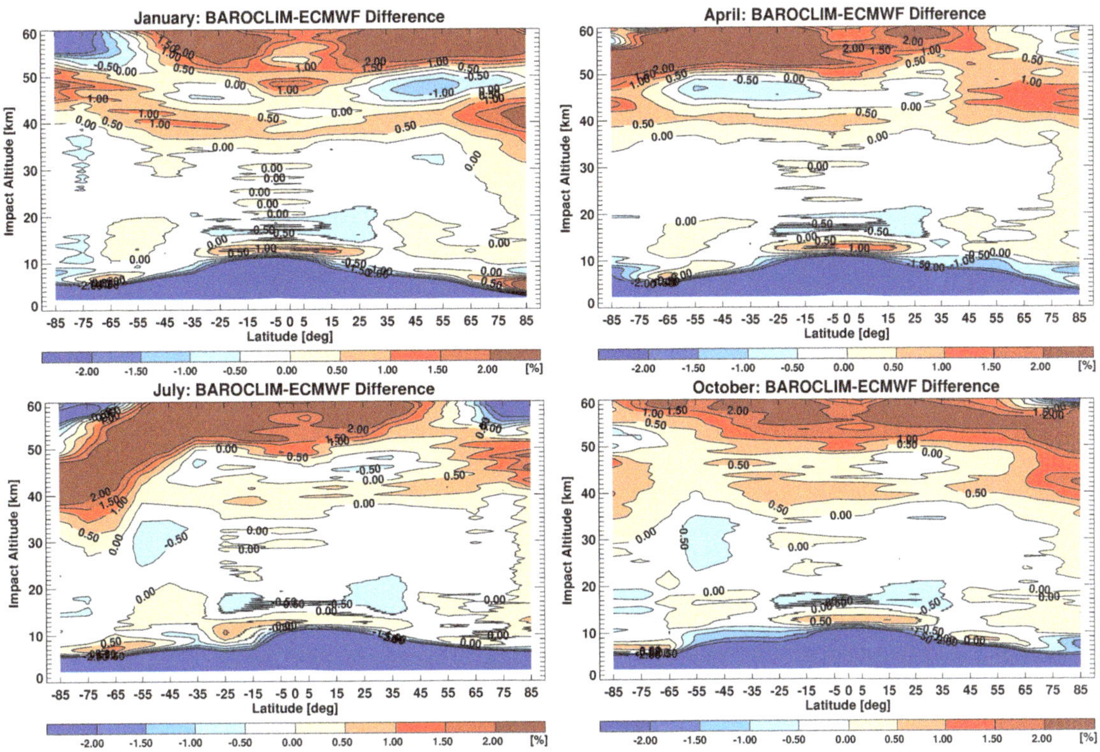

Figure 5. Systematic difference between the BAROCLIM spectral model and ECMWF analyses forward-modeled to bending angle as a function of latitude and impact altitude up to 60 km for January (top left), April (top right), July (bottom left), and October (bottom right).

Another BAROCLIM systematic error component results from the additional use of the MSIS model at low (tropospheric) and high (mesospheric and above) altitudes. Large systematic BAROCLIM errors in the troposphere are due to the absence of atmospheric water vapor in the MSIS model. For this reason BAROCLIM is not generally useful for tropospheric studies. Systematic errors from MSIS a priori information used at high altitudes (below 70 km) are assumed to be small due to the way MSIS was used (finding a bending angle profile that fits the mean RO data at high altitudes).

Finally, errors in BAROCLIM are caused by discrete sampling times and locations of RO measurements (sampling error; see, e.g., Foelsche et al., 2008; Scherllin-Pirscher et al., 2011a). The sampling error depends on the number of profiles and atmospheric variability and can be estimated from reference data that reflect true spatial and temporal variability. Using 6 to 7 years of RO data for BAROCLIM with more than 200 000 profiles per month (exception: June with 198 177 profiles), the BAROCLIM sampling error is negligible. However, since BAROCLIM is only based on measurements from 6 or 7 years, BAROCLIM might be biased relative to the long-term mean atmospheric state over 30 years. During the BAROCLIM time period, e.g., several major sudden stratospheric warming (SSW) events occurred in Northern Hemisphere winter (e.g., in January 2009 and 2010), yielding an RO climatology biased towards too high temperatures and too high atmospheric densities (i.e., too large bending angles) at northern high latitudes in these months. Other potential biases associated with the short RO record (e.g., the influence of QBO and ENSO) were estimated to be small.

4.2 Comparison to ECMWF

In December 2006 ECMWF started assimilating RO data in its operational assimilation system (Healy, 2007), which implies that ECMWF analyses and RO measurement data are not independent anymore. Healy et al. (2005) and Healy and Thépaut (2006) showed that assimilation of RO data significantly improved the forecast skill of the ECMWF operational system in the upper troposphere and lower stratosphere. Foelsche et al. (2009) found reduced biases in mean ECMWF analysis fields at least up to 30 km after ECMWF started assimilating RO. However, even though the assimilation is performed up to 50 km (von Engeln et al., 2009), a large RO observational error assumed in the assimilation at high altitudes limits the impact of RO data in this region. Systematic differences between the BAROCLIM spectral model and mean ECMWF analysis profiles therefore provide valuable information not only about the quality of BAROCLIM but also on the quality of ECMWF analyses, especially at high altitudes.

Figure 5 shows that the differences are small ($< 0.5\,\%$) in the upper troposphere and lower stratosphere region between approximately 10 and 35 km. Besides this good agreement, striking features, which are very similar in all months, are (i) large negative differences in the troposphere, (ii) a band of positive differences (mostly $< 1\,\%$) above approximately 35 km, and (iii) large positive differences ($> 2\,\%$) above 50 km.

Large negative tropospheric differences are caused by BAROCLIM being a dry-air model. Neglecting atmospheric humidity yields unrealistically small bending angles in regions where humidity is high. Positive BAROCLIM minus ECMWF analysis differences above 35 km ($< 1\,\%$) and above 50 km ($> 2\,\%$) are mainly attributable to biases in ECMWF analyses rather than to BAROCLIM. In general, these biases above 35 km might be related to the bias correction of assimilated radiances from satellite measurements, whereas the specific bias above 50 km is most likely attributable to data from AMSU-A channel 14, which are assimilated without bias correction (S. Healy, ECMWF, personal communication, October 2014).

Similar ECMWF biases at high altitudes have been found with satellite measurements from the Michelson Interferometer for Passive Atmospheric Sounding (MIPAS) instrument on the European environmental satellite ENVISAT and from the Microwave Limb Sounder (MLS) instrument on the U.S. Aura satellite (Chauhan et al., 2009). Further comparisons of ECMWF analyses and temperature data from MIPAS and Sounding of the Atmosphere using Broadband Emission Radiometry (SABER, an instrument on the U.S. TIMED (Thermosphere, Ionosphere, Mesosphere Energetics and Dynamics) satellite) also revealed a similar bias structure in ECMWF analyses above 40 km (J. Innerkofler, WEGC, personal communication, October 2014). All these comparisons consistently showed that ECMWF temperatures are too high at high altitudes, yielding a negative measurement minus ECMWF temperature bias. When pressure biases are small, a negative temperature bias corresponds to a positive refractivity bias, which in turn corresponds to a positive bending angle bias (see, e.g., Scherllin-Pirscher et al., 2011b) as shown in Fig. 5.

This comparison clearly shows that BAROCLIM is of very high quality at least up to 60 km and has the potential to validate middle-atmosphere data.

5 Use of BAROCLIM in RO profile retrievals

The intended aim of BAROCLIM was its use as a priori information in RO profile retrievals. We therefore evaluated its performance by processing occultation data from different RO missions and comparing retrieved atmospheric profiles obtained with different a priori information.

As mentioned in Sect. 2, we used level 1a RO data provided by UCAR/CDAAC for all missions and applied the WEGC OPSv5.6 retrieval to obtain ionosphere-corrected bending angles. As bending angle a priori information for statistical optimization we used BAROCLIM and MSIS profiles co-located to RO events (termed "BAROCLIM-Col" and "MSIS-Col", respectively) and BAROCLIM and MSIS profiles that best fit the ionosphere-corrected bending angle (termed "BAROCLIM-SF" and "MSIS-SF", respectively, where SF means "searched" and "fit"). This best-fit algorithm was similar to the "enhanced IGAM high-altitude retrieval scheme" described by Gobiet and Kirchengast (2004), searching for the best-fitting MSIS/BAROCLIM profile between 35 and 55 km and performing linear regression to find a multiplication factor to refine the fit to the data from 45 to 65 km.

With this approach we do not necessarily take a profile from MSIS/BAROCLIM corresponding to the latitude and season of the retrieval, but one that fits the data the best at high altitudes. Thus, with the SF approach we use MSIS/BAROCLIM as a library of different profiles representing different (average) atmospheric conditions on Earth. The approach should reduce sensitivity to biases in the climatology, although it does not guarantee that biases in the retrieved profiles are absent.

For comparison, we also included operationally retrieved OPSv5.6 profiles (Schwärz et al., 2013), which use ECMWF short-range forecasts as a priori information (termed "OPSv5.6").

To assess the performance of BAROCLIM in RO profile retrievals we calculated monthly statistics of raw ionosphere-corrected bending angles minus optimized bending angles. Since the purpose of statistical optimization is to reduce random errors, while preserving the mean, the mean difference between raw and optimized bending angles is an indicator of the quality of the background climatology.

Figure 6 shows mean results for January and July 2008 for CHAMP, GRACE-A, SAC-C, and F3C from 30 to 60 km impact altitude. Since noise of individual ionosphere-corrected raw bending angle profiles is high in the middle and upper stratosphere, difference profiles have a very large standard deviation. We did not include this information in the plot but note that it reaches approximately 15 % between 45 km (CHAMP) and 50 km (F3C).

Figure 6 shows that – while bending angle systematic differences of BAROCLIM-SF, BAROCLIM-Col, and OPSv5.6 are very close to 0 for all satellites and both months – MSIS-Col and MSIS-SF data are slightly negatively biased above 40 and 50 km, respectively. Largest differences ($> 5\,\%$ above 55 km) can be found for MSIS-Col.

Figure 7 shows how differences propagate from statistically optimized bending angle to refractivity using MSIS-SF, BAROCLIM-SF, or OPSv5.6. Since bending angle and refractivity differences shown in Fig. 7 are calculated against ECMWF analyses, zero difference does not necessarily mean that it is close to the truth (cf. Fig. 5 and the discussion there). Bending angle and refractivity systematic differences are very similar for all satellites, with differences amongst the satellites being, in general, smaller than 1.5 % up to

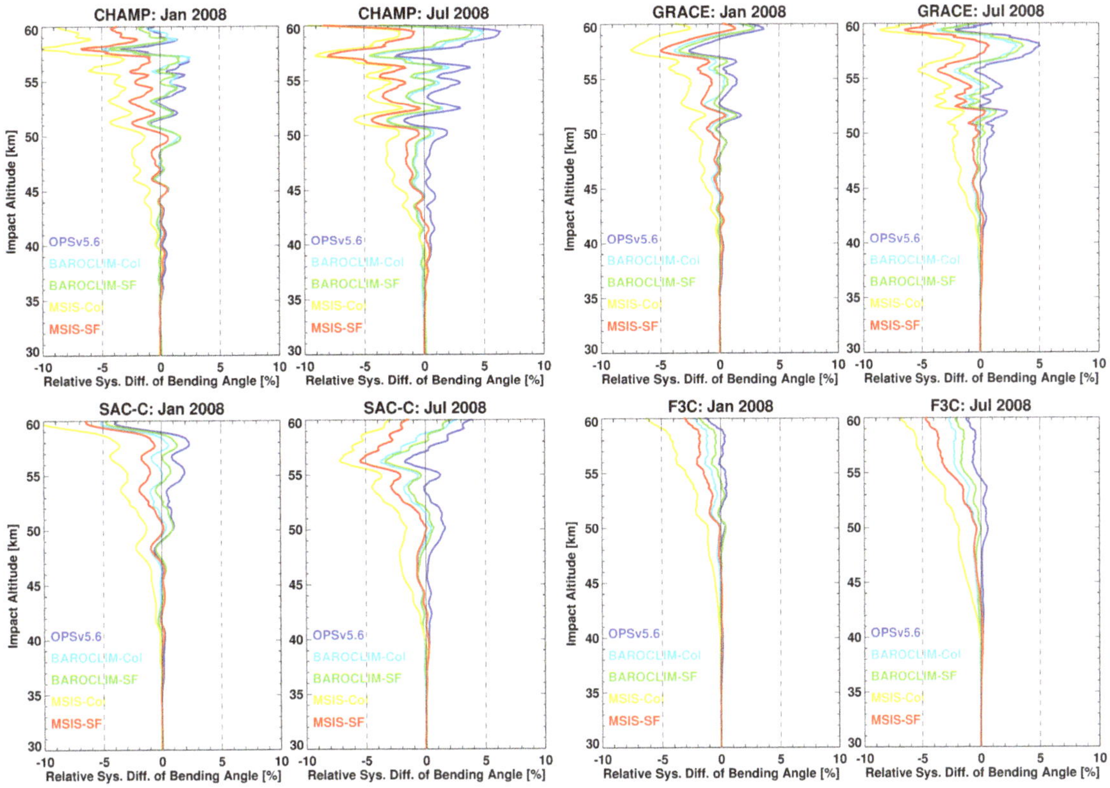

Figure 6. Global statistics of systematic difference between raw and statistically optimized RO bending angles for CHAMP, GRACE-A, SAC-C, and F3C for January 2008 and July 2008. Different colors indicate different a priori information used in the retrieval.

60 km. Comparison of the three methods reveals distinctively larger differences in bending angle and refractivity. While BAROCLIM-SF profiles are close to operationally retrieved OPSv5.6 profiles in the entire altitude range, MSIS-SF clearly performs worse compared to the other methods. Difference between MSIS-SF and ECMWF profiles reaches 3 % at approximately 55 km in bending angle and refractivity.

Contrary to systematic differences, the magnitude of the standard deviation features distinct satellite-dependent characteristics. Since CHAMP data noise is larger compared to the other satellites, OPS uses more weight of the a priori information, which results in smoother profiles and smaller standard deviation above 40 km. When comparing standard deviations of the three methods, larger standard deviations are found for the two search and fit algorithms than for OPSv5.6.

We conclude that the results using BAROCLIM seem promising, in particular when used in combination with the SF approach. As mentioned, such an approach should reduce the sensitivity to possible biases in BAROCLIM because it is then merely used as a library of different profiles representative of different (average) atmospheric conditions. The fact that BAROCLIM is based on data from only one mission (F3C) and from a limited period of time (2006 to 2012) is therefore not so important in this context; BAROCLIM can be used in this way for other RO missions in the past and in the future as long as the climate in the upper stratosphere does not change drastically in terms of global variations of bending angle.

6 Summary, conclusions, and outlook

In this study, we used radio occultation data from the F3C mission from August 2006 to December 2012 (more than 6 years of data) to compile a bending angle radio occultation climatology (BAROCLIM). After careful quality control we calculated multiyear monthly means for 10° zonal bands from the surface up to 100 km. Since mean RO profiles become noisier above 60 km (in a fractional sense) and are error dominated above 80 km, we used a priori information from the MSIS climatology modified for RO purposes and applied statistical optimization from 60 to 80 km to extend BAROCLIM into the thermosphere. We also used the MSIS model to obtain smooth bending angles in the troposphere and performed a cosine transition to mean RO profiles in the middle troposphere. This implies that BAROCLIM is a dry-air model in the troposphere. However, BAROCLIM relies on RO data from the upper troposphere up to 60 km and also contains RO-derived information higher up.

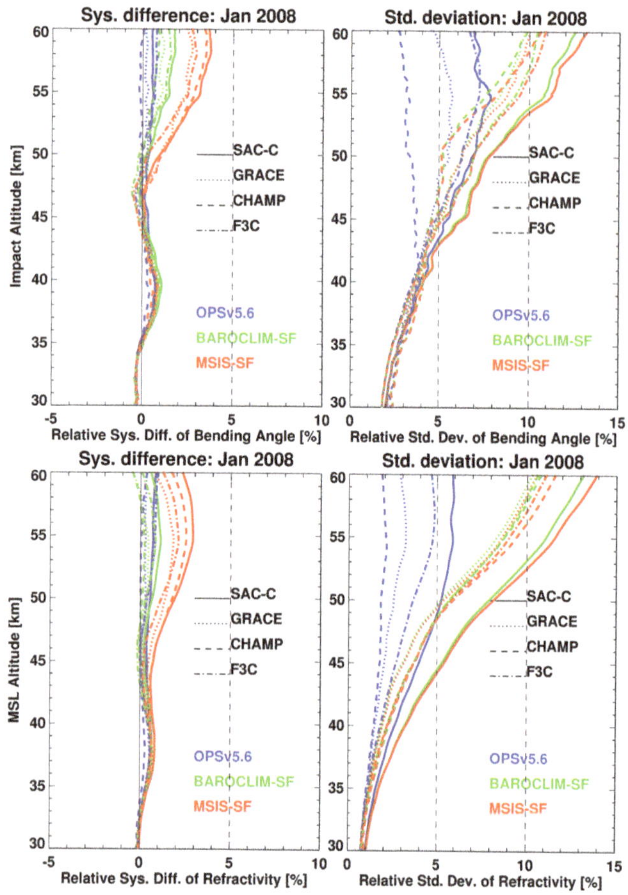

Figure 7. Global statistics of systematic difference (left) and standard deviation (right) between RO retrievals and ECMWF analyses for January 2008. The top panels show (statistically optimized) bending angle, and bottom panels refractivity. Different line types indicate retrievals for SAC-C, GRACE-A, CHAMP, and F3C. Different colors different indicate a priori information used in the retrieval.

BAROCLIM spectral coefficients and the reconstruction code, which is needed to obtain bending angles, are designed to be included in a future release of the ROPP software. This RO package can be downloaded for free after registration at http://www.romsaf.org.

We showed that BAROCLIM is of very high quality in the stratosphere and lower mesosphere, where systematic biases are small. In this altitude range differences between BAROCLIM and ECMWF analyses (forward-modeled to bending angle) rather show deficiencies in ECMWF analyses than in BAROCLIM. At 40 km, e.g., we find BAROCLIM minus ECMWF analysis bending angle differences of about 0.5 to 1.0 % and sometimes more. Above 50 km, this difference even exceeds 2 %. This is generally consistent with findings by Chauhan et al. (2009), using other types of satellite measurements.

BAROCLIM was originally developed to be used as a priori information for bending angle initialization in RO data processing. We thus evaluated BAROCLIM by comparing retrieved RO profiles initialized with different a priori information provided by BAROCLIM, MSIS, and ECMWF. These comparisons showed that RO bending angles initialized with BAROCLIM are close to raw (unoptimized) bending angles. This means that BAROCLIM-initialized bending angles preserve the mean of the raw measurements, while MSIS-initialized bending angles are slightly negatively biased. Comparison of retrieved RO profiles to ECMWF analyses also indicated that BAROCLIM outperforms MSIS. These results confirmed the capability of BAROCLIM to be used in RO profile retrievals.

The main advantage of BAROCLIM compared to the average bending angle approach proposed by Ao et al. (2012) and Gleisner and Healy (2013) is that utilization of BAROCLIM yields individual RO profiles rather than climatological fields and individual RO profiles are known to provide accurate information of, e.g., tropopause characteristics, occurrence of multiple tropopauses, or the height of the atmospheric boundary layer.

Our current BAROCLIM spectral model only includes profiles of bending angle. An important BAROCLIM update could comprise its inversion to refractivity, density, pressure, and temperature so that these parameters could be used for other applications as well.

Acknowledgements. We want to thank Gottfried Kirchengast and Josef Innerkofler (WEGC) as well as Sean Healy (ECMWF) for valuable scientific discussions. We are also grateful to UCAR/CDAAC for the provision of level 1a RO data and WEGC for the provision of level 1b RO data. Special thanks to Marc Schwärz and Johannes Fritzer (WEGC) for the contributions in OPS system development and operations. Furthermore, we thank ECMWF (Reading, UK) for providing analysis data. B. Scherllin-Pirscher and U. Foelsche were partly funded by GRAS SAF (visiting scientist project VS14) and ROM SAF (visiting scientist project VS19), and by the Austrian Science Fund (FWF) under grants P22293-N21 (BENCHCLIM) and T620-N29 (DYNOCC). S. Syndergaard and K. B. Lauritsen were supported by the ROM SAF (Radio Occultation Meteorology Satellite Application Facility), which is an operational RO processing center under EUMETSAT.

Edited by: R. Anthes

References

AIAA: Guide to Reference and Standard Atmosphere Models, ANSI/AIAA G-003B-2004, American Institute of Aeronautics and Astronautics, 2004.

Anthes, R. A.: Exploring Earth's atmosphere with radio occultation: contributions to weather, climate and space weather, Atmos. Meas. Tech., 4, 1077–1103, doi:10.5194/amt-4-1077-2011, 2011.

Anthes, R. A., Bernhardt, P. A., Chen, Y., Cucurull, L., Dymond, K. F., Ector, D., Healy, S. B., Ho, S.-P., Hunt, D. C.,

Kuo, Y.-H., Liu, H., Manning, K., McCormick, C., Meehan, T. K., Randel, W. J., Rocken, C., Schreiner, W. S., Sokolovskiy, S. V., Syndergaard, S., Thompson, D. C., Trenberth, K. E., Wee, T.-K., Yen, N. L., and Zeng, Z.: The COSMIC/FORMOSAT-3 mission: early results, B. Am. Meteorol. Soc., 89, 313–333, doi:10.1175/BAMS-89-3-313, 2008.

Ao, C. O., Mannucci, A. J., and Kursinski, E. R.: Improving GPS radio occultation stratospheric refractivity retrievals for climate benchmarking, Geophys. Res. Lett., 39, L12701, doi:10.1029/2012GL051720, 2012.

Chauhan, S., Höpfner, M., Stiller, G. P., von Clarmann, T., Funke, B., Glatthor, N., Grabowski, U., Linden, A., Kellmann, S., Milz, M., Steck, T., Fischer, H., Froidevaux, L., Lambert, A., Santee, M. L., Schwartz, M., Read, W. G., and Livesey, N. J.: MIPAS reduced spectral resolution UTLS-1 mode measurements of temperature, O_3, HNO_3, N_2O, H_2O and relative humidity over ice: retrievals and comparison to MLS, Atmos. Meas. Tech., 2, 337–353, doi:10.5194/amt-2-337-2009, 2009.

CIRA: COSPAR International Reference Atmosphere–2012, CIRA-2012, Models of the Earth's Upper Atmosphere, Technical report, chapters 1 to 3, CIRA, available at: http://spaceweather.usu.edu/files/uploads/PDF/COSPAR_INTERNATIONAL_REFERENCE_ATMOSPHERE-CHAPTER-1_328rev-01-11-08-201229.pdf (last access: July 2014), 2012.

Cucurull, L. and Derber, J. C.: Operational implementation of COSMIC observations into NCEP's global data assimilation system, Weather Forecast., 23, 702–711, doi:10.1175/2008WAF2007070.1, 2008.

Culverwell, I.: The Radio Occultation Processing Package (ROPP) – An Overview, Version 7.0 (ROPP-7 v7.0), ROM SAF CDOP-2, ROM SAF, Ref: SAF/ROM/METO/UG/ROPP/001, July 2013, available at: http://www.romsaf.org (last access: July 2014), 2013.

Danzer, J., Scherllin-Pirscher, B., and Foelsche, U.: Systematic residual ionospheric errors in radio occultation data and a potential way to minimize them, Atmos. Meas. Tech., 6, 2169–2179, doi:10.5194/amt-6-2169-2013, 2013.

Fjeldbo, G., Kliore, A. J., and Eshleman, V. R.: The neutral atmosphere of Venus as studied with the Mariner V radio occultation experiments, Astron. J., 76, 123–140, doi:10.1086/111096, 1971.

Fleming, E. L., Chandra, S., Barnett, J. J., and Corney, M.: Zonal mean temperature, pressure, zonal wind, and geopotential height as functions of latitude, COSPAR International Reference Atmosphere: 1986, Part II: Middle Atmosphere Models, Adv. Space Res., 10, 11–59, 1990.

Foelsche, U. and Scherllin-Pirscher, B.: Development of bending angle climatology from RO data, CDOP visiting scientist report 14, GRAS SAF, Ref: SAF/GRAS/DMI/REP/VS14/001, July 2012, available at: http://www.romsaf.org (last access: July 2014), 2012.

Foelsche, U., Borsche, M., Steiner, A. K., Gobiet, A., Pirscher, B., Kirchengast, G., Wickert, J., and Schmidt, T.: Observing upper troposphere-lower stratosphere climate with radio occultation data from the CHAMP satellite, Clim. Dynam., 31, 49–65, doi:10.1007/s00382-007-0337-7, 2008.

Foelsche, U., Pirscher, B., Borsche, M., Steiner, A. K., Kirchengast, G., and Rocken, C.: Climatologies based on radio occultation data from CHAMP and Formosat-3/COSMIC, in: New Horizons in Occultation Research: Studies in Atmosphere and Climate, edited by: Steiner, A. K., Pirscher, B., Foelsche, U., and Kirchengast, G., Springer, 181–194, doi:10.1007/978-3-642-00321-9_15, 2009.

Foelsche, U., Scherllin-Pirscher, B., Ladstädter, F., Steiner, A. K., and Kirchengast, G.: Refractivity and temperature climate records from multiple radio occultation satellites consistent within 0.05 %, Atmos. Meas. Tech., 4, 2007–2018, doi:10.5194/amt-4-2007-2011, 2011.

Fritzer, J., Kirchengast, G., and Pock, M.: EGOPS 5.6/DDD, End-to-End Generic Occultation Performance Simulation and Processing System Version 5.6 (EGOPS 5.6)/Detailed Design Document, WEGC-IGAM/UniGraz technical report for ESA/ESTEC no. 2/2012, doc-id: WEGC-EGOPS-2012-TR-02, issue 1.1, WEGC and IGAM, 2012.

Fujiwara, M., Polavarapu, S., and Jackson, D.: A proposal of the SPARC reanalysis/analysis intercomparison project, SPARC Newsletter, 38, 14–17, 2012.

Gleisner, H. and Healy, S. B.: A simplified approach for generating GNSS radio occultation refractivity climatologies, Atmos. Meas. Tech., 6, 121–129, doi:10.5194/amt-6-121-2013, 2013.

Gobiet, A. and Kirchengast, G.: Advancements of Global Navigation Satellite System radio occultation retrieval in the upper stratosphere for optimal climate monitoring utility, J. Geophys. Res., 109, D24110, doi:10.1029/2004JD005117, 2004.

Gobiet, A., Kirchengast, G., Manney, G. L., Borsche, M., Retscher, C., and Stiller, G.: Retrieval of temperature profiles from CHAMP for climate monitoring: intercomparison with Envisat MIPAS and GOMOS and different atmospheric analyses, Atmos. Chem. Phys., 7, 3519–3536, doi:10.5194/acp-7-3519-2007, 2007.

Gorbunov, M. E.: Canonical transform method for processing radio occultation data in the lower troposphere, Radio Sci., 37, 1076, doi:10.1029/2000RS002592, 2002.

Gorbunov, M. E. and Lauritsen, K. B.: Analysis of wave fields by Fourier integral operators and their application for radio occultations, Radio Sci., 39, RS4010, doi:10.1029/2003RS002971, 2004.

Gorbunov, M. E., Benzon, H.-H., Jensen, A. S., Lohmann, M. S., and Nielsen, A. S.: Comparative analysis of radio occultation processing approaches based on Fourier integral operators, Radio Sci., 39, RS6004, doi:10.1029/2003RS002916, 2004.

Hajj, G. A., Ao, C. O., Iijima, B. A., Kuang, D., Kursinski, E. R., Mannucci, A. J., Meehan, T. K., Romans, L. J., de la Torre Juarez, M., and Yunck, T. P.: CHAMP and SAC-C atmospheric occultation results and intercomparisons, J. Geophys. Res., 109, D06109, doi:10.1029/2003JD003909, 2004.

Healy, S.: Operational assimilation of GPS radio occultation measurements at ECMWF, ECMWF Newsletter, 111, 6–11, 2007.

Healy, S. B.: Radio occultation bending angle and impact parameter errors caused by horizontal refractive index gradients in the troposphere: a simulation study, J. Geophys. Res., 106, 11875–11889, doi:10.1029/2001JD900050, 2001.

Healy, S. B. and Thépaut, J. N.: Assimilation experiments with CHAMP GPS radio occultation measurements, Q. J. Roy. Meteor. Soc., 132, 605–623, doi:10.1256/qj.04.182, 2006.

Healy, S. B., Jupp, A. M., and Marquardt, C.: Forecast impact experiment with GPS radio occultation measurements, Geophys. Res. Lett., 32, L03804, doi:10.1029/2004GL020806, 2005.

Hedin, A. E.: A revised thermospheric model based on mass spectrometer and incoherent scatter data: MSIS-83, J. Geophys. Res., 88, 10170–10188, doi:10.1029/JA088iA12p10170, 1983.

Hedin, A. E.: MSIS-86 thermospheric model, J. Geophys. Res., 92, 4649–4662, doi:10.1029/JA092iA05p04649, 1987.

Hedin, A. E.: Extension of the MSIS thermosphere model into the middle and lower atmosphere, J. Geophys. Res., 96, 1159–1172, doi:10.1029/90JA02125, 1991.

Ho, S.-P., Hunt, D., Steiner, A. K., Mannucci, A. J., Kirchengast, G., Gleisner, H., Heise, S., von Engeln, A., Marquardt, C., Sokolovskiy, S., Schreiner, W., Scherllin-Pirscher, B., Ao, C., Wickert, J., Syndergaard, S., Lauritsen, K., Leroy, S., Kursinski, E. R., Kuo, Y.-H., Foelsche, U., Schmidt, T., and Gorbunov, M.: Reproducibility of GPS radio occultation data for climate monitoring: Profile-to-profile inter-comparison of CHAMP climate records 2002 to 2008 from six data centers, J. Geophys. Res., 117, D18111, doi:10.1029/2012JD017665, 2012.

Hocke, K., Igarashi, K., and Tsuda, T.: High-resolution profiling of layered structures in the lower stratosphere by GPS occultation, Geophys. Res. Lett., 30, 1426, doi:10.1029/2002GL016566, 2003.

Høeg, P., Hauchecorne, A., Kirchengast, G., Syndergaard, S., Belloul, B., Leitinger, R., and Rothleitner, W.: Derivation of atmospheric properties using a radio occultation technique, DMI Sci. Rep. 95-4, Danish Meteorological Institute, Copenhagen, Denmark, 1995.

Kursinski, E. R., Hajj, G. A., Schofield, J. T., Linfield, R. P., and Hardy, K. R.: Observing Earth's atmosphere with radio occultation measurements using the Global Positioning System, J. Geophys. Res., 102, 23429–23465, doi:10.1029/97JD01569, 1997.

Lohmann, M.: Application of dynamical error estimation for statistical optimization of radio occultation bending angles, Radio Sci., 40, RS3011, doi:10.1029/2004RS003117, 2005.

Melbourne, W. G., Davis, E. S., Duncan, C. B., Hajj, G. A., Hardy, K. R., Kursinski, E. R., Meehan, T. K., Young, L. E., and Yunck, T. P.: The application of spaceborne GPS to atmospheric limb sounding and global change monitoring, JPL Publication, 94–18, 147 pp., 1994.

Picone, J. M., Hedin, A. E., Drob, D. P., and Aikin, A. C.: NRLMSISE-00 empirical model of the atmosphere: statistical comparisons and scientific issues, J. Geophys. Res., 107, 1468, doi:10.1029/2002JA009430, 2002.

Press, W. H., Flannery, B. P., Teukolsky, S. A., and Vetterling, W. T.: Numerical Recipes: The Art of Scientific Computing, Cambridge University Press, New York, NY, USA, 1986.

Ramsauer, J. and Kirchengast, G.: Sensitivity of atmospheric profiles retrieved from GNSS radio occultation data to instrumental errors, Igam/ug technical report for esa/estec no. 6/2001, Institute for Geophysics, Astrophysics, and Meteorology, University of Graz, 2001.

Randel, W., Udelhofen, P., Fleming, E., Geller, M., Gelman, M., Hamilton, K., Karoly, D., Ortland, D., Pawson, S., Swinbank, R., Wu, F., Baldwin, M., Chanin, M.-L., Keckhut, P., Labitzke, K., Remsberg, E., Simmons, A., and Wu, D.: The SPARC intercomparison of middle-atmosphere climatologies, J. Climate, 17, 986–1003, doi:10.1175/1520-0442(2004)017<0986:TSIOMC>2.0.CO;2, 2004.

Ringer, M. A. and Healy, S. B.: Monitoring twenty-first century climate using GPS radio occultation bending angles, Geophys. Res. Lett., 35, L05708, doi:10.1029/2007GL032462, 2008.

Rocken, C., Anthes, R., Exner, M., Hunt, D., Sokolovskiy, S., Ware, R., Gorbunov, M., Schreiner, W., Feng, D., Herman, B., Kuo, Y.-H., , and Zuo, X.: Analysis and validation of GPS/MET data in the neutral atmosphere, J. Geophys. Res., 102, 29849–29866, doi:10.1029/97JD02400, 1997.

Rocken, C., Sokolovskiy, S., Schreiner, W., Hunt, D., Ho, S.-P., Kuo, Y.-H., and Foelsche, U.: Climate monitoring with radio occultation data: systematic error sources, presentation at the GPS RO climate applications workshop, 2008.

Rodgers, C. D.: Inverse methods for atmospheric sounding: theory and practice, in: Series on Atmospheric Oceanic and Planetary Physics, vol. 2, World Scientific Publishing Company, Singapore, 2000.

Scherllin-Pirscher, B.: Further development of BAROCLIM and implementation in ROPP, ROM SAF CDOP-2 visiting scientist report 19, ROM SAF, Ref: SAF/ROM/DMI/REP/VS19/001, June 2013, available at: http://www.romsaf.org (last access: July 2014), 2013.

Scherllin-Pirscher, B., Kirchengast, G., Steiner, A. K., Kuo, Y.-H., and Foelsche, U.: Quantifying uncertainty in climatological fields from GPS radio occultation: an empirical-analytical error model, Atmos. Meas. Tech., 4, 2019–2034, doi:10.5194/amt-4-2019-2011, 2011a.

Scherllin-Pirscher, B., Steiner, A. K., Kirchengast, G., Kuo, Y.-H., and Foelsche, U.: Empirical analysis and modeling of errors of atmospheric profiles from GPS radio occultation, Atmos. Meas. Tech., 4, 1875–1890, doi:10.5194/amt-4-1875-2011, 2011b.

Schreiner, W., Rocken, C., Sokolovskiy, S., Syndergaard, S., and Hunt, D.: Estimates of the precision of GPS radio occultations from the COSMIC/FORMOSAT-3 mission, Geophys. Res. Lett., 34, L04808, doi:10.1029/2006GL027557, 2007.

Schreiner, W., Kuo, Y.-H., Ho, S.-P., Sokolovskiy, S., and Hunt, D.: Use of GNSS Radio Occultation data for Climate Applications, world Climate Research Program Conference, 24–28 October 2011, Denver, Colorado USA, available at: http://www.cosmic.ucar.edu/groupAct/references/WCRP-2011-Schreiner-Final.pdf (last access: July 2014), 2011.

Schreiner, W. S., Rocken, C., Sokolovskiy, S., and Hunt, D.: Quality assessment of COSMIC/FORMOSAT-3 GPS radio occultation data derived from single- and double-difference atmospheric excess phase processing, GPS Solut., 14, 13–22, doi:10.1007/s10291-009-0132-5, 2009.

Schwärz, M., Scherllin-Pirscher, B., Kirchengast, G., Schwarz, J., Ladstädter, F., Fritzer, J., and Ramsauer, J.: Multi-Mission Validation by Satellite Radio Occultation, Final report for ESA/ESRIN No. 01/2013, WEGC, 2013.

Sokolovskiy, S., Rocken, C., Hunt, D., Schreiner, W., Johnson, J., Masters, D., and Esterhuizen, S.: GPS profiling of the lower troposphere from space: inversion and demodulation of the open-loop radio occultation signals, Geophys. Res. Lett., 33, L14816, doi:10.1029/2006GL026112, 2006.

Spiegel, M. R. (Ed.): Handbuch für Mathematik: Formeln und Tabellen, Theorie und Anwendung, Schaum's Outline, McGraw-Hill, Düsseldorf, New York, St. Louis, 1979.

Steiner, A. K., Lackner, B. C., Ladstädter, F., Scherllin-Pirscher, B., Foelsche, U., and Kirchengast, G.: GPS radio occultation for climate monitoring and change detection, Radio Sci., 46, RS0D24, doi:10.1029/2010RS004614, 2011.

Steiner, A. K., Hunt, D., Ho, S.-P., Kirchengast, G., Mannucci, A. J., Scherllin-Pirscher, B., Gleisner, H., von Engeln, A., Schmidt, T., Ao, C., Leroy, S. S., Kursinski, E. R., Foelsche, U., Gorbunov, M., Heise, S., Kuo, Y.-H., Lauritsen, K. B., Marquardt, C., Rocken, C., Schreiner, W., Sokolovskiy, S., Syndergaard, S., and Wickert, J.: Quantification of structural uncertainty in climate data records from GPS radio occultation, Atmos. Chem. Phys., 13, 1469–1484, doi:10.5194/acp-13-1469-2013, 2013.

Syndergaard, S.: Modeling the impact of the Earth's oblateness on the retrieval of temperature and pressure profiles from limb sounding, J. Atmos. Sol.-Terr. Phy., 60, 171–180, doi:10.1016/S1364-6826(97)00056-4, 1998.

Syndergaard, S.: Retrieval Analysis and Methodologies in Atmospheric Limb Sounding Using the GNSS Radio Occultation Technique, Danish Meteorological Institute, Copenhagen, Denmark, 1999.

von Engeln, A., Healy, S., Marquardt, C., Andres, Y., and Sancho, F.: Validation of operational GRAS radio occultation data, Geophys. Res. Lett., 36, L17809, doi:10.1029/2009GL039968, 2009.

Vorob'ev, V. V. and Krasil'nikova, T. G.: Estimation of the accuracy of the atmospheric refractive index recovery from Doppler shift measurements at frequencies used in the NAVSTAR system, Izv. Atmos. Ocean. Phy., 29, 602–609, 1994.

Zeng, Z. and Sokolovskiy, S.: Effect of sporadic E clouds on GPS radio occultation signals, Geophys. Res. Lett., 37, L18817, doi:10.1029/2010GL044561, 2010.

17

Assessment of the performance of the inter-arrival time algorithm to identify ice shattering artifacts in cloud particle probe measurements

A. Korolev[1] **and P. R. Field**[2]

[1]Environment Canada, Toronto, Canada
[2]Met Office, Exeter, UK, and University of Leeds, Leeds, UK

Correspondence to: A. Korolev (alexei.korolev@ec.gc.ca)

Abstract. Shattering presents a serious obstacle to current airborne in situ methods of characterizing the microphysical properties of ice clouds. Small shattered fragments result from the impact of natural ice crystals with the forward parts of aircraft-mounted measurement probes. The presence of these shattered fragments may result in a significant overestimation of the measured concentration of small ice crystals, contaminating the measurement of the ice particle size distribution (PSD). One method of identifying shattered particles is to use an inter-arrival time algorithm. This method is based on the assumption that shattered fragments form spatial clusters that have short inter-arrival times between particles, relative to natural particles, when they pass through the sample volume of the probe. The inter-arrival time algorithm is a successful technique for the classification of shattering artifacts and natural particles. This study assesses the limitations and efficiency of the inter-arrival time algorithm. The analysis has been performed using simultaneous measurements of two-dimensional (2-D) optical array probes with the standard and antishattering "K-tips" collected during the Airborne Icing Instrumentation Experiment (AIIE). It is shown that the efficiency of the algorithm depends on ice particle size, concentration and habit. Additional numerical simulations indicate that the effectiveness of the inter-arrival time algorithm to eliminate shattering artifacts can be significantly restricted in some cases. Improvements to the inter-arrival time algorithm are discussed. It is demonstrated that blind application of the inter-arrival time algorithm cannot filter out all shattered aggregates. To mitigate against the effects of shattering, the inter-arrival time algorithm should be used together with other means, such as antishattering tips and specially designed algorithms for segregation of shattered artifacts and natural particles.

1 Introduction

Ice particle size distributions (PSDs) are used in atmospheric models for computing both the radiative impact of ice clouds and the microphysical process rates that control ice cloud evolution. Therefore, an inaccurate representation of the ice PSD can have adverse impacts on the accuracy of climate and weather prediction models. Measurements of ice PSDs from aircraft-based observations form the basis of size distribution representations in models. Prior to entering the instrument sample volume, an ice particle may impact the probe's upstream tips or inlet and shatter into many small fragments. These shattering products can then significantly contaminate measurements of airborne particle probes.

Over three decades the abundance of small ice particles measured by particle probes in the tropospheric clouds remained an intriguing problem. Recent studies using simultaneous measurements of standard and modified probes (Korolev et al., 2011, 2013b) have unambiguously demonstrated that in many cases measurements made with standard cloud probes are adversely affected by shattering artifacts.

There are two main approaches currently used within the cloud physics community to mitigate the effect of shattering. One approach is based on modifying the inlet configuration of the cloud probes to minimize the area that deflects shattered particles towards the sample volume (Korolev et al., 2013a). This method can lead to a significant reduction in the effect of shattering but is not able to completely eradicate the problem (Korolev et al., 2011, 2013b; Lawson, 2011).

The second approach is a postprocessing methodology based on the fact that shattering products are spatially clustered. Because of close spacing, the time difference between two successive shattered fragments passing through the sample volume will be much shorter than that for naturally occurring particles. This time difference is usually referred to as "inter-arrival time".

Cooper (1977) suggested that these artifacts could be filtered out by identifying the characteristically short inter-arrival times of particles associated with these shattering products.

Field et al. (2003) used a fast forward scattering spectrometer probe (FSSP) to measure particle spacing in ice clouds. The inter-arrival time distribution in ice clouds was found to have a bimodal shape with modes at 10^{-2} and 10^{-4} s corresponding to approximately 1m and 1 cm spatial separations. The particles from the long and short inter-arrival time modes corresponded to estimated concentrations of 0.1–1 cm^{-3} and $\sim 100\,cm^{-3}$ respectively. No conclusions were drawn as to whether the latter localized clusters of high particle concentration were natural or artifacts. Assuming they were artifacts, their inter-arrival time algorithm suggested average and maximum concentration overestimates of a factor of 2 and 5 respectively.

Field et al. (2006) applied an inter-arrival time algorithm (ITA) to filter out shattering artifacts, choosing a threshold inter-arrival time in the range 10^{-4} to 10^{-5} s, depending on the instrument and the aircraft type used for the data collection, to reject the short inter-arrival mode. It was found that the OAP-2DC and CIP concentrations were reduced by up to a factor of 4 when the mass-weighted mean size exceeded 3 mm. The ice water content (IWC) estimate was reduced by up to 20–30 %, most notable in cases of narrow size distributions. It was also found that the corrected PSDs could show a reduction in particle concentrations over a wide range of sizes from 200 microns for narrow distributions up to 1000 microns for the broadest distributions that are subject to the most shattering.

It should be noted that the time separation between particle arrivals through the probe's sample volume depends on true air speed. The relevant metric, which should be used to segregate shattered artifacts and natural particles is inter-particle distance along the flight direction (Δx). It is acknowledged that the community is accustomed to referring to "inter-arrival time" Δt since this parameter is measured in the probes. In the following discussion we will be using both metrics Δx and Δt.

The ITA is now routinely used in two-dimensional (2-D) probe data processing to eliminate shattering artifacts (e.g., Baker et al., 2009; Lawson, 2011; Jackson et al., 2014). Moreover, it is recognized that the best practice for the operation of cloud probes in the presence of ice is provided by a combination of modified tips and the application of the inter-arrival time filtering together (e.g., Korolev et al., 2011; Baumgardner et al., 2012).

While it has been demonstrated that the ITA reduces the effects of shattering, its accuracy and efficiency remains poorly quantified. For instance, can the ITA alone identify all shattering artifacts? Is the efficiency of the ITA dependent upon the probe specifications such as pixel resolution, response time, sample area, and inlet configuration?

Motivation for improving our understanding of the limitations and efficiency of the ITA is threefold. Firstly, there is a need to determine if the ITA can be used to successfully reanalyze the historical data collected over the past thirty years. Secondly, an improved quantitative understanding of the limitations of the ITA will provide statements of the accuracy of the measurements of the particle number concentration, ice water content, extinction coefficient and other PSD derived parameters. Thirdly, knowledge of the efficiency of the ITA will aid in the design process of future cloud probes.

The objective of this paper is to evaluate the efficiency and limitations of ITA. A detailed description and the main assumptions underlying the algorithm are presented in Sect. 2. Section 3 considers general limitations of the algorithm. Analysis of the results of the OAP-2DC data processing using the ITA are presented in Sect. 4. In Sect. 5 statistical simulations of ice particle shattering are used to explore the limitations of the ITA. Finally, Sect. 6 provides a summary.

2 Description of the inter-arrival time algorithm

2.1 Basic assumptions

In the following discussion the term "shattering event" will be applied to the group of shattered fragments, that (a) formed as a result of impact of a single particle with the upstream tips (or inlet) of a probe, and (b) at least one particle from the group of the shattered fragments was registered by the probe. It should be noted that if the particle rebounds to the sample area without shattering, it still falls in the definition of the "shattering event".

The successful application of the ITA is predicated on two basic assumptions: (1) the maximum inter-arrival time of the shattered fragments is shorter than the minimum inter-arrival time between intact particles; (2) shattered particles are always passing through the sample volume as a group of no less than two particles.

The first assumption is a necessary condition for the complete separation of the inter-arrival time distributions $\phi_s(\Delta t)$ and $\phi_i(\Delta t)$ without overlap. Here $\phi_s(\Delta t)$ and $\phi_i(\Delta t)$ are the distribution of inter-arrival times associated with shattering events and intact particles, respectively. The absence of an overlap between $\phi_s(\Delta t)$ and $\phi_i(\Delta t)$ allows for the existence of a cut-off time τ^* such that all shattered and only shattered

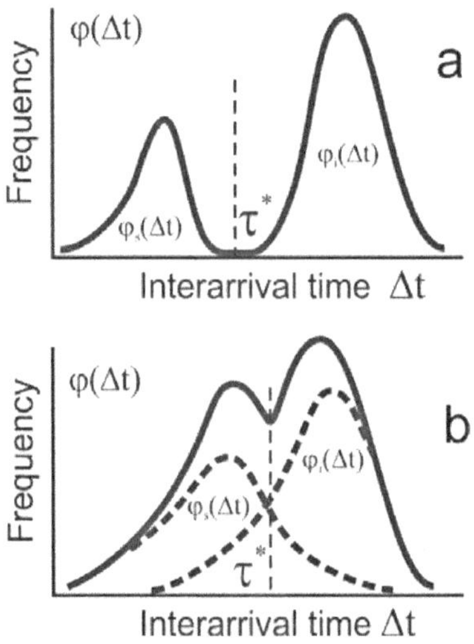

Figure 1. Conceptual diagram of the distribution of inter-arrival times $\phi(\Delta t)$ with well separated short and long inter-arrival time modes **(a)**, when shattering artifacts can be segregated from the intact particles. When the distributions of inter-arrival time associated with intact particles $\phi_{\mathrm{i}}(\Delta t)$ and shattered fragments $\phi_{\mathrm{S}}(\Delta t)$ have significant overlap, then segregation of intact particles and shattered artifacts is hindered **(b)**.

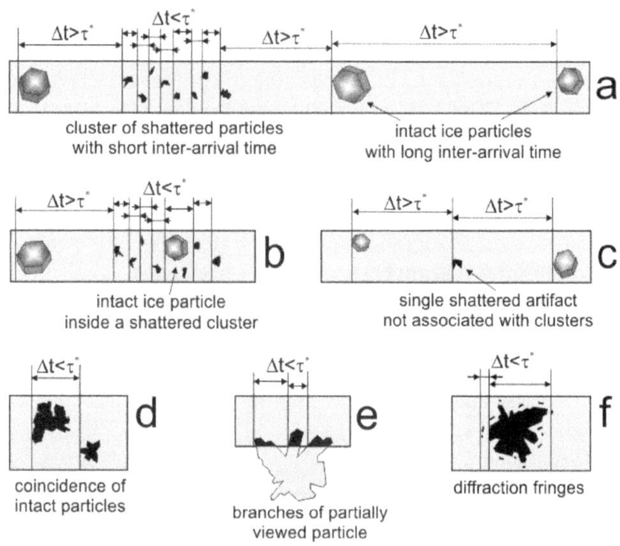

Figure 2. Conceptual diagram of **(a)** idealised spatial sequence of intact particles and shattered artifacts passing through the sample volume. Case **(c)** when the inter-arrival time algorithm may confuse shattered artifact with intact particles, and **(b, d, e, f)** when intact particles may be confused with shattering artifacts.

particles satisfy the condition $\Delta t < \tau^*$, whereas all intact particles and only intact particles are associated with $\Delta t > \tau^*$ (Fig. 1a). Given this assumption it is trivial to identify the shattered artifacts and intact particles based simply on a comparison of measured inter-arrival time Δt between two successive particles and the cut-off time τ^*.

The second assumption forms the second necessary condition to identify shattered artifacts. A minimum of two closely spaced particles is necessary to allow artifacts to be identified.

The conceptual diagram in Fig. 2a demonstrates these two basic assumptions about particle spacing that is required for the successful segregation of shattered artifacts and intact particles using an ITA. In reality, the challenges of image sampling and the statistical nature of particle spacings impose limitations on the performance of the ITA in the segregation of shattered artifacts and intact particles. As will be shown below, the first assumption cannot be satisfied due to statistical limitations, whereas the second condition is necessary but not sufficient.

2.2 Inter-arrival time algorithm

Here we describe the sequence of operations composing the basic ITA. This algorithm in its basic form will be used in the present study.

1. The inter-arrival time algorithm starts from the calculations of the distribution of inter-arrival times $\phi(\Delta t)$ as in Fig. 1. The calculation of $\phi(\Delta t)$ are performed for each averaging time interval. Similar to Field et al. (2003, 2006) the time bins in $\phi(\Delta t)$ were logarithmically spaced. The width of the time bins was optimized to trade-off accuracy of the estimation of τ^* and the statistical significance related to the number of counts in each time bin.

2. The cut-off-time τ^* was calculated for each averaging time interval as a minimum between two maxima associated with long and short time modes (Fig. 1). In cases when only one mode was present, τ^* was forced to be equal to the minimum inter-arrival time found in this averaging interval. It should be noted that the function $\phi(\Delta t)$ is a non-normalized distribution of counts in each time bin. Normalization using the bin width leads to a disappearance of the minimum between the short and long time modes in $\phi(\Delta t)$, which hinders calculation of τ^*.

3. Pairs of particles that satisfied the condition $\Delta t < \tau^*$ were identified and marked as artifacts. It is important to note that the ITA cannot identify a single-particle shattered artifact (singleton) and that the minimum number of particles identified as artifacts is two.

The value of τ^* should be calculated for each averaging time interval. As will be shown below, τ^* has a wide dynamic range and it depends on many microphysical, environmental and instrumental parameters (Sect. 4.3). The assumption, that τ^* remains constant, may result in large errors in identifying shattering artifacts.

The values of τ^* and Δt depend on the airplane speed. In the described algorithm it is assumed that the aircraft speed remained approximately constant at each averaging time interval. This assumption works well for a few seconds averaging intervals. However, the aircraft speed depending on the altitude may change by as much as a factor of two during the flight operation. This gives another reason to recalculate τ^* at each averaging time interval.

It is relevant to mention here that alternative techniques for determining τ^* were used by Field et al. (2003, 2006), Lawson (2011) and Jackson et al. (2014). These techniques were based on fitting the function $\phi_{\mathrm{i}}(\Delta t)$ by the Poisson distribution.

3 Limitations of the inter-arrival time algorithm

This section presents a list of sampling effects that demonstrate how the assumptions underlying the inter-arrival time algorithm can be contravened. These cases impose limitations on the ability of the ITA to segregate intact particles and shattering artifacts.

In the following we assume that particles are distributed randomly in space and that their spacing and hence inter-arrival time is well represented by a Poisson distribution. For the Poisson process the density function for counting one intact particle during time Δt is described by

$$\frac{\mathrm{d}P(\Delta t)}{\mathrm{d}t} = \frac{e^{-\Delta t/\tau}}{\tau}, \tag{1}$$

where $\tau = 1/nSu$ is the average inter-arrival time between intact particles passing through the probe's sample area S; n is the particle concentration; u is the sampling speed. Shattered particles detected by the probe were deflected into the sample area after the impact with the inlet. Therefore, the shattered particles have external origin, are intermittent and their distribution can be considered as independent with respect to the intact particles. Examination of the short inter-arrival time mode does indicate that these particles also appear to be characterized quite well with a Poisson distribution (e.g., Field et al., 2003, 2006; Sect. 4.3 in this paper).

3.1 Naturally occurring particles with inter-arrival times shorter than the cut-off-time interval

Two closely spaced intact particles will be identified as a shattering artifact if the inter-arrival time $\Delta t < \tau^*$ (Fig. 2d). Such cases break the first assumption in Sect. 2.1. The probability for coincidence of three or more particles falls very rapidly and has been ignored. The probability of two particles arriving within τ^* can be found from the Poisson statistics as

$$P_2\left(\Delta t < \tau^*\right) = \frac{1}{2}\left(\frac{\tau^*}{\tau}\right)^2 e^{-\frac{\tau^*}{\tau}}. \tag{2}$$

As follows from Eq. (1), the probability of such an event increases with increasing particle concentration n (decreasing τ) and increases in the cut-off time τ^*.

3.2 Coincidence of a shattering event particle and an intact particle

The probability of coincidence of the arrival time of an intact particle during the shattered event (Fig. 2b) can be estimated as

$$P\left(\Delta t < \Delta t_{\mathrm{sh}}\right) = \int_0^{\Delta t_{\mathrm{sh}}} \frac{e^{-\Delta t/\tau}}{\tau} d(\Delta t) = 1 - e^{-\Delta t_{\mathrm{sh}}/\tau}. \tag{3}$$

Here $\Delta t_{\mathrm{sh}} = \sum_i \Delta t_{\mathrm{sh}}(i) = L_{\mathrm{sh}}/u$ is the duration of the shattering event registered by the probe; $\Delta t_{\mathrm{sh}}(i)$ is the inter-arrival time between two subsequent shattered fragments registered by the probe within the shattered event; L_{sh} is the spatial length of the shattering event along the flight direction (i.e., distance between the first and last fragments in the shattering event). Equation (2) indicates that even when $\Delta t_{\mathrm{sh}} \ll \tau$, the probability of the arrival of the intact particle in the sample volume during the shattering event remains non-zero. Basically, it means that in principle it is impossible to separate all shattering artifacts and intact particles, and the functions $\phi_{\mathrm{s}}(\Delta t)$ and $\phi_{\mathrm{i}}(\Delta t)$ always overlap (Fig. 1b). The relative fraction of the overlapping area of $\phi_{\mathrm{s}}(\Delta t)$ and $\phi_{\mathrm{i}}(\Delta t)$ characterizes the frequency of misidentifying intact particles and shattering artifacts.

It is possible to attempt to correct for the removal of intact particles using Poisson statistics to estimate the fraction of intact particles rejected and then scale the remaining intact size distribution (e.g., Field et al., 2006).

3.3 Singletons: single particle shattering artifacts

A significant limitation of the ITA is related to situations when only one particle from the group of the shattered fragments is registered by the probe (Fig. 2c). Such a situation may occur when most of the shattered fragments travel outside of the sample volume, but a single fragment passes through the sample volume. It may also happen when most of the shattered fragments are smaller than the probe's detecting threshold, and only one particle exceeds the threshold and is registered by the probe. Ice particles may also rebound from the inlet without fragmentation, thus forming a single particle shattering event. Rebounding without shattering was demonstrated in Korolev et al. (2013a, Fig. 13a–d). With respect to the ITA, the single particle artifacts described above can have long inter-arrival times and are therefore indistinguishable from the natural population of intact particles. Due to the random nature of particle impact with the probe's arms and the direction of the trajectories of the rebound shattered fragments the probability of the single particle shattering events always remains non-zero. This imposes a significant and difficult-to-quantify limitation on the performance of the ITA.

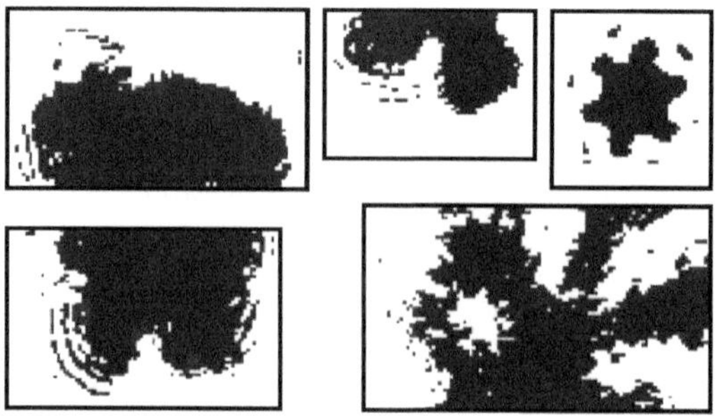

Figure 3. Examples of diffraction fringes around out-of-focus images measured by CIP at 15 μm pixel resolution. The diffraction fringes and the particle generating the fringes may be confused with shattered fragments and be rejected by the inter-arrival time algorithm.

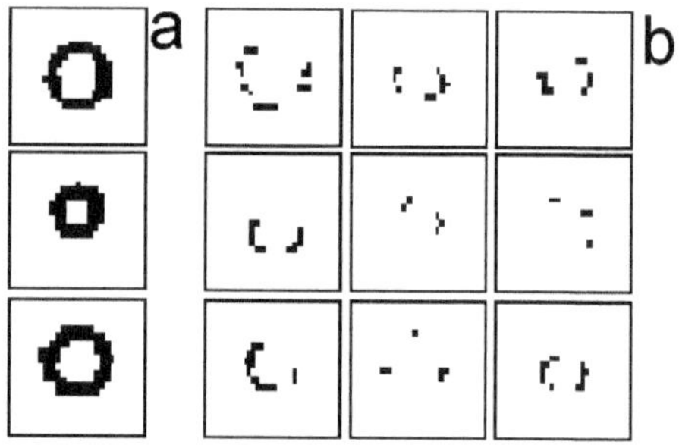

Figure 4. Examples of out-of-focus images measured by 2-D-S at 10 μm pixel resolution. **(a)** Complete circle out-of-focus images; **(b)** fragmented out-of-focus images, which were registered in two or three image frames and identified as shattering artifacts by the inter-arrival time algorithm. The fragmented out-of-focus images are related to the particles passing through the sample volume near the edge of the depth-of-field.

3.4 Partially viewed ice branches

Many ice particles develop branches extending from a few hundred micrometers up to 2–3 mm away from its center (i.e., bullet rosettes, dendrites and aggregated ice particles). The partially viewed branches of such particles could be confused with separate particles possessing short inter-arrival time (Fig. 2e), and be identified as artifacts. Rejecting images that are in contact with the edge of the array is one way to mitigate against this problem.

3.5 Diffraction fringes

Most particle imaging probes use coherent sources of light that result in the formation of diffraction fringes around the image of a particle. The binary representations of these fringes may manifest themselves as sparse disconnected pixel images that surround the main particle image. Such optical and imaging instrumentation effects may be confused with shattered fragments and result in identifying both diffraction fringes and the intact particle producing these fringes as artifacts.

For spherical particles such fringes become most pronounced when the dimensionless distance of the particles from the focal plane is close to $Z_d \approx 1.9$ (Korolev, 2007; his Fig. 9) where $Z_d = \frac{4\lambda Z}{D^2}$; λ is the wavelength; Z is the distance from the object plane; D is the particle diameter. Non-circular images produce diffraction fringes over a wider range of Z_d. The probability of imaging the diffraction fringes increases with the increasing pixel resolution. Thus, for probes with coarse pixel resolution 100–200 μm (e.g., HVPS, PIP, OAP-2DP), diffraction artifacts are quite rare, whereas for probes with 10–15 μm pixel resolution (e.g., 2-D-S, CIP) the effect of the diffraction fringes may have a significant effect on misidentification of intact particles as shattering artifacts. A few examples of diffraction fringes around the CIP out-of-focus images are shown in Fig. 3. These images were identified by the inter-arrival time algorithms as shattering artifacts and rejected. The 2-D data processing software can be tuned to return large images to the pool of accepted images. However, setting the threshold for the sizes of the accepted images is ambiguous, and it may result in accepting shattering artifacts and rejecting intact particles.

3.6 Out-of-focus fragmented images

Out-of-focus images of particles traversing the sample area near the edges of the depth-of-field, when $Z_d > 6$ may appear as disconnected images (Korolev, 2007; Fig. 7). If the out-of-focus fragmented image has a gap along the flight direction, it may be confused with a shattering artifact. Examples of the out-of-focus images, which were identified by ITA as shattered fragments, are shown in Fig. 4b. Out-of-focus images, such as images of transparent plates, quite often appear as fragmented and may also be identified by the ITA as artifacts.

4 Results of measurements of inter-particle distances

Because shattering generates particles by a very different physical process to those that occur naturally, the mode that describes the inter-arrival time distribution of these particles can be very different to that associated with the natural intact particles, which usually is well described by the Poisson distribution. This difference in the distribution of the particles manifests itself through differing inter-arrival time populations. Therefore, the distribution $\phi(\Delta t)$ can be used as one of the metrics for identifying shattering. The purpose of this section is to demonstrate the variety of $\phi(\Delta t)$ distributions and show their link to the particle size distributions. This consideration is expected to help further understanding of limitations of the ITA.

In order to reduce the effect of the air speed u on Δt, the inter-particle distance $\Delta x = \Delta t/u$ will be used instead of Δt. Accordingly, the distribution $\phi(\Delta x)$ and the cut-off-distance

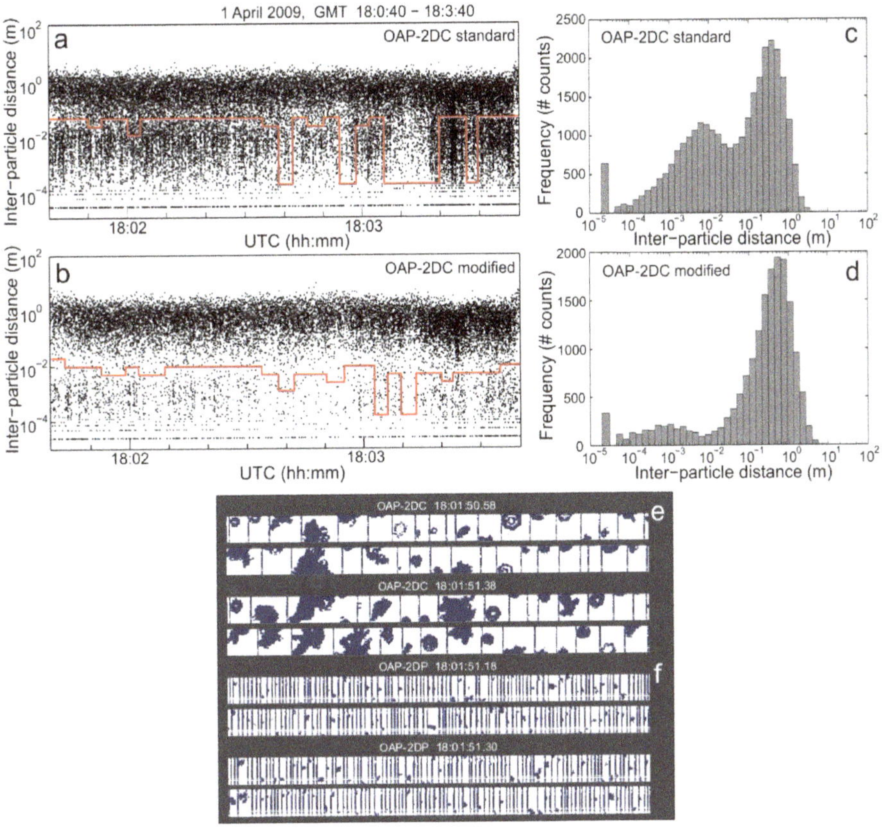

Figure 5. Comparison of inter-particle distances measured by standard **(a)** and modified **(b)** 2DC. Red lines in **(a)** and **(b)** indicate the cut-off distance χ^*. The distribution of the inter-particle distances for standard **(c)** and modified **(d)** probes. Examples of images obtained with an OAP-2DC at 25 μm pixel resolution **(e)** and an OAP-2DP at 200 μm pixel resolution **(f)**. The measurements were conducted on 1 April during an ascent through ice cloud from approximately 4600 to 5300 m in the Ottawa region. The temperature varied from -12 to $-17\,^{\circ}$C.

$\chi^* = \tau^*/u$ will be used instead of $\phi(\Delta t)$ and τ^*, respectively. We will also keep using the conventional term "inter-arrival time algorithm", although the term "inter-particle distance algorithm" would be more accurate.

4.1 Description of the data set

The data used here were collected during the Airborne Icing Instrumentation Evaluation Experiment (AIIE) flight campaign (Korolev et al., 2011, 2013b). The analysis of the inter-particle distance is focused on the measurements of two OAP-2DCs installed side-by-side in the NRC Convair-580 aircraft. Both instruments have the same pixel resolution 25 μm, optics and electronics. However, one of the probes had the standard configuration, whereas the second one had modified arms with the K-tips installed (Korolev et al., 2013a). While K-tips still shatter ice particles, it has been demonstrated that they significantly mitigate the effect of shattering on ice particle measurements. Comparing measurements made with the standard and modified OAP-2DCs before and after applying the ITA provides an opportunity to assess the efficiency of the algorithm to successfully identify and filter out shattering artifacts. The 2-D data were averaged over 5-second time intervals. For most clouds sampled during the AIIE project such averaging provided statistically significant particle numbers to estimate the function $\phi(\Delta x)$ and cut-off-distance χ^*. In the frame of this study the number of bins in $\phi(\Delta x)$ was selected to be 25. This yields a reasonable compromise between the statistical significance of number of counts in each bin and the accuracy of finding χ^*. Usually, for a typical shape of $\phi(\Delta x)$, a number of particle counts over 100 yielded an acceptable estimate of χ^* for the purposes of this work. A higher number of bins for $\phi(\Delta x)$ will require a higher number of counts, and therefore a longer averaging time.

4.2 Examples of the inter-particle distance distribution

The following three examples are based on the data collected during three different flights (1, 8 April 2009) and demonstrate how the presence of large particles and their concentration affect the inter-particle distance distribution $\phi(\Delta x)$. The first example demonstrates a moderate level of shattering and a high concentration of ice. The second example demonstrates more intense shattering and a low concentration of ice.

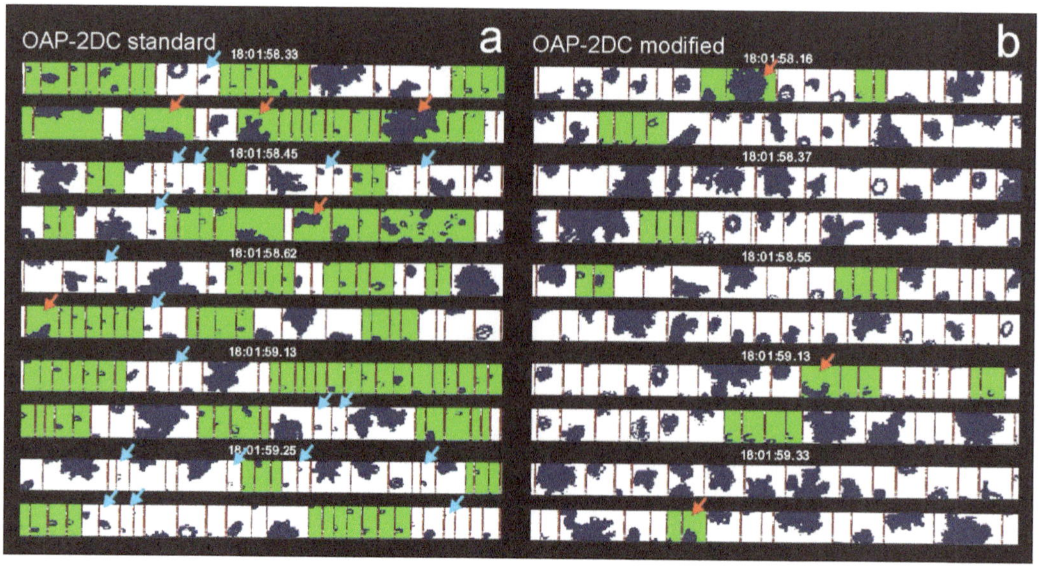

Figure 6. Examples of the results of the image rejection/acceptance processing with the inter-arrival time algorithm. The images of ice particles were simultaneously sampled by the standard (left) and modified (right) OAP-2DC during the time period shown in Fig. 5. Green backgrounds highlight the images identified by the inter-arrival time algorithm as artifacts. Images with a white background were accepted by the algorithm. As seen in **(a)**, in some cases the standard OAP-2DC rejects large particles which appear to be intact (red arrows), at the same time it accepts particles which have the features of shattered fragments (blue arrows).

The last example demonstrates a case where shattering has a negligible effect.

4.2.1 Overlapping modes

Figure 5a and c show the time series of inter-particle distances measured by the standard and modified OAP-2DC in an ice cloud. Each inter-particle distance in Fig. 5a and b is represented by a dot. The red lines indicate the cut-off-distances. As seen from these two diagrams, the density of points below the red line is greater for the standard probe (Fig. 5a) than that for the modified one (Fig. 5b). The concentration measured by the modified 2DC, and corrected with the help of the ITA, varied from 20 to 80 L^{-1}. Whereas, the uncorrected concentration measured by the standard 2DC varied from 300 to 1600 L^{-1}. After applying corrections to the standard 2DC measurements its concentration varied in the range 150 to 700 L^{-1}. The ice particle images measured by 2DC and 2DP are presented in Fig. 5e and f. Analysis of these images shows that the ensemble of ice particles was composed of two distinct habits: large spatial dendrites with sizes up to few millimeters and transparent plates with a characteristic size of a few hundred micrometers. The maximum particle size D_{max} calculated for each averaging time interval remained approximately constant and did not exceed 5 mm.

The distributions of the inter-particle distances $\phi(\Delta x)$ calculated from the standard and modified 2DC probe data are shown in Fig. 5c and d. The inter-particle distance distribution, $\phi(\Delta x)$, for the standard 2DC displays a significant overlap between the long and short inter-arrival modes (Fig. 5c). This may result in rejecting intact particles along with shattered artifacts when $\Delta x < \chi^*$, and, vice versa, accepting shattered fragments with intact particles for $\Delta x > \chi^*$.

The inter-particle distance distribution $\phi(\Delta x)$ for the modified probe has a relatively small overlap between the long and short distance modes, which suggests a better separation of shattered and intact particles. The number of particles associated with the short distance mode for the modified probe (Fig. 5d) is also reduced when compared to Fig. 5b. However, despite the larger separation between the short and long distance modes, the ITA still identifies large particles, which appear intact, as shattered fragments (e.g., Fig. 6b).

Figure 6 shows results from applying the ITA to the standard and modified OAP-2DC images sampled at the same time during one flight from the AIIE project. Images identified as shattered artifacts appear on a green background, whereas the accepted images have a white background. The number of rejected images for the standard probe (Fig. 6a) turned out to be nearly four times higher than that for the modified probe (Fig. 6b). Visual examination of the measurements made with the standard probe (Fig. 6a) reveals a number of images that we interpret as being intact (red arrows) but are rejected by the algorithm. Similar particle images can also be found in data for the modified probe (Fig. 6b), but there are far fewer images that look intact but are rejected. Some of the accepted images obtained by the standard probe we interpret as shattered artifacts (Fig. 6a, blue arrows). Shattered artifacts usually appear as elongated along the flight direction images due to the slower speed that they enter the sample volume. In many cases, the images of the shattering artifacts also have a hole in the center, that is the result of the greater likelihood of shatter products entering the sample volume closer to the arms and, consequently, far from the center of the depth-of-field. Further support for the assumption that the accepted images indicated by the blue arrows are shattered artifacts is provided by the absence of similar images recorded by the modified 2DC (Fig. 6b).

The apparently erroneous acceptance of the shattered artifacts and rejection of intact images in Fig. 6a is consistent with the large overlap between the short and long distance modes indicated in Fig. 5c. This example demonstrates that the ability of the ITA to segregate shattered artifacts and intact particles is reduced when the short and long distance modes become less separated.

One of the important goals of filtering out shattering artifacts is to obtain an accurate estimation of the PSD. Figure 7 shows the distributions of particle counts, concentration and mass calculated for all images before corrections, after corrections. These distributions were calculated for the image sizes measured along the photodiode array direction (i.e., perpendicular to the flight direction). Equivalent distributions are shown for the rejected images. During the image processing, the distribution of counts is used as a starting point for the following calculations of other distributions and bulk microphysical parameters. As seen from Fig. 7a and d for both standard and modified OAP-2DCs the number of small particles rejected by the inter-arrival algorithm is nearly two orders of magnitude greater than for the large particles. This is consistent with the concept that shattered particles are mainly composed of small fragments. Experimental studies indicate that the fragments of ice particles shattered at the aircraft speed have characteristic size from tens to hundreds of micrometers (Vidaurre and Hallett, 2009). Therefore, it is hypothesized that rejected particles larger than approximately 1 mm (Fig. 7) are likely related to the cases depicted in Fig. 2b, d, e and f.

It is interesting to note that for the modified probe the ITA corrected and uncorrected distributions agree to better than 10 % for particles larger 600 μm (in Fig. 7d–f). However, for the standard probe the separations between ITA corrected and uncorrected distributions remain approximately constant for $D > 600\,\mu$m (Fig. 7a–c) and it varies from 20 to 30 %. Despite the rejection of a large fraction of small particles (Fig. 7b) in the modified probe, the concentration of small particles ($D < 200\,\mu$m) is only reduced after correction by a factor of 2 to 3 (Fig. 7e). At the same time, the concentration of particles in the small size bins ($D < 200\,\mu$m) for the standard probe after ITA correction still remains higher than for the modified probe. This is consistent with the above conclusion that the ITA is unable to filter out all shattering artifacts. The corrected concentration of small particles ($D < 200\,\mu$m) measured by the modified probe is still high. It is difficult to conclude whether these particles are real or associated with shattering artifacts, for instance due to singletons, or other mis-sizing and concentration errors.

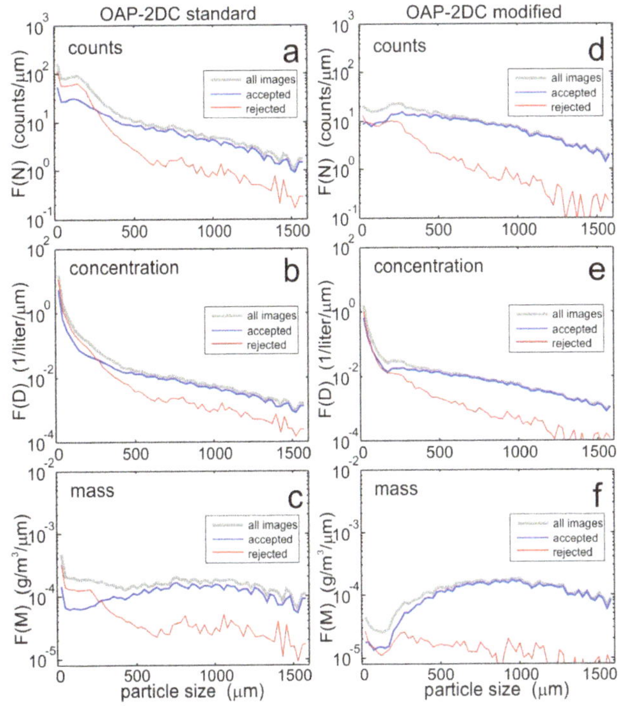

Figure 7. Distributions of particle counts **(a, d)**, concentration **(b, e)**, and mass **(c, f)** calculated for all images (grey), and accepted (blue) and rejected (red) identified by the inter-arrival time algorithm. The distributions were calculated for the standard (left) and modified (right) OAP-2DCs data collected in the cloud shown in Fig. 5.

Comparisons of corrected mass distributions for standard and modified probes in Fig. 7c and f show a large difference between them for $D < 500\,\mu$m. However, since large particles are the major contributors into the total mass, the discrepancy at the small size end of the PSD does not produce any significant effect on the bulk IWC. Integration of the mass distributions for standard and modified probes shows that IWC corrected standard is systematically higher than that for the modified OAP-2DCs. For this particular case, IWC corrected standard is approximately 20 % higher, and the mean IWC values averaged over the entire time interval is approximately 4 % higher than the modified OAP-2DC.

4.2.2 Large particles

Figure 8 shows the changes of inter-particle distances and distributions $\phi(\Delta x)$ measured during a descent through precipitating aggregates of dendrites with $D_{\max}$ ranging from 0.5 to 1.5 cm. The total number of particle counts per second in the modified probe varied from 10 to 70 and the rate of counts for the standard probe varied in the range 100 to 300. The ITA corrected concentration measured by the modified OAP-2DC varied from 0.5 to 5 L^{-1}, whereas the uncorrected concentration measured by the standard probe varied from 100 to 300 L^{-1}. After applying corrections to the standard

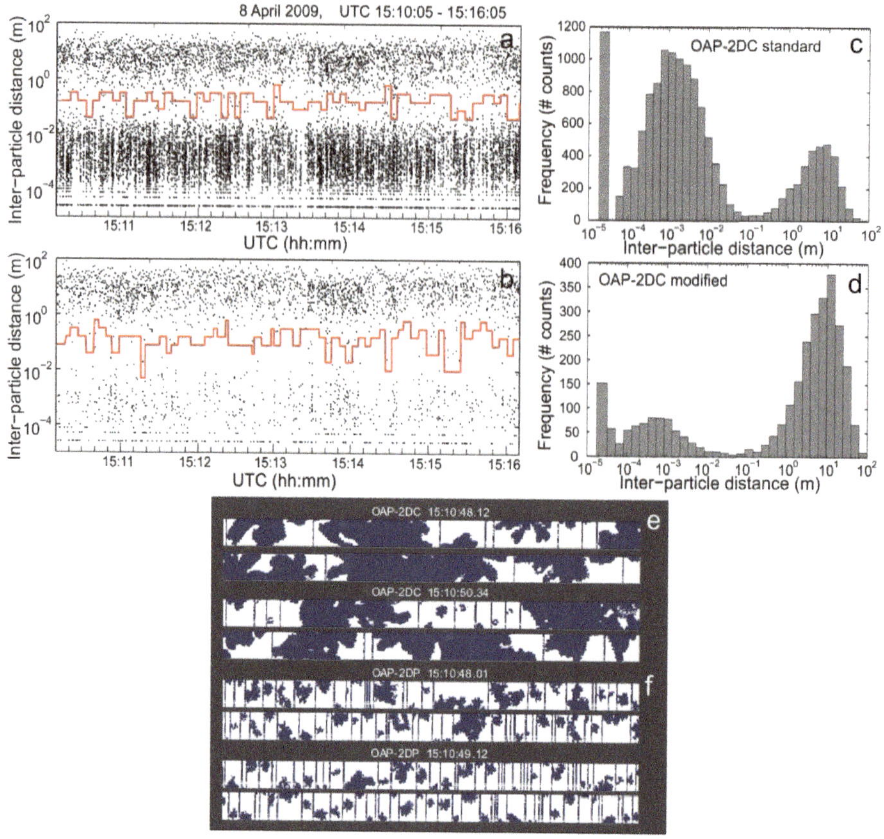

Figure 8. Comparison of inter-particle distances measured by standard **(a)** and modified **(b)** 2DC. Red lines in **(a)** and **(b)** indicate the cut-off distance χ^*. The distribution of the inter-particle distances for standard **(c)** and modified **(d)** probes. Examples of images obtained with an OAP-2DC at 25 μm pixel resolution **(e)** and an OAP-2DP at 200 μm pixel resolution **(f)**. The measurements were conducted during on 8 April (1st flight) during descent through precipitating dendrites from approximately 1300 to 500 m in the Ottawa region. The temperature varied from −8 to −2 °C.

2DC data its concentration varied in the range 1 to 10 L^{-1}. The appearance of the particle images measured by OAP-2DC and OAP-2DP are shown in Fig. 8e and f.

The distinctive feature of the inter-particle spatial distribution $\phi(\Delta x)$ calculated for the standard OAP-2DC is that it has an exceptionally high number of counts (88 %) associated with the short distance mode. This indicates that the standard probe observed mostly shattering artifacts. In contrast, for the modified probe the distribution $\phi(\Delta x)$ the number of counts in the short distance mode is smaller than for the long distance mode.

The results of segregating the intact particles and shattered artifacts for standard and modified OAP-2DC performed by the ITA are shown in Fig. 9. The measurements made with the standard probe are dominated by artifacts with very few accepted images (Fig. 9a). The particle imagery obtained with the modified probe is largely devoid of the small images typically associated with shattered fragments (Fig. 9b). An absence of small particles in subsaturated precipitating regions is consistent with the commonly accepted concept of ice formation. It is worth noting that a few small accepted images (indicated by the blue arrows) still appear in the standard 2DC imagery in Fig. 9a. It is possible that these small images are related to single fragments as in Fig. 2c and were misidentified by the ITA as intact particles.

Despite the fact that most of the images in Fig. 9b appear to be intact dendrites, many of them were rejected. Visual inspection of the rejected images in Fig. 9b indicates that nearly all of them are partially viewed images. It is likely that the partially viewed branches of dendrites (e.g., Fig. 2e) has led to the ITA confusing the partially observed closely spaced arms of the dendrite with the shattered artifacts (Sect. 3.4).

The results shown in Fig. 9 demonstrate that even for the cases where there is a good separation of the short and long distance modes (Fig. 8c, d) the ITA still misidentifies shattering artifacts and intact particles.

The distributions of particle counts, concentration and mass obtained with the standard and modified probes are shown in Fig. 10. Figure 10a and d demonstrate that the number of counts of small particles is much greater for the standard probe than for the modified probe. This is consistent with the 2-D images shown in Fig. 9. The ITA identified

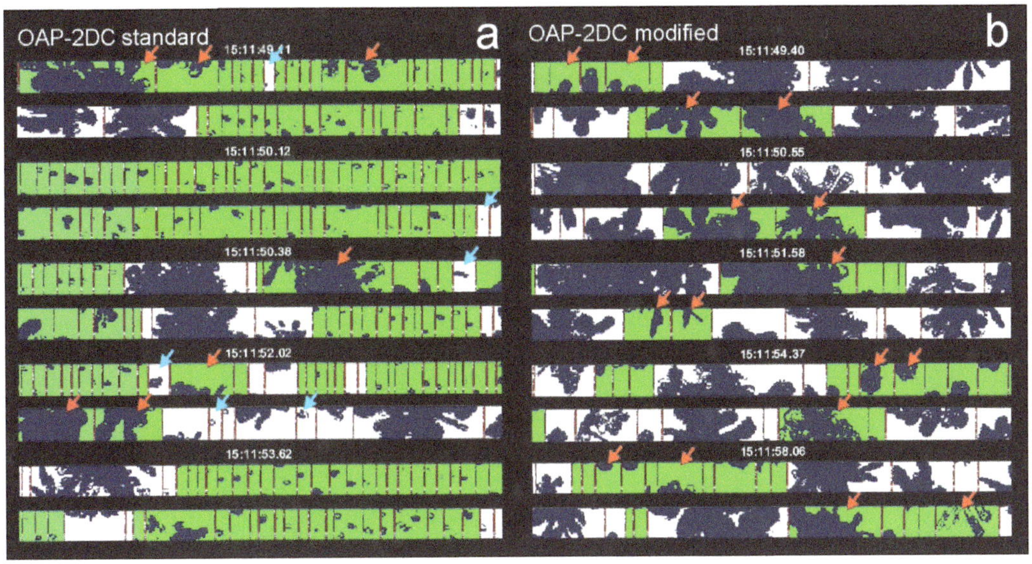

Figure 9. Examples of the results of the image rejection/acceptance processing with the inter-arrival time algorithm. The images of ice particles were simultaneously sampled by the standard (left) and modified (right) OAP-2DC during the time period shown in Fig. 8. A green background highlights the images identified by the inter-arrival time algorithm as artifacts. Images with a white background were accepted by the algorithm. As seen in **(a)**, in some cases the standard OAP-2DC rejects large particles which appear to be intact (red arrows), at the same time it accepts particles which have the features of shattered fragments (blue arrows).

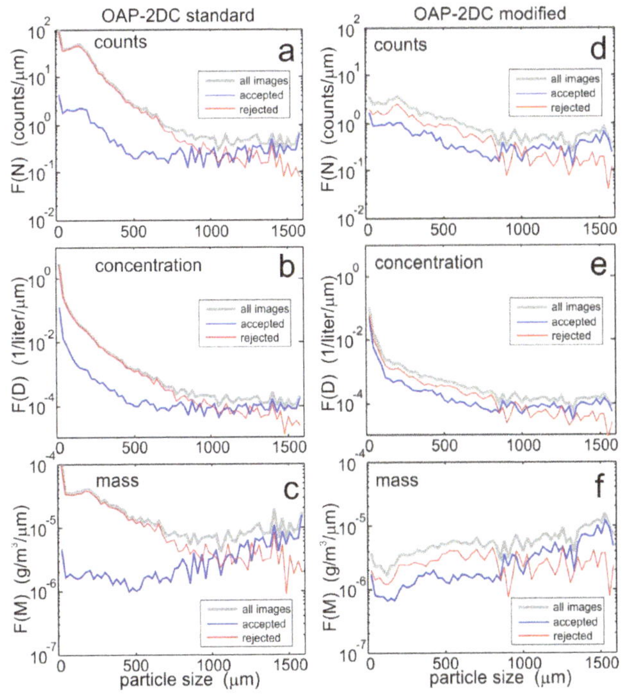

Figure 10. Distributions of particle counts **(a, d)**, concentration **(b, e)**, and mass **(c, f)** calculated for all images (grey), and accepted (blue) and rejected (red) identified by the inter-arrival time algorithm. The distributions were calculated for the standard (left) and modified (right) OAP-2DCs data collected in the cloud shown in Fig. 8.

most of the small image counts as artifacts (Fig. 10a). However, the number of small images accepted for the standard probe is still greater than the modified probe (Fig. 10d). The same conclusion applies to the concentration and mass distributions shown in Fig. 10b, c and Fig. 10e, f.

We note that the shape of rejected distributions for the modified probe is very similar to the shape of the uncorrected distributions (Fig. 10d, e, f). Furthermore, the distributions of the accepted and uncorrected images approach each other only for particle sizes $D > 1500\,\mu$m (Fig. 10a, b, c), in contrast to the case shown in Fig. 7. For both probes, the number of rejected large particles remains relatively large, and it is related to misidentifying large particles as shattered artifacts.

4.2.3 Small ice particles

The next example was obtained during a sampling of cirrus clouds at a temperature of $-35\,^{\circ}$C and altitude of 7500 m. The maximum size of particles varied from 200 to $400\,\mu$m, and did not exceed $500\,\mu$m. The particle concentration measured by the standard and modified OAP-2DCs agreed well and varied from 20 to $180\,\mathrm{L}^{-1}$.

Figure 11 shows the inter-particle spacings (Fig. 11a, b) and their distributions $\phi(\Delta x)$ for the standard and modified probes. As can be seen in Fig. 11c and d, the inter-particle spacing distribution $\phi(\Delta x)$ is monomodal for both probes. Since the ITA used here is most efficient only for the cases with bimodal distributions, for this particular case with monomodal $\phi(\Delta x)$, the fraction of rejected particles for both 2DCs is quite small and does not exceed 0.7 %. This is

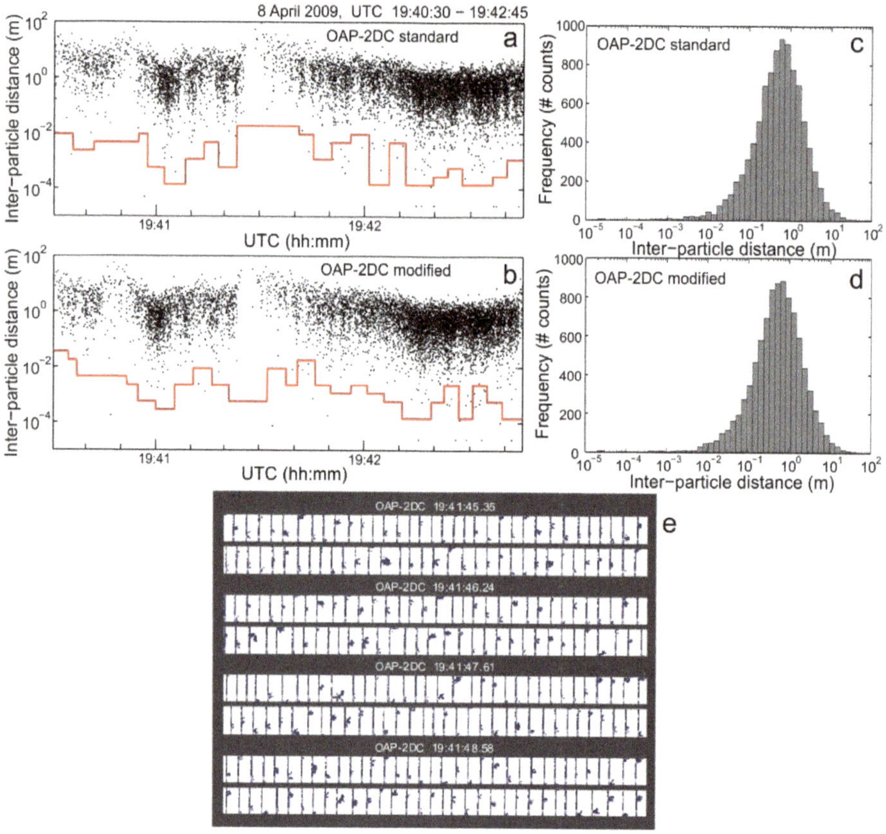

Figure 11. Same as in Fig. 8. The measurements were collected on 8 April (2nd flight) in cirrus clouds at altitude 7500 m and temperature −35 °C.

demonstrated by the small number of points that lie below the cut-off-distance line (red) in Fig. 10a and b (for which $\Delta x < \chi^*$), and that Fig. 12 shows very few images identified by the ITA as artifacts.

The distributions of particle counts, concentration and mass are shown in Fig. 13. All three distributions obtained with the standard and modified probes agree well with each other. The fraction of rejected artifacts is small and for practical purposes their effect on the size and mass distributions has negligible effect.

These results show that for this specific case there is not much difference between the measurement obtained by the standard and modified probes. This suggests that for the cases where the ice PSD is narrow ($D_{max} < 400\,\mu$m) the effect of shattering on the standard OAP-2DC measurements is quite small and can be neglected. This finding also implies that there exists a threshold size below which shattering of ice crystals does not produce any significant effect on measurements. This will likely vary according to instrumental and microphysical characteristics. Of course, even though the probes agree for this case, problems relating to mis-sizing and concentration errors for particles smaller than ~ 100 microns still exist (Korolev et al., 1998; Strapp et al., 2001).

4.3 Statistical characteristics of inter-particle distances in shattering events

Figures 5a, b, 8a, b and 11a, b contain a plethora of information about the statistical characteristics of shattering events and their effect on ice particle measurements. This data can also be used in the development of future algorithms for the data processing and numerical simulations of the shattering effects. In this section some of the statistical characteristics of shattering processes will be investigated and compared with microphysical metrics.

Figure 14 shows distributions of the number of particles N_s within each shattering event (a1–a3), distributions of the length of spatial clusters along the flight direction L_s (b1–b3), distributions of distances between shattering events L_i (c1–c3), and the number of particles between shattering events N_i (d1–d3). These distributions were calculated for the cloud regions shown in Figs. 5, 8 and 11.

The distribution $f(N_s)$ (Fig. 14a1–a3) for both standard and modified probes is well represented by an exponential function (Field et al., 2003). This suggests that the statistics of shattered fragments passing through the sample volume can be approximated by the Poisson distribution. Analysis of the other cases indicates that the slope of the distribution

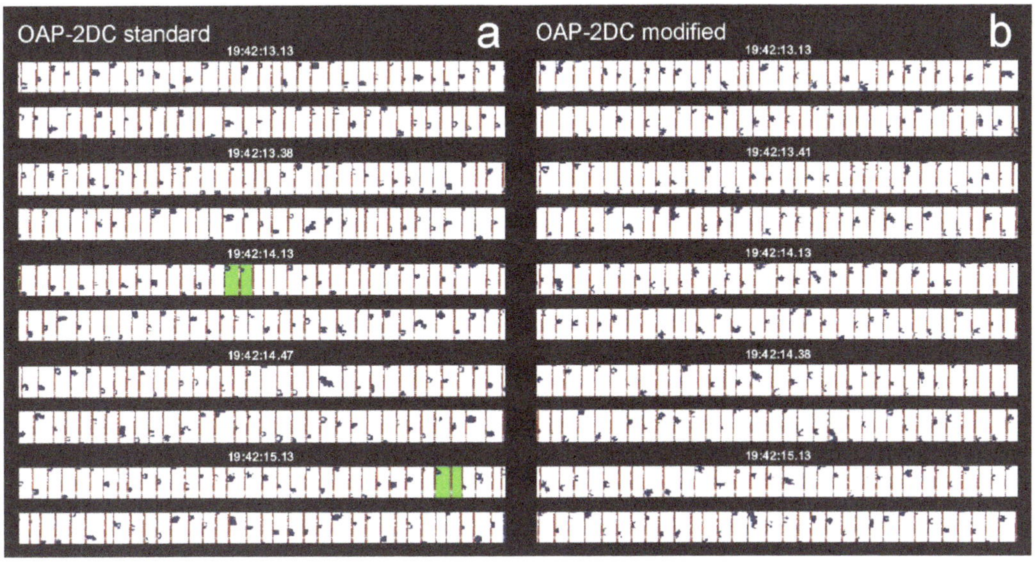

Figure 12. Same as in Fig. 6. The measurements of the images were obtained during the time period shown in Fig. 11.

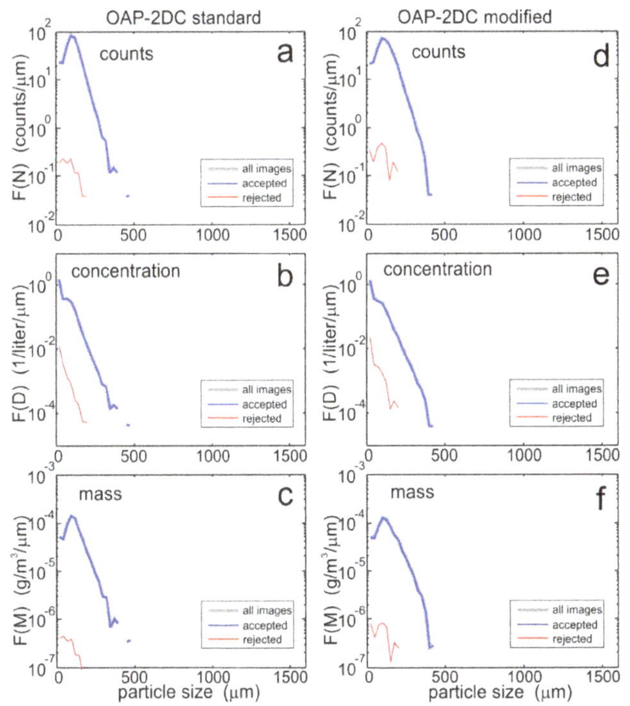

Figure 13. Distributions of particle counts **(a, d)**, concentration **(b, e)**, and mass **(c, f)** calculated for all images (grey), and accepted (blue) and rejected (red) identified by the inter-arrival time algorithm. The distributions were calculated for the standard (left) and modified (right) OAP-2DCs data collected in the cloud shown in Fig. 11.

depends on the presence of large particles in the size distribution and is correlated with D_{max} such that increasing D_{max} leads to a shallower slope for $f(N_s)$: Figure 14a2 displays the shallowest slope in $f(N_s)$ for the case with the greater D_{max}.

The distribution of $f(N_s)$ monotonically increases with decreasing N_s, having a maximum for events with two shattered fragments. Extrapolating $f(N_s)$ towards 1, i.e., shattering events with one particle, suggests that the number of singletons may be quite high. Existence of single particle shattering events presents a principal limitation of the inter-arrival algorithm, since such particles cannot be unambiguously identified as artifacts merely based on the analysis of Δx.

The maximum number of the shattered fragments in a shattering event measured by the standard OAP-2DC during the AIIE project reached 60. Whereas for the modified probe, the maximum number of fragments for this data set was found to be 14. Figure 14a1 and a2 also demonstrate that the modified OAP-2DC on average has a smaller number of fragments per shattering event, and therefore the antishattering tips can efficiently mitigate shattering.

The density function of the length of spatial clusters of shattered fragment $f(L_s)$ can also be well approximated by an exponential function (Fig. 14b1, b2). Clusters with a short length have the highest probability and they are associated with two-particle shattering events. The spatial length of the shattering clusters is presented by a cascade of scales ranging from zero to tens of centimeters. The maximum length of the shattered clusters measured by the standard and modified OAP-2DCs during the AIIE project reached 30 and 3 cm, respectively.

The density functions of distances between shattered events $f(L_i)$ are shown in Fig. 14c1–c3. The characteristic scale of L_i is determined by the concentration of particles capable of generating shattering events. For example, the case

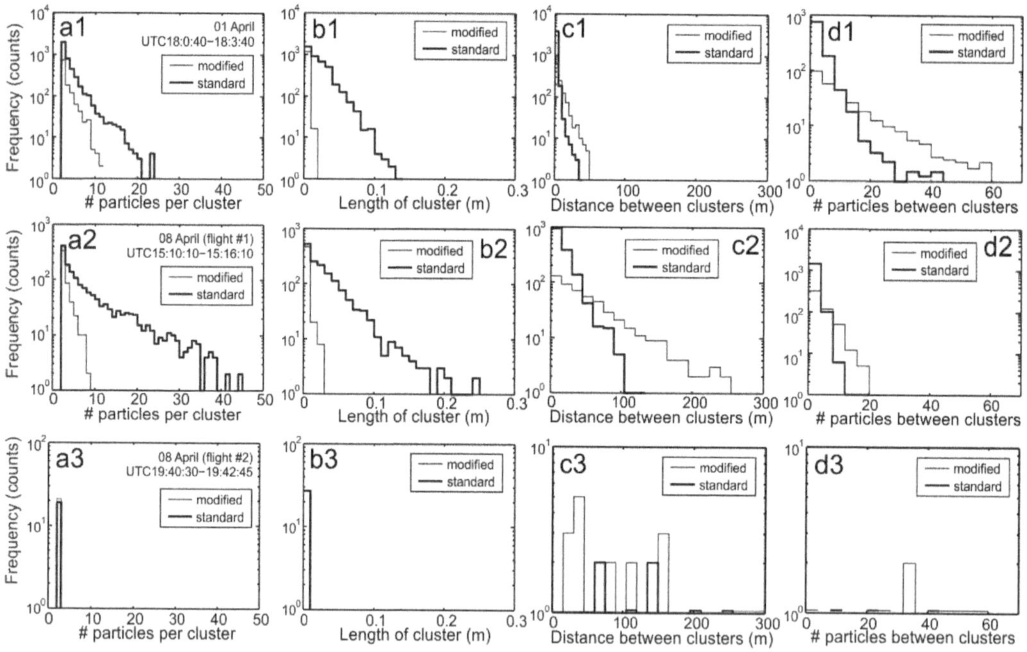

Figure 14. Distributions of number of particles per shattering event **(a1, a2, a3)**, length of the clusters of shattered artifacts along the flight direction **(b1, b2, b3)**, distance between the clusters with shattered fragments **(c1, c2, c3)**, number of intact particles between shattered events **(d1, d2, d3)** calculated for the cases shown in Fig. 5 (top row), Fig. 8 (middle row) and Fig. 11 (bottom row), respectively.

in Fig. 14c3 has the highest total concentration of ice particles. However, only 0.7 % (~ 30 counts) of them generate shattering events. As a result the $f(L_i)$ in Fig. 14c3 has a statistically insignificant distribution. As in the previous cases $f(L_i)$ can be well approximated by an exponential function and the characteristic value of L_i for modified OAP-2DC is larger than that for the standard probe (Fig. 14c1, c2).

The behavior of the function $f(N_i)$ in Fig. 14d1–d3 is very similar to that of $f(L_i)$ (Fig. 14c1, c2, c3). For the case depicted in Fig. 14d3 the low number of shattering events mean that most of the counts are outside of the scale.

Figure 15 shows the distributions of cut-off distances, $f(\chi^*)$, for the standard and modified OAP-2DCs averaged over all ice clouds sampled during the AIIE project. Because the value of χ^* depends of the size distribution of ice particles and their habits, the shape of the distribution, $f(\chi^*)$, will be determined by a combination of cloud characteristics, aircraft and instrument properties. However, the distributions of $f(\chi^*)$ in Fig. 15 allow a few conclusions to be made. Firstly, for the OAP-2DC χ^* can vary from tens of micrometers to approximately one meter. Secondly, the standard probe χ^*_{st} has a mode at approximately 10 cm, whereas the modified probe χ^*_{mdf} has a mode at approximately 2 cm.

4.4 Effect of particle sizes

The effect of particle sizes on shattering is demonstrated in Fig. 16 which shows the maximum number of fragments per shattering event N_{smax} vs. D_{max} for standard and modified probes. The N_{smax} and D_{max} were calculated for each 5 s averaging interval for the entire AIIE project. Figure 16a shows that for $D_{max} < 15$ mm, N_{smax} correlates well with D_{max} and therefore the dependence $N_{smax}(D_{max})$ can be parameterized with a linear function (Fig. 16a). However, for the modified probe the correlation coefficient between N_{smax} and D_{max} is low (0.57) and the $N_{smax}(D_{max})$ saturates at $N_{smax} \sim 15$ when $D_{max} > 5$ mm (Fig. 16b).

Figure 17 shows the maximum length of shattering clusters L_{smax} vs. D_{max} for standard and modified probes. Similar to the case in Fig. 16, L_{smax} and D_{max} were calculated for each 5 s averaging interval. A relatively high correlation coefficient between L_{smax} and D_{max} (0.78) for the standard probe allows linear parameterization of $L_{smax}(D_{max})$ (Fig. 17a). The correlation coefficient between L_{smax} and D_{max} for the modified probe is quite low (0.29), but L_{smax} is never longer than 3 cm for all D_{max} encountered.

The nearly linear increase of N_{smax} and L_{smax} with the increase of D_{max} for the standard probe (Figs. 16a, 17a) indicates a strong dependence of particle size on particle shattering.

5 Monte Carlo simulation of inter-particle distance function

The statistical parameters obtained from the shattering analysis were used to constrain Monte Carlo simulations of the effect of shattering on particle measurements. These simula-

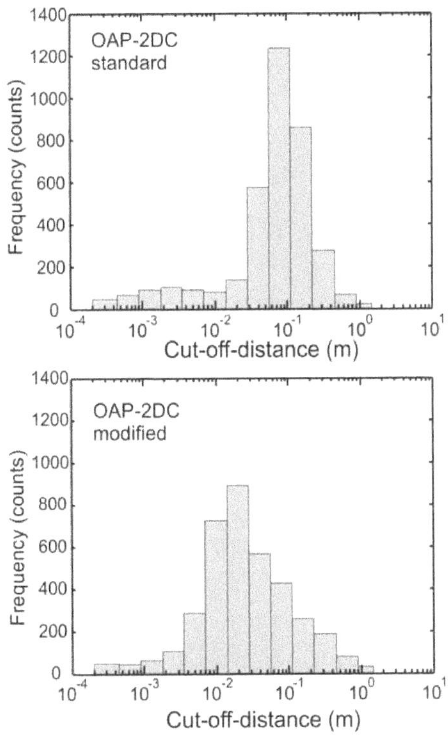

Figure 15. Distribution of cut-off distances χ^* for standard **(a)** and modified **(b)** OAP-2DCs averaged over all ice clouds sampled during the AIIE project.

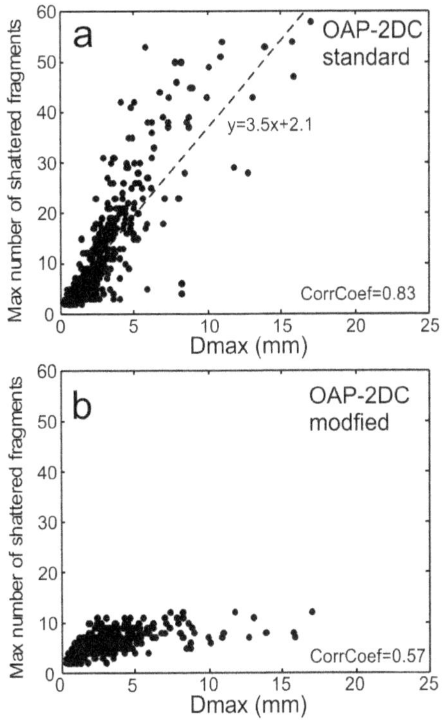

Figure 16. Scatter diagram of the maximum number of shattered fragments per event vs. maximum particle sizes for standard **(a)** and modified **(b)** OAP-2DCs for all ice clouds sampled during the AIIE project.

tions were used to understand the influence of various parameters on the shape of the inter-particle distance distribution, $\phi(\Delta x)$.

For each simulation, two fluxes of particles with the same concentration n_0 are assumed to interact with the probe. The first flux represents intact particles and passes through the sample area S_0 of the probe. The second flux represents the shattered particles and passes through a "shattering" area for the probe, S_{sh}, rebounds and subsequently passes through S_0. After passing though S_{sh}, a particle breaks down into N_{sh} fragments.

The inter-particle distance between the intact particles was simulated assuming a Poisson distribution by combining a random number generator with an exponential distribution with mean distance $\bar{\chi}_0 = \frac{n_0}{S_0}$. The inter-particle distance between the shattering events was also simulated using an exponential distribution with mean distance $\bar{\chi}_{ev} = \frac{n_0}{S_{sh}}$. Based on the results obtained in Sect. 4.3, the number of shattered fragments was simulated by a random number generator with exponential distribution with average $\bar{N}_{sh}$. The distance between the shattered fragments in each shattered event was also simulated by the exponential distribution with average $\bar{\chi}_{sh} = \frac{\bar{L}_s}{\bar{N}_{sh}}$, where $\bar{L}_s$ is the average distance of the cluster of shattered fragments. After sorting arrival times, the two flows were merged together to form a time series of intact particles mixed with the shattering artifacts.

The modeling results presented below were performed for $S_0 = 50\,\text{mm}^2$, $S_{sh} = 5\,\text{mm}^2$, $\bar{L}_s = 1$ cm and $\bar{N}_{sh} = 5$. The particle size distribution was assumed to be monodisperse.

Figure 18 shows the modeled distributions of the inter-particle distances for intact particles only, $\phi_i(\Delta x)$ (blue), shattered particles only, $\phi_s(\Delta x)$ (red), and all particles that passed through the sample volume, $\phi(\Delta x)$ (black). The distribution $\phi_i(\Delta x)$ represents the Poisson process and has a single mode, whereas $\phi_s(\Delta x)$ has two modes. The long distance mode in $\phi_s(\Delta x)$ is determined by the characteristic distance between the shattered events $\bar{\chi}_{ev}$, whereas the short distance mode is associated with the characteristic distances between the shattered fragments $\bar{\chi}_{sh}$, which passed through the sample volume. The long distance mode in $\phi_s(\Delta x)$ also includes single particle shattering events. This is what we see for the long mode of the red line (rebounders/singletons) and the blue line. Both of these are continuous processes, whereas the shorter shattering mode is intermittent conditional on a collision occurring in the shattering volume. It is important to mention that the inter-particle distribution $\phi(\Delta x)$ representing the flow of all particles cannot be presented as an algebraic superposition of $\phi_s(\Delta x)$ and $\phi_i(\Delta x)$. In this regard, the conceptual diagrams in Fig. 1 show over-simplified inter-arrival time distributions, which help understanding, but do not reflect the actual shape of $\phi_i(\Delta x)$.

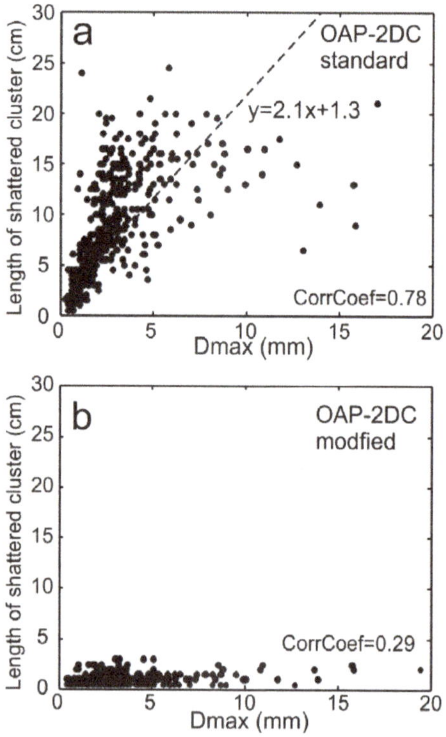

Figure 17. Scatter diagram of the length of clusters of shattered fragments vs. maximum particle sizes for standard **(a)** and modified **(b)** OAP-2DCs for all ice clouds sampled during the AIIE project.

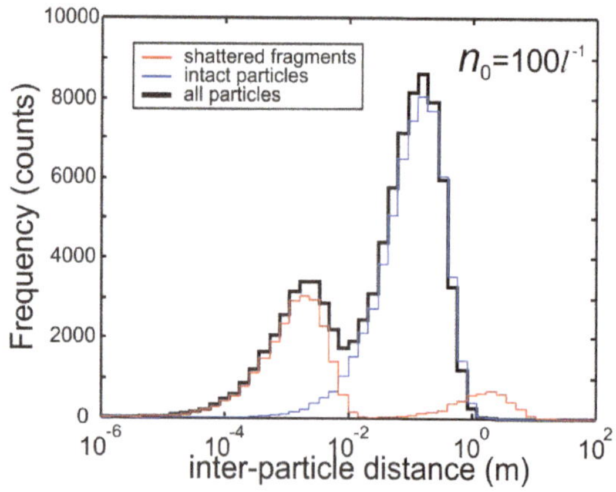

Figure 18. Results of Monte Carlo simulations of the distribution of inter-particle distance $\phi(\Delta x)$ for intact particles (blue), shattered particles (red), and all particles passed through the sample volume (black). The calculations were conducted for particle concentration $n_0 = 100\,\mathrm{L}^{-1}$.

Figure 19 shows four distributions $\phi(\Delta x)$ calculated for the particle concentrations 1, 10, 100, and $1000\,\mathrm{L}^{-1}$. Figure 19 shows that the long distance mode approaches the short distance mode, when the particle concentration n_0 increases. However, the short distance mode is insensitive to the changes of the particle concentration, and it remains at the same position, when n_0 increases (Fig. 19a, b, c). The location of the cut-off-distance χ^* appears insensitive to the changes in n_0, remaining approximately constant.

For increasing n_0 the reduced separation of the long and short distance mode results in an increased overlap of the distributions associated with these modes. As described above, increasing the overlap of these distributions reduces the efficiency of the ITA to segregate intact particles and shattered artifacts. At high concentrations the short and long distance modes merge, resulting in the vanishing of the inter-modal minimum. At that stage the ITA becomes much less efficient or even disabled (e.g., Fig. 19d).

In the above simulations the particle size distribution was assumed to be monodisperse. One of the consequences of this assumption is that all of the particles possessed the same shattering efficiency and $\bar{N}_{sh}$ remains the same for all particles. In reality, particle sizes in natural clouds are represented by broad distributions. As indicated above, $\bar{N}_{sh}$ depends on particle sizes and that for small particles with $D < 400\,\mu$m $\bar{N}_{sh} \to 0$. Therefore, the concentration n_0 should be interpreted as a concentration of the particle's contribution in the effect of shattering, but not as a total concentration, which included small ice particles which do not affect shattering. For simplicity, the effect of the particle size distribution was not included in the simulation. It should be noted that the threshold $400\,\mu$m refers to OAP-2DC. Probes with different pixel resolution, response time, and inlet configuration have different threshold sizes below which the effect of shattering becomes insignificant.

The analysis of the modeling results suggests that: (1) the number of counts in the long inter-particle distance mode is mainly controlled by the sample area S_0 and particle concentration n_0; (2) the number of counts in the short distance mode depends on the area S_{sh}, which deflects particles towards the sample volume, the characteristic number of fragments per shattering event $\bar{N}_{sh}$ and the particle concentration n_0. As demonstrated in Sect. 4, $\bar{N}_{sh}$ is a function of particle size. It should also be noted that $\bar{N}_{sh}$ depends on S_{sh}, S_0 and the spatial proximity of S_{sh} to S_0. For example, as S_0 decreases the probability of passing at least two shattered fragments ($N_{sh} \geq 2$) through S_0 diminishes too. If S_0 becomes too small, most of the shattered artifacts will be associated with single particle shattered events ($N_{sh} = 1$), that cannot be identified by the ITA. Therefore, it is anticipated that the inter-arrival time algorithm will be more efficient for OAP-2DS with $S_0 \approx 50\,\mathrm{mm}^2$, than for the FSSP with the sample area $S_0 \approx 3\,\mathrm{mm}^2$. The number of shattered fragments viewed by the probe N_{sh} also depends on spatial separation of S_{sh} to S_0. Thus, if S_{sh} is located too far from S_0 then the shattered fragments may not have sufficient momentum to travel large distances across the airflow and reach S_0. Comparisons of the measurements of standard and modified probes demonstrated that $\bar{N}_{sh}$ strongly depends upon S_{sh}. The reduced number of the shattered fragments N_{sh} for the modified probe is ex-

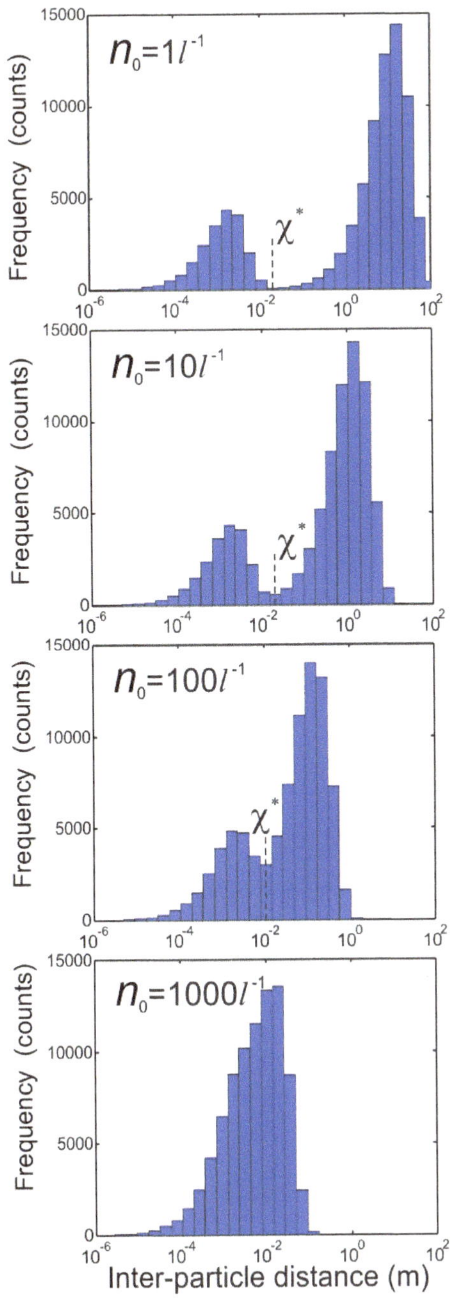

Figure 19. Results of Monte Carlo simulations of the distribution of inter-arrival time $\phi(\Delta x)$ for different particle concentrations **(a)** $1\,L^{-1}$; **(b)** $10\,L^{-1}$; **(c)** $100\,L^{-1}$; **(d)** $1000\,L^{-1}$.

plained by the fact that the antishattering tips have a significantly reduced shattering area compared to the standard tips.

6 Conclusions

Based on the analysis of data obtained with standard and modified OAP 2DC probes in a variety of ice cloud conditions and Monte Carlo simulations, the following conclusions have been obtained:

1. The inter-arrival time algorithm cannot segregate all shattering artifacts from the intact particles in principle. These limitations are imposed by the Poisson statistics of particle spatial distribution. It was demonstrated that the short inter-arrival times are not necessarily associated with shattering artifacts, and that shattered artifacts are not constrained to exhibit short inter-arrival times. In order to mitigate the effect of shattering, the inter-arrival time algorithm should be used together with other means, such as antishattering tips and special algorithms to improve its performance (e.g., reacceptance of intact particles, corrections for accepted singletons, etc.).

2. The inter-arrival time algorithm has a range of conditions when short and long inter-arrival modes are well separated and it can effectively segregate shattered artifacts and intact particles (e.g., low concentration).

3. The inter-arrival time algorithm has a number of limitations which under certain circumstances may significantly degrade its performance or even disable it. Such cases are relevant to the clouds with high particle concentration, when the spatial separation between shattered fragments Δx_s becomes comparable with the distance between intact particles Δx_i, i.e., $\Delta x_s \sim \Delta x_i$. For mixed-phase clouds it may not be possible to use the inter-arrival time algorithm with probes that have a fine pixel resolution.

4. It was found that in clouds with particles with $D_{max} < 400\,\mu m$ the effect of shattering on measurements of the standard OAP-2DC can be neglected.

5. The inter-arrival distance and number of registered shattered artifacts is well represented by an exponential function. The slope of the distribution is a function of the characteristic particle size. The number of shattered fragments correlates with particle size.

6. The results on the statistics of shattering events open the door for a statistical simulation to study the effect of shattering on measurements. These studies could potentially be useful not only for developing correction algorithms for historical data sets but also for developing recommendations on how to design probes in the future to best accommodate the ITA.

This analysis of the efficiency of the inter-arrival time algorithm is based on OAP-2DC data. Nevertheless, most of the conclusions obtained in this study can be applied to other particle probes. In this regard it is worth mentioning that the effectiveness of the inter-arrival time algorithm depends on the pixel resolution, size of the sample area, the response time of the electronics and the inlet configuration.

Acknowledgements. The work of A. Korolev was supported by Environment Canada, Transport Canada and the US Federal Aviation Administration.

Edited by: A. Lambert

References

Baker, B., Mo, Q., Lawson, R. P., O'Connor, D., and Korolev, A.: The Effects of Precipitation on Cloud Droplet Measurement Devices, J. Atmos. Ocean. Tech., 7 1404–1409, 2009.

Baumgardner, L., Avallone, A., Bansemer, S., Borrmann, P., Brown, U., Bundke, P., Chuang, Y., Cziczo, D., Field, P., Gallagher, M., Gayet, J.-F., Heymsfield, A., Korolev, A., Krämer, M., McFarquhar, G., Mertes, S., Möhler, O., Lance, S., Lawson, P., Petters, M. D., Pratt, K., Roberts, G., Rogers, D., Stetzer, O., Stith, J., Strapp, W., Twohy, C., and Wendisch, M.: In Situ, Airborne Instrumentation: Addressing and Solving Measurement Problems in Ice Clouds, B. Am. Meteorol. Soc., 93, ES29–ES34, doi:10.1175/BAMS-D-11-00123.1, 2012.

Cooper, W. A.: Cloud physics investigation by the University of Wyoming in HIPLEX 1977, Bureau of Reclamation Rep, AS, 119, 321 pp., 1977.

Field, P. R., Wood, R., Brown, P. R. A., Kaye, P. H., Hirst, E., Greenaway, R., and Smith, J. A.: Ice particle inter-arrival times measured with a Fast FSSP, J. Atmos. Ocean. Tech., 20, 249–261, 2003.

Field, P. R., Heymsfield, A. J., and Bansemer, A.: Shattering and Particle Inter-arrival Times Measured by Optical Array Probes in Ice Clouds, J. Atmos. Ocean. Tech., 23, 1357–1370, 2006.

Jackson, R. C., McFarquhar, G. M., Stith, J., Beals, M., Shaw, R. A., Jensen, J., Fugal, J., and Korolev, A.: An assessment of the impact of antishattering tips and artifact removal techniques on cloud ice size distributions measured by the 2-D Cloud Probe, J. Atmos. Ocean., Tech., 31, 2567–2590, 2014.

Korolev, A. V.: Reconstruction of the Sizes of Spherical Particles from Their Shadow Images. Part I: Theoretical Considerations, J. Atmos. Ocean. Tech., 24, 376–389, 2007.

Korolev A. V., Strapp, J. W., and Isaac, G. A.: Evaluation of accuracy of PMS Optical Array Probes, J. Atmos. Ocean. Tech., 15, 708–720, 1998.

Korolev, A., Emery, E. F., Strapp, J. W., Cober, S. G., Isaac, G. A., Wasey, M., and Marcotte, D.: Small ice particles in tropospheric clouds: fact or artifact? Airborne Icing Instrumentation Evaluation Experiment, B. Am. Meteorol. Soc., 92, 967–973, 2011.

Korolev, A., Emery, E., and Creelman, K.: Modification and tests of particle probe tips to mitigate effects of ice shattering, J. Atmos. Ocean. Tech., 30, 690–708, 2013a.

Korolev, A. V., Emery, E. F., Strapp, J. W., Cober, S. G., and Isaac, G. A.: Quantification of the effects of shattering on airborne ice particle measurements, J. Atmos. Ocean. Tech., 30, 2527–2553, 2013b.

Lawson, R. P.: Effects of ice particles shattering on the 2D-S probe, Atmos. Meas. Tech., 4, 1361–1381, doi:10.5194/amt-4-1361-2011, 2011.

Strapp, J. W., Albers, F., Reuter, A., Korolev, A. V., Maixner, U., Rashke, E., and Vukovic, Z.: Laboratory Measurements of the Response of a PMS OAP-2DC, J. Atmos. Ocean. Tech., 18, 1150–1170, 2001.

Vidaurre, G. and Hallett, J.: Particle Impact and Breakup in Aircraft Measurement, J. Atmos. Ocean. Tech., 26, 972–983, 2009.

Permissions

All chapters in this book were first published in AMT, by Copernicus Publications; hereby published with permission under the Creative Commons Attribution License or equivalent. Every chapter published in this book has been scrutinized by our experts. Their significance has been extensively debated. The topics covered herein carry significant findings which will fuel the growth of the discipline. They may even be implemented as practical applications or may be referred to as a beginning point for another development.

The contributors of this book come from diverse backgrounds, making this book a truly international effort. This book will bring forth new frontiers with its revolutionizing research information and detailed analysis of the nascent developments around the world.

We would like to thank all the contributing authors for lending their expertise to make the book truly unique. They have played a crucial role in the development of this book. Without their invaluable contributions this book wouldn't have been possible. They have made vital efforts to compile up to date information on the varied aspects of this subject to make this book a valuable addition to the collection of many professionals and students.

This book was conceptualized with the vision of imparting up-to-date information and advanced data in this field. To ensure the same, a matchless editorial board was set up. Every individual on the board went through rigorous rounds of assessment to prove their worth. After which they invested a large part of their time researching and compiling the most relevant data for our readers.

The editorial board has been involved in producing this book since its inception. They have spent rigorous hours researching and exploring the diverse topics which have resulted in the successful publishing of this book. They have passed on their knowledge of decades through this book. To expedite this challenging task, the publisher supported the team at every step. A small team of assistant editors was also appointed to further simplify the editing procedure and attain best results for the readers.

Apart from the editorial board, the designing team has also invested a significant amount of their time in understanding the subject and creating the most relevant covers. They scrutinized every image to scout for the most suitable representation of the subject and create an appropriate cover for the book.

The publishing team has been an ardent support to the editorial, designing and production team. Their endless efforts to recruit the best for this project, has resulted in the accomplishment of this book. They are a veteran in the field of academics and their pool of knowledge is as vast as their experience in printing. Their expertise and guidance has proved useful at every step. Their uncompromising quality standards have made this book an exceptional effort. Their encouragement from time to time has been an inspiration for everyone.

The publisher and the editorial board hope that this book will prove to be a valuable piece of knowledge for researchers, students, practitioners and scholars across the globe.

List of Contributors

L. B. Hande
School of Earth, Atmosphere and Environment, ARC Centre of Excellence for Climate System Science, Monash University, Melbourne, Australia

S. T. Siems
School of Earth, Atmosphere and Environment, ARC Centre of Excellence for Climate System Science, Monash University, Melbourne, Australia

M. J. Manton
School of Earth, Atmosphere and Environment, ARC Centre of Excellence for Climate System Science, Monash University, Melbourne, Australia

D. H. Lenschow
National Center for Atmospheric Research, Boulder, Colorado, USA

C. Y. Lin
Institute of Space Science, National Central University, Chungli, Taiwan

T. Matsuo
Cooperative Institute for Research in Environmental Sciences, University of Colorado, Boulder, Colorado, USA
Space Weather Prediction Center, National Oceanic and Atmospheric Administration, Boulder, Colorado, USA

J. Y. Liu
Institute of Space Science, National Central University, Chungli, Taiwan
National Space Organization, HsinChu, Taiwan

C. H. Lin
Department of Earth Sciences, National Cheng Kung University, Tainan, Taiwan

H. F. Tsai
Department of Earth Sciences, National Cheng Kung University, Tainan, Taiwan

E. A. Araujo-Pradere
Cooperative Institute for Research in Environmental Sciences, University of Colorado, Boulder, Colorado, USA

D. P. Donovan
Royal Netherlands Meteorological Institute (KNMI), P.O. Box 201, 3730 AE, De Bilt, the Netherlands

H. Klein Baltink
Royal Netherlands Meteorological Institute (KNMI), P.O. Box 201, 3730 AE, De Bilt, the Netherlands

J. S. Henzing
Netherlands Organisation for Applied Scientific Research (TNO), Princetonlaan 6, Utrecht, the Netherlands

S. R. de Roode
Technical University of Delft (TUD), Delft, the Netherlands

A. P. Siebesma
Royal Netherlands Meteorological Institute (KNMI), P.O. Box 201, 3730 AE, De Bilt, the Netherlands
Technical University of Delft (TUD), Delft, the Netherlands

K. E. Yttri
NILU – Norwegian Institute for Air Research, P.O. Box 100, 2027 Kjeller, Norway

J. Schnelle-Kreis
Helmholtz Zentrum München, Cooperation group Comprehensive Molecular Analytics, 85764 Neuherberg, Germany

W. Maenhaut
Department of Analytical Chemistry, Ghent University, 9000 Gent, Belgium

G. Abbaszade
Helmholtz Zentrum München, Cooperation group Comprehensive Molecular Analytics, 85764 Neuherberg, Germany

C. Alves
Centre for Environmental and Marine Studies, Department of Environment, University of Aveiro, 3810-193 Aveiro, Portugal

A. Bjerke
NILU – Norwegian Institute for Air Research, P.O. Box 100, 2027 Kjeller, Norway

N. Bonnier
Laboratoire des Sciences du Climat et de l'Environnement, CEA-CNRS-UVSQ, 91190 Gif-sur-Yvette, France

R. Bossi
Department of Environmental Science, Aarhus University, Frederiksborgvej 399, 4000 Roskilde, Denmark

M. Claeys
Department of Pharmaceutical Sciences, University of Antwerp, Campus Drie Eiken, 2610 Antwerp, Belgium

C. Dye
NILU – Norwegian Institute for Air Research, P.O. Box 100, 2027 Kjeller, Norway

M. Evtyugina
Centre for Environmental and Marine Studies, Department of Environment, University of Aveiro, 3810-193 Aveiro, Portugal

D. García-Gacio
Group of Analytical Chemistry (QANAP), Institute of Environment (IUMA), Department of Analytical Chemistry, Faculty of Sciences, University of A Coruña, Campus de A Coruña, 15071 A Coruña, Spain

R. Hillamo
Finnish Meteorological Institute, Atmospheric Composition Research, Aerosol Research, Erik Palménin Aukio 1, 00560, Helsinki, Finland

A. Hoffer
MTA-PE Air Chemistry Research Group, P.O. Box 158, 8201Veszprém, Hungary

M. Hyder
Center for Analysis and Synthesis, Department of Chemistry, Lund University, Sweden
Pakistan Space and Upper Atmosphere research commission (SUPARCO), Lahore, Pakistan

Y. Iinuma
Leibniz-Institut für Troposphärenforschung, Permoserstr. 15, 04138 Leipzig, Germany

J.-L. Jaffrez
Univ. Grenoble Alpes, LGGE, 38000 Grenoble, France, CNRS, LGGE, 38000 Grenoble, France

A. Kasper-Giebl
Institute of Chemical Technologies and Analytics, Vienna University of Technology, Getreidemarkt 9/164, 1060 Vienna, Austria

G. Kiss
MTA-PE Air Chemistry Research Group, P.O. Box 158, 8201Veszprém, Hungary

P. L. López-Mahia
Group of Analytical Chemistry (QANAP), Institute of Environment (IUMA), Department of Analytical Chemistry, Faculty of Sciences, University of A Coruña, Campus de A Coruña, 15071 A Coruña, Spain

C. Pio
Centre for Environmental and Marine Studies, Department of Environment, University of Aveiro, 3810-193 Aveiro, Portugal

C. Piot
Univ. Grenoble Alpes, LGGE, 38000 Grenoble, France, CNRS, LGGE, 38000 Grenoble, France
Université de Savoie, LCME, 73376 Le Bourget du lac, France

C. Ramirez-Santa-Cruz
Institute of Chemical Technologies and Analytics, Vienna University of Technology, Getreidemarkt 9/164, 1060 Vienna, Austria

J. Sciare
Laboratoire des Sciences du Climat et de l'Environnement, CEA-CNRS-UVSQ, 91190 Gif-sur-Yvette, France

K. Teinilä
Finnish Meteorological Institute, Atmospheric Composition Research, Aerosol Research, Erik Palménin Aukio 1, 00560, Helsinki, Finland

R. Vermeylen
Department of Pharmaceutical Sciences, University of Antwerp, Campus Drie Eiken, 2610 Antwerp, Belgium

A. Vicente
Centre for Environmental and Marine Studies, Department of Environment, University of Aveiro, 3810-193 Aveiro, Portugal

R. Zimmermann
Helmholtz Zentrum München, Cooperation group Comprehensive Molecular Analytics, 85764 Neuherberg, Germany

M. P. Jensen
Brookhaven National Laboratory, Upton, NY, USA

T. Toto
Brookhaven National Laboratory, Upton, NY, USA

D. Troya
Brookhaven National Laboratory, Upton, NY, USA

P. E. Ciesielski
Colorado State University, Fort Collins, CO, USA

D. Holdridge
Argonne National Laboratory, Argonne, IL, USA

J. Kyrouac
Argonne National Laboratory, Argonne, IL, USA

J. Schatz
ARM Climate Research Facility, Southern Great Plains site, Billings, OK, USA

Y. Zhang
Lawrence Livermore National Laboratory, Livermore, CA, USA

S. Xie
Lawrence Livermore National Laboratory, Livermore, CA, USA

G. M. Miles
Remote Sensing Group, STFC Rutherford Appleton Laboratory, Harwell Oxford, UK

R. Siddans
Remote Sensing Group, STFC Rutherford Appleton Laboratory, Harwell Oxford, UK

B. J. Kerridge
Remote Sensing Group, STFC Rutherford Appleton Laboratory, Harwell Oxford, UK

B. G. Latter
Remote Sensing Group, STFC Rutherford Appleton Laboratory, Harwell Oxford, UK

N. A. D. Richards
Remote Sensing Group, STFC Rutherford Appleton Laboratory, Harwell Oxford, UK
School of Earth and Environment, University of Leeds, Leeds, UK

A. Di Noia
SRON Netherlands Institute for Space Research, Sorbonnelaan 2, 3584 CA Utrecht, the Netherlands

O. P. Hasekamp
SRON Netherlands Institute for Space Research, Sorbonnelaan 2, 3584 CA Utrecht, the Netherlands

G. van Harten
Leiden Observatory, Leiden University, Niels Bohrweg 2, 2333 CA Leiden, the Netherlands

J. H. H. Rietjens
SRON Netherlands Institute for Space Research, Sorbonnelaan 2, 3584 CA Utrecht, the Netherlands

J. M. Smit
SRON Netherlands Institute for Space Research, Sorbonnelaan 2, 3584 CA Utrecht, the Netherlands

F. Snik
Leiden Observatory, Leiden University, Niels Bohrweg 2, 2333 CA Leiden, the Netherlands

J. S. Henzing
Netherlands Organisation for Applied Research (TNO), Princetonlaan 6, 3584 CB Utrecht, the Netherlands

J. de Boer
Leiden Observatory, Leiden University, Niels Bohrweg 2, 2333 CA Leiden, the Netherlands

C. U. Keller
Leiden Observatory, Leiden University, Niels Bohrweg 2, 2333 CA Leiden, the Netherlands

H. Volten
National Institute for Public Health and the Environment (RIVM), Antonie van Leeuwenhoeklaan 9, 3721 MA Bilthoven, the Netherlands

B. B. Almand-Hunter
Mechanical Engineering Department, University of Colorado, 427 UCB, Boulder, CO, 80303, USA

J. T. Walker
Air Pollution Prevention and Control Division, National Risk Management Research Laboratory, US Environmental Protection Agency, E305-2, MD-63, Research Triangle Park, NC 27711, USA

N. P. Masson
Mechanical Engineering Department, University of Colorado, 427 UCB, Boulder, CO, 80303, USA

L. Hafford
Mechanical Engineering Department, University of Colorado, 427 UCB, Boulder, CO, 80303, USA

M. P. Hannigan
Mechanical Engineering Department, University of Colorado, 427 UCB, Boulder, CO, 80303, USA

T. Fang
School of Earth and Atmospheric Sciences, Georgia Institute of Technology, Atlanta, GA, USA

V. Verma
School of Earth and Atmospheric Sciences, Georgia Institute of Technology, Atlanta, GA, USA

H. Guo
School of Earth and Atmospheric Sciences, Georgia Institute of Technology, Atlanta, GA, USA

L. E. King
School of Earth and Atmospheric Sciences, Georgia Institute of Technology, Atlanta, GA, USA

E. S. Edgerton
Atmospheric Research and Analysis, Inc., Cary, NC, USA

R. J. Weber
School of Earth and Atmospheric Sciences, Georgia Institute of Technology, Atlanta, GA, USA

C. Warneke
Cooperative Institute for Research in Environmental Sciences, Univ. of Colorado, Boulder, CO, USA
NOAA Earth System Research Laboratory, Chemical Sciences Division, Boulder, CO, USA

P. Veres
Cooperative Institute for Research in Environmental Sciences, Univ. of Colorado, Boulder, CO, USA
NOAA Earth System Research Laboratory, Chemical Sciences Division, Boulder, CO, USA

S. M. Murphy
Department of Atmospheric Science, University of Wyoming, Laramie, WY, USA

J. Soltis
Department of Atmospheric Science, University of Wyoming, Laramie, WY, USA

R. A. Field
Department of Atmospheric Science, University of Wyoming, Laramie, WY, USA

M. G. Graus
Cooperative Institute for Research in Environmental Sciences, Univ. of Colorado, Boulder, CO, USA
NOAA Earth System Research Laboratory, Chemical Sciences Division, Boulder, CO, USA
Institute of Meteorology and Geophysics, Innsbruck University, Innsbruck, Austria

A. Koss
Cooperative Institute for Research in Environmental Sciences, Univ. of Colorado, Boulder, CO, USA
NOAA Earth System Research Laboratory, Chemical Sciences Division, Boulder, CO, USA

S.-M. Li
Air Quality Process Research Division, Environment Canada, Toronto, ON, Canada

R. Li
Cooperative Institute for Research in Environmental Sciences, Univ. of Colorado, Boulder, CO, USA
NOAA Earth System Research Laboratory, Chemical Sciences Division, Boulder, CO, USA

B. Yuan
Cooperative Institute for Research in Environmental Sciences, Univ. of Colorado, Boulder, CO, USA
NOAA Earth System Research Laboratory, Chemical Sciences Division, Boulder, CO, USA

J. M. Roberts
NOAA Earth System Research Laboratory, Chemical Sciences Division, Boulder, CO, USA

J. A. de Gouw
Cooperative Institute for Research in Environmental Sciences, Univ. of Colorado, Boulder, CO, USA
NOAA Earth System Research Laboratory, Chemical Sciences Division, Boulder, CO, USA

X. Yue
COSMIC Program Office, University Corporation for Atmospheric Research, Boulder, CO, USA

W. S. Schreiner
COSMIC Program Office, University Corporation for Atmospheric Research, Boulder, CO, USA

Z. Zeng
COSMIC Program Office, University Corporation for Atmospheric Research, Boulder, CO, USA

Y.-H. Kuo
COSMIC Program Office, University Corporation for Atmospheric Research, Boulder, CO, USA

X. Xue
CAS Key Laboratory of Geospace Environment, Department of Geophysics & Planetary Sciences, University of Science & Technology of China, Hefei, Anhui, China

A. Sathe
DTU Wind Energy, Risø campus, Roskilde, Denmark

J. Mann
DTU Wind Energy, Risø campus, Roskilde, Denmark

N. Vasiljevic
DTU Wind Energy, Risø campus, Roskilde, Denmark

G. Lea
DTU Wind Energy, Risø campus, Roskilde, Denmark

L. Norin
Atmospheric Remote Sensing Unit, Research Department, Swedish Meteorological and Hydrological Institute, Norrköping, Sweden

M. Schneider
Institute for Meteorology and Climate Research (IMK-ASF), Karlsruhe Institute of Technology, Karlsruhe, Germany

Y. González
Izaña Atmospheric Research Center, Agencia Estatal de Meteorología (AEMET), Izaña, Spain

C. Dyroff
Institute for Meteorology and Climate Research (IMK-ASF), Karlsruhe Institute of Technology, Karlsruhe, Germany

E. Christner
Institute for Meteorology and Climate Research (IMK-ASF), Karlsruhe Institute of Technology, Karlsruhe, Germany

A. Wiegele
Institute for Meteorology and Climate Research (IMK-ASF), Karlsruhe Institute of Technology, Karlsruhe, Germany

S. Barthlott
Institute for Meteorology and Climate Research (IMK-ASF), Karlsruhe Institute of Technology, Karlsruhe, Germany

O. E. García
Izaña Atmospheric Research Center, Agencia Estatal de Meteorología (AEMET), Izaña, Spain

E. Sepúlveda
Izaña Atmospheric Research Center, Agencia Estatal de Meteorología (AEMET), Izaña, Spain

F. Hase
Institute for Meteorology and Climate Research (IMK-ASF), Karlsruhe Institute of Technology, Karlsruhe, Germany

J. Andrey
Area de Investigación e Instrumentación Atmosférica, INTA, Torrejón de Ardoz, Spain
CNRM-GAME, Météo France and CNRS, Toulouse, France

T. Blumenstock
Institute for Meteorology and Climate Research (IMK-ASF), Karlsruhe Institute of Technology, Karlsruhe, Germany

C. Guirado
Izaña Atmospheric Research Center, Agencia Estatal de Meteorología (AEMET), Izaña, Spain

R. Ramos
Izaña Atmospheric Research Center, Agencia Estatal de Meteorología (AEMET), Izaña, Spain

S. Rodríguez
Izaña Atmospheric Research Center, Agencia Estatal de Meteorología (AEMET), Izaña, Spain

M. J. Granados-Muñoz
Dpt. Applied Physics, Faculty of Sciences, University of Granada, Fuentenueva s/n, 18071, Granada, Spain
Andalusian Institute for Earth System Research (IISTA-CEAMA), Avda. del Mediterráneo s/n, 18006, Granada, Spain

F. Navas-Guzmán
Dpt. Applied Physics, Faculty of Sciences, University of Granada, Fuentenueva s/n, 18071, Granada, Spain
Andalusian Institute for Earth System Research (IISTA-CEAMA), Avda. del Mediterráneo s/n, 18006, Granada, Spain
Institute of Applied Physics (IAP), University of Bern, Bern, Switzerland

J. A. Bravo-Aranda
Dpt. Applied Physics, Faculty of Sciences, University of Granada, Fuentenueva s/n, 18071, Granada, Spain
Andalusian Institute for Earth System Research (IISTA-CEAMA), Avda. del Mediterráneo s/n, 18006, Granada, Spain

J. L. Guerrero-Rascado
Dpt. Applied Physics, Faculty of Sciences, University of Granada, Fuentenueva s/n, 18071, Granada, Spain
Andalusian Institute for Earth System Research (IISTA-CEAMA), Avda. del Mediterráneo s/n, 18006, Granada, Spain

H. Lyamani
Dpt. Applied Physics, Faculty of Sciences, University of Granada, Fuentenueva s/n, 18071, Granada, Spain
Andalusian Institute for Earth System Research (IISTA-CEAMA), Avda. del Mediterráneo s/n, 18006, Granada, Spain

A. Valenzuela
Dpt. Applied Physics, Faculty of Sciences, University of Granada, Fuentenueva s/n, 18071, Granada, Spain
Andalusian Institute for Earth System Research (IISTA-CEAMA), Avda. del Mediterráneo s/n, 18006, Granada, Spain

G. Titos
Dpt. Applied Physics, Faculty of Sciences, University of Granada, Fuentenueva s/n, 18071, Granada, Spain
Andalusian Institute for Earth System Research (IISTA-CEAMA), Avda. del Mediterráneo s/n, 18006, Granada, Spain

J. Fernández-Gálvez
Dpt. Applied Physics, Faculty of Sciences, University of Granada, Fuentenueva s/n, 18071, Granada, Spain
Andalusian Institute for Earth System Research (IISTA-CEAMA), Avda. del Mediterráneo s/n, 18006, Granada, Spain

L. Alados-Arboledas
Dpt. Applied Physics, Faculty of Sciences, University of Granada, Fuentenueva s/n, 18071, Granada, Spain
Andalusian Institute for Earth System Research (IISTA-CEAMA), Avda. del Mediterráneo s/n, 18006, Granada, Spain

B. Scherllin-Pirscher
Wegener Center for Climate and Global Change (WEGC) and Institute for Geophysics, Astrophysics, and Meteorology/Institute of Physics (IGAM/IP), University of Graz, Graz, Austria

S. Syndergaard
Danish Meteorological Institute, Copenhagen, Denmark

U. Foelsche
Wegener Center for Climate and Global Change (WEGC) and Institute for Geophysics, Astrophysics, and Meteorology/Institute of Physics (IGAM/IP), University of Graz, Graz, Austria

K. B. Lauritsen
Danish Meteorological Institute, Copenhagen, Denmark

A. Korolev
Environment Canada, Toronto, Canada

P. R. Field
Met Office, Exeter, UK, and University of Leeds, Leeds, UK